PRINCIPLES OF SURFACE WATER QUALITY MODELING AND CONTROL

Robert V. Thomann

John A. Mueller

Manhattan College

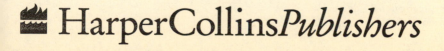 HarperCollins*Publishers*

ISBN 0-06-046677-4

90000

9 780060 466770

Sponsoring Editor: Cliff Robichaud
Project Editor: Steven Pisano
Cover Design: Wanda Lubelska Design
Text Art: RDL Artset Ltd.
Production Manager: Jeanie Berke
Production Assistant: Paula Roppolo
Compositor: Waldman Graphics, Inc.

PRINCIPLES OF SURFACE WATER QUALITY MODELING AND CONTROL

Library of Congress Cataloging-in-Publications Data

Thomann, Robert V.
    Principles of surface water quality modeling and
control.

        Includes bibliographies and index.
        1. Water quality—Mathematical models. 2. Water
quality management—Mathematical models. I. Mueller,
John A. II. Title.
TD370.T38   1987   363.7'39456   86-19510
ISBN 0-06-046677-4
Harper International Edition
International ISBN 0-06-350728-5

95 96 97 98 99   12 11 10 9 8

This book is but a small part of our lives.
For the fullness of life we have experienced,
for the listening, caring, and support,
for the love—we dedicate this small part to
our wives, Joan and Kathleen.

CONTENTS

PREFACE

For more than 60 years, since the early work by Harold Streeter and Earle Phelps on the dissolved oxygen of the Ohio River, decision makers, engineers, and scientists have continued to seek more rigorous means for assessing the effectiveness of environmental control actions.

The effectiveness of such actions is not only measured by the attainment of a water quality standard but of a concomitant expected water use associated with that water quality. The goal is to achieve such water quality and water use objectives through a control program, where presumably the benefits outweigh the costs, in some sense.

Decision makers who assess which environmental control strategies to implement are particularly wary of two possibilities:

1. Reducing waste inputs to a water body and observing little or no improvement in water quality (the environmental engineering equivalent of the bridge falling down).
2. Mandating control actions that are subsequently shown to be excessively costly, with an associated poor return in water use benefits (the environmental engineering equivalent of building a bridge to nowhere).

Mathematical modeling of water quality systems arose out of the need to address these two possibilities as the questions became more complex (e.g., the effectiveness of nutrient reduction programs on eutrophication), and the economic consequences of making a wrong decision increased markedly.

The modeling of surface water quality has a twofold result:

1. A better understanding of the mechanisms and interactions that give rise to various types of water quality behavior, such understanding to be sharpened by the formulation and testing of hypotheses of the cause–effect relationships between residual inputs and resulting water quality.
2. A more rational basis for making water quality control decisions, such a basis to include a defensible, credible, predictive framework within the larger framework of cost–benefit analysis.

This book was written to meet a very simple need—dissemination of the fundamentals and principles which underlie the mathematical modeling techniques used to analyze the quality of surface waters. In today's high-tech, computer-oriented, hardware/software-focused world, the need to understand the basic concepts of mathematical models of water quality is particularly acute, at least in our opinion. The ability to compute continues to grow exponentially, while the ability to understand, using simplified calculations, is too often still in the lag phase of growth.

We believe that, for all the apparent and obvious shortcomings and difficulties of water quality modeling, there is no viable alternative. There must be frameworks to analyze observed data in order to hypothesize what mechanisms and interactions lie "behind" the data. There must be methods for predicting the consequences of a range of potential control actions in order to rank in some way the advantages and disadvantages of various alternatives. We cannot sidestep such frameworks by a call for removals of inputs and then "watch what happens," and if "what happens" is not sufficient, call for more reductions. Such approaches, which have continued to be advocated at various times in the history of water quality, are bankrupt of understanding and are cost inefficient. Rather, it is essential that every attempt be made to assess the outcomes expected of water quality controls before the implementation of those controls, and to continue to monitor the effectiveness of the controls by both field monitoring efforts and mathematical modeling analyses.

It is our hope that this book will aid students and others who are approaching the analysis of water quality in surface waters for the first time as well as those more seasoned veterans who may occasionally need a quick glance at a formulation, approach or equation. Such a quick glance will show that the book is not a treatise, but rather a beginning; it is not exhaustive but rather selective in its choice of topics (hopefully wisely). The many seemingly distorted idealizations of "real-world" water quality situations should indicate clearly to the reader the nature of mathematical models in general, and of water quality mathematical models in particular. McFague* (1982, p. 92), in summarizing models in science, says it well:

> . . . a model in science aims at discovering that structure or set of relations in an unfamiliar area which is believed to be a genuine but partial reflection of its reality.

Three key aspects of models are then highlighted by this insightful definition:

> First, models are concerned with discovery . . .
> Second, models are concerned with behavior . . .
> Third, models are at the same time both true and untrue . . .

We are hopeful that the users of this text will enjoy the discovery of new insights, marvel at how water systems behave, and be sobered by both the reality and unreality of their calculations and reflections.

The book is divided into two broad parts:

1. An examination of the different water bodies that provide the natural setting for water quality problems:
   Chapter 1   Introduction

* McFague, S., 1982. *Metaphorical Theology, Models of God in Religious Language*, Fortress Press, Philadelphia, PA, 225 pp.

2. A review of the principal water quality problem contexts across the various water body types in Chapters 2–4:

The order of this second part reflects our perspective on the problem contexts that are most significant in both extent and impact; hence the progression from communicable disease problems, through dissolved oxygen, eutrophication, and toxic substances, to temperature.

The book has been used for a two-semester graduate environmental engineering course as well as for a one-semester senior-year civil engineering course. The selections of various chapters or topics within each chapter permits use at these different educational levels.

We are under no delusions that the work reported on in this book represents ''our work''—the material, like ourselves, is clearly a product of the many people who have contributed to the field of water quality analysis and modeling. First among these stands the work of Donald J. O'Connor, our colleague, friend, and long-time teacher at Manhattan College. To have spent the better part of our careers under his guidance, challenged by his insight and penetrating questions, has contributed greatly to our own growth and understanding. Indeed, it was his example, enthusiasm, and encouragement that led both of us, in varying ways, to water quality analysis in the first place. ''What's the mechanism?'' and ''Where's the data?'' have subsequently continued to challenge us in our own modeling efforts.

Our friend and colleague at Manhattan College, Dominic M. Di Toro, deserves special mention, too; his insightful analysis, extremely wide-reaching grasp of scientific and engineering principles, and passion for mathematical elegance have always sent us stumbling back to our own drawing boards.

Our gratefulness extends to many others, including John Connolly and Richard Winfield at Manhattan College and Thomas Gallagher, John St. John, Paul Pacquin, and William Leo of HydroQual, Inc., all of whom have stimulated us with their own experience and practical insights. We also acknowledge our past and present associations with HydoQual, Inc. (formerly Hydroscience, Inc.) from which we have drawn much practical experience, as the reader will see reflected in examples throughout the text. Stephen Chapra of Texas A&M reviewed an early manuscript and provided valuable comments for which we are grateful. The critical review and insights of Professor Robert M. Sykes, Ohio State University, are particularly appreciated since they caused the authors to look at the manuscript one more time with an even more demanding eye.

The students for whom we have had the special privilege of sharing our own incomplete knowledge are at the heart of our appreciation. Their eagerness and excitement about water quality, and their simple yet complex questions have forced us to be more critical of ourselves and have challenged us to continually repackage classic concepts for better communication.

Manuscripts get typed, proofread, and retyped again and again. This one was no exception, but it was only through the good-hearted, consistent, and steadfast determination of Eileen Lutomski that we finished at all. So our thanks and truly grateful appreciation are extended to her as well as to Madeleine Moore and Cynthia O'Donnell for their collective valiant effort and willingness to respond rapidly to our ever-changing requests for ''one more page.''

But whatever is in this book is our summary, interpretation, and collection, and we therefore freely embrace its shortcomings, errors, and faults.

ROBERT V. THOMANN
JOHN A. MUELLER

Chapter 1

Introduction

1.1 NATURE OF PROBLEM

The basic objective of the field of water quality engineering is the determination of the environmental controls that must be instituted to achieve a specific environmental quality objective. The problem arises principally from the discharge of the residues of human and natural activities that result, in some way, in an interference of a desirable use of water. What constitutes a desirable use is, of course, a matter of considerable discussion and interaction between the social-political environment and the economic ability of a given region to live with or otherwise improve its water quality.

The principal desirable uses of water are:

1. Water supply—municipal and industrial.
2. Recreational—swimming, boating, and aesthetics.
3. Fisheries—commercial and sport.
4. Ecological balance.

1.1.1 The Principal Pollution Problems, the Uses Affected, and Associated Water Quality Variables

Table 1.1 shows the major water quality problems that have been perceived through interferences with various uses of the water and subsequently confirmed through water quality sampling and analyses. The associated water use interferences are shown and the specific manifestation of that interference is indicated. For example,

1

it has long been noted that the problem of low dissolved oxygen in a stream interferes with the fish life of that stream and the manifestation of that interference is indicated by fish kills in the stream and associated aesthetic nuisances. The water quality variables associated with each problem and water use interference are also shown. The basic purpose of water quality engineering is to diagnose the type of problem shown in Table 1.1, relate that problem to the water use interference and the manifestation of that interference, and then make a judgment on which water quality variables need to be controlled and the means available for control.

> *Water quality criterion:* That concentration of a water quality measure that will meet a specific water use (USEPA, 1979).

> *Water quality standard:* The translation of a water quality criterion into a legally enforceable mass discharge or effluent limitation. Thus, the framework for assessing effluent limitations consists of two parts: (1) A use for which the water body is to be protected or designated (e.g., recreation, agriculture). (2) A numerical or qualitative pollutant concentration limit which will support that use.

TABLE 1.1   PRINCIPAL POLLUTION PROBLEMS, AFFECTED USES, AND ASSOCIATED WATER QUALITY VARIABLES

Manifestation of problem	Water use interference	Water quality problem	Water quality variables
1. Fish kills Nuisance odors, H_2S "Nuisance" organisms Radical change in ecosystem	Fishery Recreation Ecological health	Low DO (dissolved oxygen)	BOD NH_3, org N Organic solids Phytoplankton DO
2. Disease transmission Gastrointestinal disturbance, eye irritation	Water supply Recreation	High bacterial levels	Total coliform bacteria Fecal coliform bacteria Fecal streptococci Viruses
3. Tastes and odors—blue green algae Aesthetic beach nuisances—algal mats "Pea soup" Unbalanced ecosystem	Water supply Recreation Ecological health	Excessive plant growth (Eutrophication)	Nitrogen Phosphorus Phytoplankton
4. Carcinogens in water supply Fishery closed—unsafe toxic levels Ecosystem upset; mortality, reproductive impairment	Water supply Fishery Ecological health	High toxic chemical levels	Metals Radioactive substances Pesticides Herbicides Toxic product chemicals

1.1.2 The Overall Perspective

In general then, the role of the water quality engineer and scientist is to analyze water quality problems by dividing the problem into its principal components. Figure 1.1 shows such principal components as:

1. Inputs, that is, the discharge of residue into the environment from man's and nature's activities.
2. The reactions and physical transport, that is, the chemical and biological transformations, and water movement that result in different levels of water quality at different locations in time in the aquatic ecosystem.
3. The output, that is, the resulting concentration of a substance, such as dissolved oxygen or nutrients, at a particular location in the water body, during a particular time of the year or day.

The inputs are discharged into an ecological system such as a river, lake, estuary, or oceanic region. As a result of chemical, biological, and physical phenomena (such as bacterial biodegradation, chemical hydrolysis, physical sedimentation) these inputs result in a specific concentration of the substance in the given water body. Concurrently, through various mechanisms of public hearings, legislation, and evaluation, a desirable water use is being considered or has been established for the particular region of the water body under study. Such a desirable water use is translated into public health and/or ecological standards, and such standards are then compared to the concentration of the substance resulting from the discharge of the residue. This desired versus actual comparison may result in the need for an environmental engineering control, if the actual or forecasted concentration is not equal to that desired. Environmental engineering controls are then

Figure 1.1 Flow diagram—water quality engineering.

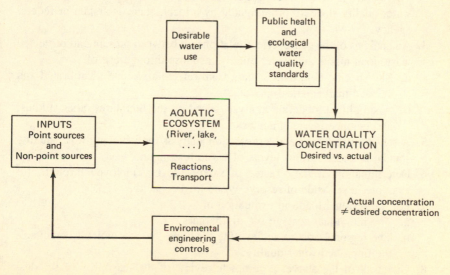

instituted on the inputs to provide the necessary reduction to reach the desired concentration. The presentation of various environmental engineering control alternatives to reach the same objective, forms a central role in the decision-making process of water quality management. This is made more specific in the application of the principles of waste load allocation.

1.2 WASTE LOAD ALLOCATION PRINCIPLES

The central problem of water quality management is the assignment of allowable discharges to a water body so that a designated water use and quality standard is met using basic principles of cost–benefit analysis. Figure 1.2 is a representation of the overall waste load allocation (WLA) problem for dissolved oxygen. There are several components to the problem, including the determination of desirable water use standards, the relationship between load and water quality and selection of projected conditions. It is generally not sufficient to simply make a scientific engineering analysis of the effect of waste load inputs on water quality. The analysis framework must also include economic impacts which, in turn, must also recognize the sociopolitical constraints that are operative in the overall problem context.

1.2.1 Steps in the WLA Process

The principal steps in the WLA process are summarized in Fig. 1.3 as:

1. A designation of a desirable water use or uses, for example, recreation, water supply, agriculture.
2. An evaluation of water quality criteria that will permit such uses.
3. The synthesis of the desirable water use and water quality criteria to a water quality standard promulgated by a local, state, interstate or federal agency.
4. An analysis of the cause–effect relationship between present and projected waste load inputs and water quality response through use of:
   a. Site-specific field data or data from related areas and a calibrated and verified mathematical model.
   b. A simplified modeling analysis based on the literature, other studies, and engineering judgment.
5. A sensitivity analysis and a projection analysis for achieving water quality standards under various levels of waste load input.
6. Determination of the "factor of safety" to be employed through, for example, a set-aside of reserve waste load capacity.
7. For the residual load, an evaluation of
   a. The individual costs to the dischargers.
   b. The regional cost to achieve the load and the concomitant benefits of the improved water quality.
8. Given all of the above, a complete review of the feasibility of the designated water use and water quality standard.

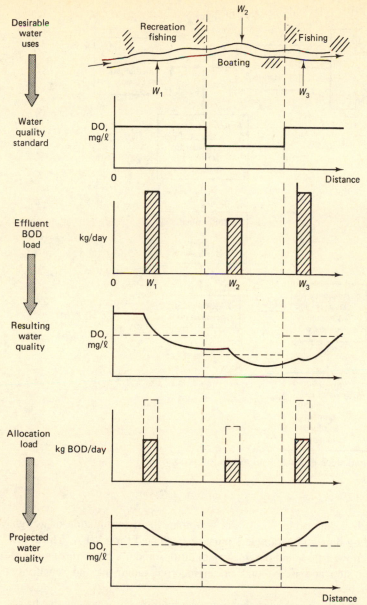

Figure 1.2 Representation of waste load allocation problem for dissolved oxygen.

9. If both are satisfactory, a promulgation of the waste load allocated to each discharger.

Within the above framework, it is assumed that a calibrated and verified water quality model is available. There are several points at which careful judgments are

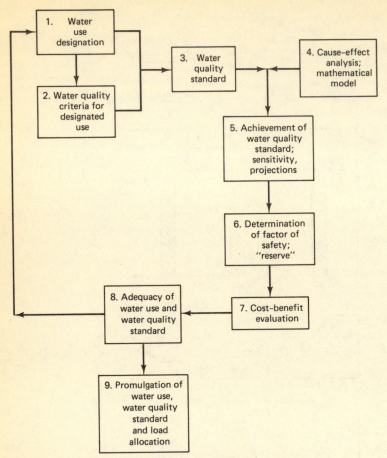

Figure 1.3   Principal steps in waste load allocation process.

required to provide a defensible WLA. For example, the determination of design conditions including flow and parameters must be evaluated for a WLA. The specification or projection of flow and parameter conditions under a given design event is a most critical step and is a blend of engineering judgment and sensitivity analysis.

There are several other issues that must be addressed to answer the basic WLA question which is "What is the permissible equitable discharge of residuals that will not exceed a water quality standard?" These questions include:

1. What does "permissible" mean? Is "permissible" in terms of maximum daily load, 7-day average load, or 30-day average load?
2. What does "exceed" mean? Does it mean "never" or 95% of the time and for all locations?
3. What are the design conditions to be used for the analysis?

4. How credible is the water quality model projection of expected responses due to the WLA, that is, what is the "accuracy" of the model calculations and how should the level of the analysis be reflected, if at all, in the WLA?

From a water quality point of view, the basic relationship between waste load input and the resulting response is given by a mathematical model of the water system. The development and applications of such a water quality model in the specific context of a WLA involves a variety of considerations including the specifications of parameters and model conditions. This relationship between input and the resulting water quality response is the principal focus of this book. Thus, the overall issues of WLA are recognized and indeed are not minimized. It is crucial, however, that the principles of water quality modeling be understood and such understanding begins with the major steps and elements of modeling.

Figure 1.4 shows the principal components of a mathematical modeling framework. The upper two steps enclosed with the dashed lines, namely, "theoretical construct" and "numerical specification" constitute what is considered a mathematical model. This is to distinguish the simple writing of equations for a model from the equally difficult task of assigning a set of representative numbers to inputs and parameters. Following this initial model specification are the steps of (a) model calibration, that is, the first "tuning" of model output to observed data and (b) the step of model verification, that is, the use of the calibrated model on a different set of water quality data. This verification data set should presumably represent a condition under a sufficiently perturbed condition (i.e., high flows, decreased temperature, changed waste input) to provide an adequate test for the model. Upon the completion of this verification or auditing step, the model would be considered verified [Fig. 1.4(a)].

The following definitions are therefore offered:

Model. A theoretical construct, together with assignment of numerical values to model parameters, incorporating some prior observations drawn from field and laboratory data, and relating external inputs or forcing functions to system variable responses.

Model Calibration. The first stage testing or tuning of a model to a set of field data, preferably a set of field data not used in the original model construction; such tuning to include a consistent and rational set of theoretically defensible parameters and inputs.

Model Verification. Subsequent testing of a calibrated model to additional field data preferably under different external conditions to further examine model validity.

The calibrated model, it should be noted, is not simply a curve-fitting exercise, but should reflect wherever possible more fundamental theoretical constructs and parameters. Thus, models that have widely varying coefficients to merely "fit" the observed data are not considered calibrated models.

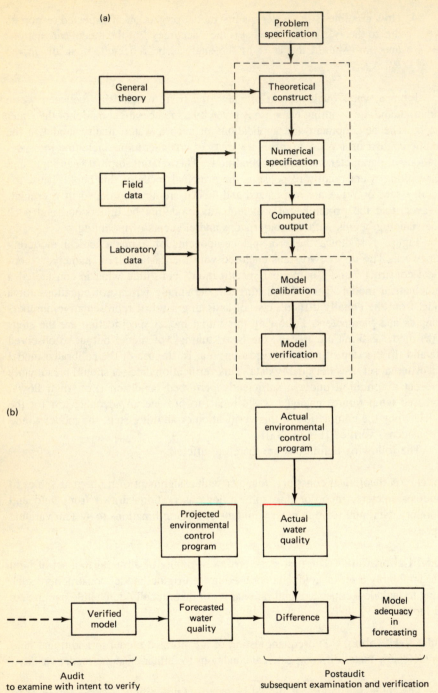

Figure 1.4 Principal components of modeling framework. (a) Steps up through verification. (b) Postaudit of models. After Thomann (1980).

8

The verified model is then often used for forecasts of expected water quality under a variety of potential scenarios. However, it is apparently rare that, following a forecast and a subsequent implementation of an environmental control program, an analysis is made of the actual ability of the model to predict water quality responses. This can be termed a ''postaudit'' of the model, as shown in Fig. 1.4. A fourth step therefore in determining model credibility is suggested as follows:

Model Postaudit.   A subsequent examination and verification of model predictive performance following implementation of an environmental control program.

Given this general review of the WLA process and the framework of mathematical modeling, the next section discusses the inputs that are common to all water bodies.

1.3 NATURE OF INPUTS

The principal inputs can be divided into two broad categories: (a) point sources and (b) non-point sources. The point sources are those inputs that are considered to have a well-defined point of discharge which, under most circumstances, is usually continuous. A discharge pipe or group of pipes can be located and identified with a particular discharger. The two principal point source groupings are: (a) municipal point sources that result in discharges of treated and partially treated sewage [with associated bacteria and organic matter, biochemical oxygen demand (BOD), nutrients, and toxic substances] and (b) industrial discharges which also result in the discharge of nutrients, BOD, and hazardous substances.

The principal non-point sources are: (1) agricultural, (b) silviculture, (c) atmospheric, (d) urban and suburban runoff, and (e) groundwater. In each case, the distinguishing feature of the nonpoint source is that the origin of the discharge is diffuse. That is, it is not possible to relate the discharge to a specific well-defined location. Furthermore, the source may enter the given river or lake via overland runoff as in the case of agriculture or through the surface of the land and water as an atmospheric input. The urban and suburban runoff may enter the water body through a large number of smaller drainage pipes not specifically designed for the carriage of wastes but for the carriage of storm runoff. In some instances in urban runoff, the discharge may be a large pipe draining a similarly large area. Other non-point sources include pollution due to groundwater infiltration, drainage from abandoned mines and construction activities, and leaching from land disposal of solid wastes.

In addition to the fact that the non-point sources result from diffuse locations, non-point sources also tend to be transient in time although not always. For example, agriculture, silviculture, and urban and suburban runoff tend to be transient resulting from flows due to precipitation at various times of the year. Other inputs such as the atmospheric input and leaching of substances out of solid waste disposal sites are more or less continuous.

1.3.1 Mass Loading

One of the more important aspects of the water quality engineering is the determination of the input mass loading; that is, the total mass of a material discharged per unit time into a specific body of water.

Figure 1.5 indicates the variability of flow and concentration for both point and non-point sources. The mass input depends on both the input flow and the input concentration and, for atmospheric inputs, includes the air-borne deposition on an areawide basis.

1.4 POINT SOURCE MASS LOADING RATES

For those defined sources with continuous flow, the input load is given by

$$W(t) = Q(t)c(t) \qquad (1.1)$$

where $c(t)$ is the concentration of the input $[M/L^3]$, $Q(t)$ is the input flow $[L^3/T]$, and $W(t)$ is the mass rate of input $[M,/T]$, all quantities occurring simultaneously

Figure 1.5 (a) Flow and concentration for continuous point source discharges (municipal and industrial wastes). (b) Flow and concentration for noncontinuous, usually non-point sources (urban runoff, agricultural drainage).

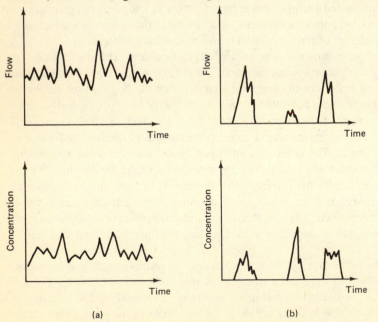

at a given time t. The units that are in widespread use in the United States are a mixture of English and metric units. Concentrations are often given in metric units, for example, mg/l but flow and mass input are in English units, for example, million gallons per day (MGD), or cubic feet per second (cfs), and lb/day, respectively. The conversion for Eq. 1.1 with these units is thus:

$$W = 8.34Qc \qquad (1.2a)$$

$$\text{lb/day} = \left(\frac{\text{lb}}{\text{MG-mg/l}}\right) \text{(MGD) (mg/l)}$$

or

$$W = 5.39Qc \qquad (1.2b)$$

$$\text{lb/day} = \left[\frac{\text{lb/day}}{\text{cfs-mg/l}}\right] \text{(cfs) (mg/l)}$$

In metric units, the flow is often given in m^3/s. Also, for almost all practical purposes:

$$1 \text{ mg/l} = 1 \text{ g/m}^3 = 10^{-3} \text{ kg/m}^3$$

Then

$$W = Qc \qquad (1.3)$$

where W is in g/s, Q in m^3/s, and c in mg/l ($= g/m^3$). Sample Problem 1.1 illustrates the continuous source loading calculation for total nitrogen and compares the result to that of an intermittent source (Section 1.6.)

It should be recognized that effluents from waste treatment plants can vary significantly over time. Although average values of various constituents can be specified, the temporal variability of the load often needs to be incorporated either directly or indirectly into the analysis. Some examples of temporal variations in effluent quality are shown in Fig. 1.6.

As indicated in Figs. 1.6(a) and 1.6(b), there may be significant diurnal variation in both flow and quality constituent, in this case, ammonia. In streams that are dominated by the effluent, these variations may be reflected in downstream water quality. Figure 1.6(c) shows the variation in BOD loading (lb/day) from a primary effluent. Inspection and also analysis of this time series indicates a significant 7-day oscillation or periodicity in the effluent. Figure 1.6(d) shows the variation in the final effluent BOD loading from an activated sludge plant and indicates a strong seasonal increase in load. In this particular case, the marked increase in the late summer is due to an additional load from a seasonal canning industry that discharges to the municipal system. These examples indicate the variability that is normally present in all effluents.

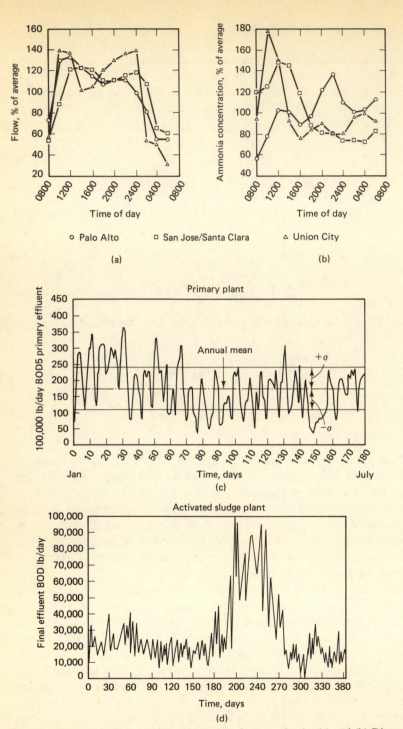

Figure 1.6 Examples of temporal variations in waste loads. (a) and (b) Diurnal variations in flow and NH₃ concentrations. From Randtke and McCarty (1977). (c) and (d) Weekly and seasonal BOD variations. From Thomann (1970). Reprinted by permission of American Society of Civil Engineers.

SAMPLE PROBLEM 1.1

DATA

A city with combined sewers extends over an area of 15 mi^2. Rainfall occurs on average every 77 hr, with an average intensity and duration of 0.055 in./hr and 6.5 hr, respectively. The permeability of the city is such that the volumetric runoff coefficient $C = 0.6$ and the average overflow total nitrogen (TN) concentration is 9 mg/l. The city treatment plant discharges 20 MGD with an effluent TN concentration of 40 mg/l. It captures on average 20% of the runoff.

PROBLEM

Compare the TN discharge from the combined sewer overflows (CSOs) with that from the city sewage treatment plant (STP).

ANALYSIS

$$W(STP) = Q(MGD) \times c(mg/l) \times 8.34 \qquad \text{[Eq. (1.2a)]}$$
$$= 20 \times 40 \times 8.34 = 6670 \text{ lb/day}$$

Runoff—Wet Periods Only (See Section 1.6)

$$Q_R = CIA \qquad \text{[Eq. (1.8)]}$$
$$Q_R = 0.6 \times 0.0555(\text{in./hr})$$
$$\times (15 \text{ mi}^2 \times 640 \text{ acres/mi}^2) = 317 \text{ cfs}$$
$$\text{overflow} = 317 \text{ cfs}[1 - 0.20(\text{STP capture})] = 254 \text{ cfs} \qquad \text{[Eq. (1.9)]}$$
$$W_R = Q_R(\text{cfs}) \times \bar{c}(mg/l) \times 5.4$$
$$W_R = 254 \times 9 \times 5.4 = 12,300 \text{ lb/day}$$

Runoff—Wet and Dry Period Average

$$W_A = W_R D/\Delta = 12,300 \times \frac{6.5}{77} = 1040 \text{ lb/day} \qquad \text{[Eq. (1.10)]}$$

Conclusion: During wet periods, the discharge of TN by the CSOs is almost twice that of the STP. Averaged over wet and dry periods, the CSO discharge is only about 15% of the point source STP.

1.5 TRIBUTARY MASS LOADING RATES

For some sources, the flow is continuously measured (as for example in the tributary to a larger river or lake), but estimates of the concentration of the water quality parameter are only available at certain intervals. If Eq. 1.3 is applied to only those times when both flow and concentration are available, the actual loading from the source may be significantly underestimated. If, for example, the concentration is not measured during times of peak runoff, a major component of the load may be missed. Therefore, some other approach is required to estimate the average load.

One approach is to plot the available concentration data as a function of the river flow, usually both as logarithms, and estimate the relationship between concentration and flow. For each flow then where the concentration was not measured, this relationship is used to estimate the concentration, and hence, the load. Thus, it is assumed in this method that

$$c = aQ^b \tag{1.4}$$

Figure 1.7 shows a typical plot of this type (for the Potomac River at Paw Paw) and illustrates the difficulty with this approach. The relationship may depend on whether the concentration was measured when the flow was rising (when bottom material is suspended and tributaries and land runoff are discharging into the river) or when the flow was declining. For the nitrite and nitrate data shown in Fig. 1.7, the scatter is large and the estimate of the proper relationship to use is somewhat judgmental. Use of this method is illustrated in Sample Problem 1.2.

Another method that has been shown to work well in estimating phosphorus loads from tributaries (Dolan et al., 1981) uses the "unbiased stratified ratio estimator." The flow record may be divided or stratified into various periods, for example, seasonal, annual, or high and low flow periods. The mean load is then

$$\overline{W}_p = \overline{Q}_p \frac{\overline{W}_c}{\overline{Q}_c}\left[\frac{1 + (1/n)(S_{QW}/\overline{Q}_c\overline{W}_c)}{1 + (1/n)(S_Q^2/\overline{Q}_c^2)}\right] \tag{1.5}$$

where $\overline{W}_p$ is the estimated average load for the period p, $\overline{Q}_p$ is the mean flow for the period, $\overline{W}_c$ is the mean daily loading for the days on which concentrations were determined, $\overline{Q}_c$ is the mean daily flow for the days on which concentrations were determined, and n is the number of days when concentrations were measured. Also,

$$S_{QW} = [1/(n - 1)]\left[\left(\sum_{i=1}^{n} Q_{ci}W_{ci}\right) - n\overline{Q}_c\overline{W}_c\right] \tag{1.6}$$

and

$$S_Q^2 = [1/(n - 1)]\left[\left(\sum_{i=1}^{n} Q_{ci}^2\right) - n\overline{Q}_c^2\right] \tag{1.7}$$

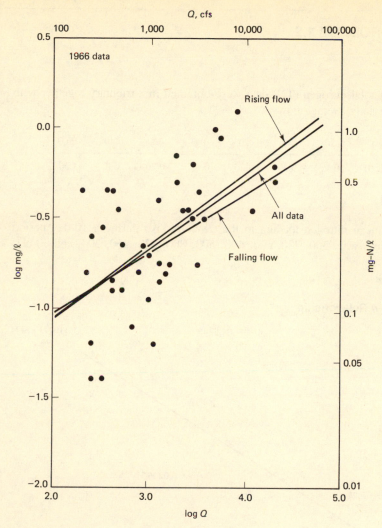

Figure 1.7 Log-log relationships for nitrate + nitrite nitrogen and river flow for the Potomac River at Paw Paw, Md. From Hydroscience (1976).

where Q_{ci} are the individually measured flows and W_{ci} is the daily loading for each day on which the concentration was measured. Dolan et al. (1981) applied several estimation procedures to a detailed sampling set of phosphorus and flow for the Grand River in the area of Grand Rapids, Michigan. They found this estimator to work best by testing various subsets of the actual data. The method is best suited where there is extensive flow data but concentration data are sparse.

SAMPLE PROBLEM 1.2

DATA

The following total nitrogen (TN) data were obtained in a tributary together with corresponding daily average flows:

Q (cfs)	20	40	70	100	300	440	1000
TN (mg/l)	0.11	0.30	0.30	0.42	0.70	1.40	1.50

PROBLEM

Estimate the mean nitrogen loading in the tributary ($\overline{W}$) during a 10-day period when daily flows were 300, 280, 600, 400, 500, 800, 620, 360, 200, and 80 cfs.

ANALYSIS

Nitrogen–Flow Relationship

$$c = aQ^b \qquad \text{[Eq. (1.4)]}$$

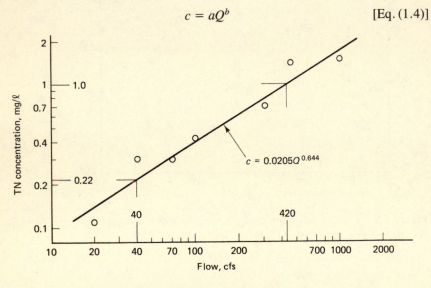

For (c, Q) of $(0.22, 40)$ and $(1.0, 420)$ on best-fit line,

$$b = \frac{\ln(c_1/c_2)}{\ln(Q_1/Q_2)} = \frac{\ln(0.22/1.0)}{\ln(40/420)} = 0.644$$

$$a = \frac{c_1}{Q_1^b} = \frac{0.22}{40^{0.644}} = 0.0205$$

$$\therefore \ c(\text{mg/l}) = 0.0205[Q(\text{cfs})]^{0.644}$$

Sample Problem 1.2 (continued)

TN Daily Mass Loadings

Day	1	2	3	4	5	6	7	8	9	10
Q(cfs)	300	280	600	400	500	800	620	360	200	80
c(mg/l)	0.81	0.77	1.26	0.97	1.12	1.52	1.29	0.91	0.62	0.34
W(10^3 lb/day)	1.31	1.16	4.08	2.10	3.02	6.57	4.32	1.77	0.67	0.15

$$\therefore \overline{\overline{W}} = \frac{\Sigma W}{n = 10} = \frac{25.15 \times 10^3}{10} = \underline{\underline{2500 \text{ lb/day}}}$$

1.6 INTERMITTENT MASS LOADING RATES

Loading for intermittent sources depends on a number of factors that may influence both the flow and the concentration. As noted earlier, the flow from urban runoff is usually highly transient resulting from variable precipitation, so at times there is no load being discharged. For discharges from combined sewers, therefore, several input loads can be estimated:

1. Equivalent annual (or other interval) loading rate (comparable to continuous load).
2. Average load discharged per event of overflow.
3. Distribution of load within an event of overflow.

Figure 1.8 shows the variability of intermittent rainfall events, volume of precipitation, resulting runoff, and loading. Data are usually available on hourly rainfall or daily totals. Note that several quantities are of importance: the volume of precipitation, the duration of the event, and the interval between events. The seasonal variation of rainfall intensity may vary by significant amounts throughout the year.

A statistical procedure has been developed (DiToro, 1979) which allows one to make an estimate of the mean load and the expected high load events with a given probability. Discussion here will be restricted to estimating average loading per event and equivalent continuous loading, both comprised of flow and concentration.

The simplest estimate of the runoff flow Q_R [L^3/T] is from the so-called rational formula:

$$Q_R = CIA \tag{1.8}$$

where I is the rainfall rate [L/T], A is the area over which the runoff will occur [L^2], and C is the runoff coefficient. The value of C depends on land use, population density, and degree of imperviousness, and ranges from 0.1 to 0.3 for population

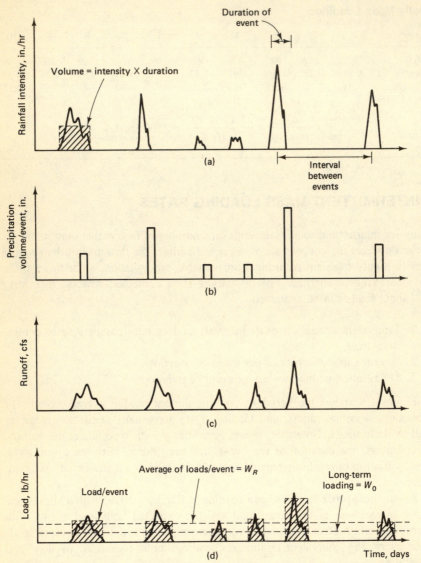

Figure 1.8 (a) Rainfall intensity events. (b) Volume. (c) Runoff. (d) Load for intermittent-type discharges.

densities of about 1 person/acre (rural) to 0.7 to 0.9 for heavy industrial and commercial areas with densities greater than 50 persons/acre. Note that if I is in./hr and A in acres, then Q_R is in acre-in./hr which is equal to about 1 cfs. Therefore,

$$Q_R(\text{cfs}) = CI(\text{in./hr})\, A(\text{acres})$$

The mean load per overflow event is given by (DiToro, 1979)

$$W_R = \bar{c} Q_R \tag{1.9}$$

where $\bar{c}$ is the average concentration during the event and Q_R is the average flow during the event, assuming that $\bar{c}$ and Q_R do not depend on each other.

The long-term average loading rate can be estimated from

$$W_A = \frac{W_R D}{\Delta} \tag{1.10}$$

where D is the average duration of storms $[T]$, and Δ is the average time between storms $[T]$. One should have at least 5 years of rain gage data to estimate these rainfall parameters. (See Sample Problem 1.1, page 13.)

Di Toro et al. (1978, 1979), and Di Toro (1979) also present an approach to estimating the exceedance frequencies of occurrence of a discharged load which is particularly appropriate for time variable water quality problems. The frequency distribution of the rainfall parameters was estimated by a gamma-type probability density function. The coefficient of variation of the rainfall intensity, v_i, was then used as the relevant statistical measure, where

$$v_i = \frac{s_{di}}{I} \tag{1.11}$$

for s_{di} as the standard deviation of the rainfall intensity and I as the average rainfall intensity.

If the rational formula for runoff Q_R is used, Eq. 1.8, then

$$v_q = v_i \tag{1.12}$$

where v_q is the coefficient of variation of the runoff flow, given by

$$v_q = s_q/Q_R \tag{1.13}$$

If the concentration and flow of the event are assumed independent, then Di Toro et al. (1978) show that

$$v_w = v_q v_c \sqrt{1 + (1/v_q)^2 + (1/v_c)^2} \tag{1.14}$$

for v_c as the coefficient of variation of the input concentration and v_w as the coefficient of variation of the input load. Note that if c is assumed a constant, (as for example through use of the averages in Table 1.3, page 21) then

$$v_w = v_q = v_i \tag{1.15}$$

As a first estimate, the variation of the loading rate is assumed as a gamma

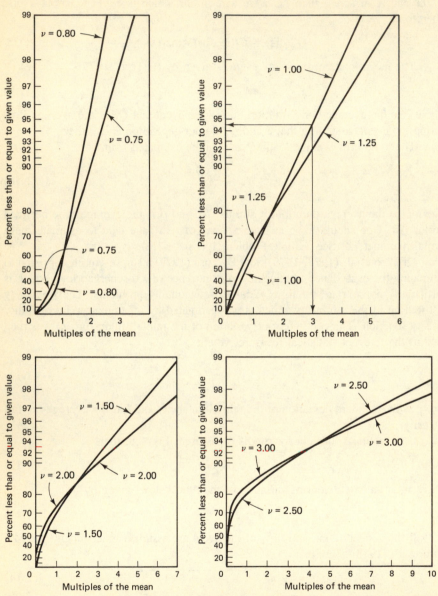

Figure 1.9 Cumulative density function for load estimation ($\nu = 0.5$ to 3.0). From DiToro et al. (1979).

TABLE 1.2 FREQUENCY OF OCCURRENCE OF
LOAD FROM INTERMITTENT LOADING

Coefficient of variation of load	Multiples of W_R % of events with load > value shown		
	50	10	5
0.5	0.9	1.3	1.7
1.0	0.5	2.4	3.2
1.5	0.3	2.5	3.7
3.0	0.1	2.6	5.2

variable. Figure 1.9 shows the frequency distribution behavior of multiples of the mean of a gamma variable as a function of v. This figure can be used to estimate the exceedance frequencies of the input load.

Thus, if W_R is calculated, and v_w estimated, Fig. 1.9 shows that for $v_w = 1.00$, about 95% of the events would be expected to be equal to or less than $3W_R$. Now the average number of storms $\bar{S}$ in an interval T is

$$\bar{S} = \frac{T}{\Delta} \qquad (1.16)$$

Therefore, 5% of the events would be expected to have discharge loading rates greater than $3W_R$. For example, suppose $\Delta = 72$ hr (between storm centers), then $\bar{S} = 120$ events for $T = 1$ yr. A 5% frequency then corresponds to about 6 events in a year where the load would be greater than $3W_R$.

Table 1.2 (from Fig. 1.9) summarizes the frequency of occurrence of the input loads for different levels of the coefficient of variation of the load.

1.7 COMPARISON OF CONTINUOUS AND INTERMITTENT WASTE SOURCES

In any given problem context, reductions in waste inputs may be required to mitigate existing water quality violations. The choice of reducing point or non-point sources will depend not only on the feasibility and economics of existing engineering controls but also on the relative magnitudes of the sources. Early identification of the major waste sources will allow concentration of available resources on the most significant inputs. To this end, published values of point and nonpoint sources may be used in a preliminary screening effort, although caution should be exercised due to the wide variability in such data. Data from areas similar to the study area should be used as much as possible.

For illustration purposes, selected values of continuous and intermittent waste input parameters are shown in Table 1.3. Note that urban inputs are typically characterized by concentrations whereas rural and atmospheric inputs are presented as areal mass rates. Calculations of longer-term average inputs, both point and distributed, are contained in Sample Problem 1.3.

TABLE 1.3 REPORTED VALUES OF SELECTED WASTE INPUT PARAMETERS IN THE UNITED STATES

Variable	Units[a]	Municipal influent[b]	CSO[c]	Urban runoff[d]	Agriculture (lb/mi²-day)[e]	Forest (lb/mi²-day)[e]	Atmosphere (lb/mi²-day)[f]
Average daily flow	gcd	125					
Total suspended solids	mg/l	300	410	610	2500	400	
CBOD5[g]	mg/l	180	170	27	40	8	
CBODU[g]	mg/l	220	240				
NBOD[g]	mg/l	220	290				
Total nitrogen	mg-N/l	50	9	2.3	15	4	8.9–18.9
Total phosphorus	mg-p/l	10	3	0.5	1.0	0.3	0.13–1.3
Total coliforms	10⁶/100 ml	30	6	0.3			
Cadmium	μg/l	1.2	10	13			0.015
Lead	μg/l	22	190	280			1.3
Chrome	μg/l	42	190	22			0.088
Copper	μg/l	159	460	110			
Zinc	μg/l	241	660	500			1.8
Total PCB	μg/l	0.9	0.3	—			0.002–0.02

[a]Units apply to municipal, CSO (combined sewer overflow), and urban runoff sources; gcd = gallons per capita per day.
[b]Thomann (1972); heavy metals and PCB, HydroQual (1982).
[c]Thomann (1972); total coli, Tetra Tech, (1977); heavy metals Di Toro et al. (1978); PCB, Hydroscience (1978).
[d]Tetra Tech (1977); heavy metals, Di Toro et al. (1978).
[e]Hydroscience (1976a).
[f]Nitrogen and phosphorus, Tetra Tech (1982); heavy metals and PCB, HydroQual (1982).
[g]CBOD5 = 5 day carbonaceous biochemical oxygen demand (CBOD); CBODU = ultimate CBOD; NBOD = nitrogenous BOD.

SAMPLE PROBLEM 1.3

DATA

Runoff from 100 mi² of agricultural lands drains to a point in a river where a city of 100,000 people is located. The city has a land area of 10 mi² and its sanitary sewers are separated from its storm drains. A sewage treatment plant discharges to the river immediately downstream of the city. The area receives an annual rainfall of 30 in. of which 30% runs off the agricultural lands and 50% drains off the more impervious city area.

PROBLEM

Using the loading data from Table 1.3, and the residual fractions cited below for the treatment plant, compare the contributions of the atmospheric, agricultural, and urban sources to annual average values of flow, CBOD5, total coliform bacteria, and lead in the river. Neglect decay mechanisms for all parameters.

ANALYSIS

Basic Data

Item	Atmospheric (atmos)	Agricultural (ag)	Runoff	STP Influent	Residual fraction
Flow	—	30% precipitation	50% precipitation	125 gcd	1.00
CBOD5	—	40 lb/mi²-day	27 mg/l	180 mg/l	0.15
Total coliform	—	100/100 ml	3 × 10⁵/100 ml	30 × 10⁶/100 ml	0.0001
Lead	1.3 lb/mi²-day	—	280 μg/l	22 μg/l	0.05

Flow Contributions

$$Q(\text{ag}) = 100 \text{ mi}^2 \times (30 \times 0.3) \text{ in./yr} \times (5280 \text{ ft})^2/\text{mi}^2$$

$$\times \frac{1 \text{ ft}}{12 \text{ in.}} \times \frac{1 \text{ yr}}{365 \text{ days}} \times \frac{1 \text{ day}}{86,400 \text{ sec}}$$

$$= 100 \text{ mi}^2 \times \left(9 \text{ in./yr} \times \frac{0.07367 \text{ cfs/mi}^2}{\text{in./yr}} = 0.663 \text{ cfs/mi}^2\right)$$

$$= \underline{\underline{66.3 \text{ cfs}}}$$

$$Q(\text{urban}) = 10 \text{ mi}^2 \times (30 \times 0.5) \text{ in./yr} \times \frac{0.0767 \text{ cfs/mi}^2}{\text{in./yr}}$$

$$= \underline{\underline{11.1 \text{ cfs}}}$$

$$Q(\text{STP}) = 100,000 \text{ cap} \times 125 \frac{\text{gal}}{\text{cap-day}} \times \frac{1 \text{ MG}}{10^6 \text{ gal}}$$

$$= \underline{\underline{12.5 \text{ MGD}}} \ (\times 1.548 \text{ cfs/MGD} = 19.4 \text{ cfs})$$

(continued)

Sample Problem 1.3 (continued)

CBOD5

$$W(\text{ag}) = 100 \text{ mi}^2 \times 40 \text{ lb/mi}^2\text{-day} = \underline{4000 \text{ lb/day}}$$

$$W(\text{urban}) = 11.1 \text{ cfs} \times 27 \text{ mg/l} \times 5.4 \frac{\text{lb/day}}{\text{cfs-mg/l}} = \underline{1620 \text{ lb/day}}$$

$$W(\text{STP}) = 12.5 \text{ MGD} \times (180 \times 0.15) \text{ mg/l} \times \frac{8.34 \text{ lb/day}}{\text{MGD-mg/l}} = \underline{2810 \text{ lb/day}}$$

Total Coliform Bacteria

$$W(\text{ag}) = 66.3 \text{ ft}^3/\text{sec} \times 100 \text{ org/100 ml} \,(1000 \text{ ml/l} \times 3.785 \text{ l/gal}$$
$$\times 7.48 \text{ gal/ft}^3 \times 86,4000 \text{ sec/day})$$

$$= 66.3 \text{ cfs} \times 100 \text{ org/100 ml} \times 2.446 \times 10^7 \frac{\text{org/day}}{\text{cfs-org/100 ml}}$$

$$= \underline{0.2 \times 10^{12} \text{ org/day}}$$

$$W(\text{urban}) = 11.1 \times 3 \times 10^5 \times 2.446 \times 10^7 = \underline{82 \times 10^{12} \text{ org/day}}$$

$$W(\text{STP}) = 19.4 \times (30 \times 10^6 \times 0.0001) \times 2.446 \times 10^7$$
$$= \underline{1.4 \times 10^{12} \text{ org/day}}$$

Lead

$$W(\text{atmos}) = 1.3 \text{ lb/mi}^2\text{-day} \times 100 \text{ mi}^2 \times 0.05(\text{resid. fract.})$$
$$= \underline{6.5 \text{ lb/day}}$$

$$W(\text{urban}) = 11.1 \text{ cfs} \times 280 \text{ } \mu g/l \times 10^{-3} \text{mg/}\mu g \times 5.4 \frac{\text{lb/day}}{\text{cfs-mg/l}} = \underline{16.8 \text{ lb/day}}$$

$$W(\text{STP}) = 12.5 \text{ MGD} \times (22 \times 0.5 \text{ } \mu g/l) \times 10^{-3} \times 8.34 = \underline{1.1 \text{ lb/day}}$$

Summary of Annual Average Contributions to River

	Q (cfs)	CBOD5 (lb/day)	Total coliform (10^{12} org/day)	Lead (lb/day)
Atmos	—	—	—	6.5
Agric	66.3	4000	0.2	—
Urban	11.1	1620	82	16.8
STP	19.4	2810	1.4	1.1
Total	96.8	8430	83.6	24.4

REFERENCES

Di Toro, D. M., J. A. Mueller, and M. J. Small, 1978. *Rainfall-Runoff and Statistical Receiving Water Models,* NYC 208 Task Report 225. Prepared by Hydroscience, Inc. for Hazen & Sawyer Engr. and NYC Department of Environmental Protection, March 1978, 271 pp.

Di Toro, D. M., 1979. *Statistics of Receiving Water Response to Runoff.* Paper presented at the Urban Stormwater and Combined Sewer Overflow Impact on Receiving Water Bodies, National Conference, University of Central Florida, Orlando, FL, Nov. 1979, 35 pp.

Di Toro, D. M., E. D. Driscoll, and R. V. Thomann, 1979. *A Statistical Method for the Assessment of Urban Stormwater.* Prepared by Hydroscience, Inc. for the U.S. Environmental Protection Agency, Office of Water Planning and Standards, Washington, D.C., May 1979, 7 Chapters + Appendix EPA 440/3-79-023.

Dolan, D. M., A. K. Yui, and R. D. Geist, 1981. "Evaluation of River Load Estimation Methods for Total Phosphorus," *J. Great Lakes Res.* **7**(3):207–214.

HydroQual Inc., 1982. "Contaminant Inputs to the Hudson–Raritan Estuary, by J. A. Mueller, T. A. Gerrish, and M. C. Casey, for National Oceanic and Atmospheric Administration (NOAA), Office of Marine Pollution Assessment. NOAA Technical Memorandum OMPA-21.

Hydroscience, Inc., 1976a. *Areawide Assessment Procedure Manual.* Prepared for U. S. Environmental Protection Agency, Office of Research and Development, Municipal Environmental Research Laboratory, Cincinnati, OH. EPA-600/9-76-014.

Hydroscience, Inc., 1976b. *Water Quality Analysis of the Potomac River.* Report to the Interstate Commission on the Potomac River Basin.

Hydroscience, Inc., Oct. 1978. *PCB Analysis,* by W. M. Leo. Prepared for Hazen & Sawyer, Engrs. and NYC Department of Water Resources. NYC 208 Task Report 317 Addendum.

Randtke, S. J., and P. L. McCarty, 1977. "Variations in Nitrogen and Organics in Wastewaters, *J. Environ. Eng. Div. Am. Soc. Civ. Eng.* **EE4**:539–550.

Tetra Tech, Inc., 1977. *Water Quality Assessment: A Screening Method for Nondesignated 208 Areas,* by S. W. Zison, K. F. Haven, and W. B. Mills. Prepared for USEPA, ERL, Athens, GA. EPA-600/9-77-023.

Tetra Tech, Inc., 1982. *A Screening Procedure for Toxic and Conventional Pollutants—Part 1,* by W. B. Mills, J. D. Dean, D. B. Porcella, S. A. Gherini, R. J. M. Hudson, W. E. Frick, G. L. Rupp, and G. L. Bowie. Prepared for USEPA, ERL, Athens, GA. EPA-600/6-82-004a.

Thomann, R. V., 1970. "Variability of Waste Treatment Plant Performance," *J. Sanit. Eng. Div. Am. Soc. Civ. Eng.* **SA3**:819–837.

Thomann, R. V., 1972. *Systems Analysis and Water Quality Mangement,* McGraw-Hill, New York, 272 pp. (Reprint: J. Williams Book Co., Oklahoma City, OK)

Thomann, R. V., 1980. "Measures of Verification," in Workshop on Verification of Water Quality Models, R. V. Thomann and T. O. Barnwell, Jr. Co-chairmen, USEPA, Athens, GA, pp 37–61. EPA-600/9-80-016.

U.S. Environmental Protection Agency, 1979. "Water Quality Criteria," *Federal Register,* Part V, 15926–15981.

PROBLEMS

1.1. (a) The allowable discharge BOD load from a municipality has been established by the state regulatory agency at 4000 lb/day. If the flow is 50 MGD, what is the allowable effluent concentration in mg/l?

(b) A lake has an area of 1000 ha. The concentration of the toxicant PCB in rainfall is 100 ng/l and the average annual rainfall is 70 cm. What is the average annual mass loading of PCB from the precipitation to the lake in mt/yr?

1.2. (a) A municipal treatment plant has a discharge flow of 36 MGD and an effluent BOD concentration of 14 mg/l. What is the discharge load in lb/day?

(b) The flow of a tributary to a large river has been measured at an average of 4.8 m^3/s. The concentration of the insecticide diflubenzuron (used for mosquito control) has been quantified in this flow at a level of 6 ng/l. What is the mass loading of this insecticide from the tributary to the large river in g/s and kg/day?

(c) An industrial plant has been given a mass loading allocation of phosphorus of 855 lb/day. The flow is estimated at 8.4 MGD. What is the permissible effluent concentration for this plant in mg/l?

1.3. A particular river has a significant effect on a reservoir through the discharge of suspended solids from the river to the reservoir. It is important to estimate the mass loading of solids to the reservoir to determine the extent of long-term sedimentation of solids in the reservoir, reducing its available volume. A brief survey over a 10-day period is shown below.

(a) Plot the flow versus time in days.

(b) and (c) Estimate the mean mass loading of solids (in kg/day) into the reservoir, using (b) log–log plot of flow vs. solids and (c) unbiased stratified ratio estimator.

Day	Flow (m^3/s)	Suspended solids concentration (mg/l)
1	1.0	10
2	1.5	
3	15.0	
4	100.0	
5	20.0	40
6	10.0	18
7	5.0	
8	2.5	20
9	1.5	
10	1.0	8

1.4. The drainage area of an urban area encompasses 1875 acres of which 1310 acres drain via a separate sewer system, and 565 acres drain via a combined sewer system (Di Toro et al., 1979). The average storm intensity for a summer period is 0.055 in./hr, the average duration of storms is 3 hr, and the average time between storms is 80 hr. The BOD of the runoff from the combined area is 111 mg/l and for the separate area is 27 mg/l. If the runoff coefficient is 0.35 for both the separate and combined areas, estimate the following quantities for the *entire area:*

(a) Mean runoff (cfs) for the summer.

(b) Mean BOD load (lb/day) during the summer storms.

(c) Long-term summer loading rate (lb/day).

1.5. For the period of June to September (122 days) of Problem 1.4, the following statistics were calculated:

	Mean	Coefficient of variation
Storm intensity	0.055 in./hr	1.55
Duration	3 hr	1.15
Time between storms	80 hr	1.15

Coefficient of variation of BOD5 concentration = 1.0. Using the same drainage areas, runoff coefficients, and mean BOD concentrations of Problem 1.4, calculate for the entire area:

(a) Average number of storms per summer.

(b) The percent frequency of storms equal to or less than 50,000 lb/day.

(c) The number of storms per summer that would have a load greater than 50,000 lb/day.

1.6. Some rainfall statistics for Denver, Colorado (Di Toro et al., 1979) are given as follows:

Month	Intensity (in./hr)	Duration (hr)	Interval between storms (hr)
March	0.02	6	90
July	0.06	2	85

An urban area in Denver is 5000 acres in size and has a population density of 20 persons/acre ($C_v = 0.7$). The BOD from the runoff from the area has been measured at 100 mg/l. For March and for July compute

(a) the mean runoff in cfs.

(b) the mean BOD load (lb/day) during a storm event.

(c) the average BOD loading rate (lb/day).

Chapter 2

Rivers and Streams

For anyone who has been thrilled by the excitement of canoeing on a river or quietly paddling on a stream, it is quite clear that the distinguishing feature of describing water quality in rivers and streams is the movement of the water, more or less rapidly, in a downstream direction. In this chapter, the basic characteristics of rivers are presented, but detailed analyses of several problem contexts such as dissolved oxygen (DO) are given in later chapters. From a water quality engineering point of view, rivers have been studied more extensively and longer than other bodies of water, probably reflecting the fact that many people live close to or interact with rivers and streams. Hydrologically then, our interest in rivers begins with the analysis of river flows. The magnitude and duration of flows coupled with the chemical quality of the waters determines, to a considerable degree, the biological characteristics of the stream. The river is an extremely rich and diverse ecosystem and any water quality analysis must recognize this diversity. The river system may therefore be considered from the physical, chemical, and biological perspectives. The principal characteristics of rivers that are of interest include:

1. Geometry: width, depth.
2. River slope, bed roughness, "tortuosity."
3. Velocity.
4. Flow.
5. Mixing chracteristics (dispersion in the river).
6. River water temperature.
7. Suspended solids and sediment transport.

For river water quality management, the important chemical characteristics are:

1. DO variations including associated effects of oxidizable nitrogen on the DO regime.
2. pH, acidity, alkalinity relationships in areas subjected to such discharges, for example, drainage from abandoned mines.
3. Total dissolved solids and chlorides in certain river systems, for example, natural salt springs in the Arkansas–White–Red River basins and TDS buildup in the Colorado River basin.
4. Chemicals that are potentially toxic.

Biological characteristics of river systems that are of special significance in water quality studies are:

1. Bacteria and viruses.
2. Fish populations.
3. Rooted aquatic plants.
4. Biological slimes; *Sphaerotilus*.

As with all water quality analyses, our objective in river water quality engineering is to recognize and quantify, as much as possible, the various interactions between river hydrology, chemistry, and biology.

2.1 RIVER HYDROLOGY AND FLOW

The study of river hydrology includes many factors of water movement in river systems including precipitation, stream flow, droughts and floods, groundwater, and sediment transport. The student is referred to the text entitled *Hydrology for Engineers* (Linsley et al., 1982) for a more in-depth treatment. The flows of the five longest rivers on each continent are found in Table 2.1.

For our purposes, the most important aspects of river hydrology are the river flow, velocity, and geometry. Each of the characteristics will be used in various ways in the water quality modeling of rivers. For reasons that will be discussed very shortly, it is probably clear that our interest in river flows will focus on those times when the flow is "low" based on the most intuitive notion of dilution. If a discharge is running into a stream, then conditions will probably be most critical during the times when there is less water in the channel. The flow at a given point in a river will depend on:

1. Watershed characteristics such as the drainage area of the river or stream basin up to the given location.
2. Geographical location of the basin.
3. Slope of the river.
4. Dams, reservoirs, or locks which may regulate flow.
5. Flow diversions into or out of the river basin.

The flow in the river can be obtained by several methods (Fig. 2.1). A direct

TABLE 2.1 CHARACTERISTICS OF THE WORLD'S LONGEST RIVERS[a]

Continent	River	Length (mi)	Drainage Area (1000 mi²)	Flow[b] (1000 cfs)
Africa	Nile	4160	1082	93
	Congo	2880	1476	1377
	Niger	2550	808	201
	Zambezi	1650	514	565
	Shebeli	1550	77	11
Asia	Yangtze	3720	705	1137
	Yenisey	3650	1011	636
	Amur	3590	792	346
	Ob-Irtysh	3360	1154	360
	Yellow	3010	297	54
Australia	Murray	2330	408	25
	Murray upper	1090	103	5.9
	Barwon	980	87	5.2
	Murrumbidgee	980	37	4.2
	Lachlan	920	33	1.5
Europe	Volga	2290	525	1360
	Danube	1770	298	227
	Dnieper	1420	195	57
	Don	1160	163	30
	N. Dvina	1160	139	120
N. America	Mississippi	3740	1247	565[c]
	Mackenzie	2630	697	343
	Yukon	1980	328	240
	St. Lawrence	1900	503	460
	Rio Grande	1880	172	2.7
S. America	Amazon	4080	2375	6180
	Plata	3030	1197	809
	Madeira	1990	463	770
	Jurua	1860	93	141
	Purus	1860	154	—

[a]*Source:* Showers, *World Facts and Figures,* Copyright © 1979 by John Wiley and Sons, Inc. Reprinted with permission.
[b]Flow at mouth.
[c]Including 112,000 cfs diversion.

measurement of river velocity $U[L/T]$ and cross-sectional area $A[L^2]$ at a specific location can give the flow as

$$Q = UA \qquad (2.1)$$

where $Q[L^3/T]$ is the river flow at that location and time. River velocities are measured either directly by current meters or indirectly by tracking the time for objects in the water to travel a given distance. Since the velocity of a river varies with width and depth due to frictional effects, the mean vertical velocity must be estimated. Linsley et al. (1982) recommend the following procedure:

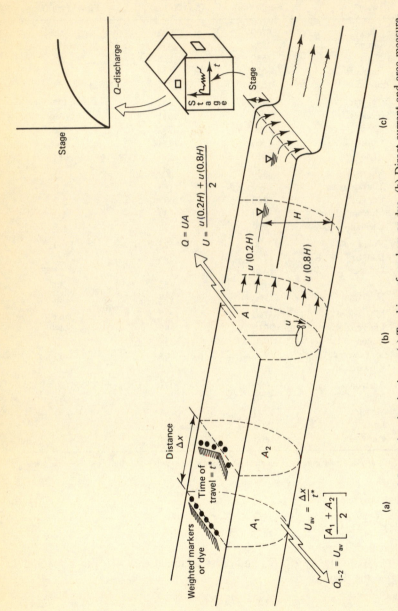

Figure 2.1 Methods of flow estimation in rivers. (a) Tracking of markers or dye. (b) Direct current and area measurement. (c) USGS gaging station and resulting stage-discharge relation.

32

1. Measure the total depth of the water (sounding or depth meter).
2. Raise the current meter to 0.8 of the total depth and measure velocity.
3. Raise the current meter to 0.2 of the total depth and measure velocity.
4. Average the velocity from the 0.8 and 0.2 depth measurements.

In shallow waters, a single velocity at 0.6 of the depth may be used. If velocity is in ft/sec and cross-sectional area in ft^2 then the flow is in ft^3/sec (cfs).

As shown in Fig. 2.1, the flow can also be obtained by estimation of the velocity over a reach of the river using weighted markers or a dye. The time of travel $t*$ between two locations on the river is measured from these tracers and since

$$U_{av} = \Delta x / t*$$

where Δx is the distance between the tracer measurement locations, the average velocity U_{av} can be obtained. With measurement of the cross-sectional areas A_1 and A_2 at the upstream and downstream stations, respectively, the average flow in the reach can be computed as

$$Q_{av} = U_{av}\left(\frac{A_1 + A_2}{2}\right)$$

The U.S. Geological Survey (USGS) has established a large number of flow gaging stations throughout the country. At each station a control section or dam is used to measure river stage (height) which, in turn, is correlated to the flow past the point. The flow is also often expressed as a discharge from a unit area of the drainage basin, that is, m^3/s-1000 km^2 (cfs/mi^2). Table 2.2 shows the average flow from the 21 water use regions in the United States shown in Fig. 2.2. The average flow ranges from 0.8 m^3/s-1000 km^2 (0.07 cfs/mi^2) in the Great Basin region to 19.1 m^3/s-1000 km^2 (1.75 cfs/mi^2) in the humid northeastern New England region. Overall, the aggregate flow of streams in the United States averages about 6.8 m^3/s-1000 km^2 (0.62 cfs/mi^2) (Piper, 1965). An example of an annual summary of daily flows produced by the USGS is contained in Table 2.3.

Within any year, of course, the flow will vary seasonally depending on location and degree of upstream regulation. Figure 2.3 shows some examples of this seasonal variation. For the Delaware River at Trenton, a spring peak in runoff results from snow melt and spring rains. A long almost exponential decrease in flow then occurs and minimum flows tend to be recorded in late summer or early fall. This is typical of largely unregulated river systems in more humid climates. For the Boise River in Idaho, the situation is markedly different and maximum values occur during the summer due to the high degree of regulation of this river for agricultural purposes.

2.1.1 Low Flow Frequency Analysis

Minimum flows are often of interest in water quality work, and questions immediately arise as to which minimum is most appropriate. A variety of minimum

TABLE 2.2 CHARACTERISTICS OF WATER USE REGIONS

Water use regions	Estimated sewered population[a] (millions)	Drainage area (1000 km²)[b]	% area in land use SMSA[c]	% area in land use Farms	% area in land use Forest	Average flow/area (m³/s·1000 km²)[d]
1. New England	8.3	153	8.2	10.9	79.2	19.1
2. Delaware–Hudson	28.7	93	36.8	22.1	41.1	17.3
3. Chesapeake	10.9	145	19.1	30.1	50.8	12.9
4. E. Great Lakes–St. Lawrence	13.0	109	32.4	29.9	37.7	12.5
5. S. Atlantic	7.5	427	11.0	35.7	53.3	10.2
6. E. Gulf	4.1	298	8.9	37.3	53.7	15.1
7. Tennessee–Cumberland	1.8	181	6.7	44.9	48.4	15.6
8. Ohio	8.8	345	11.6	53.2	35.1	12.9
9. W. Great Lakes	15.0	199	22.4	35.3	42.4	8.7
10. Hudson Bay	—	153	0	75.6	17.1	1.8
11. Upper Mississippi	4.3	492	4.7	70.5	23.7	5.8
12. Upper Missouri	2.1	1189	1.2	78.1	12.5	1.3
13. Lower Missouri	4.7	142	11.7	68.0	20.2	5.8
14. Lower Mississippi	2.8	155	7.9	46.0	46.1	13.4
15. Upper Arkansas	2.0	466	5.9	75.7	18.4	1.6
16. Lower Arkansas	0.8	298	2.8	58.5	38.7	11.6
17. W. Gulf	6.5	855	6.2	73.9	18.9	2.3
18. Colorado	1.9	632	10.8	47.1	26.1	1.2
19. Great Basin	0.7	508	0.6	24.9	27.4	0.8
20. Pacific NW	4.0	699	6.2	34.8	43.3	9.3
21. S. Pacific	22.1	290	44.3	25.2	30.4	13.1
Total or average	150.0	7829				6.8

[a] As of approx. 1970.

[b] 1000 mi² = 0.39 · (1000 km²).

[c] SMSA—Standard Metropolitan Statistical Area.

[d] cfs/mi² = 0.0914 · (m³/s·1000 km²).

Source: After NCWQ (1976).

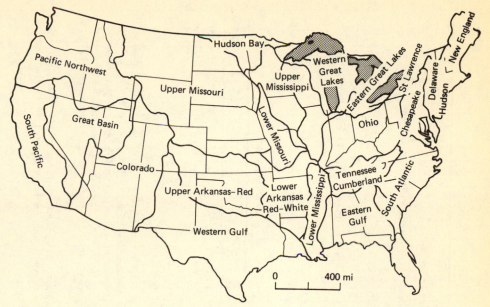

Figure 2.2 Water use regions of the United States.

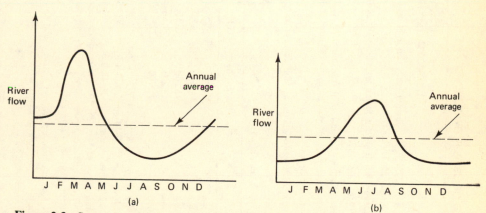

Figure 2.3 Seasonal variation of river flow. (a) Delaware River at Trenton, N. J. (b) Boise River, Twin Springs, ID.

TABLE 2.3 EXAMPLE OF USGS FLOW RECORD

Hudson River Basin

01358000 Hudson River at Green Island, N.Y.

Location: Lat 42°45′08″, long 73°41′22″, Albany County, on right bank at Green Island, just upstream from Troy lock and dam, 0.5 mile downstream from 5th branch Mohawk River.

Drainage Area: Approximately 8,090 mi^2 (including that above site of former auxiliary gage).

Period of Record: Feb. 1946–Sept. 1970.

Gage: Water-stage recorder. Datum of gage is 0.31 ft below mean sea level (U.S. Army Corps of Engineers benchmark). From July 1, 1946, to Mar. 12, 1962, auxiliary water-stage recorder on bypass channel at datum 10.59 ft higher.

Average Discharge: 24 yr, 12,520 cfs.

Extremes: Maximums and minimums (discharge in cubic feet per second and gage height in feet) for the water years 1966–1970 are contained in the following table:

| Wtr yr | Maximum | | | Minimum daily | |
	Date	Discharge	G.H.	Date	Discharge
1966	Mar. 25, 1966	50,700	20.25	July 4, 1966	1,800
1967	Apr. 4, 1967	56,900	20.59	Dec. 26, 1966	1,640
1968	Mar. 24, 1968	91,200	22.12	Sept. 2, 1968	882
1969	Apr. 23, 1969	82,200	21.86	Sept. 1, 1969	2,110
1970	Apr. 3, 1970	91,700	22.33	Oct. 20, 1969	1,990

Period of record: Maximum discharge, 181,000 cfs Dec. 31, 1948 (gage height: 27.05 ft from high-water mark in gage well), maximum daily, 141,000 cfs Dec. 31, 1948, Jan. 1, 1949; minimum daily, 882 cfs Sept. 2, 1968; minimum gage height; 13.92 ft, Sept. 2, 1946. Flood of Mar. 19, 1936, reached a stage of 29.48 ft at gage on opposite bank, from information supplied by U.S. Army Corps of Engineers (discharge, 215,000 cfs). Flood of Mar. 28, 1913, prior to construction of Sacandaga Reservoir and Troy lock and dam, reached a stage about 0.2 ft higher upstream from former dam near same site. Downstream from dams, flood in 1913 was about 3.3 ft higher than flood in 1936, from information by U.S. Army Corps of Engineers.

Remarks: Records good. Records include flow over spillway, and estimates of flow through lock and leakage through inoperative (since Nov. 30, 1960) powerplant which is located on right bank just downstream from gage. Water quality records for the water years 1966–1970 are published in reports of the U.S. Geological Survey.

Discharge, in Cubic Feet per Second, Water Year October 1965–September 1966

Day	Oct	Nov	Dec	Jan	Feb	Mar	Apr	May	Jun	Jul	Aug	Sep
1	5,040	3,430	14,300	12,100	7,250	10,600	16,000	16,000	9,050	6,260	2,200	4,420
2	11,000	5,900	12,000	13,000	6,900	16,300	14,400	17,500	9,000	5,570	3,810	4,240
3	8,990	6,570	11,500	11,600	6,600	21,200	15,300	15,000	6,880	3,300	3,960	3,500
4	5,540	6,840	11,900	12,300	7,140	20,700	13,900	14,800	6,640	1,800	3,520	4,730
5	5,320	6,270	12,300	11,900	5,990	22,100	11,800	13,200	5,100	2,690	3,940	8,630
6	6,650	6,310	12,600	11,000	6,850	39,000	10,800	13,200	4,860	4,060	2,940	8,730
7	4,750	5,140	15,000	11,100	4,030	33,300	11,000	14,800	5,700	4,450	2,150	7,230
8	6,170	4,240	11,800	11,500	6,070	26,700	12,000	14,600	7,040	4,190	2,040	7,310
9	8,630	6,630	13,400	8,660	6,090	21,900	12,500	17,500	8,840	3,510	3,570	6,570
10	9,030	9,580	13,600	6,830	7,230	19,000	11,500	18,900	10,400	2,980	3,720	5,440
11	7,770	10,400	12,300	8,780	8,850	18,400	10,100	18,200	16,400	2,850	3,980	3,260
12	8,190	9,210	8,880	8,440	16,500	18,200	8,010	15,000	11,800	4,570	4,810	2,700
13	8,650	8,980	9,240	7,670	21,500	16,800	6,940	23,400	8,980	4,400	4,160	4,510
14	8,890	7,430	10,000	7,160	27,100	14,800	11,500	25,300	9,590	4,460	3,360	4,560
15	8,180	5,970	10,900	7,570	31,300	15,600	9,250	21,400	9,650	3,430	2,380	4,980
16	8,470	9,560	10,700	7,130	23,500	16,000	9,700	18,500	11,000	3,200	3,940	5,080
17	7,640	21,800	11,000	4,090	18,800	15,800	11,900	17,800	12,300	2,160	3,950	3,960
18	5,930	22,600	10,100	6,660	15,400	16,400	12,400	17,800	11,600	1,850	4,450	2,780
19	6,890	16,700	8,790	7,060	13,100	22,600	12,400	21,400	9,820	3,990	4,540	3,020
20	7,500	13,300	6,500	8,000	9,740	27,800	14,200	36,100	7,380	4,400	3,310	4,780
21	7,230	12,400	7,400	7,680	8,190	28,800	15,000	35,900	8,250	5,600	2,760	4,870
22	6,580	10,800	7,570	7,580	9,790	24,800	18,300	29,500	8,220	4,680	3,140	8,380
23	6,420	12,200	7,960	6,960	10,700	23,100	21,700	25,600	7,430	3,450	6,660	9,720
24	9,160	11,700	7,480	4,660	10,600	28,100	24,800	21,400	7,670	2,150	11,700	8,260
25	5,840	10,600	6,940	7,370	10,600	45,400	35,300	19,200	6,460	2,050	8,450	5,990
26	7,760	8,760	12,200	7,390	10,100	38,700	30,800	14,400	4,480	3,790	6,780	5,290
27	7,820	10,800	12,300	6,900	8,110	31,200	26,500	13,500	3,960	3,580	5,730	7,170
28	7,250	21,800	10,100	5,990	7,560	24,600	22,300	13,000	6,410	3,990	3,290	6,470
29	7,000	18,000	10,400	5,470	—	20,900	21,000	12,100	6,630	4,390	2,740	6,400
30	7,000	16,500	10,700	5,050	—	18,800	17,500	8,130	6,550	3,510	4,590	5,930
31	4,950	—	10,400	4,430	—	18,200	—	7,460	—	2,580	4,660	—
Total:	226,240	320,420	330,260	252,030	325,590	715,800	468,800	570,590	248,090	113,890	131,230	168,910
Mean:	7,298	10,680	10,650	8,130	11,630	23,090	15,630	18,410	8,270	3,674	4,233	5,630
Max:	11,000	22,600	15,000	13,000	31,300	45,400	35,300	36,100	16,400	6,260	11,700	9,720
Min:	4,750	3,430	6,500	4,090	4,030	10,600	6,940	7,460	3,960	1,800	2,040	2,700

Cal yr 1965: Total: 2,828,860 Mean: 7,750 Max: 35,200 Min: 1,590
Wtr yr 1966: Total: 3,871,850 Mean: 10,610 Max: 45,400 Min: 1,800

flows can be considered; for example, minimum daily flow of record or minimum weekly or minimum monthly. Also, the frequency of occurrence of the flow must be included, that is, the minimum flow one might expect each year or every 5 or 10 years. Generally, for most design work, the minimum average 7-day flow expected to occur once every 10 yr is suggested as a design flow. The designation of these flows is given by aQb where a is the number of days used in the average and b is the interval in years over which the average is expected to occur. Thus, $7Q10$ is the minimum average 7-day flow once every 10 yr. However, several different types of such flows can be estimated from a hydrologic record. For example, the minimum average 7-day flow occurring at any time in a year can be estimated, or the minimum average flow in a given month or season can be computed.

In order to estimate a given flow sequence, the following procedure can be used.

Define T = recurrence interval in years of a given flow or flow sequence, p = probability of a flow occurring that is equal to or less than a stated flow, and N = number of years (or number of years of a given month) of record. If a minimum is to be computed for the entire year, then the record is divided into years. If a monthly flow minimum is to be computed then it is divided into the months of January, February, and so on. For each year (or month) the minimum average 7-day flow (or other sequence) is obtained, that is, the 7-day average is continually computed throughout the year (or month) and the minimum value is retained regardless of its time of occurrence during the year (or during the month). The resulting flows across all years (or across all of the individual months, e.g., across all the July flows) are then ranked according to magnitude from the smallest to the largest and the rank number in order of magnitude, m, is assigned to the flows. Then the cumulative probability of occurrence is given by

$$p = \frac{m}{N + 1} \tag{2.2}$$

and

$$T = 1/p \tag{2.3}$$

Table 2.4 shows an example of this calculation for the Schuylkill River at Philadelphia for 33 yr of record and for three flow sequences of 7-day, 30-day, and 120-day minimum average occurring at any time during the year. The flows can also be plotted on probability paper (usually log normal) from which the flows corresponding to different recurrence intervals can be estimated. The data in Table 2.4 are shown in Fig. 2.4 where a log normal plotting is used. (That is, the logarithm of the flows is plotted on the ordinate and the abscissa is scaled to a normal probability distribution). Table 2.5 shows the resulting estimates of the low flow recurrences. Therefore, the estimated average minimum 7-day flow for the Schuylkill River expected to occur once very 10 yr is 330 cfs.

The use of the minimum average flow for the entire year regardless of when it occurs may mean a rather stringent design flow since the minimum may occur in the winter during ice conditions. During that time of the year, the impact of any

TABLE 2.4 SCHUYLKILL RIVER AT PHILADELPHIA, PA.
Lowest Consecutive-Day Mean Discharge (cfs) (1932–1964) (Climatic Year)

Rank	Recurrence interval (yr)	Probability (% of time) ≤	Flow 7-day	30-day	120-day
1	34.00	2.9	292	324	434
2	17.00	5.9	300	332	495
3	11.33	8.8	314	346	516
4	8.50	11.8	336	357	548
5	6.80	14.7	349	452	600
6	5.67	17.6	380	459	651
7	4.86	20.6	389	460	683
8	4.25	23.5	407	503	709
9	3.78	26.5	434	512	777
10	3.40	29.4	438	517	777
11	3.09	32.4	461	537	855
12	2.83	35.3	473	544	873
13	2.62	38.2	495	569	920
14	2.43	41.2	502	570	1007
15	2.27	44.1	507	603	1040
16	2.13	46.9	507	611	1143
17	2.00	50.0	560	688	1149
18	1.89	52.9	577	726	1159
19	1.79	55.9	610	730	1183
20	1.70	58.8	615	751	1206
21	1.62	61.7	616	756	1373
22	1.54	64.9	623	770	1396
23	1.48	67.6	631	784	1442
24	1.42	70.4	672	788	1512
25	1.36	73.5	680	798	1537
26	1.31	76.3	682	827	1552
27	1.26	79.4	720	838	1662
28	1.21	82.6	744	1072	1756
29	1.17	85.5	760	1079	1844
30	1.13	88.5	835	1136	1994
31	1.10	90.9	860	1205	2394
32	1.06	94.3	909	1212	2407
33	1.03	97.1	1297	1571	3149

TABLE 2.5 LOW FLOW RECURRENCES, SCHUYLKILL RIVER AT
PHILADELPHIA, PA (1932–1964)

Recurrence interval (yr)	Probability of occurrence; % of time flow < shown	Mean annual low flow (cfs) 7-day	30-day	120-day
5	20	390	440	700
10	10	330	360	540
25	4	300	330	460
50	2	280	320	410

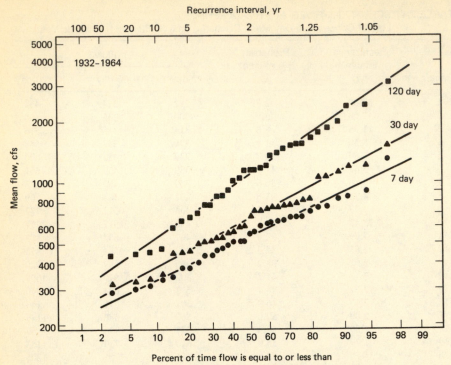

Figure 2.4 Low flow frequency curve—Schuylkill River at Philadelphia, Pa.

change in water quality may be minimal or nonexistent. Also, since minimum flows are much higher generally during various high precipitation months, the possibility exists of regulating waste discharges to a river as a function of the river flow. As a result, for some water quality management contexts, it is more desirable to analyze the flow record as noted above by months rather than across the entire year. Figure 2.5 from Reheis et al. (1982) shows the result of an analysis of flow record from the Appalachee River in Georgia where the minimum average 7-day flow by months is shown. It may be noted that the $7Q10$ from yearly records is about 50 cfs (1.4 m^3/s) but the $7Q10$ on a monthly basis is seen to vary up to a level over 300 cfs (8.5 m^3/s) in March.

2.1.2 Empirical Flow Estimation

Some attempts have been made to empirically estimate river flows of various durations directly from meteorological data, climate, and watershed characteristics. Thomas and Benson (1970) carried out an extensive statistical analysis of the relationship between river flows and watershed characteristics. Four regions were studied:

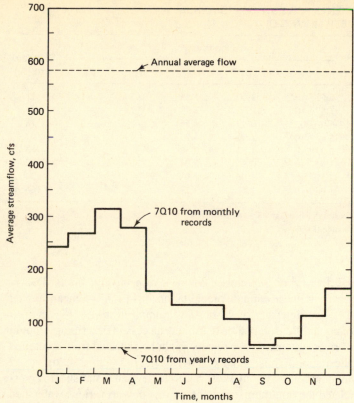

Figure 2.5 Monthly streamflow variation of Appalachee River near Buckhead, Ga. Reheis et al. (1982). Reprinted by permission of Water Pollution Control Federation.

1. Eastern—Potomac River basin.
2. Central—Missouri and Arkansas River tributary basins.
3. Southern—Louisiana Gulf coast drainage basins.
4. Western—California Central Valley (Sacramento and San Joaquin basins).

These areas were chosen to represent a variety of basin types. The results of this work indicate that the percent error in predicting drought flows (e.g. minimum average 7-day 10-yr flow) was large (>100%) and therefore such flows could only be very approximately estimated. Also, stream flow could be more accurately defined from statistical equations for the humid Eastern and Central regions than for the more arid Western and Southern Regions. Table 2.6 shows some of the results from this study for the following flow relations:

$$Q_{a,s} = a[(DA)^{b_1}](P)^{b_2}(s_n)^{b_3} \qquad \text{for Eastern, Central} \qquad (2.4)$$

$$Q_{a,s} = a[(DA)^{b_1}](P - 40)^{b_2} \qquad \text{for Southern} \qquad (2.5)$$

$$Q_{a,s} = a[(DA)^{b_1}](P)^{b_2}(I_{24,2})^{b_4} \qquad \text{for Western} \qquad (2.6)$$

TABLE 2.6 COEFFICIENTS FOR ESTIMATING ANNUAL AND
SEPTEMBER MEAN FLOW

Region	Flow index[a]	Coefficient[a]					Standard error(%)
		a	b_1	b_2	b_3	b_4	
Eastern	Q_a	$9.52 \cdot 10^{-4}$	1.00	1.59	0.3	—	11
	Q_s	$1.57 \cdot 10^{-5}$	1.06	2.90	-0.28	—	24
Central	Q_a	$6.0 \cdot 10^{-8}$	0.99	4.56	—	—	28
	Q_s	$2.45 \cdot 10^{-6}$	1.07	4.20	-1.10	—	41
Southern	Q_a	0.23	1.02	0.58	—	—	13
	Q_s	$1.52 \cdot 10^{-4}$	0.90	-2.94	—	—	62
Western	Q_a	$7.57 \cdot 10^{-4}$	1.06	1.92	—	—	34
	Q_s	$1.96 \cdot 10^{-9}$	1.50	3.29	—	1.97	165

[a]See text (Eqs. 2.4 to 2.6) for description of variables and coefficients.
Source: After Thomas and Benson (1970).

where $Q_{a,s}$ = annual flow (a) or September flow (s), respectively, in cfs; DA = drainage area in mi^2; P = mean annual precipitation in in.; s_n = mean annual snowfall in in.; and $I_{24,2}$ = intensity of 24-hr, 2-yr rainfall in in.

Although Thomas and Benson had only limited success in describing drought flows for a variety of river systems, better results might be expected from a detailed analysis of a specific region. Thus, Chang and Boyer (1977) in a study of 12 tributaries of the Monongahela River in West Virginia devised a statistical regression equation that related the minimum 7-day, 10-yr low flow to various watershed parameters. The equation provided a high degree of predictability ($r^2 = 0.999$) but as the authors point out, it should be considered strictly valid only for streams in mountainous humid regions (average elevation of the river was about 2000 ft and mean 7-day, 10 yr maximum temperature was about 88°F). This expression is

$$Q_{7,10} = 1.024 \exp[-69.0846 + 0.0388(CO) - 0.0410(CL) \\ - 3.6743(WF) + 1.8015(LA) - 0.0021(EL)] \qquad (2.7)$$

where $Q_{7,10}$ is the minimum 7-day, 10-yr flow in cfs; CO is the watershed perimeter in mi, CL is the main channel length in mi, WF is a watershed "form" factor and is equal to drainage area/(maximum watershed length)^2 (dimensionless), LA is the latitude in deg and EL is the mean elevation in ft.

It may be noted that the $7Q10$ is in general use as a design flow for water quality analyses, although there are exceptions. The choice of a minimum 7-day, 10-yr flow is, of course, somewhat arbitrary, and selection of the appropriate design flow is a subject of continuing studies (USEPA, 1984). It is also important to note that a 1 in 10 yr flow does not mean that only one of these flows will definitely occur in the next 10-yr period. Indeed, several such flows may occur in a 10-yr period or, conversely, a 10-yr period may pass without the minimum flow occurring at all. The estimated flow and recurrence interval represent an average over many 10-yr periods.

2.1.3 Morphometry–Hydraulic Geometry

The form or shape of a river, its width and depth, its "movement" as it meanders and twists and turns, collectively make up the morphometry of a river. The characteristics of interest include width, depth, slope, and the length and width of river meander (the "waves" in the course of a river). Other factors include the grain size distribution of the river bed and its floodplain. Blench (1972) provides a broad review of some of these river factors. Figure 2.6 from Blench shows the remarkable consistency in the shape of river systems as evidenced by the relationship between meander wavelength and the width of the channel.

Relationships between the flow, velocity, and depth or area of a river are desirable so that the hydraulic characteristics can essentially be specified in terms of a single variable. Leopold and Maddock (1953) have examined various rivers and developed empirical relationships between flow, velocity, depth, and width. These relationships take the form of power functions with flow as the independent variable. Thus,

$$H = aQ^b \qquad (2.8)$$
$$U = cQ^d \qquad (2.9)$$
$$B = eQ^f \qquad (2.10)$$

where H is stream depth, U is stream velocity, and B is stream width. Log–log plots of field data permit the determination of the constants and exponents in each

Figure 2.6 Meander length vs. channel width. From Blench (1972). Reprinted by permission of Academic Press, Inc.

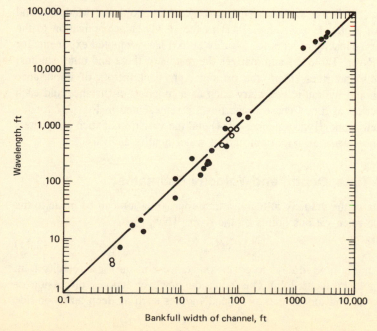

TABLE 2.7   AVERAGE VALUES OF EXPONENTS IN HYDRAULIC
GEOMETRY RELATIONSHIPS

Depth = aQ^b,   Velocity = cQ^d,   Width = eQ^f

River or river basin	Exponent		
	b	d	f
Great Plains and Southwest	0.40	0.34	0.26
Tennessee Valley	0.48	0.46	—
Scioto River, Susquehanna basin, PA	0.30	0.70	—
Willamette River, Eugene to Oregon City, OR	0.61	—	—
Potomac River (MP340-116)	0.40	0.47	0.13
Black River, NY Watertown (MP11-16)	0.1	0.7	0.20
(MP54-65)	0.4	0.4	0.20
Delaware River, below Easton, PA (MP135-185)	0.5	0.4	0.10

of the equations. The constants will, of course, vary with the size of the river basin. The average value of the exponents, however, tends to fall within a range that is relatively stable. There is a statistical distribution of the exponents which reflects the changes occurring in a river over the whole range of flows from drought to flood. If, however, for pollution control studies, the results are analyzed during critical low flow conditions, the range of interest in the values of the exponents is narrowed.

O'Connor (1962) has summarized the work of Leopold and Maddock and others (Churchill et al., 1962; Kehr, 1941) in the steady-state application to the waste assimilation capacity of streams. Worley (1963) has computed exponents for the Willamette River. Table 2.7 summarizes the results of these and other studies.

For streams and rivers then, the principal physical factors of importance include the flow and hydraulic geometry each as a function of distance and each for a specific period of time. These properties are represented in Fig. 2.7 for the Black River, New York (Hydroscience, 1974) and the variation of such parameters with river distance forms an important part of water quality analysis.

2.1.4 Travel Time, Depth, and Velocity Estimates

With an estimate of the velocity at hand, a first approximation can be made to the time of travel between various points on the river. Thus,

$$t^* = x/U \qquad (2.11)$$

where t^* is the travel time in days to cover a distance x in mi for a velocity U in mpd. (See Sample Problem 2.1.) This equation however ignores dispersion or mixing in the river and any effects of "dead" zones such as deep holes or side channel coves.

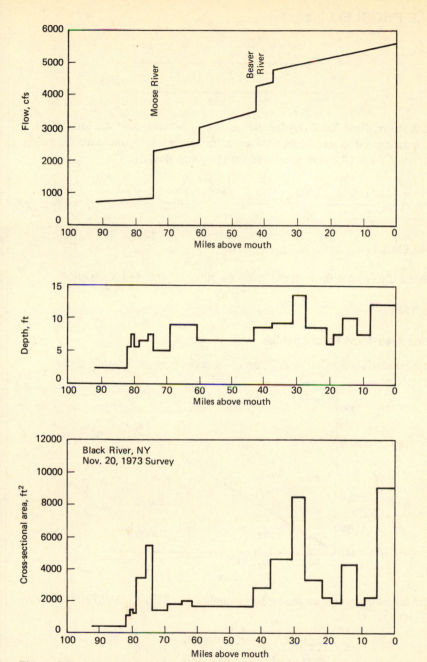

Figure 2.7   Flow and morphometric factors—depth and cross-sectional area—for Black River, NY. From Hydroscience (1974).

SAMPLE PROBLEM 2.1

DATA

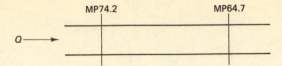

The Black River, New York (Hydroscience, 1974) between MP74.2 and MP64.7 is to be characterized as a constant flow–constant area reach. Assume the following cross-sectional areas (A) were measured for the given flows:

Q (cfs)	500	750	1300	2200	3400
A (ft^2)	680	950	1100	1600	2200

PROBLEM

Estimate the travel time through the reach for flows of 600 and 3000 cfs.

ANALYSIS

Determine Area–Flow Relationships

Assume a power function, $A = aQ^b$, and plot data above on log–log paper.

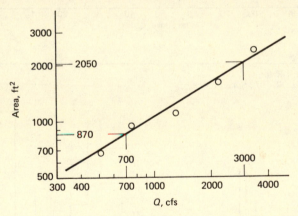

Using the following points on the best-fit straight line (2050 ft^2, 3000 cfs) and (870 ft^2, 700 cfs):

$$b = \frac{\ln(A_2/A_1)}{\ln(Q_2/Q_1)} = \frac{\ln(2050/870)}{\ln(3000/700)} = 0.589$$

$$a = \frac{A_1}{Q_1^b} = \frac{870}{700^{0.589}} = 18.3$$

and $\underline{A = 18.3\ Q^{0.589}}$ [A(ft^2), Q(cfs)]

Sample Problem 2.1 (continued)

Calculate Travel Time

For the selected flows, determine cross-sectional areas from the area–flow relationship and calculate velocities and travel times as tabulated below.

Q (cfs)	A (ft²)	U (fps)	U (mpd)	t* (days)
600	790	0.759	12.4	0.77
3000	2040	1.47	24.1	0.39

where $U = \dfrac{Q}{A}$

$U\text{(mpd)} = 16.4U\text{(fps)}$

$t^* = (74.2 - 64.7 = 9.5 \text{ mi})/U\text{(mpd)}$

Note: Although the flow increased fivefold between 600 and 3000 cfs, the velocity in the river only increased by a factor of 2 due to the increase in cross-sectional area.

For some small streams it may be very difficult to obtain a good estimate of the average depth of the water because of alternating pools and riffles. Shallow areas, bars, and lateral variation in depth may further complicate the estimation of the average depth in a reach of the stream. The flow, however, may be known from a downstream gaging station. Further, the width of most streams does not vary nearly as much as the depth. Thus, a stream may have highly variable depth, but the flow and average width can be obtained. Situations such as this can be addressed by the following procedure:

1. Using a dye or other tracer, conduct a time of travel study to estimate the average time of passage through a given reach.
2. From the time of travel study, calculate the average velocity in the reach, that is,

$$\bar{U} = x/t^*$$

3. From the flow estimate for the reach and the estimated average velocity, compute the effective average cross section for the reach, that is,

$$\bar{A} = Q/\bar{U}$$

4. With the average cross-sectional area and the average width, $\bar{B}$, estimate the average depth $\bar{H}$ as

$$\bar{H} = \bar{A}/\bar{B}$$

Estimates of stream velocity can also be obtained from the empirical open channel equation developed by Chezy in 1775 and Manning in 1890, commonly referred to as the Manning equation (Chow, 1959). This equation relates average stream velocity directly to the channel bed slope and hydraulic radius and inversely to a channel roughness coefficient:

$$U = \frac{1.49}{n} R_h^{2/3} S_0^{1/2} \tag{2.12}$$

where U = average velocity in fps
   n = roughness coefficient
   R_h = hydraulic radius in ft
   $= A \div P$
   A = cross-sectional area in ft^2
   P = wetted perimeter in a cross section, the length of contact of the water with the solid boundary in ft
   S_0 = slope of the channel bed in ft/ft

This equation is limited to small slopes ($<10°$), steady flow, and reasonably long, straight reaches where the bed slope is constant.

Values of the roughness coefficient vary considerably. For natural streams values of n range from 0.025 to 0.033 for smoothest beds, 0.045 to 0.060 for roughest beds, and from 0.075 to 0.150 for very weedy streams. Photographs of constructed and natural channels, with accompanying n values may be found in the text by Chow (1959).

As will be discussed below, water quality analyses are often conducted over a reach of a river or stream. Thus point estimates of velocity may not be suitable. Also, stream conditions may change markedly at the low flows at which water quality analyses are made. Empirical estimates of the average velocity in a reach accounting for pools, riffles, dead zones, and so on have thus been made. For example Burke (1983) has presented empirical equations based on observed data for reach velocities for different soil types in Georgia. The equations are of the form

$$U = \alpha Q^a S_0^b (DA)^c L^d \tag{2.13}$$

where U is the reach velocity (flow-through velocity) in ft/s; Q is flow in cfs; S_0 is slope in ft/mi; DA is the drainage area in mi^2; L is reach length in mi; and α, a, b, c, and d are empirical coefficients. Table 2.8 shows the coefficients as reported by Burke (1983).

2.1.5 Effect of Land Use on River Flow

Human activities can significantly alter river flows through:

1. Construction of regulating structures such as dams and reservoirs.
2. Increased agricultural uses including drainage of low lying areas or diversions through agricultural canals.
3. Changes in forest area through either increased deforestation or reforestation.
4. Increased urban land use.

TABLE 2.8 COEFFICIENTS FOR ESTIMATING REACH
VELOCITIES IN GEORGIA FOR STREAM FLOW < 100 cfs

	Equation[a] coefficients for				
Soil type	Coefficient α	Flow a	Slope b	Drainage area c	Reach length d
Atlantic Coast flatwoods and S. coastal plain province	0.181	0.190	0.000	0.000	0.219
Sand Hills province (the fall line)	0.051	0.112	0.695	0.000	0.000
S. Piedmont province	0.137	0.522	0.264	−0.333	0.348
Sand Mountain, S. Appalachian ridges and valleys, Blue Ridge provinces	0.270	0.662	0.000	−0.437	0.000

[a]$U = \alpha Q^a S_0^b (DA)^c (L)^d$ for U = fps, Q = cfs, S_0 = ft/mi; DA = mi^2, L = mi.
Source: From Burke (1983). Reprinted by permission of American Water Resources Association.

TABLE 2.9 EFFECT OF URBANIZATION ON RUNOFF OF EAST
MEADOW BROOK, LONG ISLAND

	Ratio of runoff to rainfall		
Rainfall (in.)	Preurban period (1937–1943)	Urban period (1964–1966)	Urban period/ preurban period
0.5	0.06	0.07	1.1
1.0	0.06	0.09	1.5
2.0	0.06	0.13	2.2
3.0	0.06	0.17	2.8
4.0	0.06	0.19	3.2
5.0	0.06	0.22	3.7

Source: After Seaborn (1969).

For the latter influence, Seaborn (1969) detailed the changes in runoff in a small stream on Long Island due to increased urban development. Specifically, increased flow resulted from an increase in storm sewers discharging directly to the stream. Of particular interest is the fact that the increase in runoff increased with rainfall volume as shown in Table 2.9.

2.2 DISCHARGE OF RESIDUAL MATERIAL INTO RIVERS

With the flow and hydraulic properties of the river system defined and the estimates of these properties at hand from the above, we can turn now to some first approaches to describing the discharge of residual substances into rivers and streams. Such residuals may include discharges from waste treatment plants, from combined sewer overflows, or from agricultural and urban runoff. We shall consider first the

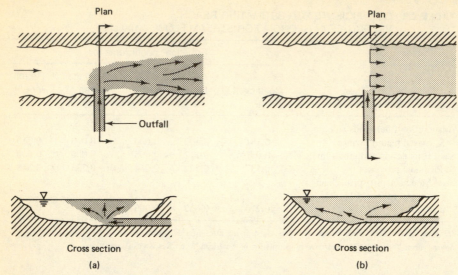

Figure 2.8   An important assumption at the outfall. (a) Typical actual situation. (b) ''Well-mixed'' assumption, laterally and vertically.

point sources, that is, those sources that enter the river from a fixed discharge point such as an effluent pipe or tributary stream.

2.2.1 Assumptions

The basic idea in describing the discharge of material into a river is to write a mass balance equation for various reaches of the river. We will begin by examining the mass balance right at the point of discharge and the first key assumption we will make is that the river is homogeneous with respect to water quality variables across the river (laterally) and with depth (vertically). This important assumption is illustrated in Fig. 2.8, where (a) indicates the actual situation from an onshore discharge. A plume often develops which gradually spreads across the river and gradually extends over the entire depth. If, of course, the discharge is from ports along the length of the pipe and across the river, then the discharge will tend to be more uniform in the cross section. We will therefore make this assumption, as shown in Fig. 2.8(b).

Consideration of how to compute the distance from an outfall to complete mixing is a separate more complicated topic. However, the order of magnitude of the distance from a single point source to the zone of complete mixing is obtained from

$$L_m = 2.6U\frac{B^2}{H} \tag{2.14}$$

for a side bank discharge, and from

$$L_m = 1.3U\frac{B^2}{H} \tag{2.15}$$

for a midstream discharge (Yotsukura, 1968). In the above,

L_m = distance from the source to the zone where the discharge has been well mixed laterally in ft

U = average stream velocity in fps

B = average stream width in ft

H = average stream depth in ft

Sample Problem 2.2 shows the calculation of the length to complete mixing for a side bank discharge.

The second key assumption to be made in this analysis of water quality in streams is that there is no mixing of water in the longitudinal downstream direction. Each element of water and its associated quality flows downstream in a unique and discrete fashion. There is no mixing of one parcel with another due to dispersion or velocity gradients. This condition is referred to as an advective system, a plug flow system, or maximum gradient system. A pulse discharge retains its identity and any spreading of the pulse is assumed negligible. In actual streams and rivers, however, true plug flow is never really reached. Lateral and vertical velocity gradients, ''dead'' zones in the river (coves, deep holes, backwater regions), produce some mixing and retardation of the material in a stream. For many purposes, however, especially for the steady state analyses discussed herein, the nondispersive assumption is a good one. In some situations, especially non-steady-state problems, even a small amount of dispersion can be important. This is discussed more fully in Section 2.3.2.

2.2.2 Mass Balance At Discharge Point

For the mass balance at the outfall, the notation of Fig. 2.9 provides the starting point. The principal statement for the mass balance is:

Mass rate of substance upstream + mass rate added by outfall
= mass rate of substance immediately downstream from outfall

Figure 2.9 Notation for mass balance at outfall.

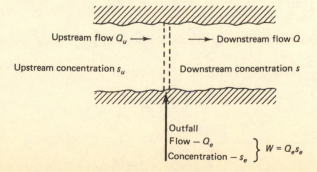

Upstream flow Q_u ⟶   ⟶ Downstream flow Q

Upstream concentration s_u Downstream concentration s

Outfall
Flow — Q_e
Concentration — s_e } $W = Q_e s_e$

SAMPLE PROBLEM 2.2

DATA

Plan

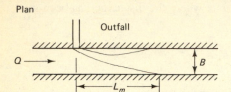

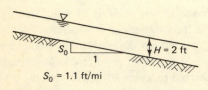

Cross section

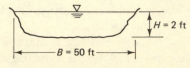

$S_0 = 1.1$ ft/mi

PROBLEM

For the clean, straight natural stream, estimate the flow and approximate the distance to complete mixing for the side-bank discharge.

ANALYSIS

Estimate Stream Velocity

Use Manning equation (2.12).

$$A = 50 \text{ ft} \times 2 \text{ ft} = 100 \text{ ft}^2$$
$$P = 2 \text{ ft} + 50 \text{ ft} + 2 \text{ ft} = 54 \text{ ft}$$
$$R_h = \frac{A}{P} = \frac{100}{54} = 1.85 \text{ ft}$$
$$S_0 = 1.1 \text{ ft/mi} = 1.1 \text{ ft}/5280 \text{ ft} = 0.000208 \text{ ft/ft}$$

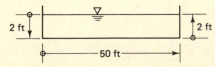

Select $n = 0.025$ appropriate for straight, clean natural channel.

$$U = \frac{1.49}{0.025} (1.85)^{2/3}(0.000208)^{1/2} = 1.30 \text{ fps}$$
$$\therefore Q = AU = 100 \text{ ft}^2 \times 1.30 \text{ fps} = \underline{\underline{130 \text{ cfs}}}$$

Sample Problem 2.2 (continued)

Use USGS Estimate (Eq. 2.14) for L_m

$$L_m = \frac{2.6UB^2}{H} = 2.6 \times 1.30 \text{ fps} \times \frac{(50 \text{ ft})^2}{2 \text{ ft}}$$

$$L_m = 4230 \text{ ft} = \underline{0.80 \text{ mi}}$$

Note: This distance is approximately 80 stream widths downstream of the outfall. The scale of a mile or more for complete mixing is typical. Use of a midstream discharge will approximately halve the distance; further reductions are possible through use of a diffuser at the outfall.

assuming complete mixing. Recalling that the mass rate is the product of flow and concentration, the mass balance is therefore given by

$$Q_u s_u + Q_e s_e = Qs \qquad (2.16)$$

or

$$Q_u s_u + W = Qs$$
$$[L^3/T][M/L^3] = [L^3/T][M/L^3]$$
$$[M/T] = [M/T]$$

where $W = Q_e s_e$ the input waste load.

A similar statement can also be made for the balance of the flows, that is, flow continuity:

$$Q_u + Q_e = Q \qquad (2.17)$$

The upstream conditions of flow and concentration (Q_u and s_u) are often known or can be measured, and typically some information is available on the effluent conditions (Q_e and s_e). For example, the discharge may be a proposed industrial treatment plant and an estimate is available of the flow and concentration of the waste substance to be expected upon completion of the plant. Interest then centers on estimating the concentration of the substance in the river at the outfall after mixing of the effluent with the upstream concentration. Therefore, we often want to solve Eq. (2.16) for the downstream concentration s or

$$s = \frac{Q_u s_u + Q_e s_e}{Q} = \frac{Q_u s_u + W}{Q}$$

$$[M/L^3] = \frac{[L^3/T][M/L^3]}{[L^3/T]} = \frac{[M/T]}{[L^3/T]} \qquad (2.18)$$

where Q is given by Eq. 2.17. The downstream concentration is thus dependent on the upstream and downstream flows and the concentrations of the upstream and effluent inputs. If the upstream concentration of the substance, s_u is zero, then

$$s = \left(\frac{Q_e}{Q}\right) s_e = \frac{W}{Q} \tag{2.19}$$

which shows that the downstream concentration is the effluent concentration reduced by the ratio of influent flow to total river flow. This is a dilution effect. If the river flow is increased, a given mass discharge W will result in a decreased concentration in the river due to the diluting effect of the increased river flow. Equations 2.18 and 2.19 contain a considerable amount of information regarding water quality engineering controls, including reduction of W and/or upstream concentration and increasing the river flow. Each of these changes will result in a reduction in the concentration of the substance (e.g., a toxic substance, bacteria, or BOD) at the outfall. But as we will see shortly, although the concentration at the outfall may be reduced due to increased river flow, the situation becomes more complex as one proceeds downstream from the point of discharge.

The mass balance also applies if the input to a river is a tributary. The flow Q_e and concentration s_e then becomes the tributary flow and concentration and Eq. 2.18 can be used to estimate the concentration of the substance after the intersection of the tributary and the main stem of the river. The assumption of complete mixing sometimes comes into question here since tributaries rarely result in immediate local mixing.

Figure 2.10 illustrates this difficulty. The Illinois River flows roughly south from the Chicago area and arrives at the Mississippi River north of St. Louis. About 89 miles from its intersection with the Mississippi, the Sangamon River joins the Illinois as shown in Fig. 2.10(a). The dissolved oxygen of the Illinois at mile 89.3 was about 2 mg/l during the 1960s across its entire width. Figure 2.10(b) shows the DO at mile 88.6 after the entrance of the Sangamon River and indicates the higher oxygenated waters of that river tend to remain on the right-hand shore (looking upstream) while the low DO waters of the Illinois River tend to remain

Figure 2.10 Example of incomplete lateral mixing of dissolved oxygen in Illinois River. (a) Location map. (b) Results of dissolved oxygen sampling. From Mills et al. (1966).

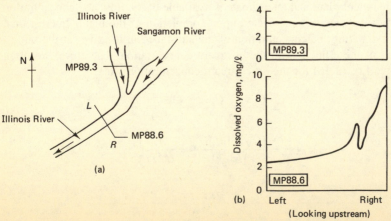

on the left side. Some mixing is beginning and other data indicate that mixing is almost complete across the stream by about mile 88 or about $1\frac{1}{2}$ mi downstream from the confluence of the Sangamon and Illinois Rivers. If one is studying the entire length of the Illinois (272 mi) then this is equivalent to complete mixing at the intersection. However, if one is studying the local situation over a scale of 5 mi in the vicinity of the Sangamon, then the assumption of complete mixing is a poor one.

2.2.3 Water Quality Downstream of Point Source

2.2.3.1 Conservative Substances   Attention will now be given to the downstream behavior of water quality. Consider first a substance that is conserved, that is, there are no losses due to chemical reactions or biochemical degradations. Such substances may include, for example, total dissolved solids, chlorides, and certain metals during times of the year when transport is in the dissolved form. The assumption of complete mixing will be retained and applied to any new entering point sources and tributaries. In order to simplify the analysis, it is also assumed that there is no change of flow between any point waste inputs or tributaries; flow into or out of the river due to groundwater effects is therefore excluded. Finally, it is assumed that the magnitudes of the waste inputs and flows are temporally invariant—the so-called steady state condition (see Section 2.5).

Because the substance is conservative, there is no change in concentration between tributaries or waste inputs. The concentration changes only at the entrances of new sources of the substance with the associated changes in flow. Therefore, Eq. 2.18 is used to compute the concentration at each tributary or waste input and the concentration remains at that level until the next tributary or waste source. The calculation is illustrated in Fig. 2.11. The schematic shown in Fig. 2.11(a) indicates

Figure 2.11   (a) Schematic of a river system. (b) Variation of conservative substance with distance.

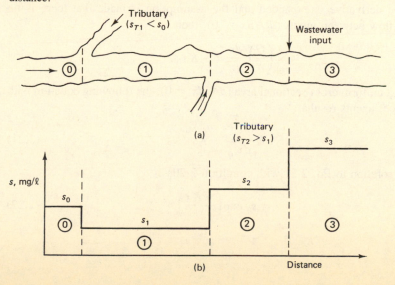

a portion of a river upstream with two tributaries and a waste input. The river is divided into three reaches and the upstream reach above the first tributary is considered as the upstream boundary reach. At the first tributary, Eq. 2.18 is applied and in this case the concentration is reduced due to the dilution effect of the tributary, where the tributary is assumed to have a negligible concentration of the substance in question. The concentration remains at the s_1 level until the next tributary where Eq. 2.18 is again applied at the entrance of the second tributary which has a concentration greater than s_1. A new level s_2 is computed and held constant (no change is possible because of the conservation assumption) until the waste input, where once again the outfall mass balance equation is applied. In this way, the complete downstream profile can easily be generated (see Sample Problem 2.3). If however, it is known or suspected that the substance is not conservative, then an additional consideration must be included.

2.2.3.2 Nonconservative Substances Assume then that the substance decays with time due perhaps to chemical reactions, bacterial degradation, radioactive decay, or perhaps settling of particulates out of the water column. Many substances exhibit decay or nonconservative behavior including oxidizable organic matter, nutrients, volatile chemicals, and bacteria. A very useful assumption is that the substance decays according to a first-order reaction, that is, the rate of loss of the substance is proportional to the concentration at any time. Then, the mass balance equation, at steady state, is a first-order linear differential equation (O'Connor, 1967),

$$\frac{1}{A}\frac{d(Qs)}{dx} = -Ks \qquad (2.20)$$

where K is the decay rate $[T^{-1}]$ of the substance.

The boundary condition is

$$s = s_0 \quad \text{at} \quad x = 0 \qquad (2.20a)$$

where s_0 is calculated from Eq. 2.18.

If the derivative is expanded and the assumption is made that there is no change of flow between inputs $(dQ/dx = 0)$, then

$$\left(\frac{Q}{A}\right)\frac{ds}{dx} = -Ks$$

Assuming a uniform cross-sectional area $(dA/dx = 0)$, the following equation with constant coefficients results:

$$U\frac{ds}{dx} = -Ks \qquad (2.21)$$

The solution to Eq. 2.21 with condition 2.20a is

$$s = s_0 \exp\left(\frac{-Kx}{U}\right) \qquad (2.22)$$

SAMPLE PROBLEM 2.3

DATA

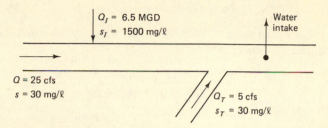

Upstream flow with a background level of chlorides, a conservative substance, of 30 mg/l is supplemented by an industrial discharge of 6.5 MGD carrying 1500 mg/l chlorides and a downstream tributary of 5 cfs with background chlorides concentration. Assume downstream tributary chlorides concentration does not vary with flow.

PROBLEM

To maintain a desired chlorides concentration of 250 mg/l at the water intake, determine: (a) the required industrial reduction or (b) the required increase in tributary flow (Q_T).

ANALYSIS

Present Conditions

$$W(\text{upstream}) = 25 \times 30 \times 5.4 \quad = 4100 \text{ lb/day}$$
$$W(\text{industrial}) = 6.5 \times 1500 \times 8.34 = 81{,}300 \text{ lb/day}$$
$$W(\text{tributary}) = 5 \times 30 \times 5.4 \quad = 800 \text{ lb/day}$$

(a) With no decay mechanisms, the allowable intake concentration is:

$$s = \frac{\Sigma W}{\Sigma Q} = 250 \text{ mg/l}$$

$$\frac{[4100 + \overline{W}(\text{industrial}) + 800]\text{lb/day}}{[25 + (6.5 \times 1.55 = 10) + 5 = 40 \text{ cfs}] \times 5.4} = 250 \text{ mg/l}$$

Allowable industrial $W = \overline{W}(\text{industrial}) = 49{,}100$ lb/day

$$\underline{\text{Percent reduction}} = \frac{81{,}300 - 49{,}100 = 32{,}200}{81{,}300} \times 100 = \underline{\underline{40\%}}$$

(continued)

Sample Problem 2.3 (continued)

This may be accomplished by decreasing the effluent chloride concentration to

$$1500 \times (1 - 0.40) = 900 \text{ mg/l}$$

(b) Assuming no industrial reduction, the tributary flow must be increased to dilute the upstream concentration.

$$s = \frac{\Sigma W}{\Sigma Q} = \frac{4100 + 81,300 + Q_T \times 30 \times 5.4}{(25 + 10 + Q_T) \times 5.4} = 250$$

$$Q_T = 32 \text{ cfs}$$

Thus, an additional release of 27 cfs into the tributary is required.

A summary of the above is shown below. Note that, at any location, $s = \Sigma W/\Sigma Q$, the summations extending to the point in question.

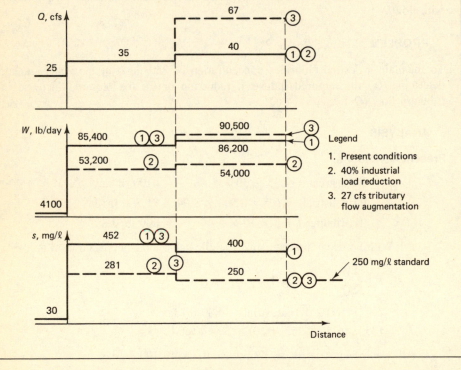

Legend

1. Present conditions
2. 40% industrial load reduction
3. 27 cfs tributary flow augmentation

or since $x/U = t^*$, the time to travel a distance x at velocity U,

$$s = s_0 \exp(-Kt^*) \tag{2.23}$$

These equations show that for a nonconservative substance decaying at a rate K, the downstream distribution of the substance will drop exponentially and asymptotically approach zero. This is in contrast to the conservative substance which changes in concentration only at points of entrance of new load and flows.

Figure 2.12(a) shows this distribution and 2.12(b) shows that a plot of ln s vs. x or t^* is a straight line, that is, from Eq. 2.23,

$$\ln s = \ln s_0 - Kt^* \tag{2.24}$$

The slope of the semilogarithmic plot is therefore equal to K. This affords the opportunity to estimate the instream loss rate K from measurements of the substance s at various downstream locations. Plotting the natural logarithm of the sampling results against the time of travel permits estimation of the rate K from the slope of the line fitted to the observed data (see Sample Problem 2.4).

Figure 2.12   Decay of nonconservative substances. (a) Cartesian coordinates. (b) Logarithmic form.

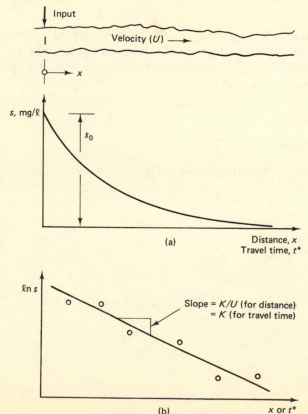

SAMPLE PROBLEM 2.4

DATA

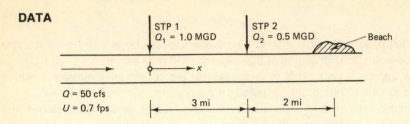

Prior to the second downstream sewage treatment plant (STP) coming on-line, during which time septic systems were used, a bacteriological survey for total coliforms was done in the river with results as follows:

x (mi)	8	16	24	32
s (MPN/100 ml)	46,500	16,700	9,000	2,800

(MPN = *m*ost *p*robable *n*umber of bacteria as determined by a multiple tube technique)

PROBLEM

If total coliform counts in the two STP effluents are 3×10^6 MPN/100 ml without disinfection, what percent kill must be achieved by disinfection to meet a beach total coliform standard of 2000 MPN/100 ml? Assume equal reductions at both plants and neglect STP flows in the flow balance.

ANALYSIS

Determine Total Coliform Rate

Assume same for both effluents; plot original river survey data on semilog paper.

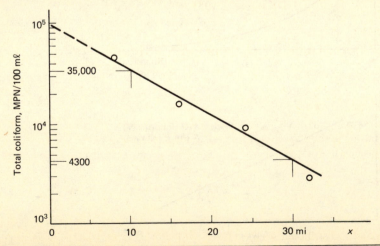

Sample Problem 2.4 (continued)

From two points on the best-fit straight line (35,000, 10 mi) and (4300, 30 mi):

$$K = \frac{-U \ln(s_2/s_1)}{x_2 - x_1} = \frac{-(0.7 \times 16.4)\text{mpd } \ln(4300/35{,}000)}{(30 - 10)\text{mi}} = \underline{1.2/\text{day}}$$

On the semilog plot note that $s_0 = 95{,}000$ MPN/100 ml, a value consistent with

$$s_{01} = \frac{W_1}{Q} = \frac{(1.0 \times 1.55)\text{cfs} \times 3 \times 10 \text{ MPN}/100\,\text{ml}}{50 \text{ cfs}} = 93{,}000 \text{ MPN}/100 \text{ ml}$$

Calculate Beach Total Coliform Count

With

$$s_{02} = \frac{W_2}{Q} = \frac{0.5W_1}{Q} = 0.5s_{01} = 46{,}500 \text{ MPN}/100 \text{ ml}$$

and

$$U = 0.7 \text{ fps} \times 16.4 = 11.5 \text{ mpd}$$

$$s = s_{01}e^{-Kx/U} + s_{02}e^{-K(x-3)/U}, \qquad x > 3$$

At the beach, $x = 5$ mi.

$$s_B = 93{,}000e^{-1.2 \times 5/11.5} + 46{,}500e^{-1.2 \times (5-3)/11.5}$$

$$s_B = 55{,}300 + 37{,}800 = 93{,}100 \text{ MPN}/100 \text{ ml} > 2000$$

$$\therefore \text{ effluent reduction} = \left(\frac{93{,}100 - 2000}{93{,}100}\right) \times 100 = 97.9\% \text{ and disinfected}$$

effluent counts $= 3 \times 10^6(1 - 0.979) = 63{,}000$ MPN/100 ml.

With the disinfection in operation, the beach total coliform count would be

$$s_B = 55{,}300 \times (1 - 0.979) + 37{,}800 \times (1 - 0.979)$$

$$s_B = 1160 + 795 = 1955 \text{ MPN}/100 \text{ ml} (\approx 2000)$$

2.2.4 Water Quality Response to Distributed Sources

Streams are often subjected to sources or sinks of a substance which are distributed along the length of the stream. An example of an external source is runoff from agricultural areas whereas oxygen demanding material distributed over the bottom of the stream exemplifies an instream, or internal, source (Fig. 2.13). Given a constant magnitude of the distributed source which originates at $x = 0$, and given a stream with reasonably constant parameters (flow, area, depth) over a given length, the mass balance equation at steady state is

$$U\frac{ds}{dx} + Ks = S_D \qquad (2.25)$$

where S_D = distributed source [$M/L^3 \cdot T$]. For external sources w [$M/L \cdot T$]

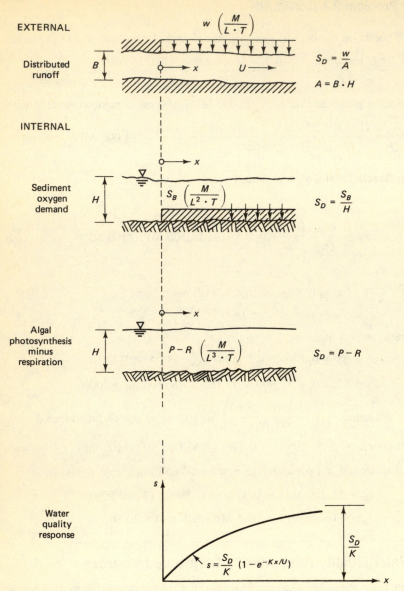

Figure 2.13 Distributed sources—types and water quality response.

$$S_D = \frac{w \cdot dx}{V = A \cdot dx} = \frac{w}{A} \qquad (2.26)$$

and for sediment oxygen demands $S_B \ [M/L^2 \cdot T]$

$$S_D = \frac{S_B \cdot B \cdot dx}{V = B \cdot H \cdot dx} = \frac{S_B}{H} \qquad (2.27)$$

Solution of the linear first-order differential equation results in

$$s = \frac{S_D}{K}(1 - e^{-Kx/U}) = \frac{S_D}{K}(1 - e^{-Kt^*}) \qquad (2.28)$$

where a boundary condition of $s = 0$ has been assumed at $x = 0$. This solution results in a zero concentration at $x = 0$ and an asymptotic approach to a value of S_D/K as x increases (Fig. 2.13). At $Kt^* = 3$, $s = 0.95(S_D/K)$ and at $Kt^* = 4.6$, $s = 0.99(S_D/K)$.

2.2.5 Effect of Spatial Flow Variations on Water Quality

Distributed flow changes in a stream can alter the spatial distributions of conservative and nonconservative parameters. If flow increases in a stream due to uncontaminated groundwater infiltration, instream concentrations will decrease due to the additional dilution. For contaminated groundwater, the instream concentration will increase—if the groundwater concentration is higher than the instream value—or decrease—if the groundwater is of lesser concentration than that instream. Decreases in flow without removal of the contaminant, such as caused by evaporation, will result in increased instream concentrations.

The general differential equation for steady state concentrations in a stream (Thomann, 1972) is

$$O - \frac{Q}{A}\frac{ds}{dx} - (v + K)s + vs_I \pm S_D \qquad (2.29)$$

where Q, A, s, v, K, s_e, and S_D are spatially variable and temporally constant and

$$v = \frac{1}{A}\frac{dQ}{dx}$$

s_I = concentration in the infiltrating flow

S_D = distributed sources or sinks $[M/L^3 \cdot T]$

The general solution of this steady state equation is given by Li (1962) as

$$s = [s_0 + I(x)] \exp[-\Phi_1(x)] \qquad (2.30)$$

where

$$\Phi_1(x) = \int_0^x \frac{K + v}{U} dx \qquad (2.30a)$$

and

$$I(x) = \int_0^x \frac{vS_I \pm S_D}{U} \exp[\Phi_1(x)] dx \qquad (2.30b)$$

Additional insight into the flow effect is obtained by specifying a spatially constant area A and decay rate K, no distributed source ($S_D = 0$), and an uncontaminated infiltrating flow ($s_I = 0$). Furthermore, assuming an exponentially increasing flow, where,

$$Q = Q_0 \exp(qx) \qquad (2.31)$$

it follows that

$$v = \frac{1}{A}\frac{dQ}{dx} = \frac{1}{A}[qQ_0 \exp(qx)]$$

or

$$v = qU_0 \exp(qx) \tag{2.32}$$

where

$$U_0 = \frac{Q_0}{A}$$

From Eqs. 2.30a and 2.30b

$$\Phi_1(x) = \int_0^x \frac{K + qU_0 \exp(qx)}{[Q_0 \exp(qx)]/A}\, dx$$

or

$$\Phi_1(x) = \frac{K}{U_0}\left[\frac{\exp(-qx) - 1}{-q}\right] + qx \tag{2.33a}$$

and

$$I(x) = 0 \tag{2.33b}$$

Substitution of Eqs. 2.33a and 2.33b into Eq. 2.30 yields

$$s = s_0 \exp\left\{\frac{K}{qU_0}[\exp(-qx) - 1] - qx\right\} \tag{2.34}$$

where s_0 is an initial point source concentration at $x = 0$ calculated from

$$s_0 = \frac{W}{Q_0} \tag{2.34a}$$

The impact of an uncontaminated infiltrating flow on the instream concentrations due to a point source is shown graphically in Fig. 2.14 for conservative and reactive substances. For the conservative substance, neglect of the inflow might lead an analyst to mistakenly conclude that some type of removal mechanism was operating in the stream when in fact dilution was the cause of decreasing concentrations. Similarly, for nonconservative substances, appropriate care in estimating q is required so as not to overestimate the reaction rate (see Sample Problem 2.5).

2.2.6 Multiple Sources—Principle of Superposition

Differential equations which govern the concentration s along a stream x are said to be linear. In a linear equation, constants or variables which multiply the concentration and its derivatives are not functions of the concentration. Products of s and its derivatives—as well as exponents of s, ds/dx, and so on—do not appear.

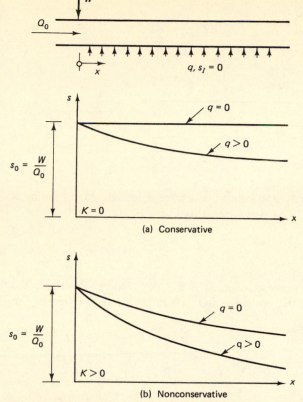

Figure 2.14   Effect of uncontaminated flow increases on stream concentrations. (a) Conservative substances. (b) Nonconservative substances.

It may be noted that the differential equations for the concentration in a stream due to point sources (Eq. 2.21) and distributed sources (Eq. 2.25) are linear.

For such linear systems, the concentration in a river or stream due to multiple point and/or distributed sources is the linear summation of the responses due to the individual sources plus the response due to any upstream boundary condition. Note that the upstream boundary condition is treated as an additional source since for all other point and distributed sources a boundary condition of zero is assumed. As an example, for a stream with constant flow and velocity subjected to a point source W at $x = 0$ as well as an infinitely long distributed source originating at $x = 0$, the water quality response would be the sum of Eqs. 2.23 and 2.28 as follows:

$$s = s_0 \, e^{-Kt^*} + \frac{S_D}{K} \, (1 - e^{-Kt^*}) \tag{2.35}$$

where $s_0 = W/Q$ and the upstream boundary concentration is assumed equal to zero. Note also that the decay rates of the material discharged by the point source

SAMPLE PROBLEM 2.5

DATA

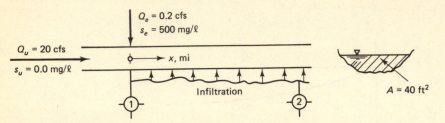

A flow gage immediately upstream of a discharge records 20 cfs during a low flow period and a gage 20 mi downstream indicates 28 cfs.

PROBLEM

If the decay rate of the substance discharged is 0.2/day, base e, determine the concentration distribution if the upstream and infiltrating flows are free of the substance.

ANALYSIS

Hydraulic Parameters

Assuming an exponential increase in flow,

$$Q = Q_0 \exp(qx) \qquad \text{[Eq. (2.31)]}$$

Letting (2) be downstream ($x = 20$ mi) and (1) upstream ($x = 0$), Eq. 2.31 can be rearranged as

$$q = \frac{\ln(Q_2/Q_1)}{x_2 - x_1} = \frac{\ln(28/20.2)}{20 - 0} = 0.0163/\text{mi}$$

The initial stream velocity at $x = 0$ is

$$U_0 = \frac{Q_0}{A} = \frac{20.2 \text{ cfs}}{40 \text{ ft}^2} = 0.505 \text{ fps} = 8.28 \text{ mpd}$$

Sample Problem 2.5 (continued)

Concentration Profiles

$$s_0 = \frac{W}{Q_0} = \frac{0.2 \text{ cfs} \times 500 \text{ mg/l}}{20.2 \text{ cfs}} = 4.95 \text{ mg/l} \qquad [\text{Eq. } (2.34\text{a})]$$

(a) Including both decay and dilution:

$$s = s_0 \exp\left\{ \frac{K}{qU_0} \left[\exp(-qx) - 1 \right] - qx \right\} \qquad [\text{Eq. } (2.34)]$$

$$s = 4.95 \exp\left\{ \frac{0.2/\text{day}}{0.0163/\text{mi} \times 8.28 \, (\text{mi/day})} \right.$$

$$\left. \times \left[\exp(-0.0163x) - 1 \right] - 0.0163x \right\}$$

$$s = 4.95 \exp\{1.48[\exp(-0.0163x) - 1] - 0.0163x\}$$

(b) If dilution were not included, a predicted profile would be given by Eq. 2.22 as

$$s = s_0 \exp\left(-\frac{Kx}{U} \right) = 4.95 \exp\left(-\frac{0.2x}{8.28} \right)$$

(c) If instream data were used to extract a decay rate, and the diluting flow was not recognized, then a value of $K = 0.36/\text{day}$ may be wrongly used.

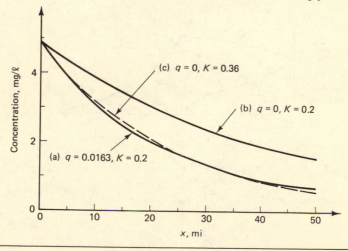

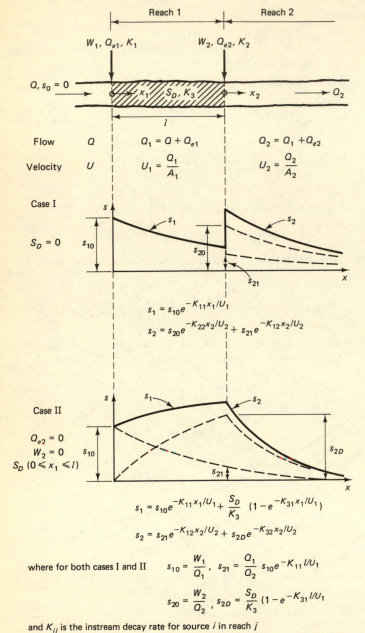

where for both cases I and II $\quad s_{10} = \dfrac{W_1}{Q_1}$, $\quad s_{21} = \dfrac{Q_1}{Q_2}\, s_{10} e^{-K_{11} l/U_1}$

$$s_{20} = \dfrac{W_2}{Q_2}, \quad s_{2D} = \dfrac{S_D}{K_3}\left(1 - e^{-K_{31} l/U_1}\right)$$

and K_{ij} is the instream decay rate for source i in reach j

Figure 2.15 Examples of water quality responses to multiple sources.

and the distributed source are assumed equal in the stream, that is, K(point) $=$ K(distributed) $= K$.

For streams with multiple sources in which flows and velocities vary, but are constant for a given length (reach), solutions are presented in Fig. 2.15 for two point sources (Case I) and a point source plus a distributed source (Case II). Dashed lines represent the responses due to individual sources and solid lines depict the total concentration. The decay rates of all sources are assumed to be different and the upstream boundary is assumed equal to zero. In Case I, note the decrease in the response due to W_1 at $x = l$ due to the effect of dilution ($Q_2 > Q_1$). In Case II, when the distributed source terminates at $x = l$, the response exponentially decays downstream according to Eq. 2.23 with a decay rate appropriate for that source (K_3).

2.3 TIME VARIABLE ANALYSIS

For some problem contexts, it is important to be able to describe the time variable behavior of water quality in a river downstream of an outfall or tributary input. Such problems include describing the downstream transport of a peak in a waste-water discharge load, an accidental spill of a chemical, or the day-to-day variation in water quality due to day-to-day changes in waste load inputs.

2.3.1 Nondispersive Streams

The basic principle of the time variable response in a river or stream can be quickly seen by making the initial assumption that there is no mixing in the longitudinal direction. It can be recalled that this assumption has already been made in the steady state analysis of the earlier Section 2.2.

If there is no mixing of the water parcels that are moving downstream, then each parcel does not interact with the parcel in front of it or behind it. Each parcel of water retains its identity and hence as noted earlier this type of condition is called plug flow. Figure 2.16(a) illustrates the process for a discharge pulse at time $t = 0$. The "slice" of water passing by the discharge pipe (assumed to mix completely in the cross section, as before) receives the pulse discharge and the resulting concentration in the slice is simply the mass of the discharge (i.e., the pulse discharge rate W times the time interval of the pulse Δt) divided by the volume of water in the slice over the interval Δt.

At some time interval later, say t_1^*, the slice of the river that contains the discharged pulse is now a distance x_1 downstream where $x_1 = t_1^* \cdot U$. If a person were therefore stationed at location x_1, the pulse would be seen for the period Δt only. There would be an equal concentration in the slice of water as it passed by and there would be sharp edges to the pulse in the river. If the decay were to occur in the slice of river water, then the concentration in the slice would be reduced proportional to the travel time t^*, as illustrated in Fig. 2.16(b).

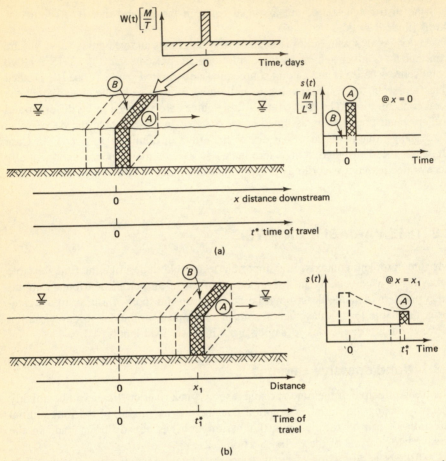

Figure 2.16 Schematic of water quality response in a nondispersive, advective stream due to a "pulse" input. (a) At time $t = 0$. (b) At time $t = t_1$.

The basic differential equation with constant coefficients in space and time that describes the concentration $s(x,t)$ in this advective nondispersive stream is

$$\frac{\partial s}{\partial t} = -U \frac{\partial s}{\partial x} - Ks \qquad (2.36)$$

for boundary condition $s = s_0(t)$ at $x = 0$

$$s_0(t) = \frac{W(t)}{Q}$$

where

for zero upstream conditions. In this simplified situation only the waste load is assumed to be varying in time. The solution to this equation is

$$s(x,t) = \frac{W(t - t^*)}{Q} \exp\left(\frac{-Kx}{U}\right) \qquad (2.37)$$

where $t*$ is the time of travel to location x. This latter equation is the mathematical expression for the translation of any variation in $W(t)$ in the downstream direction at a velocity U. Thus $W(t - t*)$ means that the time varying input at the outfall is translated downstream a distance equal to $U \cdot t*$. The concentration is then given by the dilution due to the flow and the loss of mass due to decay.

2.3.2 Time Variable Analysis: Effect of Dispersion

The preceding discussion has assumed that, although there is complete mixing from side to side and top to bottom in the river, there is no mixing as we proceed down the length of the river. In any real river, however, there is some mixing that occurs along the length of the river due principally to the horizontal and vertical gradients of velocity. In addition, coves, embayments, and river channel changes and twists add further to this mixing. The phenomenon is called longitudinal dispersion. Fischer et al. (1979) provide a full and excellent treatment of the mixing processes in rivers.

If a water quality analysis is to be conducted over a "long" distance and a short time interval of discharge (as, for example, a spill of a substance), then longitudinal dispersion must be considered. The mass balance equation for constant cross-sectional area, river flow, and dispersion and no other inputs of s except at the outfall is

$$\frac{\partial s}{\partial t} = -U \frac{\partial s}{\partial x} + E_x \frac{\partial^2 s}{\partial x^2} - Ks \qquad (2.38)$$

where E_x is the longitudinal dispersion coefficient $[L^2/T]$.

2.3.2.1 Instantaneous Input
If an instantaneous spill of material occurs, of mass M, then the solution to Eq. 2.38 is

$$s(x,t) = \frac{M}{2A\sqrt{\pi E_x t}} \exp\left[\frac{-(x - Ut)^2}{4E_x t} - Kt \right] \qquad (2.39)$$

This response may be recognized as the mathematical form of the normal probability density function with a mean of Ut and a variance, $\sigma^2 = 4Et$. Therefore, if x assumes the role of the independent variable and t is fixed, a plot of $s(x,t_i)$ as a function of x is symmetrical and bell-shaped around the peak concentration. The spread of the curve is a measure of the dispersion phenomenon incorporated in the coefficient, E_x. Figure 2.17(a) shows the travel of this response function for variable x and different times. Note that the spread increases as time increases and that the overall peak decays. Since the "spread" equivalent to the standard deviation in the normal probability density is $\sigma = 2\sqrt{E_x t}$, low values of E_x (small mixing) approach plug flow while larger values of E_x result in increased spreading due to increased mixing. For fixed x, however, and variable time, the impulsive response is skewed. Figure 2.17(b) shows the nonsymmetrical distribution when x is fixed and time is variable. These results can be compared to the stream response functions for zero dispersion shown in Fig. 2.16 where the impulse input retains its shape

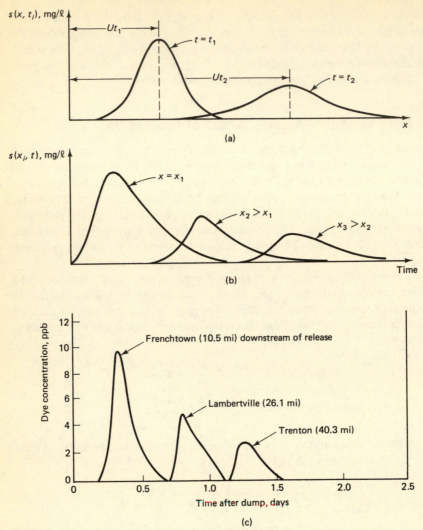

Figure 2.17   Dispersive stream quality response to pulse input. (a) As a function of distance at times t_1 and t_2 (note symmetry). (b) Response over time at distance x_1 and x_2 (note asymmetry). (c) Response of dye dump as measured in Delaware River. From Hydroscience (1975).

as it progresses downstream. In contrast, stream dispersion spreads the response temporally and spatially, as shown in Figs. 2.17(b) and (c).

Equation 2.39 has proved useful for estimating the dispersion coefficient from field studies as discussed below. The impulsive response function can then be used to estimate the transient effects due to inputs such as BOD from storm water overflow and batch discharges of municipal and industrial wastes.

2.3.2.2 Continuous Input over an Interval of Time   For some studies, it is necessary to calculate the response in the stream as a function of distance and time where the input "jumps" from zero to a fixed constant and steady value for a period of time and then drops back to zero. This is called a "rectangular input" and is representative of the input that results from the discharge of dye into a stream for measuring the reaeration capacity of a stream. Rathbun (1979), using the work of O'Loughlin and Bowmer (1975) and Rose (1977), discusses the water quality solution for this type of input for use in dye and reaeration studies.

For constant coefficients, the general solution of Eq. 2.38 with a steady input is given as

$$s(x,t) = \frac{s_0}{2} \exp\left(-\frac{Kx}{U}\right)\left[\operatorname{erf}\frac{x - U(t - \tau)(1 + \eta)}{\sqrt{4E_x(t - \tau)}} - \operatorname{erf}\frac{x - Ut(1 + \eta)}{\sqrt{4E_x t}}\right] \quad (2.40)$$

where s_0 is the concentration of the substance after mixing of the input over the cross section, τ is the time interval of the input, η is a dimensionless number given by

$$\eta = \frac{KE_x}{U^2} \quad (2.41)$$

and erf (ϕ) is the error function of the argument ϕ given by

$$\operatorname{erf} \phi = \frac{2}{\sqrt{\pi}} \int_0^{\phi} e^{-\xi^2} \, d\xi$$

A tabulation of the erf ϕ is given in Table 2.10. Note that $\operatorname{erf}(-\phi) = -\operatorname{erf}(\phi)$.

TABLE 2.10 VALUES OF THE ERROR FUNCTION FOR USE IN EQ. 2.40

ϕ	erf ϕ	ϕ	erf ϕ
0	0	1.0	0.842701
0.05	0.056372	1.1	0.880205
0.1	0.112463	1.2	0.910314
0.15	0.167996	1.3	0.934008
0.2	0.222703	1.4	0.952285
0.25	0.276326	1.5	0.966105
0.3	0.328627	1.6	0.976348
0.35	0.379382	1.7	0.983790
0.4	0.428392	1.8	0.989091
0.45	0.475482	1.9	0.992790
0.5	0.520500	2.0	0.995322
0.55	0.563323	2.1	0.997021
0.6	0.603856	2.2	0.998137
0.65	0.642029	2.3	0.998857
0.7	0.677801	2.4	0.999311
0.75	0.711156	2.5	0.999593
0.8	0.742101	2.6	0.999764
0.85	0.770668	2.7	0.999866
0.9	0.796908	2.8	0.999925
0.95	0.820891	2.9	0.999959
		3.0	0.999978

Source: Carslaw and Jaeger (1959). Reprinted by permission of Oxford University Press.

If the time of the input, τ, is small relative to the travel time (a condition usually met in dye studies or relatively short spills), then O'Loughlin and Bowner (1975) and Rose (1977) have shown that the time of travel of the peak concentration at a distance x is

$$t_p = \frac{x + U\tau(1 + \eta)}{U(1 + \eta)} \tag{2.42}$$

and the peak concentration (s_p) at x and t_p is then

$$s_p = \frac{s_0}{2} \exp\left(-\frac{Kx}{U}\right)\left[\operatorname{erf}\frac{x - U(t_p - \tau)(1 + \eta)}{\sqrt{4E_x(t_p - \tau)}} - \operatorname{erf}\frac{x - Ut_p(1 + \eta)}{\sqrt{4E_xt_p}}\right] \tag{2.43}$$

Figure 2.18 shows the form of the solution.

Figure 2.18   Dispersive stream quality response to (a) a square wave input load; (b) time response at a distance x_1 downstream; and (c) time response at a further distance x_2 downstream.

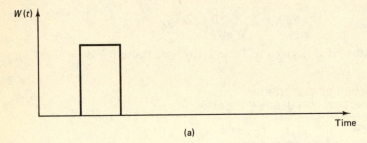

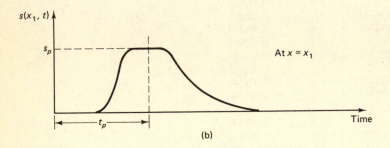

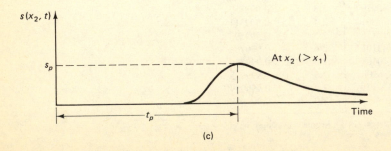

Thomann (1973) has presented a mathematical treatment of the conditions under which longitudinal dispersion is significant for varying waste load inputs (not pulse discharges). The significant dimensionless number that describes the relative effect of the longitudinal mixing effect for a nonconservative substance is given by the quantity η in Eq. 2.41.

For upland feeder streams, $\eta \approx < 0.01$; for main drainage rivers, $\eta \approx 0.01 - 0.5$, and for large rivers, $\eta \approx 0.5 - 1.0$. It is shown that for the upland feeder streams, the effect of longitudinal dispersion is generally not significant when waste sources are varying with periods of oscillation of 1–7 days and longer. As we proceed however to the main drainage rivers, that is, η increasing, then the effect of longitudinal dispersion becomes significant and generally cannot be neglected in time variable studies.

2.3.2.3 Estimation of River Dispersion Coefficient   The longitudinal dispersion coefficient for rivers and streams has been a matter of considerable investigation. A variety of theoretical and empirical relationships has been proposed. In addition one can release dye into the river, measure the resulting downstream dye distribution curves, and estimate the dispersion coefficient.

Fischer et al. (1979) suggest the following for an estimate of the dispersion coefficient in real streams,

$$E_x = 3.4 \cdot 10^{-5} \frac{U^2 B^2}{HU^*} \tag{2.44}$$

where E_x is in mi^2/day (smpd), U is the mean river velocity in fps, B is mean width in ft, H is mean depth in ft, U^* is the river shear velocity in fps given by

$$U^* = \sqrt{gHS}$$

for $g = 32$ ft/sec^2 and S as the river slope in ft/ft. Fischer et al. (1979) note that Eq. 2.44 is approximate only and does not explicitly account for "dead zones" in the river.

McQuivey and Keefer (1974) evaluated 18 streams and 40 time of travel studies for rivers with flows ranging from 35 to 33,000 cfs and slopes from 0.00015 to 0.0098 ft/ft. They proposed the following for conditions where the Froude number ($F = U/\sqrt{gH}$) is less than 0.5:

$$E_x = 1.8 \cdot 10^{-4} \frac{Q}{S_0 B} \tag{2.45}$$

for E_x in mi^2/day, Q as the steady state base flow in cfs, S_0 the bed slope in ft/ft, and B the mean width in ft.

If a dye study is conducted, then the dispersion coefficient can be estimated from the concentration–time data obtained at an upstream station and a downstream station. Fischer (1968) describes the "method of moments" that can be used to estimate the dispersion in the reach. Thus,

$$E_x = \frac{U^2}{2} \frac{\sigma_{td}^2 - \sigma_{tu}^2}{\bar{t}_d - \bar{t}_u} \tag{2.46}$$

where σ^2 is the variance of the concentration–time curve (extrapolated to 1% of the peak dye concentration), $\bar{t}$ is the time of travel to the centroid of the curve, and the subscripts d and u refer to downstream and upstream, respectively. The variances are computed from

$$\sigma^2 = \frac{\int_0^{t_{0.01}} st^2 \, dt}{\int_0^{t_{0.01}} s \, dt} - (\bar{t})^2 \tag{2.47}$$

and the centroid times are computed from

$$\bar{t} = \frac{\int_0^{t_{0.01}} st \, dt}{\int_0^{t_{0.01}} s \, dt} \tag{2.48}$$

where $t_{0.01}$ is the time at which the concentration has decreased to 1% of the peak concentration. Finally, a simple procedure is given by evaluating Eq. 2.48 for the peak concentration, that is, at $x = Ut$. Then for the approximate instantaneous discharge of a dye, the peak concentration is given by

$$s_p = \frac{M}{2A\sqrt{\pi E_x t_p}} \tag{2.49}$$

for $t_p = x/U$.

A plot, therefore, of the observed peak concentrations versus $1/\sqrt{t_p}$ provides an overall estimate of E_x (see Sample Problem 2.6).

2.3.3 Planning of Dye Studies

The design and successful execution of a dye study requires some advance planning. Dye studies are often conducted to obtain an estimate of the time of travel through a given reach of the river. The planning requires some a priori estimate of stream hydrology; flow and velocity. The USGS (1970) has provided some guidelines for estimating the volume of dye required, the approximate time to the peak concentration, and the duration of the dye cloud. The USGS (1970) suggests the following for the volume of dye:

$$V = 3.4 \cdot 10^{-4} \left(\frac{Q_m x}{U} \right)^{0.93} s_p \tag{2.50}$$

where V is the volume of Rhodamine WT-20% dye in liters required to produce a peak concentration of s_p in μg/l at a distance downstream x in mi from the injection point to the sampling point, and Q_m and U are the maximum discharge in cfs and average velocity expected in the reach. The weight of the dye mixture is

$$W_{\text{dye}}(\text{lb}) = 2.62V(\text{liters}) \tag{2.51}$$

The time to the peak concentration is estimated from a first estimate of the time of travel

$$t_p = 1.47 \frac{x}{U}$$

for x in mi, U in fps, and t_p in hr. Figure 2.19 from the USGS (1970) then provides the expected duration of the dye cloud and suggested sampling intervals.

SAMPLE PROBLEM 2.6

DATA

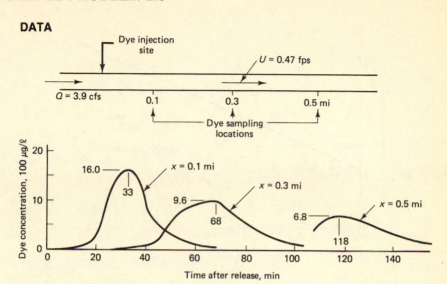

A mass of 1.21 lb of dye is rapidly introduced into the stream and the dye is monitored by a flourometer at $x = 0.1$, 0.3, and 0.5 mi downstream. Assume decay is negligible.

PROBLEM

Estimate the longitudinal dispersion coefficient in the stream.

ANALYSIS

The peak concentrations will be used to estimate E_x, where

$$s_p = \frac{M}{2A\sqrt{\pi E_x t_p}} = \text{slope} \cdot \frac{1}{\sqrt{t_p}} \qquad \text{[Eq. (2.49)]}$$

and

$$\text{Slope} = \frac{M}{2A\sqrt{\pi E_x}}$$

Quantity	x (mi)		
	0.1	0.3	0.5
s_p (100 µg/l)	16	9.6	6.8
t_p (min)	33	68	118
$1/\sqrt{t_p}$ (min$^{-0.5}$)	0.174	0.121	0.0

(continued)

Sample Problem 2.6 (continued)

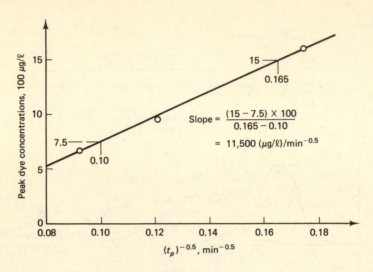

From the plot,

Slope $= 11,500 \ (\mu g/l)/min^{-0.5} \times (60 \ sec/min)^{0.5} = 89,100 \ (\mu g/l)/sec^{-0.5}$

with $A = Q/U = 3.9/0.47 = 8.3 \ ft^2$,

$$\frac{M}{2A\sqrt{\pi}\sqrt{E_x}} = \frac{1.21 \ lb \times (10^9 \ \mu g/l)/(lb/lb)}{2 \times 8.3 \ ft^2 \times \sqrt{\pi} \times \sqrt{E_x} \ ft^2/sec \times 62.4 \ lb/ft^3} = \frac{657,000}{\sqrt{E_x}} = 89,100$$

$$\underline{\underline{E_x = (657,000/89,100)^2 = 54 \ ft^2/sec}}$$

If an estimate of the dispersion in the equations given previously (Eqs. 2.44 and 2.45) is made then Eq. 2.49 for the concentration can also be used to estimate the required mass of dye to obtain a given peak concentration s_p at a time t_p.

2.4 ENGINEERING CONTROLS

There are several points at which the water quality in a system such as shown in Fig. 2.12 can be controlled. Equations 2.22 and 2.18 provide the "control point" to achieve a desired water quality objective. In the determination of a specific water quality standard which may be mandated by state requirements, the value of s in Eq. 2.22 is specified. The standard may be spatially variable and may allow for a mixing zone. In order to achieve a given value of s, the value of s_0 the initial concentration can be controlled. Reducing s_0 results in a parallel reduction in concentration throughout the stream as shown in Fig. 2.20(a).

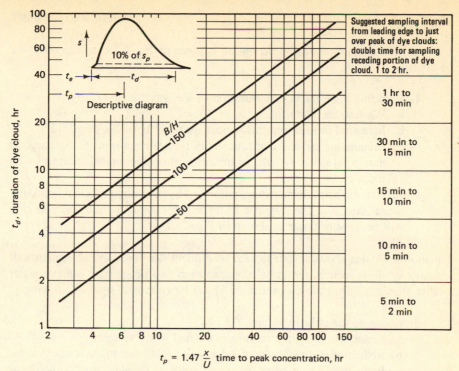

Suggested sampling interval
from leading edge to just
over peak of dye clouds:
double time for sampling
receding portion of dye
cloud. 1 to 2 hr.

1 hr to
30 min

30 min to
15 min

15 min to
10 min

10 min to
5 min

5 min to
2 min

$t_p = 1.47 \dfrac{x}{U}$ time to peak concentration, hr

Figure 2.19 Duration of dye cloud as a function of travel time to peak, and average channel width–depth ratio (B/H). Relationship is approximate and should be used only for planning initial sampling schedules and is subject to field revision. From USGS (1970).

Figure 2.20 Water quality responses to engineering contol. (a) Point source reduction. (b) Upstream flow increase. (c) Increased decay rate.

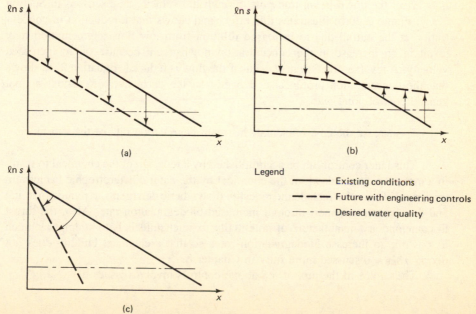

Legend

—————— Existing conditions

– – – – – Future with engineering controls

— · — · — Desired water quality

79

The initial concentration at the outfall is given by Eq. 2.18 and can therefore be controlled by

1. Reducing the effluent concentration of the waste input (s_e) by:
   a. Wastewater treatment.
   b. Industrial in-plant process control and/or "housekeeping."
   c. Eliminating effluent constituents by pretreatment prior to discharge to municipal sewer systems or by different product manufacturing for an industry.
2. Reducing the upstream concentration (s_u) by upstream point and non-point source controls. The response to these two controls on the concentration will be directly linear as shown in Fig. 2.20(a).

Equation 2.18 also shows that reduction of effluent flow and/or augmentation of stream river flow reduces the initial concentration and hence may achieve water quality standards. Thus the concentration s_0 can be controlled by the following.

3. Reducing the effluent volume (Q_e) by:
   a. Reduction in infiltration into municipal sewer systems.
   b. Reduction of direct industrial discharge volumes into the sewer system.
   c. Reduction, for industry, of waste volumes through process modifications.
4. Increasing the upstream flow (Q_u) by low flow augmentation, that is releases from upstream reservoir storage or from diversions from nearby bodies of water.

However, these latter two controls on effluent flow and upstream river flow also may affect the downstream transport through the velocity as discussed in Section 2.1.3. Figure 2.20(b) illustrates this effect and shows that a reduction in concentration at the outfall due to increased dilution from low flow augmentation may result in an increase in the concentration downstream because of an increased velocity. A reverse situation may occur if the flow is reduced. Equation 2.22 shows that the concentration profile also depends on the decay rate, K. Thus, a final general control point is to:

5. Increase the environmental, instream degradation rate of the substance.

This latter control can be accomplished by a redesign of the chemical to result in a more rapid breakdown of the chemical by the natural heterotrophic bacteria in the stream. Examples include the redesign of synthetic detergents to reduce foaming and downstream transport through increased biodegradation rate. Also, the thrust in contemporary manufacture of potentially toxic chemicals is to attempt as much as possible to increase biodegradation rates so that a chemical buildup does not occur. This is discussed more fully in Chapter 8.

The choice of the mix of the above controls involves issues of:

1. The costs of the controls, both locally, regionally, and nationally.
2. The expected benefits of the resulting water quality in terms of water use.
3. The technological bounds (e.g., available storage for low flow augmentation) on the controls.

In the chapters to follow, these general principles of engineering control will be examined for a variety of problem contexts and situations.

2.5 DERIVATION OF STEADY STATE STREAM EQUATION

The first-order differential equation for a steady state distribution of a reactive substance in a stream or river was previously presented as Eq. 2.20. This equation is based on the conservation of mass principle expressed as a balance of the time rates of mass flowing into and out of a differential water element and reacting away in the element, as shown in Fig. 2.21. It is assumed that the geometry, flow, and kinetic coefficient are temporally constant. Using the advective and reactive components developed in this figure,

$$\frac{\partial M}{\partial t} = V \frac{\partial s}{\partial t} = -\frac{\partial}{\partial x}(Qs)\,\Delta x - KsV \qquad (2.52)$$

Figure 2.21 Definition of stream mass balance components.

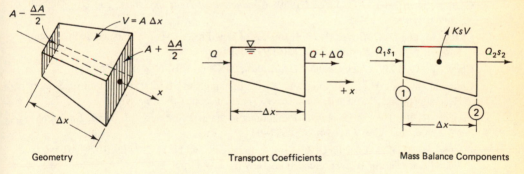

Geometry   Transport Coefficients   Mass Balance Components

Advection

$$\frac{\partial M}{\partial t} = Q_1 s_1 - Q_2 s_2 = Qs - (Q + \Delta Q)(s + \Delta s)$$

with $\Delta Q = \frac{\partial Q}{\partial x}\,\Delta x$ and $\Delta s = \frac{\partial s}{\partial x}\,\Delta x$, and neglecting higher-order terms:

$$\frac{\partial M}{\partial t} = -\frac{\partial Q}{\partial x}\,\Delta x \cdot s - Q\,\frac{\partial s}{\partial x}\cdot\Delta x = -\left[\frac{\partial}{\partial x}(Qs)\right]\Delta x$$

Reaction

Assume first-order kinetics, temporally constant volume.

$$\frac{\partial M}{\partial t} = V \frac{\partial s}{\partial t} = V\left[\frac{\partial}{\partial t}(s = s_0 e^{-Kt}) = -Ks_0 e^{-Kt} = -Ks\right] = -KsV$$

Dividing both sides by $V = A \Delta x$

$$\frac{\partial s}{\partial t} = -\frac{1}{A}\frac{\partial}{\partial x}(Qs) - Ks$$

which is the time variable equation for a reactive substance in a stream with temporally constant coefficients.

For steady state conditions, $\partial s/\partial t = 0$, and

$$0 = -\frac{1}{A}\frac{\partial}{\partial x}(Qs) - Ks \qquad (2.53)$$

This is the steady state stream equation for a reactive substance previously presented as Eq. 2.20. A detailed derivation of the general three-dimensional advective–diffusion equation is found in Fischer et al. (1979), together with a discussion of the phenomena affecting the riverine dispersion coefficient.

REFERENCES

Blench, T., 1972. *Morphometric Changes in River Ecology and Man,* Ray T. Oglesby et al., Eds., Academic, New York, pp. 287–308.

Burke, R., 1983. "Velocity Equation for Water Quality Modeling in Georgia," *Water Resourc. Bull.,* American Water Resources Association **19**(2):271–276.

Carslaw, H. S. and J. C. Jaeger, 1959. *Conduction of Heat in Solids,* 2nd ed., Oxford University Press, p 485.

Chang, M. and D. C. Boyer, 1977. Estimates of Low Flows Using Watershed and Climatic Parameters, *Water Resour. Res.,* **13**(6):997–1001.

Chow, V. T., 1959. *Open-Channel Hydraulics,* McGraw-Hill, New York, 680 pp.

Churchill, M. A., H. L. Elmore, and R. A. Buckingham, 1962. "Prediction of Stream Reaeration Rates," *J. Sanit. Eng. Div., Proc. Am. Soc. Civ. Eng.* **SA4**:1.

Fischer, H. B., 1968. "Dispersion Predictions in Natural Streams," *J. Sanit. Eng. Div., Proc. Am. Soc. Civ. Eng.* **94**(SA5):927–944.

Fischer, H. B., E. J. List, R. C. Y. Koh, J. Imberger, and N. H. Brooks, 1979. *Mixing in Inland and Coastal Waters,* Academic, New York, 483 pp.

Hydroscience, Inc., 1974. *Water Quality Analysis of the Black River.* Prepared for USEPA, Region V, Enforcement Division, Great Lakes Initiative Contract Program Dec. 1974. EPA-905/9-74-009.

Hydroscience, 1975. Time-Variable Analyses and Related Studies of the Upper Delaware River—Port Jervis to Trenton, Delaware River Basin Committee, West Trenton, NJ, 164 pp. + 5 appendixes.

Kehr, R. W., W. C. Purdy, J. B. Lackey, O. R. Placak, and W. E. Burns, 1941. "A Study of the Pollution and Natural Purification of the Scioto River," *Public Health Bull.* No. 276, U.S. Public Health Service, Washington, D.C. 153 pp.

Leopold, L. B. and T. Maddock, 1953. *The Hydraulic Geometry of Stream Channels and Some Physiographic Implications,* Geological Survey Professional Paper 252, Washington, D.C.

Li, W. H., 1962. "Unsteady Dissolved Oxygen Sag in a Stream," *J. Sanit. Eng. Div., Proc. Am. Soc. Civ. Eng.* **88**(SA3):75–85.

Linsley, R. K. Jr., M. A. Kohler, and J. L. H. Paulhus, 1982. *Hydrology for Engineers,* 3rd ed., McGraw-Hill, New York, 508 pp.

McQuivey, R. S. and T. N. Keefer, 1974. Simple Method for Predicting Dispersion in Streams, *J. Environ, Eng. Div., Proc. Am. Soc. Civ. Eng.* **100**(EE4):997–1011.

Mills, H. B., W. C. Starrett, and F. C. Bellrose, 1966. "Man's Effect on the Fish and Wildlife of the Illinois River," *Ill. Nat. Hist. Surv. Biol. Notes,* No. 57, Urbana, IL, 24 pp.

National Commission on Water Quality, 1976. Water Quality Analysis and Environmental Impact, Assessment of P. L. 92-500: Technical Volume, Chapter III-E, New Q, Washington, D.C., Harold Allen, Program Leader.

O'Connor, D. J., 1962. "The Effect of Stream Flow on Waste Assimilation Capacity." Presented at 17th Purdue Industrial Waste Conference, Lafayette, IN, May 1962.

O'Connor, D. J., 1967. "The Temporal and Spatial Variation of Dissolved Oxygen in Streams," *Water Resour. Res.* **3**(1):65–79.

O'Loughlin, E. M. and K. H. Bowmer, 1975. "Dilution and Decay of Aquatic Herbicides in Flowing Channels," *J. Hydrol.* **26**:217–235.

Piper, A. M. 1965. "Has the United States Enough Water?," Geological Survey WS Paper 1797, U.S. Department of the Interior, U.S. Government Printing Office, Washington, D.C. 27 + III + 3 plates.

Rathbun, R. E., 1979. Estimating the Gas and Dye Quantities for Modified Tracer Technique Measurements of Stream Reaeration Coefficients, U.S. Geological Survey, Water Res. Invest., 79–27, NSTL Sta. Miss., 42 pp.

Reheis, H. F., J. C. Dozier, D. M. Word, and J. R. Holland, 1982. Treatment Costs Savings Through Monthly Variable Effluent Limits, *J. Water Pollut. Control Fed.* **54**(8):1224–1230.

Rose, D. A., 1977. Dilution and Decay of Aquatic Herbicides in Flowing Channels— Comments, *J. Hydrol.* **32**:399–400.

Seaborn, G. E., 1969. Effects of Urban Development on Direct Runoff to East Meadow Brook, Nassau County, Long Island, New York, Geological Survey Professional Paper 627-B, U.S. Government Printing Office, Washington, D.C., 14 pp. + plate.

Showers, V., 1979. *World Facts and Figures,* Wiley, New York.

Thomann, R. V., 1972. *Systems Analysis and Water Quality Management,* McGraw-Hill, New York, 286 pp. (Reprint: J. Williams Book Co., Oklahoma City, OK.)

Thomann, R. V., 1973. "Effect of Longitudinal Dispersion on Dynamic Water Quality Response of Streams and Rivers," *Water Resour. Res.* **9**(2):355–366.

Thomas, D. M. and M. A. Benson, 1970. Generalization of Streamflow Characteristics From Drainage-Basin Characteristics. U.S. Geological Survey WS Paper 1975, U.S. Government Printing Office, Washington, D.C. 55 pp.

U.S. Environmental Protection Agency, 1984. Draft Report of Technical Guidance Manual for Performing Waste Load Allocations, Book VI: Selecting Design Conditions, Part A, Stream Design Flow; Office of Water Regulation & Standards, Monitoring & Data Support Division, Monitoring Branch, Washington, D.C.

U.S. Geological Survey, 1970. Measurement of Time of Travel and Dispersion by Dye Tracing, Preliminary Report, Book 3, Chapter A9 in: *Techniques of Water Resources Investigations,* by F. A. Kilpatrick, L. A. Martens, and J. F. Wilson.

U.S. Geological Survey, undated. Surface Water Supply of the United States, 1966–1970. Part 1, North Atlantic Slope Basins, Vol. 2, Basins from New York to Delaware. Water-Supply Paper 2102.

Worley, J. L., 1963. A System Analysis Method for Water Quality Management by Flow Augmentation. Thesis submitted to Oregon State University, Master of Science, June, 1963, 136 pp. + plate.

Yotsukura, 1968. As Referenced in Preliminary report *Techniques of Water Resources Investigations of the U.S. Geological Survey,* Measurement of Time of Travel and Dispersion by Dye Tracing, Book 3, Chapter A9, by F. A. Kilpatrick, L. A. Martens, and J. F. Wilson, 1970.

PROBLEMS

2.1. Given the following minimum 7-day mean discharges (cfs) for the Delaware River at Trenton:

2021	4012	1669	2721
1697	1871	1309	1351
2570	1951	6574	3213
2762	3404	1836	1639
1614	1781	3297	2793
4604	2000	1627	2050
2796	2563	2844	2990
3259	3397	1670	1622
1661	1493	3231	2899
2000	2119	1340	1560
2390	4536	2244	1737
1423	2773	3947	1890
3214	1951	2019	2691

(a) Plot a low flow frequency curve on log normal probability paper.
(b) Estimate the 1 in 10 yr 7-day minimum average flow.

2.2. A midstream outfall discharges to a reach of stream which has been excavated to form a trapezoidal cross section and lined with concrete ($n = 0.012$). Estimate the flow in the stream for the dimensions below and approximate the distance to complete mixing of the discharge. Assume an effective width equal to that at mid-depth. $S_0 = 0.5$ ft/mi.

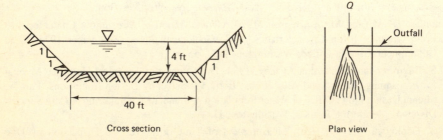

Cross section   Plan view

2.3. A stream reach between MP200 and MP125 has a velocity–flow power relationship,

$U = cQ^d$, with the exponent $d = 0.35$, U in fps, and Q in cfs. What flow is required to halve the travel time through the reach associated with a flow of 600 cfs?

2.4. The following data were collected on the color of the Prairie River and St. Lawrence River in the vicinity of Montreal.

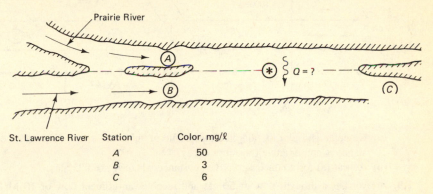

Station	Color, mg/ℓ
A	50
B	3
C	6

The flow of the Prairie River just before its entrance to the St. Lawrence River was 30,000 cfs and the flow of the St. Lawrence at Montreal was 100,000 cfs.

(a) What would be the concentration at point C if there was complete mixing between the two rivers?

(b) Using the color data, estimate the flow (cfs) that was transported from the Prairie to the St. Lawrence at the point marked ⊛.

2.5. A river receives a discharge of 10 MGD at a concentration of 150 mg/l. The river flow upstream is 20 MGD at zero concentration. For 15 mi downstream, the velocity is 10 mpd. A region of slow moving water is then encountered for the next 20 mi where the velocity drops to 2 mpd. If the decay rate of the substance is 0.2/day, what is the concentration at the point 35 mi downstream from the outfall?

2.6. For the Potomac River at Little Falls, just north of Washington, D.C., the average annual flow is 10,500 cfs. During this flow, measurements were made as follows: depth $= 10.22$ ft, velocity $= 0.81$ fps, width $= 1262$ ft. The respective exponents in the hydraulic geometry relationships are $b = 0.35$, $d = 0.55$, $f = 0.1$. For the 7-day, once in 10-yr flow of 980 cfs, estimate the depth, width, area, and velocity.

2.7. Given the following configuration and a substance that is conservative.

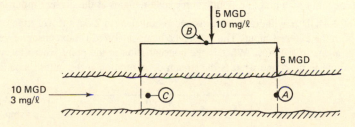

(a) What is the mass rate of the substance in lb/day passing downstream of point A?

(b) What is the concentration in mg/l at point A?

(c) Point B?

(d) Point C?

2.8. Given the following configuration.

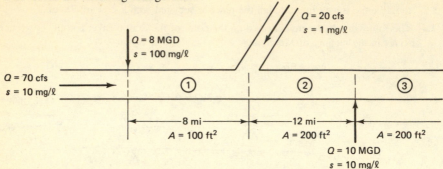

(a) Compute and plot the longitudinal profile of substance s assuming that the substance is conservative (does not decay).

(b) Repeat (a) for a nonconservative substance where $K = 0.25/\text{day}$.

2.9. A river has a drought flow of 50 cfs and receives an effluent flow of 10 MGD as shown below.

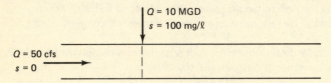

The substance s decays in the river at $0.1/\text{day}$. The velocity downstream of the outfall is related to river flow by the following equation,

$$U = 0.407 Q^{0.6}$$

where $U = $ velocity in mpd and Q is the river flow in cfs. It is required that the concentration of s be equal to or less than 15 mg/l beginning at a location 10 mi downstream from the outfall. Two methods of control are proposed: point source reduction or increased upstream river flow.

(a) What is the % reduction required from the point source load input to meet the desired water quality?

(b) What flow increase (in cfs) in the upstream flow is necessary to meet the desired water quality assuming the point source load remains at the present input level?

2.10. A flow of 3 MGD is withdrawn at a distance equivalent to 2 days travel time downstream from the external input. The substance being analyzed is nonconservative and decays at a rate of $0.1/\text{day}$. What is the concentration at the point where the water is pumped out in mg/l and what is the stream concentration at the outfall location in mg/l?

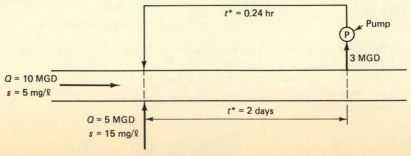

2.11. Suppose a waste source to a river claims to be discharging 5 cfs with an effluent concentration of 10 mg/l as shown below.

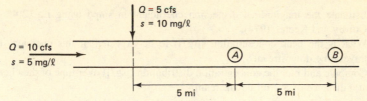

Data collected at A gave a concentration of 5 mg/l and at B a concentration of 2 mg/l. Based on this information from the river, is the discharger's claim true? If you think not, what is your estimate of the discharge concentration? The substance is nonconservative and the river velocity is 5 mpd.

2.12. Using Eq. 2.25 as a starting point, determine the equation for the instream concentration of a conservative distributed source with negligible associated inflow. Assume the upstream concentration is equal to zero.

2.13. The concentration of BOD, an oxygen demanding parameter, is 300 mg/l in a runoff of 5 cfs. This flow is distributed by overflow pipes along a 1-mi length of stream. Assuming an average flow of 55 cfs for the entire stream, calculate and plot the instream BOD profile for a distance 10 mi downstream. BOD decays at a rate of 0.4/day, base e, for this stream.

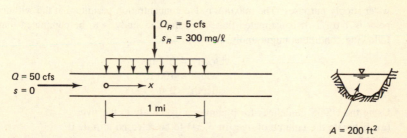

2.14. The river velocity is 5 mpd and is approximately independent of flow for the range considered here. The waste load W has been discharging 5,000 lb BOD/day for a long time. It is known that the BOD in the stream decays with a rate of 0.1/day. At 12 midnight October 1, the BOD was reduced to 1000 lb BOD/day and maintained at that level. Assuming no dispersion or mixing in the river:
(a) Calculate and plot the BOD concentration with time at $x = 0$ and $x = 10$ mi.
(b) Calculate and plot the spatial profile from $x = 0$ to $x = 10$ mi on midnight October 1, 2, 3, and 4.

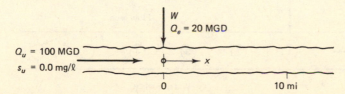

2.15. A dump of 20 lb of a conservative chemical was made accidently into a river. The release was over a short period of time and can be approximated as an instantaneous release. The river had the following characteristics during the dump:

$$U = 0.5 \text{ fps}, \qquad B = 100 \text{ ft}, \qquad H = 5 \text{ ft}$$
$$S_0 = 1.5 \text{ ft/mi}$$

(a) Estimate the longitudinal dispersion coefficient in smpd using Eq. 2.45 (McQuivey and Keefer 1974).

(b) What is the peak concentration (μg/l) to be expected of this chemical 1-day travel time downstream?

(c) Compute and plot the concentration distribution (μg/l) over time of the chemical at a distance of 8.2 mi downstream.

2.16. A dye study to be conducted on a river where the primary study reach extends for 10 mi as shown below.

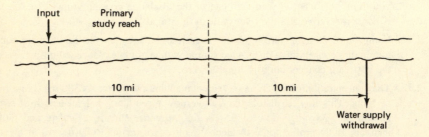

However, 10 mi below the primary study area, water is withdrawn from the river for water supply purposes. The maximum dye concentration permitted at the withdrawal point is 1 μg/l. It is estimated that when the dye study is to be conducted that the following conditions might apply:

$$\text{velocity} = 0.2 \text{ fps}$$
$$\text{width} = 100 \text{ ft}$$
$$\text{depth} = 2 \text{ ft}$$

Using the USGS guidelines for planning a dye study in the river:

(a) Estimate the volume of dye (in liters) to be released at mile 0.

(b) Plot the peak concentrations as a function of distance.

(c) Plot the arrival times of the leading edge, the peak, and the trailing edge of the dye as a function of distance.

(d) What time of day would you release the dye?

(e) Describe the sampling schedule (i.e., locations, length of sampling, and interval of sampling) that you would recommend for this dye study.

2.17. A point source W discharges a substance to the stream at $x = 0$.

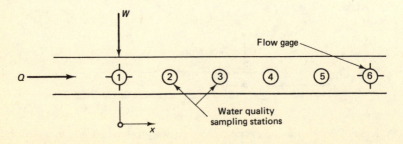

Station	1	2	3	4	5	6
x (mi)	0	5	10	15	20	25
Q (cfs)	30					50
s (mg/l)	20.0	18.5	16.0	14.7	13.5	11.9

Assuming an exponential increase in flow and uncontaminated infiltrating water, calculate the decay rate of the substance introduced using the data above. The stream has a constant cross-sectional area of 25 ft². Assume that $s_0 = 20.0$ mg/l accurately reflects the well-mixed concentration at $x = 0$.

2.18. A nondispersive (no mixing) river situation has the following input:

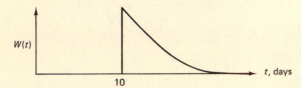

described by

$$W(t) = 0 \qquad \text{for } t < 10 \text{ days}$$
$$= 5000 \exp[-0.4(t - 10)] \qquad \text{for } t \geq 10 \text{ days}$$

Conditions are $Q = 100$ MGD, $U = 10$ mpd, $K = 0.1/\text{day}$, $W(t)$ in lb/day.

(a) Calculate and sketch the response profile of concentration (mg/l) at the outfall.

(b) Calculate and sketch the response profile of concentration (mg/l) at $x = 20$ mi downstream of the outfall.

Chapter 3

Estuaries, Bays, and Harbors

The region between the free flowing river and the ocean is a fascinating, diverse, and complex water system: the coastal regime of estuaries, bays and harbors. The ebb and flow of the tides, the incursion of salinity from the ocean, the influx of nutrients from the upstream drainage all contribute to the generation of a unique aquatic ecosystem. The estuarine and wetland regions are considered to be crucial to the maintenance of major fish stocks such as the striped bass and blue fish which to varying degrees utilize the estuarine areas as spawning and nursery grounds.

The movement of the tides into and out of estuaries, the associated density effects created by the incursion of salinity, are of particular importance in describing the water quality of such bodies of water. Many major cities are located along estuaries primarily as a result of the historical need for ready access to national and international commerce routes. For many years, such cities discharged large quantities of waste but, because of the large volumes of estuaries, effects were not immediately felt. Later however, especially in the 1950s, the load on estuaries became very great, quality deteriorated rapidly, and great interest centered on the analysis of water quality in estuaries. Figure 3.1 shows pictorially the various stages of a river in its movement from the upstream region to the ocean. Several distinct zones can be defined. The *tidal river* is that region of a river where there is some current reversal but sea salts have not penetrated to this region so that the tidal river is still ''fresh.'' The *estuary* is the ''drowned'' part of a river system due to incursion of the ocean landward with marked current reversal and brackish due to the saline water. (Note that if a river discharges to a large lake such as one of the Great Lakes, a condition similar to that in an estuary can be created through the incursion of lake water up into the mouth of the river.)

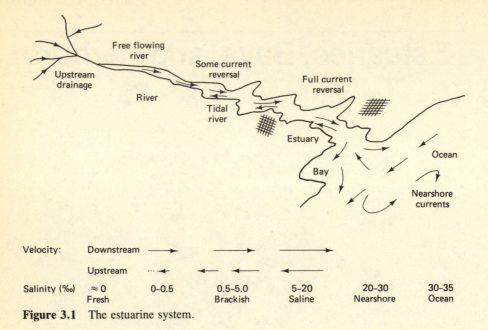

Figure 3.1   The estuarine system.

3.1 PHYSICAL ASPECTS OF ESTUARIES: ESTUARINE HYDROLOGY, TIDES, AND TIDAL CURRENTS

Dwellers by their shores have always been fascinated by the movement of water into and out of estuaries and bays along coastal regions. No coastline is without tides and over the many centuries of observation, a great degree of regularity in the vertical and horizontal motion of water along the coast has been noticed.

Tides are the movement of water above and below a datum plane, usually mean sea level. Tidal currents are the associated horizontal movement of the water into and out of an estuary. Tides and tidal currents are due to the attractive force of the moon and sun on the waters of the earth. There is a "pulling and tugging" which raises the water at certain locations and lowers it at other locations. These motions occur on a more or less regular cyclical basis reflecting the regularity of the lunar and solar cycles. Detailed discussion on tidal mechanics are given by Defant (1958), Neumann and Pierson (1966), and Ippen (1966). Measurement of tidal height or stage is easily accomplished through mechanisms such as shown in Fig. 3.2. A small stilling well is constructed to dampen the high-frequency oscillations of the water surface due to variable winds. A recorder is attached to a float which rises and falls with the water level in the well to trace out the tidal stage oscillation. The measurement of tidal currents is accomplished with current meters similar to those used in river current measurements except that current directions must also be determined. Figure 3.3 shows a typical oscillation of the tidal velocity and, in many locations, is approximately sinusoidal although there are many other oscillations that may be superimposed to result in a variety of cyclical patterns.

Table 3.1 from Defant (1958) shows the most important components of the

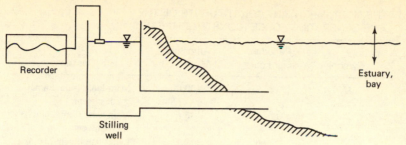

Figure 3.2 Measurement of tidal stage.

tide-producing forces and reflects the various combinations of lunar and solar positions relative to earth and the resulting variations in tidal amplitude. The semidiurnal tide, designated M_2, is the main lunar constituent. As shown in Table 3.1, there are also diurnal tides with periods close to 24 hr and longer period tides of about a fortnight (14 days). The combination of these solar and lunar forces results in typical tidal stage (height) oscillations such as shown in Fig. 3.4 (p. 95). Tides are also generated in lakes and seas produced principally by winds blowing across the lake surface and "piling up" the water which in turn sets the lake into an oscillatory motion or seiche. The approximately regular motion of the lake results in a motion in lake tributaries similar to estuarine tides. The National Oceanic and Atmospheric Administration (NOAA) publishes annually tidal height and current predictions, samples of which are shown in Table 3.2 (p. 96).

   Tidal excursion is the approximate distance a particle will travel along the main axis of an estuary in going from low to high water or vice-versa. For the principal M_2 tide, the excursion distance is given by

$$x_{te} = \frac{2}{\pi}(\overline{U}_{\max})\left(\frac{T_{m2}}{2}\right)$$

(3.1)

Figure 3.3 Typical oscillation of tidal velocity.

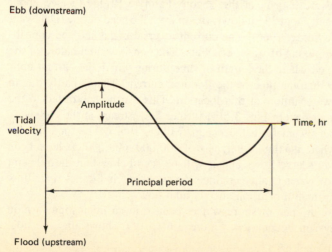

TABLE 3.1   THE MOST IMPORTANT CONSTITUENTS OF THE
             TIDE-GENERATING FORCE (CONSTITUENT TIDES)[a]

	Symbol	Period in solar hours	Amplitude $M_2 = 100$	Description
Semidiurnal tides	M_2	12.42	100.00	Main lunar (semidiurnal) constituent
	S_2	12.00	46.6	Main solar (semidiurnal) constituent
	N_2	12.66	19.1	Lunar constituent due to monthly variation in moon's distance
	K_2	11.97	12.7	Soli-lunar constituent due to changes in declination of sun and moon throughout their orbit of cycle
Diurnal tides	K_1	23.93	58.4	Soli-lunar constituent
	O_1	25.82	41.5	Main lunar (diurnal) constituent
	P_1	24.07	19.3	Main solar (diurnal) constituent
Long-period tides	M_1	327.86 (13.7 days)	17.2	Moon's fortnightly constituent

[a]From Defant (1958). Reprinted by permission of the University of Michigan Press.

where $\overline{U}_{max}$ is the average maximum tidal velocity, and T_{m2} is the period of the M_2 tide. Thus, if $\overline{U}_{max}$ is about 3 fps (91.4 cm/s), a typical value, and $T_{m2} = 12.42$ hr, then the average tidal excursion is about 8 mi (13 km).

The *tidal flow* is the total volume of water passing a given point in the estuary over time.

Figure 3.5(a) (p. 98) shows a typical tidal current pattern for East Coast estuaries, for the Delaware Estuary at the Tacony-Palmyra Bridge located about 107 mi from the ocean and about 25 mi downstream from the head of tide (Delaware Estuary Comprehensive Study, 1966). The current reversal is marked and contributes substantially to the mixing of waste effluents. The nonsinusoidal nature of the reversal can also be noted where the length of time during which the current ebbs (downstream) is longer than the time of the flooding current. This is partly due to the effect of the freshwater inflow at this location. This can be contrasted to the more symmetrical current pattern Fig. 3.5(b) further downstream at the Delaware Memorial Bridge, about 69 mi from the ocean (Miller, 1962). The total flow is shown in Fig. 3.5(c) where maximum values x of 400,000–600,000 cfs have been measured. These are very large flows and it will be seen below that the relevant variable for a variety of water quality engineering problems is the net river flow over a tidal cycle rather than this instantaneous tidal flow.

The tidal currents in open offshore waters behave in an interesting fashion due to the lack of physical boundaries. Figure 3.6 (p. 99) shows that the tidal

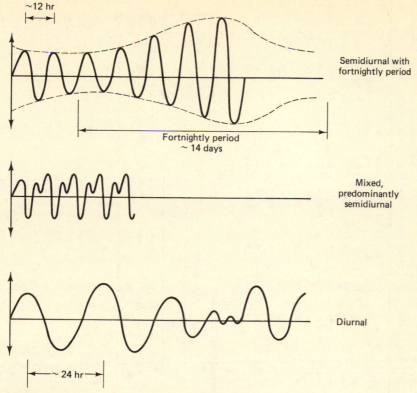

Figure 3.4 Some patterns of tidal stage oscillations.

current tends to move about a point in a rotary-type current. This type of current therefore will tend to move any wastes discharged offshore in an elliptical motion on which may be superimposed a net current drift. Figure 3.7 shows the currents as measured in a region about 34 mi southeast of the coast of New Jersey–Delaware. The various currents at the three different depths show a marked change in direction. At the surface, the current is to the NW but at 140 ft, the current is to the south. The current structure in offshore waters is therefore quite complex and is of particular significance in the transport of wastes discharged at sea.

3.1.1 Net Estuarine Flow

An important characteristic of estuarine hydrology is the net flow through the estuary over a tidal cycle or a given number of cycles. This is the flow that over a period of several days or weeks flushes material out of the estuary and is a significant parameter in the estimation of the distribution of estuarine water quality.

If the estuary is well mixed from top to bottom and side to side (i.e., no significant gradients in velocity), then the net flow at any location in the estuary is approximately equal to the sum of the upstream external flow inputs to the estuary, assuming no other significant net hydrologic inputs or losses. This is so

TABLE 3.2 SAMPLE TIDAL STAGES AND CURRENTS FROM NOAA

New York (The Battery), N.Y., 1984—Times and Heights of High and Low Waters

January

Day	Time	ft	m	Day	Time	ft	m
1	0033	-0.4	-0.1	16	0617	4.8	1.5
Su	0656	5.0	1.5	M	1254	-0.5	-0.2
	1317	-0.5	-0.2		1845	3.8	1.2
	1920	3.8	1.2				
2	0120	-0.4	-0.1	17	0050	-0.4	-0.1
M	0739	5.0	1.5	Tu	0705	5.1	1.6
	1405	-0.5	-0.2		1345	-0.8	-0.2
	2005	3.8	1.2		1936	4.0	1.2
3	0205	-0.3	-0.1	18	0142	-0.6	-0.2
Tu	0821	4.9	1.5	W	0754	5.3	1.6
	1448	-0.5	-0.2		1431	-1.0	-0.3
	2048	3.7	1.1		2026	4.2	1.3

February

Day	Time	ft	m	Day	Time	ft	m
1	0149	-0.2	-0.1	16	0129	-0.8	-0.2
W	0804	4.7	1.4	Th	0739	5.4	1.6
	1426	-0.5	-0.2		1410	-1.1	-0.3
	2030	3.8	1.2		2009	4.6	1.4
2	0232	-0.2	-0.1	17	0221	-1.0	-0.3
Th	0843	4.6	1.4	F	0831	5.5	1.7
	1504	-0.5	-0.2		1456	-1.3	-0.4
	2110	3.8	1.2		2100	4.9	1.5
3	0309	-0.2	-0.1	18	0310	-1.2	-0.4
F	0921	4.5	1.4	Sa	0921	5.4	1.6
	1541	-0.5	-0.2		1541	-1.4	-0.4
	2150	3.8	1.2		2152	5.0	1.5

March

Day	Time	ft	m	Day	Time	ft	m
1	0130	-0.1	0.0	16	0113	-0.8	-0.2
Th	0742	4.6	1.4	F	0723	5.4	1.6
	1400	-0.4	-0.1		1344	-1.1	-0.3
	2005	4.1	1.2		1949	5.2	1.6
2	0211	-0.2	-0.1	17	0205	-1.1	-0.3
F	0818	4.6	1.4	Sa	0810	5.4	1.6
	1437	-0.4	-0.1		1430	-1.2	-0.4
	2041	4.2	1.3		2037	5.4	1.6
3	0248	-0.3	-0.1	18	0254	-1.2	-0.4
Sa	0854	4.5	1.4	Su	0901	5.4	1.6
	1512	-0.4	-0.1		1515	-1.2	-0.4
	2116	4.2	1.3		2126	5.5	1.7

Tidal Differences and Other Constants

No.	Place	Position Lat. (°N)	Position Long. (°W)	Differences Time High Water (h. m)	Differences Time Low Water (h. m)	Height High Water (ft)	Height Low Water (ft)	Ranges Mean (ft)	Ranges Spring (ft)	Mean tide level (ft)
	Hudson River (8)									
1513	Jersey City, Con Rail RR, Ferry, N. J.	40 43	74 02	+0 11	+0 10	-0.2	0.0	4.4	5.3	2.2
1515	New York, Desbrosses Street	40 43	74 01	+0 14	+0 13	-0.2	0.0	4.4	5.3	2.2
1517	New York, Chelsea Docks	40 45	74 01	+0 21	+0 19	-0.3	0.0	4.3	5.2	2.1
1519	Hoboken, Castle Point, N. J.	40 45	74 01	+0 21	+0 19	-0.3	0.0	4.3	5.2	2.1
1521	Weehawken, Days Point, N. J.	40 46	74 01	+0 28	+0 26	-0.4	0.0	4.2	5.0	2.1
1523	New York, Union Stock Yards	40 47	74 00	+0 31	+0 29	-0.4	0.0	4.2	5.0	2.1
1525	New York, 130th Street	40 49	73 58	+0 41	+0 38	-0.6	0.0	4.0	4.8	2.0
1527	George Washington Bridge	40 51	73 57	+0 50	+0 46	-0.7	0.0	3.9	4.6	1.9
1529	Spuyten Duyvil, west of RR bridge	40 53	73 56	+1 02	+0 56	-0.8	0.0	3.8	4.5	1.9
1531	Yonkers	40 56	73 54	+1 13	+1 13	-0.9	0.0	3.7	4.4	1.8

THE NARROWS, New York Harbor, New York, 1984—F-Flood, Dir. 340° True E-Ebb, Dir. 160° True

January

Day	Slack water time (h.m.)	Maximum current Time (h.m.)	Vel. (knots)
1 Su	0224	0528	2.1F
	0822	1137	2.1E
	1522	1802	1.4F
	2026	2340	1.9E
2 M	0311	0609	2.1F
	0907	1226	2.1E
	1608	1839	1.4F
	2112		
3 Tu	0356	0031	1.9E
	0950	0644	2.0F
	1652	1312	2.1E
	2158	1916	1.4F

Day	Slack water time (h.m.)	Maximum current Time (h.m.)	Vel. (knots)
16 M	0155	0445	2.0F
	0757	1111	2.1E
	1503	1717	1.4F
	2003	2317	2.0E
16 Tu	0224	0532	2.2F
	0845	1203	2.3E
	1548	1802	1.6F
	2053		
18 W	0333	0009	2.1E
	0933	0619	2.3F
	1633	1254	2.4E
	2144	1849	1.7F

February

Day	Slack water time (h.m.)	Maximum current Time (h.m.)	Vel. (knots)
1 W	0339	0009	1.8E
	0928	0632	1.9F
	1628	1249	2.1E
	2138	1900	1.4F
2 Th	0422	0056	1.8E
	1008	0702	1.9F
	1707	1330	2.1E
	2221	1930	1.4F
3 F	0502	0139	1.8E
	1047	0733	1.8F
	1745	1407	2.1E
	2304	2002	1.5F

Day	Slack water time (h.m.)	Maximum current Time (h.m.)	Vel. (knots)
16 Th	0319	0601	2.3F
	0912	1229	2.5E
	1606	1829	1.9F
	2126		
17 F	0409	0045	2.4E
	1001	0650	2.4F
	1650	1320	2.6E
	2219	1916	2.1F
18 Sa	0459	0137	2.5E
	1050	0739	2.4F
	1734	1406	2.6E
	2312	2005	2.1F

Current Differences and Other Constants, 1984

No.	Place — NEW YORK HARBOR, Upper Bay. Time meridian, 75°W	Position Lat. ° ′ N	Position Long. ° ′ W	Meter depth (ft)	Time differences Min. before flood (h.m.)	Time differences Flood (h.m.)	Time differences Min. before ebb (h.m.)	Time differences Ebb (h.m.)	Speed ratios Flood	Speed ratios Ebb	Minimum before flood knots	Minimum before flood deg.	Maximum flood knots	Maximum flood deg.	Minimum before ebb knots	Minimum before ebb deg.	Maximum ebb knots	Maximum ebb deg.
3846	Bay Ridge, west of	40 37.9	74 03.4	22	−0 11	+0 20	+0 42	+0 59	0.8	0.7	0.1	104	1.4	354	0.0	− −	1.5	185
3851	Bay Ridge Channel	40 39.3	74 01.9	15	−0 58	−1 26	+0 04	−1 17	0.6	0.3	0.0	− −	1.0	032	0.1	125	0.7	212
	do	40 39.3	74 01.9	36	−1 35	−2 36	−0 50	−0 09	0.4	0.2	0.0	− −	0.6	037	0.0	− −	0.4	225
3856	Red Hook Channel	40 40.0	74 01.2		−1 03	−0 44	−0 08	−0 30	0.6	0.4	0.0	− −	1.0	353	0.0	− −	0.7	170
3861	Robbins Reef Light, east of	40 39.45	74 03.48		+0 16	+0 16	+0 02	+0 24	0.8	0.8	0.0	− −	1.3	016	0.0	− −	1.6	204
3866	Red Hook, 1 mile east of	40 40.5	74 02.5		+0 41	+1 06	+0 47	+0 52	0.8	1.2	0.0	− −	1.3	024	0.0	− −	2.3	206
3871	Statue of Liberty, east of	40 41.4	74 01.8		+0 57	+0 58	+0 56	+0 59	0.8	1.0	0.0	− −	1.4	031	0.0	− −	1.9	205

(b) Tidal Currents (NOAA), 1984b

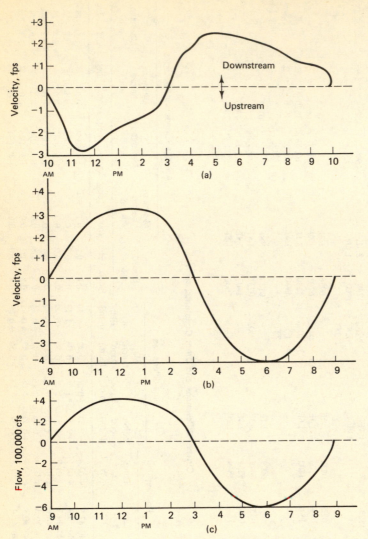

Figure 3.5 Current velocities in Delaware Estuary at (a) Tacony-Palmyra Bridge, May 11, 1964; (b) Delaware Memorial Bridge, August 21, 1957 (from Miller, 1962); (c) flow at Delaware Memorial Bridge, August 21, 1957. (From Miller, 1962).

since it is known that the estuary is not overflowing due to the flow inputs. Therefore, this flow must, on balance, be leaving the estuary at any cross section. Figure 3.8 shows this for a typical estuary that has tidal flow oscillating in one direction. The estimated net flow (Q_n) at point A in the estuary is given from the sum of the upstream flow from the up-basin drainage area, the flow from the major well-defined tributary, the flow from the waste input, and the incremental flow between these latter point flow inputs due to diffuse drainage from small tributaries and local runoff. The net flow is, of course, also a statistical variable and input flow probability distributions can be computed as described in Chapter 2.

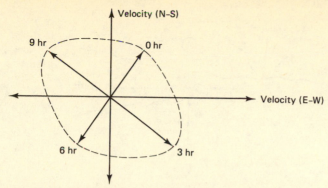

Figure 3.6 Typical rotary tide for offshore waters.

The sum of the upstream flow estimate of the net estuarine flow does not include such effects as wind driven flows or unknown groundwater inputs and losses and other flow sources and sinks. In some instances, when accurate current meter data are available in a cross section to estimate the tidal flow over time, an additional estimate can be made of the net flow. Figure 3.9 shows the notation. For the ebb and flow portions of the discharge, and assuming a sinusoidal variation, the flow can be written as (*LMS Eng.*, 1977)

$$Q = q_e \sin \left[\frac{\pi(t + \theta)}{T_e} \right] \qquad \text{for} \quad 0 \le t' \le T_e \qquad (3.2a)$$

$$Q = q_f \sin \left[\pi + \frac{\pi(t + \theta - T_e)}{\tau - T_e} \right] \qquad \text{for} \quad T_e \le t' \le \tau \qquad (3.2b)$$

Figure 3.7 Tidal currents off Delaware Bay, August 1970, 34 mi southeast of mouth of bay.

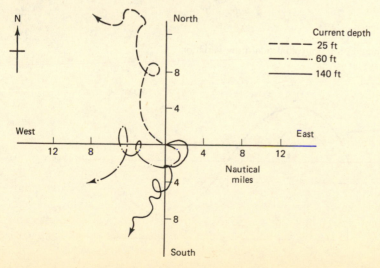

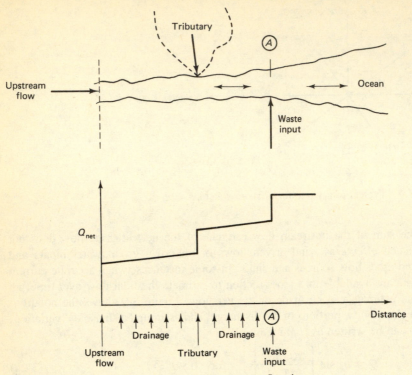

Figure 3.8 Net estuarine flow as sum of upstream flow inputs.

where Q is the tidal flow $[L^3/T]$, q_e and q_f are the amplitudes of the tidal flow $[L^3/T]$ during the ebb and flow cycles, respectively, θ is the phase lag $[T]$ (i.e., the time from a reference point in the estuary to the point at which the measurement is made), T_e and T_f are the durations $[T]$ of ebb and flood, respectively, and

$$\tau = T_e + T_f \tag{3.2c}$$

Figure 3.9 Notation for sinusoidal tidal flow variation.

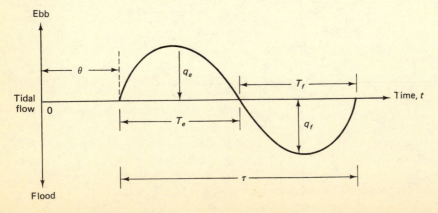

and

$$t' = t + \theta \tag{3.2d}$$

If Eqs. 3.2a and 3.2b are averaged over time (i.e., the integral is taken and divided by τ), then for Q_n as the net freshwater flow,

$$Q_n = \frac{2}{\pi} \frac{q_e T_e - q_f T_f}{\tau} \quad [L^3/T] \tag{3.3}$$

As the time of ebb flow and magnitude of the ebb flow (q_e) increases, Q_n increases reflecting the greater volume of flow that must pass through the cross section at ebb. Note that Q_n can be negative indicating an upstream flow which can, of course, persist for only a relatively brief transient time period. Equation 3.3 should be used with caution since, as indicated previously, the difference in tidal flow especially toward the mouths of estuaries is between two large numbers and therefore subject to significant error.

3.1.2 Vertical Stratification and Estuarine Circulation

The fully developed estuarine region in many instances has a complicated circulation pattern resulting from the vertical stratification of freshwater inflow "riding over" the more saline ocean water. The differences in salinity and temperature in different regions of an estuary or bay results in differences in density with associated resulting complex circulation patterns. It is important then to first examine the relationship between temperature, salinity, and the density of water.

3.1.2.1 Temperature, Salinity, Density Relationships
The great dissolving power of water in its pure state is a direct result of its dissociation power. This in turn is related to the dielectric constant, ϵ. This quantity reflects the relative difference in electric intensity of a given medium or substance compared to a vacuum. Therefore, consider a condenser which has a capacitance c_0 for a vacuum between two charged plates. The c ($= \epsilon c_0$) is the capacitance for a given medium. For air, $\epsilon = 1.0006$, for glass $\epsilon = 5.7$, for water $\epsilon = 81$ (Neumann and Pierson, 1966).

In liquid water, molecules tend to group together. This grouping is called polymerization and the groups are called polymers. Therefore, H_2O is monohydral, $(H_2O)_2$ is bihydral, and $(H_2O)_3$ is trihydral. The relative proportions of different forms depend on the temperature. Generally, the degree of polymerization decreases with increasing temperature.

Further, water arranges itself in different structures and can actually have three different forms or arrangements of molecules. The first form is a tetrahedral form and in this form water occupies the largest volume. The water can then have a form of a quartzite lattice structure and then finally a dense structure where the molecules are packed closely together. With increasing temperature, the relative proportions shift from the first form to the third form. This is the opposite effect of what one would normally expect from the general thermal expansion of a liquid.

The net effect of these properties of water is that the density of water depends in a special way on the water temperature as shown in Fig. 3.10. The density

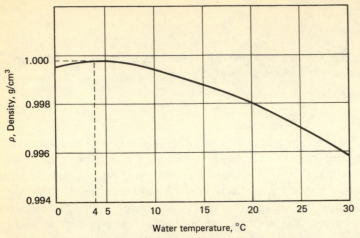

Figure 3.10   Variation of water density with temperature.

assumes its maximum value at 4°C and decreases with increasing temperature. In addition to the vertical stratification resulting in estuaries and bays from this density effect, this phenomenon is also of particular interest in the study of the vertical stratification in lakes (see Chapter 4).

The density of water also depends on the salt content or salinity of the water. Neumann and Pierson (1966) provide an excellent discussion of this dependence and the interaction with temperature. The density increases with increasing salinity and pressure. For most water quality modeling problems, pressure is not a significant variable in the density interaction. The variation of density with temperature and salinity is given by the equation of state (Crowley, 1968) as

$$\rho_{s,t,0} = 1 + \{10^{-3}[(28.14 - 0.0735T - 0.00469T^2)$$
$$+ (0.802 - 0.002T)(S - 35)]\} \tag{3.4}$$

where $\rho_{s,t,0}$ is the density at salinity S in parts per thousand (ppt), temperature T in °C and atmospheric pressure. For some studies in hydrodynamic circulation, it may be necessary to know the density to at least five decimal places and in order to facilitate computation, the quantity *sigma-t* has been defined as

$$\sigma_t = (\rho_{s,t,0} - 1) \cdot 10^3 \tag{3.5}$$

Thus, if $\rho_{s,t,0} = 1.02478$, then $\sigma_t = 24.78$, which is approximately the value for the open ocean. The temperature at which the density is a maximum can be computed from

$$T_{\rho\text{max}} = 3.975 - 0.2168S + 0.000128S^2 \tag{3.6}$$

for S in ppt (‰) and $T_{\rho\text{max}}$ in °C. Therefore, for salinity of about 30‰, (approximately offshore salinity), the temperature at which the density is a maximum drops from about 4 to about −2.4°C.

3.1.2.2 Two-Layer Estuarine Circulation   The preceding density effects, the relative amount of freshwater inflow, and the relationship of the estuary or bay to the

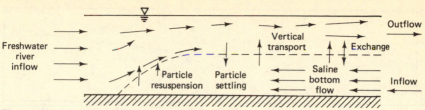

Figure 3.11 Schematic of two-dimensional flow in estuary.

open ocean all result in a complicated two-layered flow in some regions of tidal water bodies. Figure 3.11 indicates the general circulation. Postma (1967), Nichols and Poor (1967), O'Connor and Lung (1981), and Officer (1983) discuss this estuarine circulation in detail.

The principal feature to be noted is that the two-layer flow pattern consists of a net seaward transport in a surface layer and a net landward transport in a bottom layer of the estuary. The surface flow is fresh to brackish and the bottom flow is more saline and may generally be reduced in temperature. The more saline, colder, and thus more dense bottom water results in a vertical stratification that can have a significant effect on dissolved oxygen and nutrient cycling.

The landward transport of bottom water and the seaward transport of freshwater results in vertical velocities in the estuary and a convergence zone in the vicinity of the limit of saltwater intrusion. The suspended solids in the estuary are transported downstream in the surface waters, settle into the bottom waters, and then may be recylced upstream to the convergence area. In that vicinity, a "turbidity maximum" may occur. This is a region of the estuary where suspended solids concentrations are elevated due to the complex horizontal and vertical velocities and local resuspension of bottom sediments. Figure 3.12 shows such a solids maximum for the Potomac Estuary. Note that for this case it was necessary to estimate the contribution of the phytoplankton to the total solids. After that contribution was subtracted from the total solids, the estimated inorganic solids show a clear turbidity maximum in the vicinity of the limit of the saltwater intrusion. The transport of particulate and dissolved nutrients can also result in an intensification or "nutrient trap" in estuaries.

The two-layered estuarine circulation should be carefully analyzed in any actual estuarine problem and the degree of stratification should be assessed. If such stratification is not significant, as in the tidal river areas, then the following water quality analysis is appropriate. However, should vertical stratification be judged to be significant then a more complex two-layered analysis must be conducted (see for example, O'Connor and Lung, 1981). Section 3.6 below on finite difference approaches to the estuarine problem explores this area further.

3.2 DISTRIBUTION OF WATER QUALITY
IN ESTUARIES

Estimating the time and spatial behavior of water quality in estuaries is complicated by the effects of tidal motion as described above. The upstream and downstream

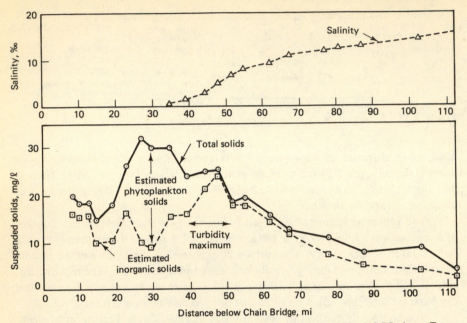

Figure 3.12 Turbidity maximum, Potomac Estuary, August 1977. USGS data. From Thomann and Fitzpatrick (1982).

currents produce substantial variations of water quality at certain points in the estuary and the calculation of such variation is indeed a complicated problem. Some simplifications can, however, be made which provide some remarkably useful results in estimating the distribution of estuarine water quality. The simplifications can be summarized through the following assumptions:

1. Estuary is one-dimensional.
2. Water quality is described as a type of average condition over a number of tidal cycles.
3. Area, flow, and reaction rates are constant with distance.
4. Estuary is in a steady state condition.

A water body is considered to be a one-dimensional estuary when it is subjected to tidal reversals (i.e., reversals in direction of the water velocity) and where only the longitudinal gradient of a particular water quality parameter is dominant.

In order to construct a model of a nonconservative variable, a mass balance around an element of the estuary can be performed. Before this is actually attempted, however, it is useful to reorient one's view of the time scale under examination. Instead of thinking in terms of real time as the scale of interest, the final equation can be simplified by considering the time scale to be composed of tidal cycle units. Thus, the mass balance and the subsequent model equations will be thought of as representing the water quality of the estuary over a sequence of tidal cycles. In other words, the models given here will not attempt to describe variations in water quality within a tidal cycle, but only from one tidal cycle to the next. Models that follow the time and space distribution of a variable from hour to hour can also be constructed and, for some problem contexts (e.g., highly transient storm water overflows), may be necessary. These models usually require large amounts of computer time for simulations. A considerable amount of insight into estuarine behavior can be obtained, however, by averaging tidal conditions with full recognition of the importance of within-tide variations.

When the system is viewed on a tidal cycle basis, the tidal reversal phenomenon introduces a relatively large amount of mixing. This can be visualized by considering the instantaneous discharge of a colored dye into an estuary at a time of zero velocity, say slack water before the ebb tide. This is illustrated in Fig. 3.13. As ebb tide begins, the dye proceeds downstream, and gradually is dispersed due to lateral and vertical velocity gradients. At the end of the ebb tide, the entire mass will halt and begin to return upstream on the flood tide. In the process, different particles will proceed at varying velocities. The mass of dye (now further spread out) will then return upstream toward the point of injection but the centroid of the mass will usually not return to the exact location of the injection point. The difference between the centroid of mass at the end of the tidal cycle is a measure of the time it takes to "flush" the estuary and reflects the discharge of freshwater into the upper end of the estuary.

In addition to this mixing, density differences in the more saline portion of the estuary may further add to the mixing process. This occurs because of the

Figure 3.13   Illustration of tidal mixing. (a) Dye "slug" at time zero. (b) At end of ebb tide. (c) At end of tidal cycle.

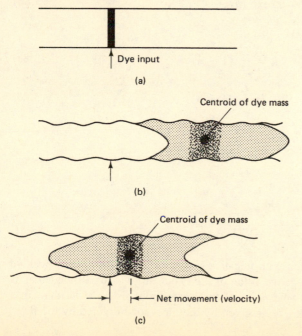

tendency for liquids of unequal density to induce water movements as the heavier water settles and lighter water rises. This overall phenomenon of mixing due to the temporal variation of tidal velocity, lateral and vertical gradients in velocity, and density differences is termed tidal dispersion. This dispersive mechanism is central to the analysis of estuarine water quality.

3.2.1 Water Quality Due to Point Sources

3.2.1.1 Nonconservative Substances

Under the assumptions given above, the following mass balance equations can be derived, see Section 3.5, (O'Connor, 1960, 1962, and 1965; Thomann, 1972) for an infinitely long, constant parameter, estuary with a point source, reactive waste input at $x = 0$,

$$E \frac{d^2s}{dx^2} - U \frac{ds}{dx} - Ks = 0 \tag{3.7}$$

for

$$s = s_0 \quad \text{at} \quad x = 0 \tag{3.7a}$$
$$s = 0 \quad \text{at} \quad x = \pm \infty \tag{3.7b}$$

where E is a tidal dispersion coefficient $[L^2/T]$ (a mixing coefficient which incorporates tidal flow oscillations, density effects, and other lateral and vertical velocity gradients) and $U = Q/A$, the net nontidal velocity, where $Q = Q_n$. The solution to Eq. 3.7 with conditions 3.7a and 3.7b is

$$s = s_0 \exp(j_1 x) \quad x \leq 0 \tag{3.8a}$$
$$= s_0 \exp(j_2 x) \quad x \geq 0 \tag{3.8b}$$

where

$$j_1 = \frac{U}{2E}(1 + \alpha) \tag{3.8c}$$

$$j_2 = \frac{U}{2E}(1 - \alpha) \tag{3.8d}$$

and

$$s_0 = \frac{W}{Q\alpha} \tag{3.8e}$$

where

$$\alpha = \sqrt{1 + \frac{4KE}{U^2}} \tag{3.8f}$$

The coefficient j_1 is associated with negative values of x and the j_2 with positive values of x, resulting in a negative value of the exponent of e for all values of x. The concentration at $x = 0$, the point of the waste input, is s_0. Note that, in a well-mixed river with no tides, s_0 is given by the input load divided by the flow, whereas in the estuary the concentration is reduced further by the coefficient α due to the transport of the substance upstream and downstream because of tidal mixing.

If the net flow is zero, then $U = 0$ and a purely dispersive system results with

$$j_1 = j_2 = \sqrt{K/E} \tag{3.9a}$$

$$s = s_0 \exp(-\sqrt{Kx^2/E}) \tag{3.9b}$$

and

$$s_0 = \frac{W}{2A\sqrt{KE}} \tag{3.9c}$$

The so-called estuary number KE/U^2 strongly controls the character of the solution (O'Connor, 1960; Dobbins, 1964; Kahlig, 1979). As KE/U^2 approaches zero, steady state concentration profiles begin to resemble those in a stream and, as KE/U^2 becomes large, profiles approach those in a purely dispersive estuary with no net flow.

3.2.1.2 Conservative Substances
For a conservative variable such as salinity or chlorides in the estuary, $K = 0$ and the mass balance equation is

$$E \frac{d^2s}{dx^2} - U \frac{ds}{dx} = 0 \tag{3.10}$$

with

$$s = 0 \quad \text{at} \quad x = -\infty \tag{3.10a}$$
$$s = s_0 \quad \text{at} \quad x = 0 \tag{3.10b}$$

The solution is

$$s = s_0 \exp\left(\frac{Ux}{E}\right) \qquad x \leq 0 \tag{3.11a}$$

$$s = s_0 \qquad x \geq 0 \tag{3.11b}$$

$$s_0 = \frac{W}{Q} \tag{3.11c}$$

Equation 3.11c is a rather interesting result since it indicates that for a discharge of a conservative variable into an estuary with all its complex mixing, the concentration at steady-state downstream of the input of the conservative substance is simply W/Q—the same result as for the river. This is to be expected since the variable is conservative and, therefore, the total mass of the discharge must pass through a downstream section or else some mass would have been lost—violating the conservative assumption. Sample Problem 3.1 illustrates the application of the estuary equations for conservative substances.

Figure 3.14 shows the water quality distributions for a nonconservative variable ($K \neq 0$) and for the conservative variable case ($K = 0$). Analyses of profiles

SAMPLE PROBLEM 3.1

DATA

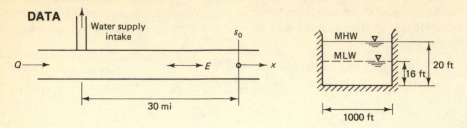

For a low flow period in the estuary, the dispersion coefficient is 5 smpd, the tidal excursion is 8 mi, and the concentration of chlorides—a conservative substance—is 18,000 mg/l at $x = 0$ (s_0).

PROBLEM

Determine the minimum value of the freshwater flow Q so that water may be withdrawn at the intake with a chloride concentration equal to or less than 250 mg/l. The magnitude of the flow diversion is negligible compared with Q.

ANALYSIS

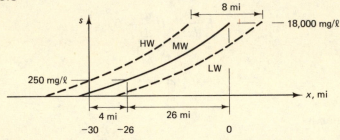

Since the MW concentration of 250 mg/l will translate 4 mi upstream on a flood tide, the location at which the 250 mg/l is required is not at $x = -30$ mi but $x = -26$ mi.

$$s = 250 = s_0 e^{Ux/E} = 18,000\, e^{U(-26)/5}$$

$$\ln \frac{250}{18,000} = -5.2U, \qquad U = 0.822 \text{ mpd}$$

$$U = 0.822 \text{ mpd}/16.4 \text{ (mpd/fps)} = 0.0501 \text{ fps}$$

With the MW cross-sectional area equal to the width times the MW depth,

$$A = 1000 \text{ ft} \times \left(\frac{20 + 16}{2} = 18 \text{ ft} \right) = 18,000 \text{ ft}^2$$

and

$$Q = UA = 0.0501 \text{ fps} \times 18,000 \text{ ft}^2 = \underline{900 \text{ cfs}}$$

Note: Both E and s_0 are functions of Q. The above conditions would be representative of a protracted (several month) low flow period.

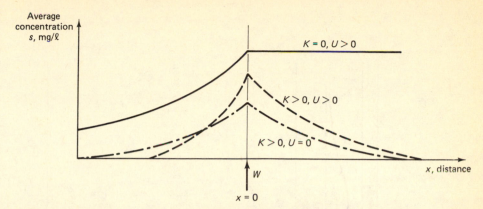

Figure 3.14 Distribution of concentration for idealized estuary for $K = 0$ and $K > 0$.

of this type provide a first approximation to the distribution of water quality in estuarine systems.

3.2.2 Water Quality Due to Distributed Sources

As with streams and rivers (Section 2.2.4), estuaries may be subjected to external and internal sources of a substance. Thus, for a reactive water quality constituent distributed along the length of a constant parameter estuary (Q, E, A, H, and K spatially constant), the differential equation is

$$E \frac{d^2s}{dx^2} - U \frac{ds}{dx} - Ks = S_D \qquad (3.12)$$

The spatially constant distributed loading (S_D) is assumed to begin at $x = 0$ and terminate at $x = a$ and the following boundary conditions are used:

$$s = 0 \quad \text{at} \quad x = -\infty \qquad (3.12a)$$
$$s = 0 \quad \text{at} \quad x = +\infty \qquad (3.12b)$$

Solution of the above differential equation for the "loaded" reach ($0 \leq x \leq a$) involves use of two "unloaded" estuarine reaches ($x < 0$ and $x > a$) on either side of the "loaded" reach in order to impose the boundary conditions. Consequently, there are three solution regions:

$$s_1, \quad \text{for} \quad x \leq 0$$
$$s_2, \quad \text{for} \quad 0 \leq x \leq a$$
$$s_3, \quad \text{for} \quad x \geq a$$

The solutions for all regions are summarized in Fig. 3.15 for two cases: $Q > 0$, the advective–dispersive case (Eqs. 3.13 through 3.15), and $Q = 0$, the purely dispersive case (Eqs. 3.16 through 3.18). Values of S_D, the distributed loading [$M/L^3 \cdot T$], are calculated for both external and internal sources the same as shown in Fig. 2.13.

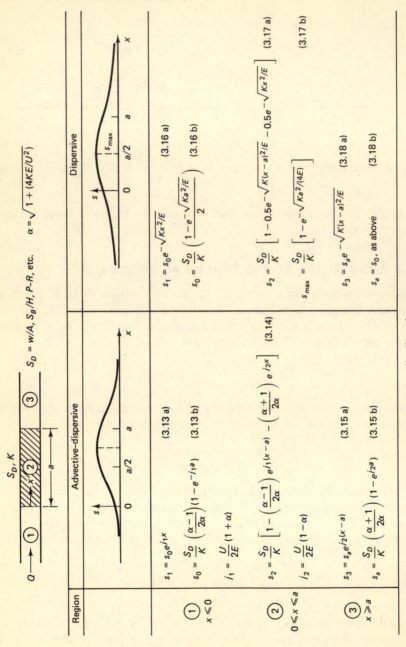

Figure 3.15 Estuarine water quality due to reactive distributed sources.

$S_D = w/A,\ S_B/H,\ P\text{-}R,\ \text{etc.}$ $\alpha = \sqrt{1 + (4KE/U^2)}$

Region	Advective-dispersive		Dispersive	
① $x \leq 0$	$s_1 = s_0 e^{i_1 x}$ (3.13 a) $s_0 = \dfrac{S_D}{K}\left(\dfrac{\alpha-1}{2\alpha}\right)(1-e^{-i_1 a})$ (3.13 b) $i_1 = \dfrac{U}{2E}(1+\alpha)$		$s_1 = s_0 e^{-\sqrt{Kx^2/E}}$ (3.16 a) $s_0 = \dfrac{S_D}{K}\left(\dfrac{1-e^{-\sqrt{Ka^2/E}}}{2}\right)$ (3.16 b)	
② $0 < x \leq a$	$s_2 = \dfrac{S_D}{K}\left[1 - \left(\dfrac{\alpha-1}{2\alpha}\right)e^{i_1(x-a)} - \left(\dfrac{\alpha+1}{2\alpha}\right)e^{i_2 x}\right]$ (3.14) $i_2 = \dfrac{U}{2E}(1-\alpha)$		$s_2 = \dfrac{S_D}{K}\left[1 - 0.5e^{-\sqrt{K(x-a)^2/E}} - 0.5e^{-\sqrt{Kx^2/E}}\right]$ (3.17 a) $s_{max} = \dfrac{S_D}{K}\left[1 - e^{-\sqrt{Ka^2/(4E)}}\right]$ (3.17 b)	
③ $x \geq a$	$s_3 = s_a e^{i_2(x-a)}$ (3.15 a) $s_a = \dfrac{S_D}{K}\left(\dfrac{\alpha+1}{2\alpha}\right)(1-e^{i_2 a})$ (3.15 b)		$s_3 = s_a e^{-\sqrt{K(x-a)^2/E}}$ (3.18 a) $s_a = s_0,\ \text{as above}$ (3.18 b)	

3.2.3 Multiple Sources

The differential equations for the distribution of reactive (Eqs. 3.7, 3.12) and conservative (Eq. 3.10) substances are linear, second-order equations. As such, the principle of superposition discussed previously (Section 2.2.6) applies to estuaries. An example of this principle is found in Fig. 3.16 where two point sources discharge a given substance into an estuary which has a constant flow and dispersion coefficient. Note that this presumes that the flows from the point sources are small compared to the upstream flow; if these flows are significant, the solutions given are not applicable. The substances discharged by each point source are assumed to have different reaction rates in the estuary—a value of K_1 for the upstream source and K_2 for the downstream discharge. This necessitates calculating two values of j_1 and j_2: j_{11} and j_{21} are the appropriate exponents in Eqs. 3.8a and 3.8b for the point source at the upstream location and j_{12} and j_{22} apply to the downstream source. Dashed lines in Fig. 3.16 represent the individual responses to each source and the solid line is the composite response. Boundary conditions are assumed equal to zero at $x = +\infty$ and $x = -\infty$, as usual. Sample Problem 3.2 illustrates multiple

Figure 3.16   Estuarine response to multiple point sources.

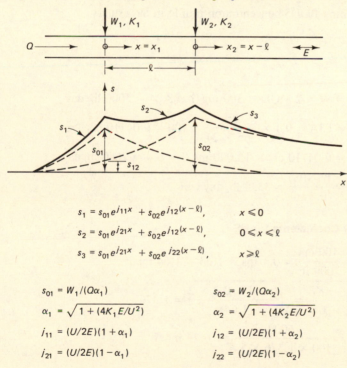

$$s_1 = s_{01}e^{j_{11}x} + s_{02}e^{j_{12}(x-\ell)}, \qquad x \leqslant 0$$
$$s_2 = s_{01}e^{j_{21}x} + s_{02}e^{j_{12}(x-\ell)}, \qquad 0 \leqslant x \leqslant \ell$$
$$s_3 = s_{01}e^{j_{21}x} + s_{02}e^{j_{22}(x-\ell)}, \qquad x \geqslant \ell$$

$$s_{01} = W_1/(Q\alpha_1) \qquad\qquad s_{02} = W_2/(Q\alpha_2)$$
$$\alpha_1 = \sqrt{1 + (4K_1E/U^2)} \qquad \alpha_2 = \sqrt{1 + (4K_2E/U^2)}$$
$$j_{11} = (U/2E)(1+\alpha_1) \qquad j_{12} = (U/2E)(1+\alpha_2)$$
$$j_{21} = (U/2E)(1-\alpha_1) \qquad j_{22} = (U/2E)(1-\alpha_2)$$

Note: $s_{02}e^{j_{12}(x-\ell)} = e^{j_{12}x} \cdot s_{02}e^{j_{12}(-\ell)} = s_{12}e^{j_{12}x}$, where s_{12} is the concentration at location 1 due to the discharge of material at location 2 at a distance of $x_2 = -\ell$ upstream of the second load.

SAMPLE PROBLEM 3.2

DATA

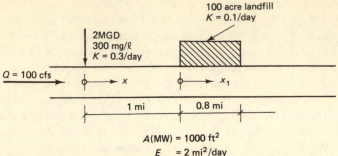

A point source ($K = 0.3$/day) and a more refractory ($K = 0.1$/day) distributed source (landfill) of BOD5 enter the estuary. For the distributed source, assume that 36 in. of rain falls annually, that the effective runoff coefficient is 0.1 and that the average BOD5 concentration in the runoff is 12000 mg/l.

PROBLEM

Determine the resulting BOD5 concentration profile in the estuary.

ANALYSIS

Calculate Loads

$$W(PS) = W = 2 \text{ MGD} \times 300 \text{ mg/l} \times 8.34 = 5000 \text{ lb/day}$$

$$Q(NPS) = CIA = 0.1 \times \frac{36}{365 \times 24} \times 100 = 0.04110 \text{ cfs}$$

$$W(NPS) = 0.04110 \text{ cfs} \times 12{,}000 \text{ mg/l} \times 5.4 = 2660 \text{ lb/day}$$

$$w = \frac{W(NPS)}{0.8 \text{ mi}} = 3330 \text{ lb/day} \cdot \text{mi}$$

Determine Estuary Coefficients for PS

$$U = \frac{Q}{A} = \frac{100 \text{ cfs}}{1000 \text{ ft}^2} = 0.1 \text{ fps} \times 16.4 = 1.64 \text{ mpd}$$

$$\alpha = \sqrt{1 + \frac{4KE}{U^2}} = \sqrt{\frac{1 + 4 \times 0.3 \times 2}{(1.64)^2}} = 1.376$$

$$s_0 = \frac{W}{Q\alpha} = \frac{5000}{100 \times 1.376 \times 5.4} = 6.73 \text{ mg/l}$$

$$j_1 = \frac{U}{2E}(1 + \alpha) = \left(\frac{1.64}{2} \times 1.64\right)(1 + 1.376) = 0.974/\text{mi}$$

$$j_2 = \frac{U}{2E}(1 - \alpha) = \left(\frac{1.64}{2} \times 1.64\right)(1 - 1.376) = -0.154/\text{mi}$$

Sample Problem 3.2 (continued)

**Determine Coefficients for NPS
(see Fig. 3.15)**

$$a = 0.8 \text{ mi}$$

$$\alpha = \sqrt{1 + 4 \times 0.1 \times \frac{2}{(1.64)^2}} = 1.139$$

$$S_D = \frac{w}{A} = \frac{3330 \text{ lb/day} \cdot \text{mi}}{1000 \text{ ft}^2 \times 5280 \text{ ft/mi} \times 62.4 \text{ lb/ft}^3} = 10.1 \text{ mg/l} \cdot \text{day}$$

$$j_1 = \frac{1.64}{2 \times 2} \times (1 + 1.139) = 0.877/\text{mi}$$

$$j_2 = \frac{1.64}{2 \times 2} \times (1 - 1.139) = -0.0570/\text{mi}$$

$$s_0 = \frac{S_D}{K}\left(\frac{\alpha - 1}{2\alpha}\right)\left(1 - e^{-j_1 a}\right) = \frac{10.1}{0.1}\left(\frac{1.139 - 1}{2 \times 1.139}\right)(1 - e^{-0.877 \times 0.8})$$

$$= 3.11 \text{ mg/l}$$

$$s_a = \frac{S_D}{K}\left(\frac{\alpha + 1}{2\alpha}\right)(1 - e^{-j_2 a}) = \frac{10.1}{0.1}\left(\frac{1.139 + 1}{2 \times 1.139}\right)(1 - e^{-0.0570 \times 0.8})$$

$$= 4.22 \text{ mg/l}$$

Estuarine BOD5 Concentration Equations

See Eqs. 3.8a and 3.8b and Fig. 3.15.

For $x \leq 0$

$$s = 6.73e^{0.974x} + 3.11e^{0.877x_1}, \qquad \text{where } x_1 = x - 1$$

For $0 \leq x \leq 1 \text{ mi}$

$$s = 6.73e^{-0.154x} + 3.11e^{0.877x_1}$$

For $1 \leq x \leq 1.8 \text{ mi}$

$$s = 6.73e^{-0.154x} + s(\text{NPS})$$

$$s(\text{NPS}) = \frac{S_D}{K}\left[1 - \left(\frac{\alpha - 1}{2\alpha}\right)e^{j_1(x_1 - a)} - \left(\frac{\alpha + 1}{2\alpha}\right)e^{j_2 x_1}\right]$$

$$= \frac{10.1}{0.1}\left[1 - \left(\frac{1.139 - 1}{2 \times 1.139}\right)e^{0.877(x_1 - 0.8)} - \left(\frac{1.139 + 1}{2 \times 1.139}\right)e^{-0.0570x_1}\right]$$

$$s(\text{NPS}) = 101[1 - 0.0610e^{0.877(x_1 - 0.8)} - 0.939e^{-0.0570x_1}]$$

For $x \geq 1.8 \text{ mi}$

$$s = 6.73e^{-0.154x} + 4.22e^{-0.0570(x_1 - 0.8)}$$

(continued)

Sample Problem 3.2 (continued)

Resulting Concentration Profiles

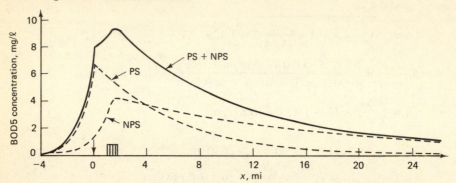

Notice the exponential decay of the PS concentrations upstream and downstream of the discharge at $x = 0$. The NPS peaks within the loaded length and exponentially decays on either side at a slower rate than the PS.

inputs for an estuary with point and distributed sources of a nonconservative parameter.

3.3 ESTIMATION OF TIDAL DISPERSION COEFFICIENT

The principal additional parameter introduced in the estuarine water quality equation (Eq. 3.7) is the tidal dispersion coefficient, E. Several approaches can be used to estimate this parameter:

1. Use of estuarine salinity as a tracer of the tidal mixing.
2. For tidal river freshwater portions of estuaries, discharge of dye as tracer substance.
3. Use of hydrodynamic theory incorporating velocity shear and salt diffusion mechanisms.

The first approach makes use of Eq. 3.11a with $x = 0$ at the ocean boundary or some other convenient boundary where the salinity can be specified. In all cases, the coordinate x is positive in the direction of net flow, that is, oceanward. Data are obtained at various distances upstream into the estuary and plotted on a semilogarithmic scale. Thus, Eq. 3.11a is written as

$$s = s_0 \exp(Ux/E)$$

or

$$\ln \frac{s}{s_0} = \frac{U}{E} x \qquad (3.19)$$

The slope U/E, as shown by Eq. 3.19, permits estimation of the tidal dispersion coefficient if U, the net advective velocity is known. An example is shown in Fig. 3.17. Alternately, selection of two points (s_1, x_1) and (s_2, x_2) on the straight line which best fits the salinity data on the semilog plot permits calculation of the dispersion coefficient from

$$E = \frac{U(x_2 - x_1)}{\ln (s_2/s_1)} \qquad (3.20)$$

Figure 3.17   Salinity distribution in Hudson Estuary.

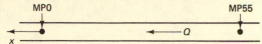

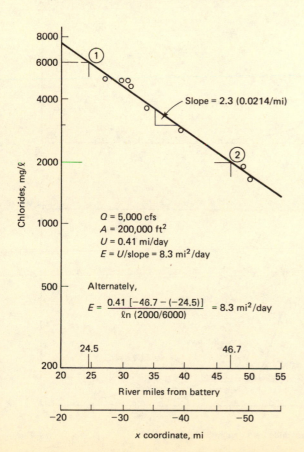

The second approach requires the discharge of a tracer substance such as dye or radioactive tracers. In recent years, the use of dyes, such as Rhodamine, has become more widely applied because of ease of handling and relatively low cost. In this technique, dye is discharged as a "slug" at a particular stage of tide and followed over several tidal cycles. The spread of the dye cloud over time is a measure of the tidal dispersion coefficient. The equation used is a time variable equation and is given by Krenkel and Orlob (1962) and Diachishin (1963) as

$$s = \frac{M}{A\sqrt{2\pi}\,\sigma_x} \exp\left[-\frac{1}{2}\left(\frac{x - Ut}{\sigma_x}\right)^2 \right] \tag{3.21}$$

Figure 3.18 Dye distributions—Wicomico River Estuary, July 22, 1974. From Salas and Thomann (1975). Data from Maryland DEC.

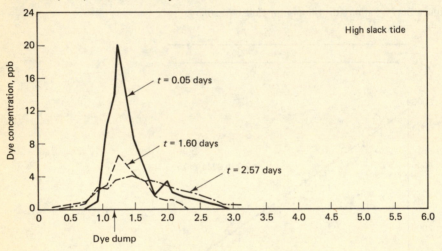

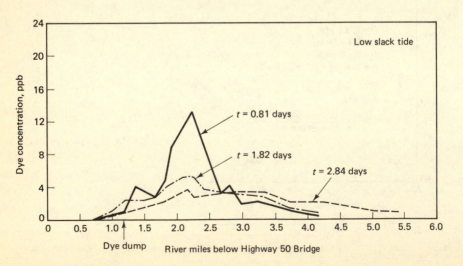

where M is the mass of dye discharged $[M]$ and

$$\sigma_x = \sqrt{2Et}$$

It is to be noted that this equation is identical with that of Eq. 2.39 (for streams) in the case of a conservative substance ($K = 0$). For the mean water, steady state estuarine model as contrasted with streams, time is measured in tidal cycles and sampling is conducted from tidal cycle to tidal cycle—as discussed below. In addition, the magnitude of the dispersion coefficient is typically 1–10 smpd (mi^2/day) compared with stream values of 0.01–0.1 smpd. The results of a dye study in the Wicomico River—an estuary of Chesapeake Bay—are shown in Fig. 3.18 (Salas and Thomann, 1975). Note that the peaks of the high water slack samples ($t = \frac{1}{2}$, $1\frac{1}{2}$, and $2\frac{1}{2}$ tidal cycles after the dump) are approximately at the location of the dye dump and the low water peaks are at a location approximately 1 mi downstream of the dump. The lack of movement of the peaks downstream indicates that this estuary has very little freshwater flow and is predominately a dispersive system. An example of the analysis of a dye profile—similar to those in the Wicomico River estuary—to obtain a dispersion coefficient is contained in Sample Problem 3.3. Dispersion coefficients for selected estuaries are shown in Table 3.3.

TABLE 3.3 ESTUARINE LONGITUDINAL DISPERSION COEFFICIENTS

Estuary	Flow (cfs)	Dispersion coefficient (smpd)	Reference
Hudson River, NY	5,000	20	Hydroscience (1971)
East River, NY	0	10	Hydroscience (1971)
Cooper River, SC	10,000	30	Hydroscience (1971)
South River, NJ	23	5	Hydroscience (1971)
Houston Ship Channel, TX	900	27	Hydroscience (1971)
Cape Fear River, NC	1,000	2–10	Hydroscience (1971)
Compton Creek, NJ	10	1	Hydroscience (1971)
Wappinger and Fishkill Creek, NY	2	0.5–1	Hydroscience (1971)
River Foyle, N. Ireland	250	5	Hydroscience (1971)
Delaware River, upper		2–7	Thomann (1972)
Delaware River, lower		7–11	Thomann (1972)
Potomac River, upper		0.6–6	Thomann (1972)
Potomac River, lower		6–10	Thomann (1972)
Narrows of Mercey		4.4–12	Tetra-Tech (1978)
San Francisco Bay, southern		0.6–6	Tetra-Tech (1978)
San Francisco Bay, northern		1.5–62	Tetra-Tech (1978)
Rio Quayas, Ecuador		25	Fischer et al. (1979)
Thames River, England, low flow		1.8–2.8	Fischer et al. (1979)
Thames River, England, high flow		11	Fischer et al. (1979)

SAMPLE PROBLEM 3.3

DATA

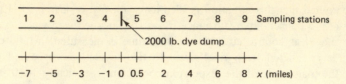

Two-thousand pounds of a conservative tracer are rapidly distributed across the width of an estuary at slack water before flood ($t = 0$). One tidal cycle later ($t = 12.42$ hr) a sampling crew simultaneously takes water samples at the nine stations indicated above. Resulting tracer concentrations in nanograms per liter (10^{-6} mg/l) are:

Station	1	2	3	4	5	6	7	8	9
s (µg/l)	0.9	1.8	3.0	3.7	4.0	3.9	2.6	1.8	0.8

The cross-sectional area at mean water is 140,000 ft² and freshwater flow is 7000 cfs, values typical of summer conditions in the Hudson River off the west side of Manhattan, New York.

PROBLEM

Estimate the dispersion coefficient.

ANALYSIS

Net freshwater velocity

$$U = \frac{Q}{A} = \frac{7000}{140,000} = 0.05 \text{ fps} \times 16.4 \frac{\text{mpd}}{\text{fps}} = 0.82 \text{ mpd}$$

With the tracer decay rate (K) equal to zero and with $t = 12.42$ hr $= 0.518$ days, concentrations can be estimated from Eq. 3.21.

$$s = \frac{M}{2A\sqrt{\pi E_x t}} \exp\left[-\frac{(x - Ut)^2}{4E_x t} \right] \qquad \text{[Eq. (3.21)]}$$

A plot of ln s vs. $(x - Ut)^2$ should yield a straight line, the slope of which equals $[-(4E_x t)^{-1}]$.

Station	1	2	3	4	5	6	7	8	9
$(x - 0.425)^2$ (mi²)	55.1	20.4	11.7	2.0	0.01	2.5	12.8	31.1	57.4

where $0.425 = Ut = 0.82$ mpd $\times 0.518$ days.

Sample Problem 3.3 (continued)

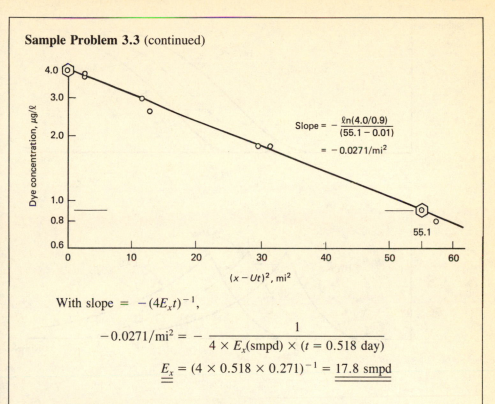

With slope $= -(4E_x t)^{-1}$,

$$-0.0271/\text{mi}^2 = -\frac{1}{4 \times E_x(\text{smpd}) \times (t = 0.518 \text{ day})}$$

$$E_x = (4 \times 0.518 \times 0.271)^{-1} = \underline{\underline{17.8 \text{ smpd}}}$$

Note: The analysis could be strengthened by obtaining another set of samples on the next tidal cycle at $t = 24.84$ hr. In that case, samples could be located symmetrically about the peak which would be at $x = Ut = 0.82$ mpd $\times$ (24.84/24)days $= 0.85$ mi.

3.4 SLACK TIDE SAMPLING

For some estuaries, where the movement of the tidal wave is progressive up the estuary, sampling of the estuary at the same stage of the tide may be possible. For example, suppose a sample is taken at a particular location at say high water slack (HWS) just at the turn of the tide. If a sampling boat is available that can move up the estuary faster than the movement of the tidal wave (about 20–30 mph) then the sampling crew can reach the next upstream station before HWS reaches that point. A sample can then be taken at each successive upstream station at HWS and a synoptic picture is obtained of the variation of the water quality at a fixed stage of the tide. If the system is a stationary wave system (where HWS occurs almost simultaneously everywhere) then a large sampling crew is necessary and slack sampling may be impractical.

Figure 3.19 shows the HWS and low water slack (LWS) profiles for salinity. As seen, a translation of the salinity results from the tidal oscillation. The inter-

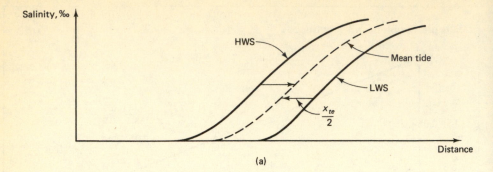

(a)

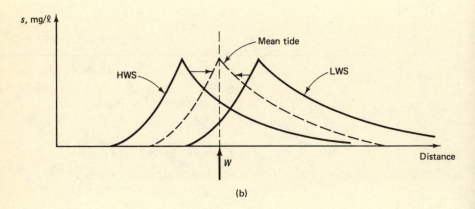

(b)

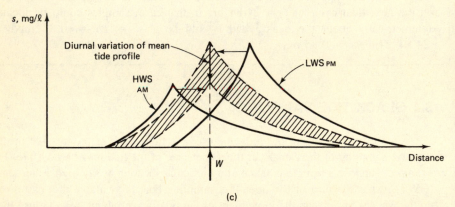

(c)

Figure 3.19 Distribution at slack tide and mean tide. (a) Salinity. (b) Nonconservative variable. (c) Nonconservative variable with diurnal input.

pretation of Eq. 3.11 can be seen from Fig. 3.19(a). If the HWS and LWS are each translated one-half of the tidal excursion distance, then an estimate is obtained of the mean tide profile, as is calculated by Eq. 3.11. Figure 3.19(b) shows a similar interpretation of Eq. 3.8 for the discharge of a nonconservative substance. Finally, Fig. 3.19(c) shows the case where the discharge may be varying with a diurnal period resulting in unequal profiles at the HWS and LWS times.

Sampling of the estuary at slack water can therefore be useful in certain cases for delineating water quality profiles in estuaries. It need only be remembered that a translation of the slack profile should be made to obtain an estimate of the mean tide profile which forms the basis of the preceding equations.

3.5 DERIVATION OF ESTUARY EQUATION

The second-order, linear differential equation for the mean water (MW) estuary water quality concentrations, Eq. 3.7, is applicable to a particular case of a more general one-dimensional estuarine setting. In the general case, flows, cross-sectional areas, dispersion coefficients, and kinetic rates may vary along the length of the estuary as indicated in Fig. 3.20. All estuary coefficients are assumed temporally constant for the MW model and intratidal variations are excluded. The components of a mass balance around a small volume element are shown in this figure to be net advection and net dispersion across the upstream and downstream cross-sectional areas and a loss of mass due to decay. The change in mass in the volume element with respect to time is then equal to the sum of these three components:

$$V\frac{\partial s}{\partial t} = -Q\frac{\partial s}{\partial x}\Delta x - \Delta Q \cdot s + \frac{\partial}{\partial x}\left(EA\frac{\partial s}{\partial x}\right)\Delta x - KsV \qquad (3.22)$$

Dividing both sides by $V = A\Delta x$ and letting Δx approach zero,

$$\frac{\partial s}{\partial t} = -\frac{1}{A}\left(Q\frac{\partial s}{\partial x} + s\frac{\partial Q}{\partial x}\right) + \frac{1}{A}\frac{\partial}{\partial x}\left(EA\frac{\partial s}{\partial x}\right) - Ks$$

Noting that

$$\frac{\partial}{\partial x}(Qs) = Q\frac{\partial s}{\partial x} + s\frac{\partial Q}{\partial x}$$

$$\frac{\partial s}{\partial t} = \frac{1}{A}\frac{\partial}{\partial x}\left(EA\frac{\partial s}{\partial x}\right) - \frac{1}{A}\frac{\partial}{\partial x}(Qs) - Ks \qquad (3.23)$$

For steady state conditions, where s is constant in time, the general one-dimensional steady-state differential equation is

$$0 = \frac{1}{A}\frac{d}{dx}\left(EA\frac{ds}{dx}\right) - \frac{1}{A}\frac{d}{dx}(Qs) - Ks \qquad (3.24)$$

Finally, if the estuary has spatially invariant flows, cross-sectional areas, dispersion coefficients, and kinetic rates, the following steady state, constant parameter equation results:

$$0 = E\frac{d^2s}{dx^2} - U\frac{ds}{dx} - Ks \qquad (3.25)$$

This is the equation previously presented as Eq. 3.7. A detailed derivation of the general three-dimensional advective–diffusion equation is found in Fisher et al. (1979), together with a disscussion of the phenomena incorporated in the estuarine dispersion coefficients.

Geometry

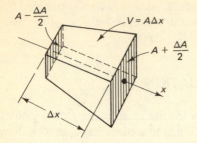

Transport coefficients

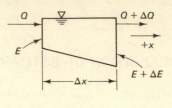

Mass Balance Components per Unit Time

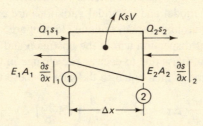

Advection

$$\frac{\partial M}{\partial t} = Q_1 s_1 - Q_2 s_2 = Qs - (Q + \Delta Q)\left(s + \frac{\partial s}{\partial x} \Delta x\right) = -Q \frac{\partial s}{\partial x} \Delta x - \Delta Q \cdot s$$

Dispersion

$$\frac{\partial M}{\partial t} = E_2 A_2 \left.\frac{\partial s}{\partial x}\right|_2 - E_1 A_1 \left.\frac{\partial s}{\partial x}\right|_1$$

$$= (E + \Delta E)(A + \Delta A)\left(\frac{\partial s}{\partial x} + \frac{\partial}{\partial x}\left(\frac{\partial s}{\partial x}\right)\Delta x\right) - EA \frac{\partial s}{\partial x} = \frac{\partial}{\partial x}\left(EA \frac{\partial s}{\partial x}\right)\Delta x$$

Reaction

Assume V = constant in time, first-order kinetics

$$\frac{\partial M}{\partial t} = V \frac{\partial s}{\partial t} = V\left[\frac{\partial}{\partial t}(s = s_0 e^{-Kt}) = -K s_0 e^{-Kt} = -Ks\right] = -KsV$$

Note: All estuary coefficients are assumed temporally
        constant for the MW model. Intratidal variations
        are not included.

Figure 3.20 Definition of estuarine mass balance components.

3.5.1 Mass Balance for Finite Length of Estuary

Mass balances can be written for any length of the estuary, for any location. Assuming a mass balance is desired about a point source for distances x_1 upstream of the load and x_2 downstream, where $x = 0$ is located at the point source, the magnitude of the point source would be shown as an input in Fig. 3.20. The net flux of mass due to transport across the upstream and downstream faces is then added to the input W and the mass reacting away as follows:

$$0 = \text{mass in} - \text{mass out}$$

$$0 = W + Q_1 s_1 + E_2 A_2 \left. \frac{ds}{dx} \right|_2$$

$$- \left(Q_2 s_2 + E_1 A_1 \left. \frac{ds}{dx} \right|_1 + \int_V Ks \, dV \right) \tag{3.26}$$

where

$$\int_V Ks \, dV = KA \int_{x_1}^{x_2} s \, dx$$

Thus, by evaluating concentrations, s_1 and s_2, and concentration gradients, $ds/dx|_1$ and $ds/dx|_2$, at the upstream and downstream faces of the desired reach and calculating the integrated mass decaying between these locations, the relative magnitudes of all components of the mass balance can be determined, as shown in Sample Problem 3.4.

3.6 MODELING "REAL" ESTUARIES— FINITE SEGMENT (FINITE DIFFERENCES)

The preceding analyses have assumed an ideal estuary, that is, an estuary with constant cross-sectional area, flow, depth, and reaction rates. Such estuaries exist only rarely in nature, but the general solution to even such idealized conditions illustrates the important principles of estuarine water quality analyses. Nevertheless there are many problem contexts where such drastic simplifications are not appropriate. A simple procedure is therefore necessary to approach more complicated real estuaries. Such a procedure involves dividing the estuary into a series of reaches, or segments, and writing the mass balance equation for each reach assuming the gradient is not significant within the reach. The development given here is a simplified version of the more extensive treatment given in Section 3.7. The approach is essentially a numerical finite difference approximation to Eq. 3.24 for the one-dimensional case. (The case of bays and harbors and vertical stratification is considered in Sections 3.6.4 and 3.6.5.)

3.6.1 The One-Dimensional Finite Segment Estuary

Figure 3.21 shows a typical estuary with increasing area as one approaches the ocean or the bay to which the estuary is a tributary. Also, the average depth of the estuary usually varies along the length of the estuary with occasional deep holes and shallow areas. In this development, it is assumed that mixing processes keep the estuary vertically and laterally (across the estuary) well mixed. Gradients are therefore only along the axis of the estuary. Several waste sources may be discharging to the estuary at various points and tributaries of freshwater may also be

SAMPLE PROBLEM 3.4

DATA

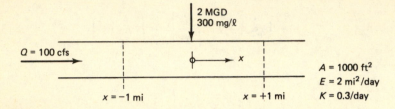

- 2 MGD
- 300 mg/ℓ
- $Q = 100$ cfs
- x
- $x = -1$ mi
- $x = +1$ mi
- $A = 1000$ ft^2
- $E = 2$ mi^2/day
- $K = 0.3$/day

PROBLEM

Given the estuary in Sample Problem 3.2 and only considering the point source, demonstrate that a mass balance exists for the reach between $x = \pm 1$ mi.

ANALYSIS

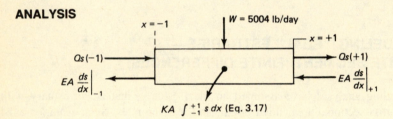

$x = -1$
$W = 5004$ lb/day
$x = +1$
$Qs(-1)$
$Qs(+1)$
$EA \left. \dfrac{ds}{dx} \right|_{-1}$
$EA \left. \dfrac{ds}{dx} \right|_{+1}$
$KA \int_{-1}^{+1} s\, dx$ (Eq. 3.17)

From Sample Problem 3.2, due to the point source alone

$$s = 6.734e^{0.9742x} \qquad \text{for} \quad x \leq 0$$
$$s = 6.734e^{-0.1542x} \qquad \text{for} \quad x \geq 0$$

Advective Transport

Thus,

$$s(-1) = 6.734e^{0.9742 \times (-1)} = 2.542 \text{ mg/l}$$
$$s(+1) = 6.734e^{-0.1542 \times 1} = 5.772 \text{ mg/l}$$

and

$$Qs(-1) = 100 \times 2.542 \times 5.391 = 1370 \text{ lb/day}$$
$$Qs(+1) = 100 \times 5.772 \times 5.391 = 3112 \text{ lb/day}$$

Dispersive Transport

$$\left. \frac{ds}{dx} \right|_{-1} = 0.9742/\text{mi} \times 6.734e^{0.9742 \times (-1)}\text{mg/l} = 2.476 \text{ mg/l·mi}$$

Sample Problem 3.4 (continued)

$$\left.\frac{ds}{dx}\right|_{+1} = -0.1542 \times 6.734 e^{-0.1542 \times (1)} = -0.8900 \text{ mg/l·mi}$$

$$EA \left.\frac{ds}{dx}\right|_{-1} = 2 \text{ mi smpd} \times 1000 \text{ ft}^2 \times 2.476 \text{ mg/l·mi} \times 5280 \text{ ft/mi}$$

$$\times 1 \text{ day/86,400}s \times 5.391$$

$$= 658.9 \times 2.476 = 1631 \text{ lb/day}$$

$$EA \left.\frac{ds}{dx}\right|_{+1} = 658.9 \times (-0.8900) = -586 \text{ lb/day (downstream)}$$

Decay

$$\int_{-1}^{+1} Ks \, dV = KA \int_{-1}^{0} s \, dx + \int_{0}^{1} s \, dx$$

$$\int_{-1}^{1} s \, dx = 6.734 \int_{-1}^{0} e^{0.9742x} \, dx + \int_{0}^{1} e^{-0.1542x} \, dx$$

$$= 6.734 \left[\frac{1}{0.9742} \left. (e^{0.09742x}) \right|_{-1}^{0} + \frac{1}{-0.1542} \left. (e^{-0.1542x}) \right|_{0}^{1} \right]$$

$$= 6.734 \left[\frac{1 - e^{-0.9742}}{0.9742} + \frac{e^{-0.1542} - 1}{-0.1542} \right] = 10.55 \text{ (mg/l)} \cdot \text{mi}$$

$$\therefore \int_{-1}^{1} Ks \, dV = 0.3/\text{day} \times 1000 \text{ ft}^2 \times 10.55 \text{ (mg/l)} \cdot \text{mi} \times 5280 \text{ ft/mi}$$

$$\times 1 \text{ day/86,400}s \times 5.391 = 1042 \text{ lb/day}$$

Summary (lb/day)

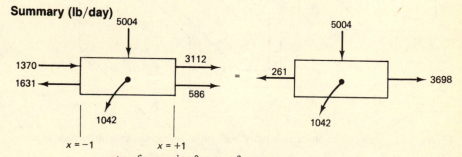

$$\text{rate of mass in} \overset{?}{=} \text{rate of mass out}$$

$$1370 + 5004 \overset{?}{=} 1631 + 1042 + 3112 + 586$$

$$6374 \approx 6371$$

Note: *Net* transport of mass is upstream across the cross section at $x = -1$ and downstream at $x = +1$. At $x = \pm\infty$, these net fluxes are zero and

$$W = \int_{-\infty}^{+\infty} Ks \, dV$$

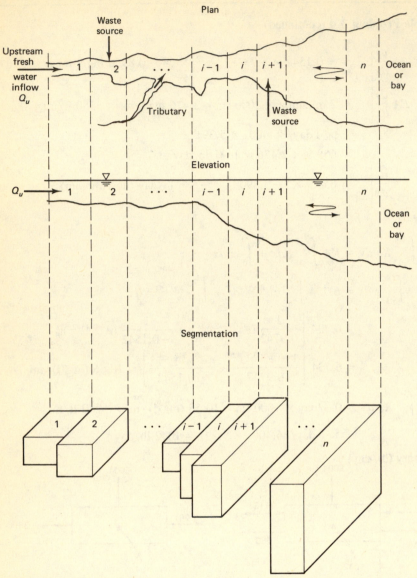

Figure 3.21 Representation of an estuary with finite segments (finite difference approximation).

discharging to the estuary. Figure 3.21 shows the division of the estuary into a series of finite reaches or segments. The concentration of a substance s $[M/L^3]$ is to be calculated in the center of each segment. The question of how many segments to use, that is, the length of each segment will be addressed more specifically later. At this stage, it is logical to use a segment length that captures approximately the principal gradients of the substance, s. With the computational power available today, one need not be overly concerned with the number of segments, but as a general guideline for estuaries, segment lengths of 1–2 mi would normally provide a good representation of the actual estuary. The segments do not need to be of a constant length, but for convenience here, equal lengths will be assumed. As noted in Fig. 3.21, the segments are numbered from 1 to n with the internal segments designated, $i - 1$, i, and $i + 1$.

3.6.1.1 The Mass Balance around Segment i

The procedure for developing the basic mass balance equation around segment i is actually quite simple. Four components are necessary: (1) transport of s due to advective flow, (2) mass transport due to tidal dispersion and density mixing, (3) loss of mass due to decay, and (4) any external sources or sinks of s. Figure 3.22 shows the notation.

First, consider the mass per unit time of s transferred into and out of segment i due to the net advective flow. The mass input $[M/T]$ from the upstream segment is simply

$$+Q_{i-1,i}s_{i-1,i} = +Q_{i-1,i}s_{i-1} \tag{i}$$
$$[L^3/T][M/L^3] = [M/T]$$

where $s_{i-1,i}$ is the concentration at the interface between $i - 1$ and i. This concentration can be approximated most simply by just using the concentration in the center of the segment $i - 1$, that is, s_{i-1} (This approximation actually constitutes a "backward differencing" scheme for the numerical approximation of the basic differential equation. Other approximations can be used (3.7.1) but the essence of the approach remains the same.) The advective mass transport leaving the segment i is then

$$-Q_{i,i+1}s_{i,i+1} = -Q_{i,i+1}s_i \tag{ii}$$

where the same approximation to the interfacial concentration has been applied and the negative sign indicates mass leaving the segment.

The mass exchange due to dispersion and mixing processes simply states that such mass transport is directly proportional to the concentration difference between adjacent segments. Thus for the dispersive exchange of mass with the upstream

Figure 3.22 Notation for ith segment showing advective flow, tidal dispersion, and density mixing and decay of the substance s_i.

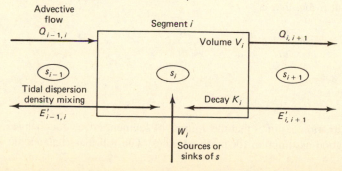

segment, one writes

$$+E'_{i,i-1}(s_{i-1} - s_i) \tag{iii}$$
$$[L^3/T][M/L^3] = [M/T]$$

where $E'_{i-1,i}$ is termed the bulk dispersion coefficient with dimensions $[L^3/T]$. In the downstream direction, a similar expression is used, that is,

$$+E'_{i+1,i}(s_{i+1} - s_i) \tag{iv}$$

In these dispersive quantities, if the upstream concentration s_{i-1} is greater than s_i, then mass is transferred from $i - 1$ to i. If, however, s_i is greater than s_{i-1} then mass is transferred upstream by mixing processes from i to $i - 1$. Similar comments apply in the downstream direction. If the concentrations are equal, then no mass exchange occurs. The bulk dispersion coefficient E' and the previously discussed tidal dispersion coefficient E are related as follows (Thomann, 1972):

$$E'_{i,i-1} = \frac{E_{i,i-1}A_{i,i-1}}{\overline{\Delta x}_{i,i-1}} \tag{3.27}$$

where $A_{i,i-1}$ is the cross-sectional area between i and $i - 1$ and $\overline{\Delta x}_{i,i-1}$ is the average length of the adjacent segments. A similar expression is used for i, $i + 1$.

The loss of mass due to first-order decay processes is

$$-K_i V_i s_i \tag{v}$$
$$[1/T][L^3][M/L^3] = [M/T]$$

where K_i is the decay coefficient $[1/T]$ for the ith segment and V_i is the volume of segment i $[L^3]$. Finally, sources or sinks of the substances s in segment i are designated

$$\pm W_i \tag{vi}$$
$$[M/T]$$

The sum of the mass quantities (i) through (vi) must equal the rate of change of the mass in segment i, M_i, that is

$$\frac{dM_i}{dt} = \frac{d(V_i s_i)}{dt} \approx V_i \frac{ds_i}{dt}$$

for

$$M_i = V_i s_i$$

and where the approximation assumes that the volume of segment i does not change with time, usually a good assumption for most estuaries. The full mass balance is then

$$V_i \frac{ds_i}{dt} = Q_{i-1,i}s_{i-1} - Q_{i,i+1}s_i + E'_{i-1,i}(s_{i-1} - s_i) + E'_{i+1,i}(s_{i+1} - s_i)$$

Net advective mass
transport

Net dispersive mass
transport

$$-K_i V_i s_i \qquad \pm W_i \qquad (3.28)$$

Mass loss
due to decay

Sources and sinks

This equation is a numerical approximation to Eq. 3.23 which clearly shows the four components of the mass transport around segment i. It is a first-order, ordinary, time-dependent, linear differential equation. If the estuary is now assumed to be at steady state, that is, all inputs, flow, exchanges, and reaction rates are temporally constant then

$$V_i \frac{ds_i}{dt} = 0$$

and Eq. 3.28 becomes a simple linear algebraic equation. There will be n such equations, one for each segment.

Therefore, the calculation of the concentration of the substance s in each segment of an estuary with spatially variable area, depth, flow, and reaction coefficients, is simply the solution of n simultaneous linear algebraic equations, an easy task given contemporary computer capability.

Special attention must be paid to the first and last segments, the "boundary" segments. A mass balance for the first segment gives at steady state

$$0 = Q_{u1}s_u - Q_{12}s_1 + E'_{21}(s_2 - s_1) - V_1 K_1 s_1 \pm W_1 \qquad (3.29)$$

where Q_{u1} is the upstream flow discharging into the first segment and s_u is the upstream input concentration, assumed to be known. This equation assumes no exchange between the first segment and the upstream river entering the estuary (i.e., a fall line or a dam).

A mass balance for the last, nth segment gives

$$0 = Q_{n-1,n}s_{n-1} - Q_{nb}s_n + E'_{n-1,n}(s_{n-1} - s_n)$$
$$+ E'_{nb}(s_b - s_n) - V_n K_n s_n \pm W_n \qquad (3.30)$$

where Q_{nb} is the net flow from segment n to the bay (or ocean) and s_b is the bay or ocean concentration of s, assumed known as the downstream boundary concentration.

Algebraic manipulation of the above equations will provide additional insight. Returning to Eq. 3.28 with $ds_i/dt = 0$ for steady state, if all terms in s_{i-1}, s_i and s_{i+1} are grouped on the left-hand side, one obtains

$$(-Q_{i-1,i} - E'_{i-1,i})s_{i-1} + (Q_{i,i+1} + E'_{i-1,i} + E'_{i,i+1}$$
$$+ V_i K_i)s_i + (-E'_{i,i+1})s_{i+1} = W_i \qquad (3.31)$$

Letting

$$a_{i,i-1} = -Q_{i-1,i} - E'_{i-1,i} \tag{3.32a}$$

$$a_{i,i} = Q_{i,i+1} + E'_{i-1,i} + E'_{i,i+1} + V_i K_i \tag{3.32b}$$

$$a_{i,i+1} = -E'_{i,i+1} \tag{3.32c}$$

one obtains

$$a_{i,i-1}s_{i-1} + a_{i,i}s_i + a_{i,i+1}s_{i+1} = W_i \tag{3.33}$$

as the general equation for the ith segment. The dimensions of the a_{ij} coefficients are $[L^3/T]$.

For the first segment, (see Eq. 3.29), following a similar procedure, the equation is

$$a_{11}s_1 + a_{12}s_2 = W_1 + Q_{u1}s_u$$

or

$$a_{11}s_1 + a_{12}s_2 = W'_1 \tag{3.34}$$

since s_u must be specified. For the nth segment (Eq. 3.30),

$$a_{n,n-1}s_{n-1} + a_{n,n}s_n = W_n + E'_{nb}s_b$$

or

$$a_{n,n-1}s_{n-1} + a_{n,n}s_n = W'_n \tag{3.35}$$

The complete set of equations to be solved is now given by

$$
\begin{aligned}
a_{11}s_1 + a_{12}s_2 +\quad 0\quad + \cdots \qquad\qquad + 0 &= W'_1 \\
a_{21}s_1 + a_{22}s_2 + a_{23}s_3 +\quad 0\quad + \cdots \qquad + 0 &= W_2 \\
0\quad + a_{32}s_2 + a_{33}s_3 + a_{34}s_4 +\quad 0\quad + \cdots + 0 &= W_3 \\
\vdots \qquad\qquad \vdots \qquad\qquad\qquad \vdots \qquad\qquad \\
0\quad + \qquad \cdots\quad + 0\quad + a_{n,n-1}s_{n-1} + a_{nn}s_n\quad &= W'_n
\end{aligned} \tag{3.36}
$$

where the a_{ij} coefficients are given by Eqs. 3.32.

The full set of n equations to be solved is now seen in Eqs. 3.36. For application of these equations, it is assumed that the a_{ij} coefficients are known or can be estimated from the observed data. With given input loads on the right-hand side, it is an easy matter to solve for the concentrations, s_i $(i = 1 \ldots n)$ in each segment, as illustrated in Sample Problem 3.5.

3.6.1.2 Steady State Response Matrix A further understanding of how the concentration of a finite segment estuary responds to external inputs is obtained by

SAMPLE PROBLEM 3.5

DATA

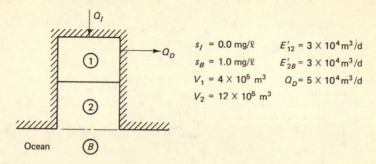

s_I = 0.0 mg/ℓ E'_{12} = 3 × 10⁴ m³/d
s_B = 1.0 mg/ℓ E'_{2B} = 3 × 10⁴ m³/d
V_1 = 4 × 10⁵ m³ Q_D= 5 × 10⁴ m³/d
V_2 = 12 × 10⁵ m³

A flow diversion (Q_D) exists in seg. 1 of the embayment.

PROBLEM

(a) If the inflow (Q_I) equals $1 \times 10^4 \text{m}^3/\text{day}$, and the substance is conservative, determine the concentration in the diversion (c_D).

(b) Repeat (a) with $Q_I = Q_D = 5 \times 10^4 \text{ m}^3/\text{day}$.

(c) For $Q_I = 1 \times 10^4 \text{ m}^3/\text{day}$ and a nonconservative substance with $K = 0.1/\text{day}$, determine c_D.

ANALYSIS

Use backward differences in all cases. $c_D = c_1$, since seg. 1 is assumed completely mixed.

(a)

Water Balances, m³/day

Mass Balances, kg/day

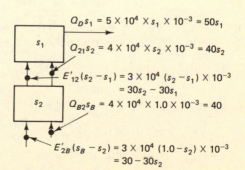

$Q_D s_1 = 5 \times 10^4 \times s_1 \times 10^{-3} = 50 s_1$

$Q_{21} s_2 = 4 \times 10^4 \times s_2 \times 10^{-3} = 40 s_2$

$E'_{12}(s_2 - s_1) = 3 \times 10^4 (s_2 - s_1) \times 10^{-3}$
$= 30 s_2 - 30 s_1$

$Q_{B2} s_B = 4 \times 10^4 \times 1.0 \times 10^{-3} = 40$

$E'_{2B}(s_B - s_2) = 3 \times 10^4 (1.0 - s_2) \times 10^{-3}$
$= 30 - 30 s_2$

(continued)

Sample Problem 3.5 (continued)

With $M_{in}(+)$ and $M_{out}(-)$, mass balances for the segments are:

seg. 1: $\quad -50c_1 + 40c_2 + (30c_2 - 30c_1) = 0$
seg. 2: $\quad -40c_2 - (30c_2 - 30c_1) + 40 + (30 - 30c_2) = 0$

$$\text{from which } c_1 = 0.8305 \text{ mg/l and } c_2 = 0.9491 \text{ mg/l}$$

$$\text{and } c_D = c_1 = 0.8305 \text{ mg/l}$$

(b)

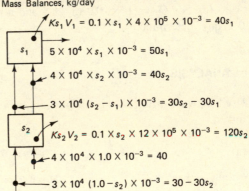

seg. 1: $\quad -50c_1 + (30c_2 - 30c_1) = 0$
seg. 2: $\quad -(30c_2 - 30c_1) + (30 - 30c_2) = 0$

$$c_1 = 0.2308 \text{ mg/l}, \qquad c_2 = 0.6154 \text{ mg/l}, \qquad \underline{c_D = 0.2308 \text{ mg/l}}$$

(c) Water Balances, m^3/day   Mass Balances, kg/day

seg. 1: $\quad -40c_1 - 50c_1 + 40c_2 + (30c_2 - 30c_1) = 0$
seg. 2: $\quad -40c_2 - (30c_2 - 30c_1) - 120c_2 + 40 + (30 - 30c_2) = 0$

$$c_1 = 0.2016 \text{ mg/l}, \qquad c_2 = 0.3457 \text{ mg/l}, \qquad \underline{c_D = 0.2016 \text{ mg/l}}$$

Note:

$$m^3/\text{day} \times \text{mg/l} \times (10^{-6} \text{ kg/mg} \times 10^3 \text{ l/m}^3 = 10^{-3}) = \text{kg/day}$$

Final mass balances, after substituting resulting concentrations, are as follows (all values kg/day):

132

Sample Problem 3.5 (continued)

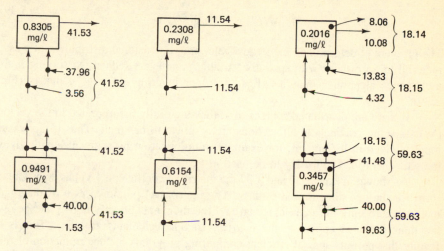

In part (a) note that the mass entering seg. 2—primarily by advection— exits seg. 1 in the diversion flow. In part (b) the amount is significantly reduced since boundary advection is absent. Finally, in part (c) notice that the mass across the boundary is greater than any other case due to the greater dispersive flux caused by the increased concentration gradient at the boundary. Also note that the mass entering the embayment (59.63 kg/day) equals the mass leaving in the diversion (10.08 kg/day) plus the masses reacted away (41.48 + 8.06 = 49.56 kg/day).

expressing Eqs. 3.36 in matrix form. Thus,

$$
\begin{bmatrix}
a_{11} & a_{12} & 0 & \cdot & \cdot & \cdot & & 0 \\
a_{21} & a_{22} & a_{23} & 0 & \cdot & \cdot & \cdot & 0 \\
0 & a_{32} & a_{33} & a_{34} & 0 & \cdot & \cdot & 0 \\
\cdot \\
\cdot \\
\cdot \\
0 & \cdot & \cdot & \cdot & 0 & a_{n,n-1} & & a_{nn}
\end{bmatrix}
\begin{pmatrix}
s_1 \\ s_2 \\ s_3 \\ \cdot \\ \cdot \\ \cdot \\ s_n
\end{pmatrix}
=
\begin{pmatrix}
W_1' \\ W_2 \\ W_3 \\ \cdot \\ \cdot \\ \cdot \\ W_n'
\end{pmatrix}
\qquad (3.37)
$$

or

$$[A](s) = (W)$$
$$[L^3/T][M/L^3] = [M/T] \qquad (3.38)$$

where $[A]$ is an $n \times n$ matrix and (s) and (W) are $n \times 1$ vectors. The solution vector (s) is then obtained formally by inversion of the A matrix, that is

$$(s) = [A]^{-1}(W)$$
$$[M/L^3] = [T/L^3][M/T]$$

(3.39)

The problem of determining the steady state, one-dimensional distribution of waste material in an estuary with spatially variable parameters reduces to solving n simultaneous algebraic equations (Eqs. 3.36) or inverting an $n \times n$ matrix (Eq. 3.39).

If there are distributed sources of s, they are easily incorporated in the vector (W) since the gradients in all sections are assumed to be small. Thus if the waste input is given in lb/day/mi, for example, multiplication by segment length gives the equivalent load in lb/day to be used in Eq. 3.36.

As shown in Eqs. 3.36 and 3.37 the matrix of coefficients [A] has a particular form for the one-dimensional estuary. The matrix [A] is known as a tri-diagonal matrix where only the main diagonal and the diagonals directly above and below the main diagonal appear in the matrix. All other elements are zero. This is a feature that permits special efficient computing programs for determination of the inverse $[A]^{-1}$. But in any event, modern computing capabilities permit rapid solution of the simultaneous equations or the matrix inverse.

However, the generation of the matrix inverse provides additional understanding of the system response. In solving either Eq. 3.36 or 3.39 care must be taken to ensure that a consistent set of units is used throughout. More specifically, if Q and E' are in MGD, K in 1/day and V in MG, then the dimensions of [A] are MGD and the dimensions of $[A]^{-1}$ are day/MG. If the input vector (W) is in lb/day, then $(1/8.34)[A]^{-1}$ has dimensions (mg/l output)/(lb/day input). The output solution vector is then

$$(s) = \frac{1}{8.34}[A]^{-1}(W)$$

$$mg/1 = \left(\frac{mg/1}{lb/day}\right)(lb/day)$$

The inverse has a special meaning which can be seen from the following line of discussion. Let the load input vector (W) in Eq. 3.39 be a vector that has a unit input of 1 lb/day. Thus

$$(W) = \begin{pmatrix} 1 \\ 0 \\ 0 \\ . \\ . \\ . \\ 0 \end{pmatrix}$$

Equation (3.39) then becomes in expanded form:

$$
\begin{pmatrix} s_1 \\ s_2 \\ s_3 \\ \cdot \\ \cdot \\ \cdot \\ s_n \end{pmatrix}
=
\begin{bmatrix} a'_{11} & a'_{12} & a'_{13} & \cdots & a'_{1n} \\ a'_{21} & a'_{22} & & & \\ a'_{31} & a'_{32} & & & \\ \cdot & \cdot & & & \cdot \\ \cdot & \cdot & & & \cdot \\ \cdot & \cdot & & & \cdot \\ a'_{n1} & a'_{n2} & & & a'_{nn} \end{bmatrix}
\begin{pmatrix} 1 \\ 0 \\ 0 \\ \cdot \\ \cdot \\ \cdot \\ 0 \end{pmatrix}
\qquad (3.40)
$$

where a'_{ij} are the elements of the inverse matrix (units mg/l per lb/day). Recalling that matrix multiplication is given by the load column times the rows of the matrix, the result of the multiplication on the right-hand side using column times row is

$$
\begin{pmatrix} s_1 \\ s_2 \\ \cdot \\ \cdot \\ \cdot \\ s_n \end{pmatrix}
=
\begin{pmatrix} a'_{11} \\ a'_{21} \\ \cdot \\ \cdot \\ \cdot \\ a'_{n1} \end{pmatrix}
\qquad (3.41)
$$

$$
\text{(mg/l)} \qquad \frac{\text{mg/l}}{\text{lb/day}} \cdot 1 \text{ lb/day}
$$

This equation indicates that the concentrations in every segment resulting from a unit load of 1 lb/day into seg. 1, is the first column of the inverse matrix. Similarly, if a unit load of 1 lb/day is input to the second segment, it can be seen that the second column of the inverse would represent the concentrations in all segments due to a unit load into the seg. 2. In general then, the columns of the inverse matrix represent the concentrations in all segments due to a unit load into the segment number of the column.

It is interesting to note that the rows of the inverse represent the concentration response in a given segment due to a unit load in every other segment. Thus, much insight can be obtained into the relative effects of individual waste sources on the water quality of the estuary by examining the "unit responses" of a discharge into one segment on the water quality of all other segments. Multiplication of the unit responses by the actual discharged load and summing over all of the input loads gives the total response. This is identical to the multiple input case for the ideal estuary discussed previously in Section 3.2.3. In this case, however, the responses can be computed for the real estuary where flows, areas, depths, and reaction rates may be varying with distance.

3.6.2 Estimation of Tidal Dispersion Coefficients for the One-Dimensional Finite Segment Model

Section 3.3 discussed the estimation of the tidal dispersion coefficient for the idealized case. It was indicated there that first approximations can be obtained for

E from the slope of the salinity data (Eq. 3.20 and Fig. 3.17). For the variable density case of the finite segment model, such a procedure is useful for a first approximation but may need to be further refined for the actual estuary. Two approaches may then be used.

3.6.2.1 Trial and Error Approach

With initial estimates of the tidal dispersion coefficient and known freshwater flows, the salinity (or chloride) concentration in each segment can be calculated using the finite segment estuary model. This calculation is made using Eqs. 3.36 with the right-hand-side load vector given by

$$(\mathbf{W}) = \begin{pmatrix} 0 \\ 0 \\ \cdot \\ \cdot \\ \cdot \\ E'_{nb}s_b \end{pmatrix} \tag{3.42}$$

where s_b is the boundary (ocean or bay) salinity concentration which must be assigned from data. For the freshwater tidal river portion of the estuary, some other tracer (e.g., dye) must be used. In that case, a time variable analysis must be conducted. The calculated concentrations are then compared to the observed data and, if the comparison is judged unsatisfactory, adjustment of the tidal dispersion coefficient can be made together with a new calculation of the longitudinal salinity distribution. This procedure is repeated until a sufficiently satisfactory comparison between the observed and computed values is obtained. Figure 3.23 illustrates this procedure and as seen several trials may be necessary. What is "acceptable" may vary from analyst to analyst. In any event a type of smoothing of the data is indicated where the trial and error procedure does not attempt to go through the

Figure 3.23 Hypothetical illustration of trial and error procedure for estimating tidal dispersion coefficients in a finite segment model.

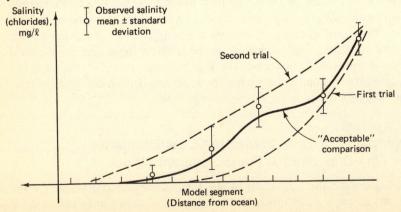

mean of every data point but rather allows for some deviations between observed and calculated values to reflect random behavior in the observed data.

The difficulties with this aproach are that it can be time consuming and no two analysts will necessarily arrive at the same set of tidal dispersion coefficients. The advantage is that the approach inherently incorporates the analyst's scientific and engineering judgment in interpreting the fluctuations in the observed data.

3.6.2.2 Direct Calculation of E from Salinity Data

The second procedure is a direct approach to calculating the dispersion coefficients. In this procedure, the E' coefficients are calculated directly from the salinity data and the known freshwater flows. This approach has been also called "the fraction of freshwater method" and is discussed fully in Officer (1976). For n segments, there are $n + 1$ coefficients that must be computed if dispersion is allowed across the boundary of the first segment. If however, the first segment is placed at the head of the estuary where upstream mixing is small or zero (as at a dam), then there are n values of E' and n equations. For these estimates of E', the interfacial salinities can be used as in the previous mass balance equation. Thus the salinity equation for the estimate of E' is

$$0 = (Qs)_{i-1,i} - (Qs)_{i,i+1} + E'_{i-1,i}(s_{i-1} - s_i) + E'_{i,i+1}(s_{i+1} - s_i) \quad (3.43)$$

If all the salinity equations are added, using the salinity input vector given by Eq. 3.42, then the result is that

$$(Q_{nb} + E'_{nb})s_{nb} = E'_{nb}s_b$$

or

$$Q_{nb}s_{nb} = E'_{nb}(s_b - s_{nb}) \quad (3.44)$$

where s_{nb} is the salinity at the interface between the nth segment and the boundary which simply says that at steady state the flux of salt entering the last segment from the boundary through dispersion, $E'_{nb}(s_b - s_{nb})$, must be balanced by the salt flux leaving the last segment to the boundary via advection, $Q_{nb}s_{nb}$. If this balance did not occur, then a nonsteady buildup or loss of salinity would occur. The boundary dispersion can then be simply calculated as

$$E'_{nb} = \frac{Q_{nb}s_{nb}}{s_b - s_{nb}} \quad (3.45)$$

The other E' coefficients are then calculated from each equation in a straightforward manner, working back up the estuary from the lower boundary segment, that is,

$$E'_{i-1,i} = \frac{E'_{i,i+1}(s_{i+1} - s_i) + [(Qs)_{i-1,i} - (Qs)_{i,i+1}]}{s_i - s_{i-1}} \quad (3.46)$$

$$i = n, n - 1, n - 2, \ldots, 1 \quad \text{and} \quad n + 1 = b$$

The values of E are then computed from Eq. 3.27 as (see Sample Problem 3.6)

$$E_{i-1,i} = \frac{E'_{i-1,i}\overline{\Delta x}_{i-1,i}}{A_{i-1,i}} \quad (3.47)$$

The difficulty with this procedure is in the specification of the values of the salinity or chlorides for each segment and interface since such closely spaced data are usually not available. Therefore the uncertainty of the acceptability of the trial and error procedure is translated in this approach to the uncertainty of specifying the observed salinity profile. This disadvantage can be overcome by a mathematical smoothing of the observed data and interpolated values computed for each segment. Practically, the analyst may smooth the data by eye, estimate the segment salinities graphically, and then substitute them into Eq. 3.45 and 3.46 for E'. Sample Problem 3.6 illustrates the procedure for the Wicomico Estuary in Maryland.

3.6.3 Guideline for Estimating Segment Length for One-Dimensional Estuary

In dividing a real estuary into a series of well-mixed segments, typically smaller segments are chosen in the vicinity of large point sources, to properly capture spatial concentration gradients, and larger segments are selected outside the area of influence of the load. If the segments are made larger in the vicinity of the load, the average concentration in the well-mixed segment decreases below that of the concentration which exists at the discharge location. The following analysis (Mueller, 1976) develops a guideline for controlling the difference between the concentration from a finite difference solution (s_0) and the concentration at the discharge location (l_0).

For simplicity an infinitely long estuary with spatially constant values of the cross-sectional area (A), net flow (Q), dispersion coefficient (E), and substance decay rate (K) is selected for analysis. Due to a single point source discharge, (at $x = 0$), the analytical solution of the estuarine concentration at the outfall is calculated from Eq. 3.8e:

$$l_0 = \frac{W}{Q\alpha} = \frac{W}{Q\sqrt{1 + 4KE/U^2}} \tag{3.48}$$

A finite difference model of the estuary is then prepared with all segments of equal length (Δx), with segments numbered $n = 0$ at $x = 0$ (centered about the discharge), $n = -1, -2, -3$, and so on upstream and $n = +1, +2, +3$, and so on downstream of the loaded segment [Fig. 3.24(a)].

Using the solution of a second-order difference equation with constant coefficients (Miller, 1960) and imposing boundary conditions of zero at $n = \pm \infty$, it can be shown that the concentrations in the well-mixed segments (s_n) can be expressed analytically as

$$s_n = s_0(\rho_l)^n, \qquad n \leq 0 \tag{3.49a}$$
$$s_n = s_0(\rho_r)^n, \qquad n \geq 0 \tag{3.49b}$$

where, for a backward differencing scheme,

$$\frac{\rho_l}{\rho_r} = \left[2\left(\frac{\tau}{\eta} + \frac{\tau^2}{\eta} \right) + 1 \right] \pm 2\sqrt{ \left(\frac{\tau}{\eta} + \frac{\tau^2}{\eta} \right)^2 + \frac{\tau^2}{\eta} } \tag{3.50}$$

SAMPLE PROBLEM 3.6

DATA

The Wicomico Estuary is a small tributary of the Chesapeake Bay receiving the discharge from Salisbury, Maryland (Salas and Thomann, 1975). Salinity data were obtained and an analysis is to be made of the tidal dispersion. The Wicomico has been segmented into 10 segments as shown.

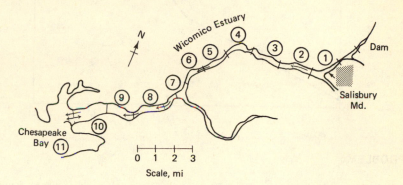

The cross-sectional area and depth are variable along the length of the estuary. The data for the geometry are shown below.

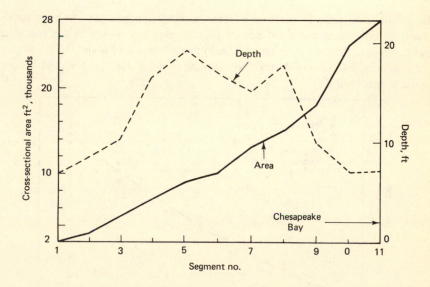

The freshwater flow is also variable along the length of the estuary as shown below.

(continued)

Sample Problem 3.6 (continued)

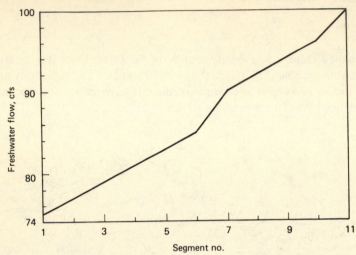

PROBLEM

Using the method of Section 3.6.2.2, calculate dispersion coefficients for each segment interface from the salinity data cited below.

ANALYSIS

The salinity data were plotted and a linear interpolation was used to estimate the salinity at the center and interfaces of each segment as shown in the following figure. Because the salinity reached "background" levels of about 0.2‰ by seg. 4, the salinity data cannot be used to estimate E and a dye study should be conducted in that region.

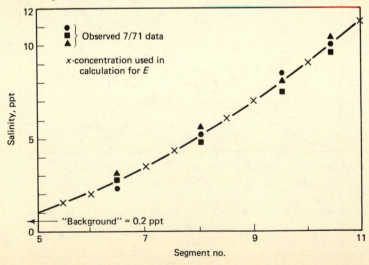

Sample Problem 3.6 (continued)

The input values for the estimate of E are given in the following table.

Mile point	Seg. no.	Interface	Area @ interface (10^4 ft^2)	Flow @ interface (cfs)	Salinity (‰) Center[a]	Salinity (‰) Interface
0–2	10	10–Bay	2.8	100	9.0	10.1
2–4	9	9–10	2.5	96	7.0	8.0
4–6	8	8–9	1.8	94	5.2	6.1
6–8	7	7–8	1.5	92	3.5	4.35
8–10	6	6–7	1.3	90	2.0	2.75
10–12	5	5–6	1.0	85	1.0	1.50
12–14	4	4–5	0.9	83	0.2	0.60
14–16	3	3–4	0.7	81	(0.2)	(0.2)
16–18	2	2–3	0.5	79	(0.2)	(0.2)
18–20	1	1–2	0.3	77	(0.2)	(0.2)

[a]Bay = 11.3‰.

Using Eqs. 3.45 through 3.47, the tidal dispersion coefficient can be computed for each segment.
Thus, from Eq. 3.45,

$$E'_{10,B} = \frac{(Qs)_{10,B}}{s_B - s_{10}} = \frac{100(10.1)}{11.3 - 9} = \underline{\underline{439.1 \text{ cfs}}}$$

and from Eq. 3.47,

$$E_{10,B} = \frac{E'_{10,B}\,\Delta x}{A_{10,B}} = 439.1\,\frac{\text{ft}^3}{\text{sec}}\frac{2\,\text{mi}}{2.8 \times 10^4\,\text{ft}^2}\,5280\,\text{ft/mi} = \underline{\underline{165.6\,\text{ft}^2/\text{sec}}}$$

or $E_{10,B} = 165.6\,\text{ft}^2/\text{sec} \cdot 0.003099\,\dfrac{\text{smpd}}{\text{ft}^2/\text{sec}} = 0.51\,\text{smpd}$

Similarly, from Eq. 3.46,

$$E'_{9,10} = \frac{E'_{10,B}(s_B - s_{10}) + (Qs)_{9,10} - (Qs)_{10,B}}{s_{10} - s_9}$$

$$= \frac{439.1(11.3 - 9) + (96 \cdot 8) - (100 \cdot 10.1)}{9 - 7} = \underline{\underline{384 \text{ cfs}}}$$

and from Eq. 3.47,

$$E_{9,10} = \frac{E'_{9,10}\,\Delta x}{A_{9,10}} = \underline{\underline{162.2\,\text{ft}^2/\text{sec} = 0.50\,\text{smpd}}}$$

(continued)

Sample Problem 3.6 (continued)

The complete calculation gave the following:

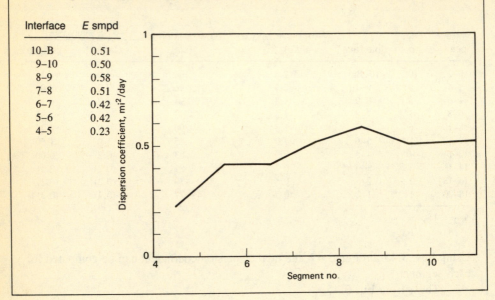

Interface	E smpd
10–B	0.51
9–10	0.50
8–9	0.58
7–8	0.51
6–7	0.42
5–6	0.42
4–5	0.23

$$\tau = K\frac{\Delta x}{U} \qquad (3.51)$$

$$\eta = \frac{4KE}{U^2} \qquad (3.52)$$

In Eq. 3.50, the positive sign before the radical is used for ρ_l and the negative sign for ρ_r. Note that the segment length (Δx) is contained in the expression for the variable τ in Eq. 3.51.

By taking a mass balance around segment zero, using Eqs. 3.49, the concentration in that segment is determined to be

$$s_0 = \frac{W}{Q\sqrt{(1 + \tau)^2 + \eta}} \qquad (3.53)$$

Substituting $\eta = 4KE/U^2$ (Eq. 3.52) in Eq. 3.53, the concentration at $x = 0$ is

$$l_0 = \frac{W}{Q\sqrt{1 + \eta}} \qquad (3.54)$$

Thus, as τ approaches zero, that is, the segment size Δx approaches zero, the finite difference solution (s_0) approaches the estuary concentration at $x = 0$ (l_0). Note that s_0 is always less than l_0 since its denominator is greater than that for $l_0 (\tau > 0)$.

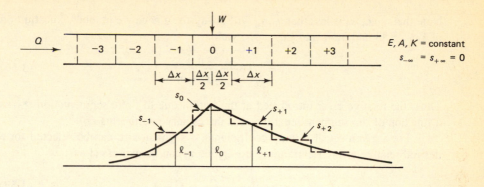

$$\ell_0 = W/Q\alpha = (W/Q) \div \sqrt{1+\eta}$$
$$s_0 = (W/Q) \div \sqrt{(1+\tau)^2 + \eta}$$
$$r_0 = (s_0 - \ell_0)/\ell_0 = (s_0/\ell_0) - 1$$
$$\tau = K\Delta x/U, \ \eta = 4KE/U^2$$

(a)

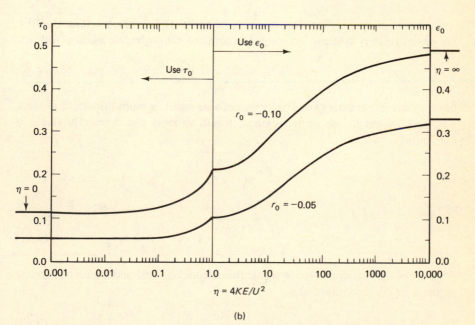

(b)

Figure 3.24 Relative error at $x = 0$ in constant parameter estuary for a backward differencing scheme. From Mueller (1976). Reprinted by permission of HydroQual, Inc.

The relative error of the finite difference solution in segment zero is defined as

$$r_0 = \frac{s_0 - l_0}{l_0} = \frac{s_0}{l_0} - 1 \tag{3.55}$$

Note that, since s_0 is less than l_0, r_0 will always be a negative number. Substitution of Eqs. 3.53 and 3.54 into Eq. 3.55, yields

$$r_0 = \sqrt{\frac{1 + \eta}{(1 + \tau_0)^2 + \eta}} - 1 \tag{3.56}$$

Thus, the relative error introduced at the outfall due to finite segmentation (r_0) is a function of the segment length τ_0 and the estuary coefficients (η).

For a given set of estuary coefficients, a segment size can be selected for a desired relative error. Rearranging Eq. 3.56, τ_0 can be expressed as

$$\tau_0 = \sqrt{\frac{1 + \eta}{(1 + r_0)^2} - \eta} - 1 \tag{3.57}$$

With τ_0 known for a desired relative error (e.g., $r_0 = -0.05$ for an error of 5%), the segment size is calculated from Eq. 3.51 as

$$\Delta x = \tau_0 \frac{U}{K} \tag{3.58}$$

A plot of τ vs. η for relative errors of 5 and 10% appears in Fig. 3.24(b). For more advective systems, η approaches zero and τ_0 may be calculated as follows:

$$\tau_0 = \frac{1}{(1 + r_0)} - 1 \tag{3.59}$$

In cases where net flow Q and velocity become small, a more dispersive system, τ is inconvenient to use as the segment length variable and a new variable ϵ is introduced.

$$\epsilon = \sqrt{\frac{K(\Delta x)^2}{4E}} \tag{3.60}$$

and

$$\epsilon_0 = \sqrt{\frac{1 + (1/\eta)}{(1 + r_0)^2} - 1} - \frac{1}{\sqrt{\eta}} \tag{3.61}$$

After choosing a relative error, ϵ is calculated from Eq. 3.61 and the corresponding segment length determined as

$$\Delta x = \sqrt{\frac{4E\epsilon_0^2}{K}} \tag{3.62}$$

For a purely dispersive case, ($U = 0$ or $\eta = \infty$), Eq. 3.61 reduces to

$$\epsilon_0 = \sqrt{\frac{1}{(1 + r_0)^2} - 1} \tag{3.63}$$

In the plot of Fig. 3.24(b), Eq. 3.57 involving τ is used for $\eta \leq 1$ and Eq. 3.61 involving ϵ is used for $\eta \geq 1$. Note that $\tau = \epsilon\sqrt{\eta}$. Sample Problem 3.7 illustrates the use of this chart in selecting an appropriate segment size.

SAMPLE PROBLEM 3.7

DATA

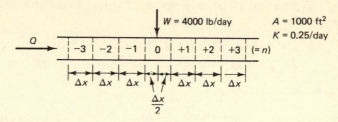

A uniformly segmented estuary is subjected to a single point source at $x = 0$.

PROBLEM

For a backward differencing scheme, and a desired maximum absolute error of 5% in the discharge segment ($n = 0$), determine the maximum segment length for the following conditions:

(a) $Q = 50$ cfs, $E = 5.0$ smpd,   (b) $Q = 200$ cfs, $E = 7.5$ smpd

Consider the flow in the waste source to be negligible.

ANALYSIS

	Q (cfs)	U^a (mpd)	E (smpd)	η^b	$\tau_0{}^c$ or ϵ_0	Δx^d (mi)	$l_0{}^e$ (mg/l)	$s_0{}^f$ (mg/l)
(a)	50	0.82	5.0	7.44	$\epsilon_0 = 0.140$	1.25	5.108	4.853
(b)	200	3.28	7.5	0.697	$\tau_0 = 0.0878$	1.15	2.848	2.706

$^a U = (Q/A) \times 16.4.$
$^b \eta = 4KE/U^2.$
^{c}Fig. 3.24 ($r_0 = -0.05$) or Eqs. 3.57 and 3.61.
^{d}Eqs. 3.58 and 3.62.
$^e l_0 = W/(Q \times \sqrt{1 + \eta} \times 5.39).$
$^f s_0 = 0.95\ l_0.$

From the tabulated solutions above, uniform segment lengths of 1.25 and 1.15 mi are required for conditions (a) and (b), respectively.

Spatial Variation of Relative Error

Although the relative error in the maximum concentration region ($x = 0$ or $n = 0$) is restricted to 5%, it is interesting to note the spatial variation of relative error.

(continued)

Sample Problem 3.7 (continued)

Using condition (b) as an example, the analytic solutions at the segment centers are:

$$l = 2.848 \exp(0.504 \times 1.15 \times n), \qquad x \le 0$$
$$l = 2.848 \exp(-0.0662 \times 1.15 \times n), \qquad x \ge 0$$

[Eq. (3.8)]

The finite difference solutions are:

$$s = 2.706 \times 1.619^n, \qquad n \le 0$$
$$s = 2.706 \times 0.929^n, \qquad n \ge 0$$

[Eq. (3.49)]

Resulting relative errors are calculated from

$$r = \frac{s}{l} - 1$$

and are plotted below, together with the analytical solution.

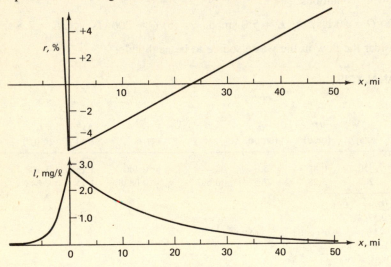

On the upstream side of the discharge, in the region of sharp concentration gradients, the relative errors are all positive ($+4.8$, $+15.4$, and $+27.6\%$ in segments $n = -1$, -2, and -3, respectively). Thus, the finite segment model overestimates concentrations in this region.

Downstream of the discharge, negative relative errors gradually decrease from -5% at $x = 0$ to 0.0% at approximately $x = +23$ mi. Thereafter, relative errors are positive, approximately $+5\%$ at $x = +47$ mi. Thus, the finite difference model—with segments 1.15 mi in length—underestimates concentrations immediately downstream of the load then overestimates thereafter in the lower concentration region.

3.6.4 Two-Dimensional Analysis—Lateral

There are many problem situations where estuaries may include tidal embayments, or wide bays and harbors, which exhibit gradients in water quality in both the longitudinal and lateral directions. The general case of multidimensional water quality problems can be derived but the notation becomes somewhat obscure. A more direct approach to the technique can be made by considering a simple example. It will be seen that the inclusion of such lateral gradients is quite easy and follows the general principles outlined for the preceding one-dimensional case.

Consider Fig. 3.25, an estuary with a tidal embayment where for simplicity and illustrative purposes, the estuary has been divided into five segments and the embayment considered as the sixth segment. A tributary enters the embayment which tidally exchanges with the main estuary.

The principles of mass balance around each segment are applied in a manner similar to the one-dimensional case. For segs. 1, 2, 4, and 5, in Fig. 3.25, the equations are the same as in Eqs. 3.31 through 3.33. Segment 3 must now include the exchange and flow from seg. 6 and a new equation must be written for seg. 6.

For seg. 3, the steady state equation is

$$
\overset{(1)}{0 = Q_{23}s_2 - Q_{34}s_3 + \overbrace{Q_{63}s_6} + E'_{23}(s_2 - s_3) + E'_{43}(s_4 - s_3)} \\
+ \underbrace{E'_{63}(s_6 - s_3)}_{(2)} - V_3 K_3 s_3 + W_3
$$

$$(3.64)$$

The two marked terms are: (1) the net advection of the concentration in the embayment, s_6 to seg. 3, and (2) the net tidal mass exchange between the embayment and the estuary proper. Also, note that

$$Q_{34} = Q_{23} + Q_{63}$$

by continuity.

Figure 3.25 Illustration of a lateral two-dimensional estuary with a tidal embayment.

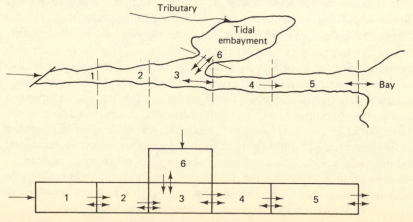

For seg. 6, the mass balance equation is

$$0 = Q_T s_T - Q_{63} s_6 + E'_{36}(s_3 - s_6) - V_6 K_6 s_6 + W_6 \qquad (3.65)$$

where Q_T and s_T are the flow and concentration, respectively, of the tributary inflow into the embayment. The second term is the mass loss from 6 due to net advection to the main estuary. The third term is the net exchange of mass from or to the main estuary due to tidal mixing. The fourth term is the loss due to decay in the embayment and the last term is any direct input of waste into the embayment which can incorporate the load from the tributary.

There are now six equations in six unknowns (the concentrations in each segment). The two-dimensional problem is therefore again simply the solution of linear algebraic equations. The matrix of coefficients and the vectors for the example in Fig. 3.25 can be shown to have the following form

$$\begin{bmatrix} a_{11} & a_{12} & 0 & 0 & 0 & 0 \\ a_{21} & a_{22} & a_{23} & 0 & 0 & 0 \\ 0 & a_{32} & a_{33} & a_{34} & 0 & a_{36} \\ 0 & 0 & a_{43} & a_{44} & a_{45} & 0 \\ 0 & 0 & 0 & a_{54} & a_{55} & 0 \\ 0 & 0 & a_{63} & 0 & 0 & a_{66} \end{bmatrix} \begin{pmatrix} s_1 \\ s_2 \\ s_3 \\ s_4 \\ s_5 \\ s_6 \end{pmatrix} = \begin{pmatrix} W_1 \\ W_2 \\ W_3 \\ W_4 \\ W_5 \\ W_6 \end{pmatrix} \qquad (3.66)$$

where

$$a_{33} = Q_{23} + E'_{23} + E'_{43} + E'_{63} + V_3 K_3 \qquad (3.67a)$$

$$a_{36} = -Q_{63} - E'_{63} \qquad (3.67b)$$

$$a_{63} = -E'_{36} \qquad (3.67c)$$

$$a_{66} = Q_{63} + E'_{36} + V_6 K_6 \qquad (3.67d)$$

Comparison of Eq. 3.66 with Eq. 3.37 shows that the effect of adding the embayment is merely to add terms off the tri-diagonal form, but that does not add any difficulty in solving the equations. The meaning of the inverse matrix is the same, that is, the columns of the inverse represent the response in all segments due to a load into the given column segment.

For the simple system of Fig. 3.25, the tidal exchange coefficients can also be obtained in a manner similar to that given for the one-dimensional case. Therefore, the steady state two-dimensional case of lateral and longitudinal variations in water quality is really no more complex conceptually and computationally than the one-dimensional case. However, as the system increases in the complexity of the transport then the principal difficulty is in estimating the net advective horizontal flows and dispersions. One direct and practical way is to utilize some estimate of the net transport from previous hydrodynamic analyses and then use the salinity or other tracer to estimate the dispersion coefficients through a trial and error procedure. Figure 3.26 shows the results of such an analysis for Boston Harbor.

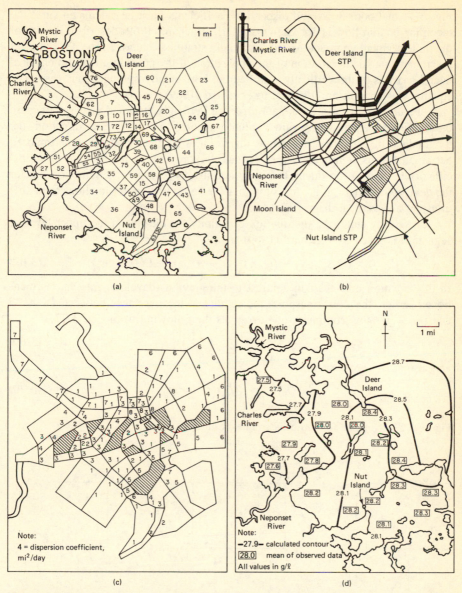

Figure 3.26 Boston Harbor two-dimensional water quality model. (a) Segmentation into 70 elements. (b) Assumed flow routing. (c) Estimated dispersion coefficients. (d) Observed vs. calculated salinity. From Hydroscience (1970).

3.6.5 Two-Dimensional Analysis—Vertical

The two-layered estuarine circulation has been discussed previously in a qualitative manner (see Section 3.1.2.2 and Fig. 3.11). The finite segment approach is easily applied to the problem as well for calculating the transport parameters. Note from

Fig. 3.11 that both flows and dispersion coefficients must be calculated since the incoming bottom flow is not known a priori. Pritchard (1969) described the use of the finite segment approach for these partially stratified estuaries. The development given here is a slight variation of the application of the preceding two-dimensional lateral analysis to the vertically stratified but horizontally homogeneous case. The equations given here are identical to Pritchard (1969).

Figure 3.27 shows the notation for the vertically stratified estuary. Because the density driven flows are now explicitly included in the vertical direction, the longitudinal dispersion is neglected, that is, it is assumed that the superimposed net circulation is significantly greater than the longitudinal mixing due to the oscillatory motion of the tides. This may not always be a good assumption but for the development here, it greatly simplifies obtaining first approximations to the net transport and vertical dispersion.

At steady state, the distribution of salinity for the salt in the upper layer of segment i is given by Pritchard (1969) as

$$0 = (Q_u s_u)_{i-1,i} - (Q_u s_u)_{i,i+1} + (Q_v s_l)_i + E_{lu}(s_{li} - s_{ui}) \qquad (3.68)$$

In this equation note that, in contrast to the previous developments and as mentioned earlier, the flow transports the salinity at the interfaces between $i - 1$, i and i, $i + 1$. Now continuity of flow gives the following from Fig. 3.27,

$$(Q_u)_{i-1,i} + Q_{vi} = (Q_u)_{i,i+1} \qquad (3.69)$$

and also at any cross section

$$(Q_u)_{i-1,i} - (Q_l)_{i,i-1} = Q_{Ri} \qquad (3.70)$$

Figure 3.27 Notation for vertically stratified estuary.

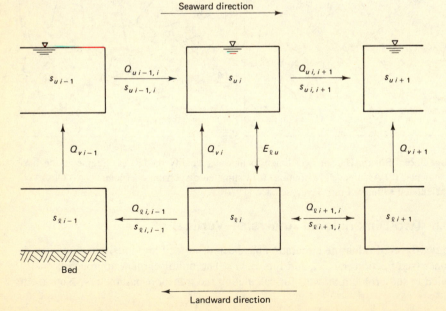

where Q_{Ri} is the freshwater flow entering section i. Finally, if all the equations surrounding segment i are added, then the necessary result emerges that the transport of salt across an interface in the bottom layer must be balanced by a transport of salt in the top layer. Thus,

$$(Q_u s_u)_{i-1,i} = (Q_l s_l)_{i,i-1} \tag{3.71}$$

Substituting Eq. 3.70 into Eq. 3.71 gives the following:

$$(Q_l)_{i,i-1} = Q_{Ri} \frac{(s_u)_{i-1,i}}{(s_l)_{i,i-1} - (s_u)_{i-1,i}} \tag{3.72}$$

for the bottom flow and

$$(Q_u)_{i-1,i} = Q_{Ri} \frac{(s_l)_{i,i-1}}{(s_l)_{i,i-1} - (s_u)_{i-1,i}} \tag{3.73}$$

for the upper flow.

   With the estimated salinity at the interfaces of the segments, the latter two equations permit calculation of the net surface and bottom advective flows. Equation 3.69 can then be used for the vertical flow and Eq. 3.68 for the vertical bulk dispersion. All of the advective transport and vertical dispersion can therefore be uniquely calculated from the salinity data.

3.7 POSITIVITY AND NUMERICAL DISPERSION OF STEADY STATE FINITE SEGMENT MODELS

3.7.1 Mass Balance Equation as a Finite Difference Approximation

The governing equation for calculating concentrations in a finite segment of an estuary, Eq. 3.28, was derived using the mass balance principle. In this section, Eq. 3.28 will be shown (Thomann, 1973) to be a special case of a finite difference approximation of the differential equation, Eq. 3.23. In Fig. 3.28 a one-dimensional estuary is segmented into three equal length reaches, of length Δx. The continuous concentration distribution, $s(x)$, is shown below together with finite segment concentrations at the midpoints of the segments, s_{i-1}, s_i, and s_{i+1}. Concentrations at the interfaces between the segments are also shown, where $s_{i-1,i}$ and $s_{i,i+1}$ are the interfacial concentrations of segment i at its upstream and downstream faces, respectively.

   The slope of the continuous concentration distribution at the center of segment i is approximated as the difference of its interfacial concentrations divided by the segment length. Thus,

$$\left.\frac{\partial s}{\partial x}\right|_i = \frac{s_{i,i+1} - s_{i-1,i}}{\Delta x} \tag{3.74}$$

Similarly, the spatial rate of change of the advective mass rate, Qs — the second

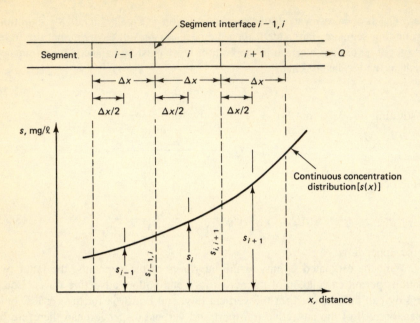

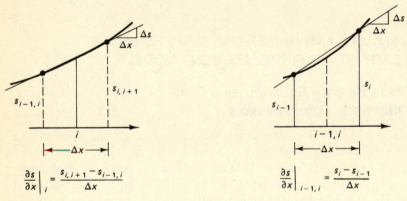

Figure 3.28 Graphical description of finite difference approximations.

term in Eq. 3.23, is given by

$$\left.\frac{\partial(Qs)}{\partial x}\right|_i = \frac{(Qs)_{i,i+1} - (Qs)_{i-1,i}}{\Delta x} \qquad (3.75a)$$

where

$$(Qs)_{i,i+1} = Q_{i,i+1}s_{i,i+1} \qquad (3.75b)$$

and

$$(Qs)_{i-1,i} = Q_{i-1,i}s_{i-1,i} \qquad (3.75c)$$

Note that $Q_{i,i+1}$ and $s_{i,i+1}$ are the flow and concentration at the downstream inter-face of segment i and $Q_{i-1,i}$ and $s_{i-1,i}$ are similar quantities at the upstream face.

The interfacial concentration $s_{i,i+1}$ is related to the concentrations in the adjoining segments, that is, s_i and s_{i+1}. Assume the relationship is linear, of the following form

$$s_{i,i+1} = \alpha_{i,i+1}s_i + \beta_{i,i+1}s_{i+1} \tag{3.76}$$

where α (a weighting value) is the fraction of the upstream segment's concentration at the interface, β is the fraction of the downstream segment's concentration at interface i, $i + 1$, and

$$\beta_{i,i+1} = 1 - \alpha_{i,i+1} \tag{3.77}$$

Substituting Eq. 3.76 into Eq. 3.75a, the finite difference approximation of the spatial derivative of the advective mass balance term is

$$\left.\frac{\partial(Qs)}{\partial_x}\right|_i = \frac{Q_{i,i+1}(\alpha_{i,i+1}s_i + \beta_{i,i+1}s_{i+1}) - Q_{i-1,i}(\alpha_{i-1,i}s_{i-1} + \beta_{i-1,i}s_i)}{\Delta x}$$

$$(3.78)$$

This is an "arbitrary" difference approximation to the derivative. Two special cases are of interest. If $\alpha = 1$, then $\beta = 0$ and Eq. 3.78 becomes

$$\left.\frac{\partial(Qs)}{\partial x}\right|_i = \frac{Q_{i,i+1}s_i - Q_{i-1,i}s_{i-1}}{\Delta x} \tag{3.79}$$

This is the "backward" difference approximation to the spatial derivative. In this case, mass is transported into segment i by the upstream interfacial flow $Q_{i-1,i}$ and the upstream segment's concentration s_{i-1}. Transport out of segment i is by $Q_{i,i+1}$ with segment i's concentration s_i. This backward difference approximation is most appropriate for those cases where advective transport predominates over dispersive transport—in which case downstream interfacial concentrations should be expected to equal those at the centers of upstream segments.

The second special case is when $\alpha = 0.5$ and $\beta = 0.5$, a "central" differencing scheme. In this case, interfacial concentrations are the averages of those at the centers of adjoining segments. This differencing scheme is appropriate when dispersive transport dominates and interfacial concentrations are not significantly influenced by advective flows in the downstream direction.

Dispersive transport in the differential equation (Eq. 3.23) may be represented by the following finite difference approximation:

$$\left.\frac{\partial}{\partial x}\left(EA\frac{\partial s}{\partial x}\right)\right|_i = \frac{\left(EA\dfrac{\partial s}{\partial x}\right)_{i,i+1} - \left(EA\dfrac{\partial s}{\partial x}\right)_{i-1,i}}{\Delta x} \tag{3.80a}$$

where

$$\left(EA\frac{\partial s}{\partial x}\right)_{i,i+1} = E_{i,i+1}A_{i,i+1}\left.\frac{\partial s}{\partial x}\right|_{i,i+1} \tag{3.80b}$$

and

$$\left(EA\frac{\partial s}{\partial x}\right)_{i-1,i} = E_{i-1,i}A_{i-1,i}\frac{\partial s}{\partial x}\bigg|_{i-1,i} \tag{3.80c}$$

Note that the concentration derivatives are to be evaluated at the interfaces, not at the centers of segments as done previously for the advective term. Resulting finite difference approximations are

$$\left(EA\frac{\partial s}{\partial x}\right)_{i,i+1} = E_{i,i+1}A_{i,i+1}\left(\frac{s_{i+1} - s_i}{\Delta x}\right) \tag{3.81a}$$

$$\left(EA\frac{\partial s}{\partial x}\right)_{i-1,i} = E_{i-1,i}A_{i-1,i}\left(\frac{s_i - s_{i-1}}{\Delta x}\right) \tag{3.81b}$$

Substitution of $E' = EA/\Delta x$ in Eq. 3.81 and subsequent substitution into Eq. 3.80a result in the following approximation for the dispersive transport term:

$$\frac{\partial}{\partial x}\left(EA\frac{\partial s}{\partial x}\right)_i = \frac{E'_{i,i+1}(s_{i+1} - s_i) - E'_{i-1,i}(s_i - s_{i-1})}{\Delta x} \tag{3.82}$$

Note that the approximations to the dispersive fluxes involve the concentrations at the centers of adjoining segments and do not depend on interfacial concentrations. Thus, these fluxes are independent of the weighting values α and β.

Multiplying the differential equation (Eq. 3.23) by the volume, $V = A\,\Delta x$, and substituting the advective and dispersive approximations, Eqs. 3.78 and 3.82, the resulting equation is

$$V_i\frac{\partial s_i}{\partial t} = -\frac{A\,\Delta x}{A}\left[\frac{Q_{i,i+1}(\alpha_{i,i+1}s_i + \beta_{i,i+1}s_{i+1}) - Q_{i-1,i}(\alpha_{i-1,i}s_{i-1} + \beta_{i-1,i}s_i)}{\Delta x}\right]$$
$$+ \frac{A\,\Delta x}{A}\left[\frac{E'_{i,i+1}(s_{i+1} - s_i) - E'_{i-1,i}(s_i - s_{i-1})}{\Delta x}\right]$$
$$- K_is_iV_i \pm W_i$$

which reduces to

$$V_i\frac{\partial s_i}{\partial t} = Q_{i-1,i}(\alpha_{i-1,i}s_{i-1} + \beta_{i-1,i}s_i) - Q_{i,i+1}(\alpha_{i,i+1}s_i + \beta_{i,i+1}s_{i+1})$$
$$+ E'_{i,i+1}(s_{i+1} - s_i) - E'_{i-1,i}(s_i - s_{i-1})$$
$$- K_is_iV_i \pm W_i \tag{3.83}$$

where $K_is_iV_i$ is the reactive loss rate and W_i is any other source or sink of mass in segment i. Use of backward differences ($\alpha = 1$, $\beta = 0$) in Eq. 3.83 reduces this difference approximation to

$$V_i\frac{\partial s_i}{\partial t} = Q_{i-1,i}s_{i-1} - Q_{i,i+1}s_i + E'_{i,i+1}(s_{i+1} - s_i)$$
$$- E'_{i-1,i}(s_i - s_{i-1}) - K_is_iV_i \pm W_i \tag{3.84}$$

which is identical with the mass balance equation previously derived as Eq. 3.28. In conclusion, a mass balance equation for a finite segment is, in actuality, an equation obtained from the governing mass balance differential equation with finite difference approximations substituted for the derivatives. This underlies the common use of "finite differences" for "finite segments."

It may be noted (Thomann, 1972) that common practice is to set a lower limit on the weighting value alpha of one-half for one-dimensional estuaries. This ensures that equal weighting is given to adjoining segment concentrations in calculating interfacial concentrations as the system becomes predominantly dispersive. The typical range of α is thus

$$0.5 \le \alpha \le 1 \tag{3.85}$$

3.7.2 Solution Positivity

Whenever a source of a substance is introduced in the water column, that is, W_i is positive, the estuary responds with an increase in concentration, that is, s_i is positive. Thus, a physical constraint on the solution vector (**s**) is that all concentrations must be positive. For the one-dimensional estuary (Thomann, 1972), where the ith segment has only two adjoining segments ($i - 1$ upstream and $i + 1$ downstream), the **A** matrix is a tri-diagonal matrix (Eq. 3.83) with an upper diagonal having elements $a_{i,i+1}$, a main diagonal with elements $a_{i,i}$, and a lower diagonal with elements $a_{i-1,i}$. It can be shown (Varga, 1962) that the elements on the main diagonal of the **A** matrix are positive and the off-diagonal elements must be negative in order that the solution vector be positive. For a backward differencing scheme, Eqs. 3.32 clearly show that the upper diagonal is negative (Eq. 3.32c), the main diagonal is positive (Eq. 3.32b), and the lower diagonal (Eq. 3.32a) is negative in all cases. For arbitrary differences, the elements of the **A** matrix are written as

$$a_{i,i+1} = Q_{i,i+1}\beta_{i,i+1} - E'_{i,i+1} \tag{3.86a}$$

$$a_{i,i} = Q_{i,i+1}\alpha_{i,i+1} - Q_{i-1,i}\beta_{i-1,i} + E'_{i-1,i} + E'_{i,i+1} + V_i K_i \tag{3.86b}$$

$$a_{i-1,i} = -Q_{i-1,i}\alpha_{i-1,i} - E'_{i-1,i} \tag{3.86c}$$

The lower off-diagonal elements (Eq. 3.86c) are unconditionally negative, whereas the upper off-diagonal (Eq. 3.86a) may be positive or negative. Thus, negativity of the upper off-diagonal elements is assured when

$$E'_{i,i+1} > Q_{i,i+1}\beta_{i,i+1}$$

or

$$E'_{i,i+1} > Q_{i,i+1}(1 - \alpha_{i,i+1}) \tag{3.87}$$

Dividing by $Q_{i,i+1}$ and subtracting 1 from both sides of Eq. 3.87,

$$-1 + \frac{E'_{i,i+1}}{Q_{i,i+1}} > -\alpha_{i,i+1}$$

and, therefore,

$$\alpha_{i,i+1} > 1 - \frac{E'_{i,i+1}}{Q_{i,i+1}} \tag{3.88}$$

Use of Eq. 3.88 will, therefore, assure that resulting concentrations are positive. Substituting for E' (Eq. 3.27), and deleting interfacial subscripts,

$$\alpha > 1 - \frac{EA/\Delta x}{Q}$$

or

$$\alpha > 1 - \frac{E}{U \Delta x}$$

Solving for Δx,

$$\Delta x < \frac{E}{U(1 - \alpha)} \tag{3.89}$$

For central differences ($\alpha = \frac{1}{2}$), Eq. 3.89 becomes

$$\Delta x < 2\frac{E}{U} \tag{3.90}$$

For backward differences, with $\alpha = 1$, Eq. 3.89 is unconditionally fulfilled ($\Delta x < \infty$).

It is interesting to note that the equations used in developing relative error (Section 3.6.3) yield the same restrictions on segment size as indicated in Eqs. 3.89 and 3.90. The equation for the calculated finite segment concentration in the region downstream of the load is

$$s = s_0(\rho_r)^n \qquad n \geq 0 \tag{3.49b}$$

For positive concentrations, the base ρ_r must be positive since n may be odd. Using arbitrary differences, the base ρ_r may be shown to be given by the following expression:

$$\rho_r = \frac{a}{1 - 4\beta\tau/\eta} - \sqrt{\left(\frac{a}{1 - 4\beta\tau/\eta}\right)^2 - \frac{1 + 4\alpha\tau/\eta}{1 - 4\beta\tau/\eta}} \tag{3.91a}$$

where

$$a = \frac{2\tau^2}{\eta} + \frac{2(2\alpha - 1)\tau}{\eta} + 1 \tag{3.91b}$$

As long as $\frac{1}{2} \leq \alpha \leq 1$ (Eq. 3.85), the value of a is positive. To ensure that ρ_r is positive, the denominator of the first term of Eq. 3.91a must be positive, resulting in the constraint

$$4\beta\tau/\eta < 1 \tag{3.92}$$

Substituting $\beta = 1 - \alpha$, $\tau = K \Delta x/U$, and $\eta = 4 KE/U^2$,

$$(1 - \alpha)\frac{U \Delta x}{E} < 1$$

or

$$\Delta x < \frac{E}{U(1 - \alpha)}$$

which is identical with Eq. 3.89.

In summary, use of a backward differencing scheme ensures unconditional positivity of resulting segment concentrations, whereas arbitrary and central differences mandate adoption of segment size restrictions (Eqs. 3.89, 3.90) to ensure positive concentrations.

3.7.3 Numerical Dispersion

Use of difference approximations in place of continuous spatial derivatives in the mass balance differential equation introduces a degree of error in the solution. The nature of this error may be understood (Thomann, 1972, 1973) by expanding the lowest order derivative in a Taylor series for a one-dimensional, constant parameter estuary. As an example, for a backward differencing scheme, a Taylor series expansion of s_{i-1} about s_i can be shown (Torrance, 1968) to yield

$$\left.\frac{\partial s}{\partial x}\right|_i = \frac{s_i - s_{i-1}}{\Delta x} + \left[\frac{\Delta x}{2}\left(\frac{\partial^2 s}{\partial x^2}\right)_i - \frac{(\Delta x)^2}{6}\left(\frac{\partial^3 s}{\partial x^3}\right)_i + \cdots\right]$$

The first term is recognized (Eq. 3.79) as the backward difference approximation to the spatial concentration derivative. The terms in the brackets are seen to be neglected and introduce a truncation error. Taking the first term in the brackets as a measure of this error and substituting into the steady state mass balance equation (Eq. 3.7), the solution of the difference equation is equivalent to solution of the equation given by (Torrance, 1968)

$$\left(E + \frac{\Delta x}{2}|U|\right)\frac{\partial^2 s}{\partial x^2} - U\frac{\partial s}{\partial x} - Ks = 0 \qquad (3.93)$$

where the second value in the dispersive term is referred to as "numerical dispersion." Thus, the effect of neglecting higher order terms in the derivative expansion is to introduce an additional undesired dispersion. For an arbitrary differencing scheme, the numerical dispersion may be written as

$$E_{\text{num}} = U \Delta x(\alpha - 0.5) \qquad (3.94)$$

For backward differences ($\alpha = 1$),

$$E_{\text{num}} = U\frac{\Delta x}{2} \qquad (3.95)$$

for central differences ($\alpha = 0.5$), E_{num} equals zero. Thus, use of backward differences introduces the largest numerical dispersion, arbitrary differences introduce less, and central differences cause no numerical dispersion (all for steady state).

Common practice has been to select the segment size so that the numerical dispersion is some small fraction of the actual dispersion coefficient. It was felt that, in so doing, reasonably uniform control of errors would result. For more advective systems, with higher net velocities (U), this results in smaller segment sizes, as compared with more dispersive systems. Using examples (Mueller, 1976) of segment sizes required for a constant relative error of 5% at the outfall for predominantly advective to predominantly dispersive systems, the ratio E_{num}/E is shown in Fig. 3.29 (Thomann, 1983) to be much higher for more advective systems ($\eta = 0.01$) than for the dispersive systems ($\eta = 100$). Consequently, a somewhat higher ratio of E_{num}/E may be used in more advective systems and a lower ratio in dispersive systems to insure equal relative errors at discharge locations.

For steady state models, backward differencing schemes provide unconditionally positive solutions but introduce maximum numerical dispersion. Segment size limitations are required for arbitrary and central differences to ensure positivity. On the other hand, the latter schemes have less (arbitrary differences) or no (central differences) numerical dispersion. Regardless of what spatial grid is selected, errors in the solutions result due to the approximations used for the first and second spatial derivatives of concentration. Estimates of these errors may be made for simpler systems using the relative error techniques of Section 3.6.3. For more complex systems, numerical tests of the impact of varying segment lengths must be performed, especially in areas of high concentration gradients.

Figure 3.29 Relationship between normalized numerical dispersion and estuary number. From Thomann (1983).

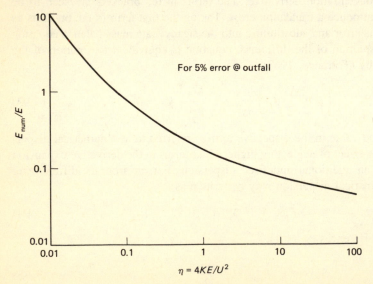

3.8 TIME VARIABLE ASPECTS OF ESTUARIES

Time variable concentrations due to instantaneous inputs of mass have been discussed in Section 2.3.2.1 and illustrative examples given in Sample Problems 2.6 and 3.3 for advective–dispersive systems. Equation 2.40 presents the temporal and spatial variation of concentration due to a constant input of mass over a relatively short time interval (Section 2.3.2.2). For an estuary, another problem of interest is determining the time period required for the estuary to come to a steady state concentration profile after the introduction of a new continuous source. This is the so-called time variable buildup calculation.

3.8.1 Time Variable Buildup

The analysis is performed on a one-dimensional, infinitely long, constant parameter estuary initially uncontaminated with the substance to be introduced. At $t = 0$, the new discharge is introduced at $x = 0$ at a constant rate, $W[M/T]$, and it continues at that magnitude indefinitely thereafter. Resulting concentrations (Thomann, 1972) are given by

$$s(x, t) = \frac{W}{2A\sqrt{\pi E}} \int_0^t t^{-1/2} \exp\left[\frac{-(x - Ut)^2}{4Et} - Kt\right] dt \qquad (3.96)$$

Equation 3.96 indicates that equilibrium is reached after a length of time governed by the system parameters, K, A, E, and U. The step function response is also a function of the location in the estuary. This is to be expected since if one directs attention to a point many miles downstream from the input sources, the effect of W must first travel to that point and then begin to build up to its new equilibrium value. Figure 3.30 contrasts the response due to a unit step for the

Figure 3.30 Unit step functions for Green River, Washington and Delaware Estuary at Philadelphia, PA. From Thomann (1972).

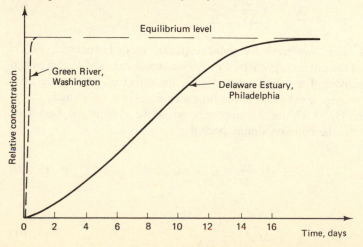

Green River in Washington and the Delaware Estuary in Philadelphia. For the Green River, the response is at a location a short distance downstream from the outfall while for the Delaware case, the response is for a station located at the outfall. In contrast to the river case (small dispersion) where equilibrium is reached almost instantaneously, equilibrium time for an estuary is more closely related to that of the completely mixed system. Times of 15–30 days and longer to reach 95% equilibrium for estuaries are not uncommon. An example of the spatial buildup in the Hudson River estuary at the Battery, New York City, is given in Sample Problem 3.8.

3.8.2 Stability and Numerical Dispersion of Finite Segment Models

For time variable finite segment models, a common practice (Thomann, 1973) has been to select a spatial grid first, that is, fix Δx using the positivity criterion discussed in Section 3.7.2, and then select a time step (Δt). It can be shown that in order to maintain stability for an explicit time differencing, the main diagonal term must be all positive. This leads to the following general stability criterion (for constant coefficients)(Kent, 1960):

$$1 + U\frac{\Delta t}{\Delta x}(\beta - \alpha) - \frac{2E\,\Delta t}{\Delta x^2} - K\,\Delta t > 0 \tag{3.97}$$

For central differences, ($\alpha = \beta = \frac{1}{2}$), this criterion becomes

$$2E\frac{\Delta t}{\Delta x^2} + K\,\Delta t < 1 \tag{3.98}$$

and for $K = 0$, a conservative substance,

$$\frac{\Delta t}{\Delta x^2} < \frac{1}{2E} \tag{3.99}$$

This is the usual expression given for stability (Kent, 1960; Leendertse, 1971). Note however, that the criterion given by Eq. 3.99 assumes central differences and conservative substances. For a fixed spatial grid, the effect of introducing first-order decay is to require a reduction in the time step necessary for stability.

Returning to Eq. 3.97, for a completely advective system and backward differences ($\alpha = 1$), the criterion simply becomes

$$1 - U\frac{\Delta t}{\Delta x} - K\,\Delta t > 0 \tag{3.100}$$

or

$$\Delta t < \frac{\Delta x}{U + K\,\Delta x} \tag{3.101}$$

SAMPLE PROBLEM 3.8

DATA

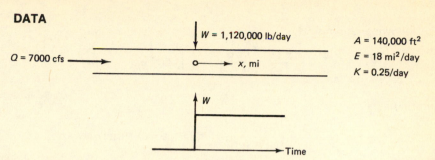

$W = 1,120,000$ lb/day

$Q = 7000$ cfs → x, mi

$A = 140,000$ ft^2
$E = 18$ mi^2/day
$K = 0.25$/day

A new discharge of 1.12 million lb/day is introduced to the estuary with coefficients typical of the Hudson River at the Battery in New York City.

PROBLEM

Determine the time to steady state at locations at 10-mi intervals upstream and downstream of the discharge. Plot equilibrium concentrations as well as concentrations at $t = 0.5, 1, 2,$ and 4 days (approximately 1, 2, 4, and 8 tidal cycles after the discharge begins) for the length of the estuary.

ANALYSIS

Time to Steady State

The time to equilibrium will be judged by comparing the time variable concentration $s(x, t)$ with the equilibrium value for the corresponding location $s(x, \infty)$. When the ratio of $s(x, t)/s(x, \infty)$ approaches unity, equilibrium has been attained.

$$s(x, t) = \frac{W}{2A\sqrt{\pi E}} \int_0^t f(t)\, dt \qquad \text{[Eq. (3.96)]}$$

where

$$f(t) = t^{-1/2} \exp\left[\frac{-(x - Ut)^2}{4Et} - Kt\right]$$

$$s(x, \infty) = s_0 \exp(hx) \qquad \text{[Eq. (3.8a,b)]}$$

where

$$s_0 = \frac{W}{Q\alpha} \qquad \text{[Eq. (3.8e)]}$$

and

$$h = j_1 = \frac{U}{2E}(1 + \alpha), \qquad x \leq 0 \qquad \text{[Eq. (3.8c)]}$$

$$h = j_2 = \frac{U}{2E}(1 - \alpha), \qquad x \geq 0 \qquad \text{[Eq. (3.8d)]}$$

(continued)

Sample Problem 3.8 (continued)

The following ratio results:

$$\frac{s(x,t)}{s(x,\infty)} = \frac{U\alpha}{2\sqrt{\pi E}} \cdot \frac{1}{e^{hx}} \int_0^t f(t)\, dt$$

Substituting

$$U = \frac{Q}{A} = \frac{7000 \text{ cfs}}{140,000 \text{ ft}^2} = 0.05 \text{ fps} = 0.82 \text{ mi/day}$$

$$E = 18 \text{ smpd} \quad \text{and} \quad \alpha = 5.27$$

$$\frac{s(x,t)}{s(x,\infty)} = 0.287 e^{-hx} \int_0^t f(t)\, dt$$

where

$$h = 0.143/\text{mi} \qquad \text{for} \quad x \leq 0$$

and

$$h = -0.0973/\text{mi} \qquad \text{for} \quad x \geq 0$$

Evaluation of the integral was accomplished by calculating $f(t)$ at short intervals ($t = 0.025$ day) and using Simpson's rule. In the first time step, $f(t)$ approaches infinity. The procedure adopted was to neglect the area under the curve from $t = 0$ to $t = 0.025$ day, generate the asymptote by making $t(\text{max}) = 80$ days, and add the difference between unity and the asymptote to all preceding values of the integral.

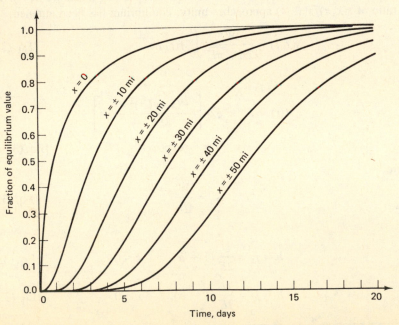

Sample Problem 3.8 (continued)

Results of the analysis are shown above. At $x = 0$, the discharge site, 95% of the equilibrium concentration is attained by day 7.5, 15 tidal cycles after discharge began. The 95% level is attained at $t = 11, 14.5, 17.5, 20.5$, and 23.5 days for $x = 10, 20, 30, 40$, and 50 mi, respectively. From the analysis results, the same fraction of the equilibrium concentration occurs at the same time at points symmetric about the source. Thus, in the plot, the same curve applies to $x = +10$ mi and $x = -10$ mi, and so on. Note the increasing lag times before the concentrations begin to build up (approximately 4 days for $x = \pm 50$ mi) as $|x|$ increases.

Spatial Profiles

The steady state concentration at $x = 0$ is

$$s_0 = \frac{W}{Q\alpha} = \frac{1,120,000 \text{ lb/day}}{7000 \text{ cfs} \times 5.27 \times 5.391} = 5.63 \text{ mg/l}$$

$$\therefore s(x, \infty) = 5.63 \exp(hx)$$

Absolute spatial concentrations were obtained by multiplying the equilibrium value by the fraction of equilibrium determined above.

$$s(x,t) = s(x,\infty)\left[\frac{s(x,t)}{s(x,\infty)}\right]$$

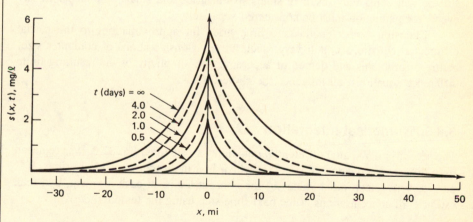

In the resulting profiles above, note that the shape of the steady state profile is maintained at all times with more curvature upstream of the discharge and less downstream.

For $K = 0$,

$$U \, \Delta t < \Delta x \tag{3.102}$$

or

$$\Delta t < \frac{\Delta x}{U} \tag{3.103}$$

In physical terms, Eq. 3.103 requires that the time step selected be less than the residence time $(\Delta x / U)$ of any segment.

As indicated previously in Section 3.7.3, numerical dispersion is introduced by backward and aribtrary spatial differencing schemes. In time variable problems, numerical dispersion is also introduced by the explicit time differencing (Bella and Grenney, 1970). Therefore, for an arbitrary spatial differencing scheme and forward explicit time differencing, the dispersion introduced is given by

$$E_{\text{num}} = U \, \Delta x \left[\left(\alpha - \frac{1}{2} \right) - \frac{U \, \Delta t}{2 \, \Delta x} \right] \tag{3.104}$$

For time variable problems, the issue of selecting appropriate space and time scales is a little clearer—at least for explicit temporal schemes. If central spatial differences are used ($\alpha = \frac{1}{2}$), the numerical dispersion due to spatial truncation is eliminated and if enough additional dispersion is added to the original equation (i.e., equal to $U^2 \, \Delta t / 2$) then the solution will not have any numerical dispersion. However, the segment size and time step constraints of Eqs. 3.89 and 3.97 must still be met. This may result in small spatial grids and small time steps. In any event, each situation must be considered separately.

Choosing proper spatial and time grids for approximations to the partial differential equations is still very much an art. Considerations of computer size, nature of problems and degree of accuracy, and simplicity of the resulting finite difference equations, all influence the choice.

3.8.3. Numerical Integration

The second-order spatial differential equation for estuaries, Eq. 3.23, was replaced by a difference equation (Eq. 3.28), which is a first-order equation in time. It is necessary to select initial values of the concentrations in each segment, and then estimate all concentrations at the next time step using the temporal derivative

$$V_i \frac{ds_i}{dt} = f(t, s_{i-1}, s_i, s_{i+1}) \tag{3.105}$$

where the function f is the right side of Eq. 3.28.

As an example, the computational procedure given by Jeglic (1966) begins from a set of initial values at time zero and in all segments, $i = 1, \ldots, n$. Numerical integration is utilized to advance the solution one time step forward. After each step is completed, the new values s_i become the initial values of the next step. The process is continued until the entire time span that is originally specified has been completed. Truncation errors are controlled by a halving and doubling procedure which, while sacrificing some computational time, does not require the user to specify an integration interval. This numerical procedure can be accomplished using a fourth-order Runge–Kutta approximation (Jeglic, 1966) although other types of numerical integration schemes are also appropriate. Equation 3.105 can be integrated as follows

$$(s_i)_{t+1} = (s_i)_t + \tfrac{1}{6}(r_{0,i} + 2r_{1,i} + 2r_{2,i} + r_{3,i}) \tag{3.106}$$

where

$$r_{0,i} = hf[t_t, (s_{i-1})_t, (s_i)_t, (s_{i+1})_t]$$

$$r_{1,i} = hf[t_t + \tfrac{1}{2}h, (s_{i-1})_t + \tfrac{1}{2}r_{0,i-1}, (s_i)_t + \tfrac{1}{2}r_{0,i}, (s_{i+1})_t + \tfrac{1}{2}r_{0,i+1}]$$

$$r_{2,i} = hf[t_t + \tfrac{1}{2}h, (s_{i-1})_t + \tfrac{1}{2}r_{1,i-1}, (s_i)_t + \tfrac{1}{2}r_{1,i}, (s_{i+1})_t + \tfrac{1}{2}r_{1,i+1}]$$

$$r_{3,i} = hf[t_t + h, (s_{i-1})_t + \tfrac{1}{2}r_{2,i-1}, (s_i)_t + \tfrac{1}{2}r_{2,i}, (s_{i+1})_t + \tfrac{1}{2}r_{2,i+1}]$$

for the time interval of integration (Δt) given by h.

With initial ($t = 0$) conditions known in all sections and boundary and forcing functions specified, Eq. 3.106 provides the numerical scheme for computing s as a function of discrete time, $t = 0, h, 2h$, and so on.

REFERENCES

Bella, D. A., and W. J., Grenney, 1970. Finite-Difference Convection Errors, *J. Sanit. Eng. Div. Proc. Am. Soc, Civ. Eng.* **96** (SA6): 1361–1375.

Crowley, W. P., 1968. A Global Numerical Ocean Model: Part 1, *J. Comp. Phys.* **3**: 111–147.

Defant, A., 1958. *Ebb and Flow, the Tides of Earth, Air and Water*, University of Michigan Press, Ann Arbor, MI, 121 pp.

Delaware Estuary Comprehensive Study, July, 1966. Preliminary Report and Findings, Department of the Interior, Federal Water Pollution Control Administration, Philadelphia, Pa., 94 pp. + xvii and Appendix.

Diachishin, A., 1963. Dye Dispersion Studies, *J. Sanit. Eng. Div. Proc. Am. Soc. Civ. Eng.* **89** (SA1), Proc. Paper 3386, 29–49.

Dobbins, W. E., 1964. BOD and Oxygen Relationships in Streams. *J. Sanit. Eng. Div, Proc. Am. Soc. Civ. Eng* **90** (SA3): 53–78.

Fischer, H. B., E. J. List, R. C. Y. Koh, J. Imberger, and N. H. Brooks, 1979. *Mixing in Inland and Coastal Waters*, Academic P, New York.

Hydroscience, Inc. 1970. Development of Water Quality Model of Boston Harbor, Interim Report Comm. of Mass. Water Res. Comm. Boston, MA, 18 pp. + Appendix.

Hydroscience, Inc. 1971. Simplified Mathematical Modeling of Water Quality, for the Mitre Corporation and the U.S. Environmental Protection Agency, Water Programs, Washington, D. C., March 1971.

Ippen, A., 1966. *Estuary and Coastline Hydrodynamics*, McGraw-Hill, New York, 744 pp. + xvii.

Jeglic, J. 1966. DECS III, Mathematical Simulation of the Estuarine Behavior, Analysis Memo. No. 1032, Gen. Elec. Co., Phil, PA, Dec. 1966, 5 chap. + Appendix.

Kahlig, P., 1979. One-Dimensional Transient Model for Short-Term Prediction of Downstream Pollution in Rivers, *Water Research*, **13**: 1311–1316.

Kent, R., 1960. Diffusion in a Sectionally Homogeneous Estuary, *J. Sanit. Eng. Div Proc. Am. Soc. Civ. Eng.* **86** (SA2): 15–47.

Krenkel, P. A., and G. T. Orlob, 1962. Turbulent Diffusion and the Reaeration Coefficient, *J. Sanit. Eng. Div. Proc. Am. Soc. Civ. Eng.* **88** (SA2): 53.

Leendertse, J. J., 1971. Digital Techniques: Finite Differences Chap. VI, Solution Techniques, in Estuarine Modeling: An Assessment, prepared for National Coastal Pollution Research Program, Water Quality Office, U.S. Environmental Protection Agency by Tracor, Inc., Austin, TX, pp. 277–302.

Miller, E.G., 1962. Observations of Tidal Flow in the Delaware River, U.S. Department of the Interior, USGS, U.S. Government Printing Office, Washington, D.C., 26 pp.

Miller, K. S., 1960. *An Introduction to the Calculus of Finite Differences and Difference Equations*, Henry Holt, New York.

Mueller, J. A., 1976. Accuracy of Steady State Finite Difference Solutions, Technical Memorandum, Hydroscience, Inc., 23 pp.

Neumann, G., and W. J. Pierson, 1966. *Principles of Physical Oceanography*, Prentice-Hall, Englewood Cliffs, NJ, 545 pp + xii.

Nichols, M., and G. Poor, 1967. Sediment transport in a Coastal Plain Estuary, *J. Waterways Harbors Div. Proc. Am. Soc. Civ. Eng.* **93**: 83–95.

NOAA, 1984a. Tide Tables: 1984 High- and Low-Water Predictions, East Coast of North and South America, Including Greenland, U.S. Dept. of Commerce, Washington, D.C.

NOAA. 1984b. Tidal Current Tables 1984, Atlantic Coast of North America; U.S. Dept. of Commerce, Washington, D.C.

O'Connor, D. J., 1960. Oxygen Balance of an Estuary, *J Sanit. Eng. Div.*, Proc. *Am. Soc. Civ. Eng.* **86** (SA3): 35–55.

O'Connor, D. J., 1962. Organic Pollution of New York Harbor—Theoretical Considerations. *J. Water Pollut. Control Fed.* **34** (9): 905–919.

O'Connor, D. J. 1965. Estuarine Distribution of Nonconservative Substances. *J. Sanit. Engr. Div., Proc. ASCE* **91**(2): 23–42.

O'Connor, D. J., and W. Lung, 1981. Suspended Solids Analsyis of Estuarine Systems. *J. Environ. Eng. Div. Proc. Am. Soc. Civ. Eng.* **107**: 101–120.

Officer, C. B., 1976. *Physical Oceanography of Estuaries (and Associated Coastal Waters)*, Wiley, New York.

Officer, C. B., 1983. 2. Physics of Estuarine Circulation, B. H. Ketchum, Ed., *Ecosystems of the World:26, Estuaries and Enclosed Seas*. Elsevier, Amsterdam, pp. 15–41.

Postma, H., 1967. Sediment Transport and Sedimentation in the Estuarine Environment, G. H. Lauff, Ed. *Estuaries*, American Association for the Advancement of Science, Washington, D.C., pp. 158–179.

Pritchard, D. W., 1969. Dispersion and Flushing of Pollutants in Estuaries, *J. Hydraulics Div. Proc. Am Soc. Civ. Eng.* **95**(HY2):115–124.

Salas, H. J., and R. V. Thomann, 1975. The Chesapeake Bay Waste Load Allocation Study, by Hydroscience, Inc. for Maryland Department of Natural Resources, Water Resources Administration, Annapolis, MD, 287 pp.

Tetra Tech, Inc., 1978. Rates, Constants, and Kinetics Formulations in Surface Water Quality Modeling, for the U.S. Environmental Protection Agency, Environmental Research Laboratory, Athens, GA, Dec. 1978. EPA-600/3-78-105.

Thomann, R. V., 1972. *Systems Analysis and Water Quality Management*, McGraw-Hill, New York, 286 pp. (Reprint: J. Williams, Oklahoma City, OK.)

Thomann, R. V., 1973. Time Variable Water Quality Models, Estuaries, Harbors and Off-Shore Waters; in Mathematical Modeling of Natural Systems, Manhattan College Summer Institute in Water Pollution Control.

Thomann, R. V., 1983. Finite Difference Approach to Estuary Water Quality Analysis, in Quality Models of Natural Water Systems, Twenty-Eighth Summer Institute in Water Pollution Control, Manhattan College.

Thomann, R. V., and J. F. Fitzpatrick, 1982. Calibration and Verification of a Mathematical Model of the Eutrophication of the Potomac Estuary; report by Hydroqual, Inc., Mahwah, NJ to DES, Dist. Col., p. 500 + Appendix.

Torrance, K. E., 1968. Comparison of Finite-Difference Computations of Natural Convection, *J. Res., Nat. Bur. Stds* **72B** (4):281–301.

Varga, R. S. 1962. *Matrix Iterative Analysis*, Prentice-Hall, Englewood Cliffs, NJ, 322 pp. + xi.

PROBLEMS

3.1. At a location in the Hudson River about 90 mi from the Battery, the maximum flow during ebb tide is 135,000 cfs and during flood tide is 120,000 cfs. The phase lag is 3.45 hr referenced to the Battery. Assume that the length of the tidal period is 12.4 hr and the ebb period is 6.8 hr. Estimate the net nontidal flow (in cfs).

3.2. For the Back River, an estuary of Chespeake Bay, salinity was measured at the mouth of the river (entrance to Chesapeake Bay) at 3.9‰ during August, 1975. At 5.5 km upstream, the salinity was 2.5‰ and at 8.6 km, salinity was 0.32‰. If the net nontidal velocity is about 3 cm/s, what is the tidal dispersion coefficient in m^2/s?

3.3. An estuary with no net flow receives an input of 10,000 kg/day. The cross-sectional area of the estuary is 5000 m^2, the tidal dispersion coefficient is $5 \cdot 10^6$ m^2/day and the decay rate is 0.1/day. What inflow (m^3/s) would be required to be discharged to the estuary so that the concentration at the outfall after mixing is reduced by 50% from the concentration at the outfall with no flow in the estuary?

3.4. Given the following estuary.

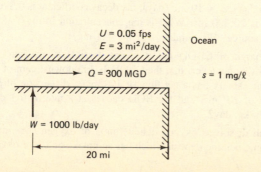

If the substance is conservative, compute and plot the profile of concentration (in mg/l) with distance—both upstream and downstream of the discharge. Note that the local oceanic boundary concentration is affected by the estuary discharge as well as other sources external to the estuary.

3.5. For an estuary with a flow of 1000 cfs, a cross-sectional area of 50,000 ft², a tidal dispersion coefficient of 5 smpd and a decay coefficient of 0.1/day:
   (a) Compute and plot the concentration in μg/l for a 1000-lb/day input.
   (b) How far downstream must another 1000 lb/day source go so as not to violate a maximum level of 50 μg/l?

3.6. An estuary has a net flow of 800 cfs, tidal dispersion of 5 smpd, net velocity of 0.2 mpd. If a conservative variable is to be discharged at mile 0, what is the allowable discharge load in lb/day such that a concentration of 8 mg/l is obtained 21 mi upstream from the input?

3.7. The following water quality data were collected in an estuary.

	Distance (mi)	Concentration (mg/l)
	−15	0.43
	−10	1.22
	−5	3.50
Discharge location	0	10.00
	5	4.49
	10	2.02
	15	0.91

It is also known that the input at mile 0 is 372,000 lb/day, the net estuarine flow is 1000 cfs and the average tidal cross-sectional area is 10^5 ft². If the substance has been tested to have a decay coefficient of 0.1/day, what is the tidal dispersion for this estuary in ft²/sec?

3.8. Given the following estuary with *zero* net flow (only dispersion):

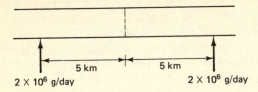

The tidal dispersion is 150 m²/s (1.3×10^7 m²/day), the decay coefficient is 0.2/day and the cross-sectional area is 20,000 m². What is the concentration in mg/l at the point 5 km in between each source? (Remember 1 g/m³ = 1 mg/l).

3.9. An estuary receives a discharge of CBOD but it is not possible to measure the discharge mass rate. It is known, however, that the CBOD 10 km downstream of the discharge is 0.15 mg/l. If the estuary has a net flow of 100 m³/s, dispersion of 300 m²/s, a cross-sectional area of 10^4 m², and the BOD decays at a rate of 0.1 day, what is the mass discharge in kg/day?

3.10. (a) Determine the maximum dissolved oxygen deficit (mg/l of oxygen consumed) by oxidation of the organic benthic deposit which has an uptake of 2 g of oxygen per day per m² of bottom area. The reaeration coefficient (K_a) is a decay, or removal rate, of the DO deficit.

(b) Determine the DO deficits at $x = -1$ mi, 0, and $+3$ mi.

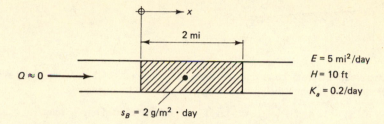

$s_B = 2$ g/m² · day

3.11. A discharge of 1000 lb/day of a conservative substance is made into the estuary at $x = 0$. Using the methods employed in Sample Problem 3.4, determine the *net* flux of the material (advective plus dispersive) across cross sections located at $x = -2$ mi, -1 mi, and $+2$ mi. The cross-sectional area at mean water is 800 ft².

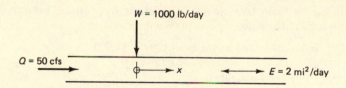

3.12. Assume for the Wicomico estuary described in Sample Problem 3.6 that the following salinity values are representative of a high flow period.

Segment interface	Flow (cfs)	Salinity (‰)	Segment no.	Salinity (‰)
10–Bay	500	8	Chesapeake Bay	10
9–10	490	4.5	10	6
8–9	480	2.25	9	3
7–8	470	1.0	8	1.5
6–7	460	0.3	7	0.5
5–6	430	0.1	6	0.1 (background)
4–5	420		5	
3–4	410		4	
2–3	400		3	
1–2	390		2	
0–1	380	0.1	1	0.1

(a) Plot the salinity values vs. estuary distance (each segment is 2 mi long).
(b) Using the cross-sectional area of Sample Problem 3.6 calculate the dispersion coefficients in smpd for the segment interfaces: 10–bay, 9–10, 8–9.

3.13. A harbor of a large bay receives a discharge at one end as shown. The harbor is divided into two segments for simplicity.

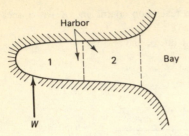

A conservative substance is released into seg. 1 at a rate of 1 kg/day. There is no advective flow through the system. A survey gave the following results:

$$s_1 = 12\,\text{mg/l}, \qquad s_2 = 5\,\text{mg/l}, \qquad \text{concentration of bay} = 1\,\text{mg/l}$$

(a) What are the dispersive exchange coefficients due to mixing of the bay and the harbor, E'_{12} and $E'_{2,\text{bay}}$ in m^3/day?

(b) What is the input to the bay from this harbor in kg/day? That is, what is the dispersive flux across the interface between seg. 2 and the bay?

3.14. As a first rough approximation, the lower 50 mi of the Potomac Estuary may be assumed to be composed of two well-mixed segments of water—one above the other— with flow entering the lower layer from the Chesapeake Bay. Mixing between the two layers also occurs. The following conditions are appropriate:

inflow of fresh water to upper layer = 11,000 cfs
salinity of inflow = 0
salinity of upper layer = 10‰
salinity of lower layer = 12‰
salinity of bottom bay water = 14‰

Assuming that horizontal dispersion is not significant, but including vertical mixing:

(a) What is the flow entering the lower layer of the estuary from the bay in cfs?

(b) What is the surface outflow from the upper layer to the bay in cfs?

(c) If the surface area is $7.68 \cdot 10^8 m^2$ and each segment depth = 3 m, what is the vertical dispersion coefficient in cm^2/s?

3.15. Organic phosphorus is discharged into an estuary that has been divided into three segments. The particulate fraction of the organic phosphorus settles at a rate of 1 ft/day and the dissolved fraction of the organic phosphorus hydrolyzes to inorganic phosphorus at a rate of 0.2/day. The depth of each of the segments is 10 ft and each volume in 100×10^6 gal. A discharge load of 1000 lbs org. P/day enters the first segment and 600 lb/day enters the third segment. The following steady state response matrix for the organic phosphorus has been obtained. (units are mg/l per lb/day):

$$\begin{bmatrix} 0.002 & 0.0005 & 0.0001 \\ 0.001 & 0.002 & 0.0005 \\ 0.0005 & 0.001 & 0.002 \end{bmatrix}$$

(a) What is the concentration of organic phosphorus in each segment in mg/l?

(b) How many lb/day of organic phosphorus settles out of seg. 2?

(c) What is the mass flux (lb/day) of the decomposition (hydrolysis) of organic phosphorus to inorganic phosphorus in seg. 3?

(d) Suppose the *observed* concentration of organic phosphorus in seg. 1 is 3 mg/l. How much additional load of organic phosphorus into 1 is necessary in pounds per day to match this observed value?

3.16. A conservative substance enters via a tidal channel into a large shallow embayment that is completely enclosed as follows.

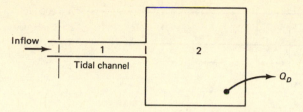

The embayment evaporates water because of the high temperature and a diversion of flow (Q_D) is made from the embayment to maintain a steady volume in the embayment. The conditions are as follows:

$$\text{inflow} = 100\,\text{m}^3/\text{s} \qquad E'_{12} = 100\,\text{m}^3/\text{s} \qquad V_1 = V_2 = 10^8\,\text{m}^3$$
$$\text{concentration of inflow} = 10\,\text{mg/l} \qquad E'_{01} = 0$$

(a) What is the concentration in segs. 1 and 2 if the evaporation from the embayment is zero?

(b) What is the concentration in segs. 1 and 2 if the evaporation from the embayment is 20 m³/s?

(c) How much mass is transferred due to dispersive exchange between segs. 1 and 2 in g/s under (a)?

(d) How much mass is transferred due to dispersive exchange between segs. 1 and 2 in g/s under (b)?

3.17. The estuary has a water withdrawal in seg. 2.

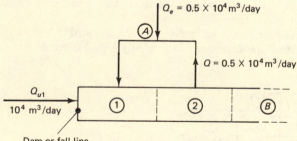

$$E'_{12} = E'_{2B} = 2 \cdot 10^4\,\text{m}^3/\text{day}, \qquad E'_{01} = 0, \qquad V_1 = V_2 = 1.5 \cdot 10^5\,\text{m}^3$$
$$K_1 = K_2 = 0.1/\text{day}, \qquad s_B = 10\,\text{g/m}^3, \qquad s_0 = 0$$

(a) What is the allowable discharge in g/day from source A so that the concentration is 4.0 mg/l in seg. 1 and 4.8 mg/l in seg. 2?

(b) It can be calculated that the concentration due just to the load input from A is 2.4 mg/l in seg. 1 and 1.28 in seg. 2. At what boundary concentration does the allowable load from source A drop to zero under the condition of maintaining a maximum concentration of 4.0 mg/l in seg. 1 and 4.8 mg/l is seg. 2?

3.18. For the uniformly segmented estuary with a single point source, determine the segment size for relative errors in the discharge segment of 1, 2, 3, 5, 7, 10, 15, and 20%. Plot resulting finite difference concentrations (s_0) vs. segment size (abscissa) and indicate the analytical solution (l_0) on the plot.

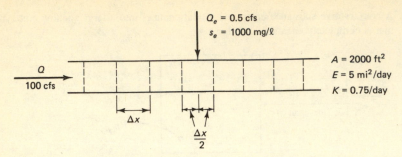

3.19. The following three segments are a schematic representation of a particular body of water.

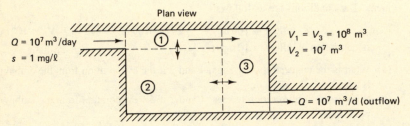

The incoming flow is routed via a deep channel from segs 1 to 3. The dispersion between these two segments is small compared to the flow. However, both segs. 1 and 3 exchange dispersively with seg. 2. The variable s is nonconservative and decays at 0.1/day.

(a) If it is estimated that the bulk dispersion between segs. 1 and 2 and between segs. 3 and 2 is equal to 10^7 m³/day, set up the basic equations for solution of this problem in matrix form.

(b) Suppose the concentration at the outflow has been measured at 0.26 mg/l. Is the original estimate of the dispersion of 10^7 m³/day a good estimate?

3.20. Given:

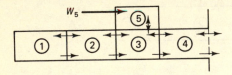

and solution vector in mg/1,

$$\begin{pmatrix} 0.5 \\ 0.8 \\ 1.0 \\ 0.9 \\ 3.0 \end{pmatrix} \qquad \begin{aligned} V_5 &= 10^4 \, \text{m}^3 \\ K_5 &= 0.1/\text{day} \\ W_5 &= 10 \, \text{kg/day} \\ Q_{53} &= 0 \end{aligned}$$

What is the exchange E'_{35} in m³/day?

Chapter 4

Lakes

A major portion of our water-based recreational activities centers about the thousands of lakes, reservoirs, and other small, relatively quiescent bodies of water. In addition, these waters serve as a source of water for municipal and industrial use, including water released from reservoirs for agricultural purposes, water quality control, and fisheries management. The ecosystems and quality of lakes throughout the world are therefore of primary concern in water quality management. Indeed, for those who have drifted across a quiet lake at sundown, half attempting to fish, but marveling at the beauty and complexity of the scene, the study of the water quality of lakes is of particular interest.

Lakes and reservoirs vary from small ponds and dams to the magnificent and monumental large lakes of the world such as Lake Superior, one of the Great Lakes, and Lake Baikal in the Soviet Union, the deepest lake in the world (1620 m, 5310 ft). The ecosystems supported by this broad range of water bodies vary from the very attractive local sport fishes such as bass and perch to the large top predators of both sport and commercial value such as the lake trout and landlocked and migratory salmon. Lake Baikal, for example, is the habitat for a fresh water seal, an aquatic animal of significant ecological symbolism (Kochov, 1962).

Limnology is the study of the physical, chemical, and biological behavior of lakes. Recreation, sport fishing (and for the larger lakes, commercial fishing), and water supply for municipal and industrial uses are all intimately related to the quality of these water bodies. The distinguishing physical features of lakes include relatively low flow-through velocities and development of significant vertical gradients in temperature and other water quality variables. Lakes therefore often become sinks for nutrients, toxicants, and other substances in incoming rivers. As a result, eutrophication is one of the more significant water quality problems of lakes and is the topic of a detailed treatment in Chapter 7. The problems of toxic sub-

173

stances in lakes are discussed in Chapter 8. In this chapter, a general overview is given of the properties of lake systems and the basic tools needed for modeling of lake water quality are presented.

Additional references in limnology and lake processes include Lerman (1978), Wetzel (1975), and the classical great works of Hutchinson (1957, 1967). These references should be consulted for more in-depth treatment of the background of physical, chemical, and biological processes in lakes. Also, Reckhow and Chapra (1983) and Chapra and Reckhow (1983) provide detailed analyses of lake water quality models and mechanisms.

4.1 PHYSICAL AND HYDROLOGIC CHARACTERISTICS

The principal physical features of a lake are: length, depth, area (both of the water surface and of the drainage area), and volume. The overall physical relationships for a lake can be summarized in area–depth and volume–depth curves. A typical example for Lake Michigan is shown in Fig. 4.1 where as indicated the relationships characterize the area and volume as a function of the depth of the lake. Table 4.1 shows some of the characteristics of the world's largest lakes.

The relationship between the flow out of a lake or reservoir and the volume is also an important characteristic. The ratio of the volume to the flow represents the hydraulic detention time, that is, the time it would take to empty out the lake or reservoir if all inputs of water to the lake ceased. The hydraulic detention time t_d is given by

$$t_d = \frac{V}{Q} \tag{4.1}$$

Figure 4.1 Variation of area and volume with depth for Lake Michigan. From Mortimer (1976).

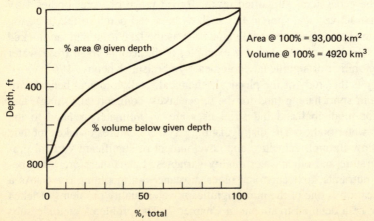

TABLE 4.1 LAKES WITH LARGEST SURFACE AREA BY CONTINENT

Continent	Lake	Area (mi^2)	Elevation (ft)	Maximum depth (ft)
Africa	Victoria	24,300	3,721	279
	Tanganyika	12,350	2,539	4,825
	Malawi	8,680	1,558	2,316
	Chad	4,000	787	36
	Rudolph	2,470	1,401	240
Asia	Caspian (Sea)	143,240	−91	3,363
	Aral (Sea)	24,900	174	220
	Baikal	12,160	1,496	5,315
	Balkhash	6,560	17,000	87
	Tonle Sap	1,040	—	39
Australia	Eyre	2,970	−39	4
	Torrens	2,230	98	—
	Gairdner	1,840	112	—
	Frome	930	160	4
	Taupo	230	1,172	522
Europe	Ladoga	7,000	13	755
	Onega	3,750	108	394
	Vanern	2,160	144	322
	Peipus	1,660	98	49
	Vattern	740	289	420
N. America	Superior	31,760	602	1,333
	Huron	23,000	581	750
	Michigan	22,400	581	923
	Great Bear	12,100	512	1,356
	Great Slave	11,030	513	2,015
S. America	Maracaibo	5,020	0	197
	Patos	3,920	0	15
	Titicaca	3,100	12,497	997
	Merim	1,150	0	33
	Buenos Aires	860	712	—

Source: Showers (1979). From "World Facts and Figures," copyright © 1979 by John Wiley & Sons, Inc. Reprinted with permission.

for a volume $V[L^3]$ and outflow $Q[L^3/T]$. Bartsch and Gakstatter (1978) have compiled the hydraulic detention times as a function of the ratio of lake drainage area to lake surface area for northern U.S. lakes and reservoirs. Their results are shown in Fig. 4.2 which indicates a range of detention times from 1 day to about 6000 days (16 years). The empirical relationship from their work is given by

$$\log_{10} t_d = 4.077 - 1.177 \log_{10} \frac{DA}{SA} \qquad (4.2)$$

where t_d is the detention time in days, DA is the drainage area of the lake, and SA is the surface area of the lake. The lake drainage area is a measure of the inflow and the lake surface area is a measure of the volume. A long detention time does not necessarily indicate a large lake. The ratio of the volume to flow determines

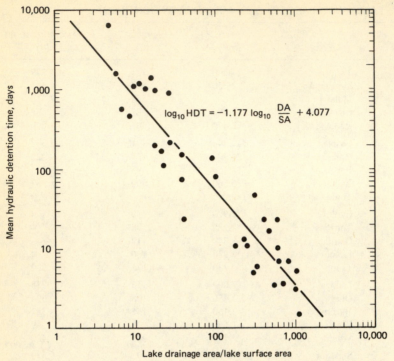

$$\log_{10} \text{HDT} = -1.177 \log_{10} \frac{DA}{SA} + 4.077$$

Figure 4.2 Relationship between mean hydraulic detention time and the ratio of the drainage area to lake surface area for a number of northern U.S. lakes and reservoirs. From Bartsch and Gakstatter (1978).

the average residence time in a lake. Thus, a small lake with a small flow may still have a long detention time.

As with rivers and estuaries, an understanding of the water balance and circulation of lakes is of considerable importance in water quality analysis and engineering. Linsley et al. (1958) provide a general review of hydrologic analyses of lakes and reservoirs. A general and simple hydrologic balance equation for a given body of water is

$$\frac{dV}{dt} = Q_{in} - Q + PA_s - E_v A_s \tag{4.3}$$

where V = lake volume [L^3], A_s = lake surface area [L^2], Q_{in} and Q [L^3/T] represent net flows into and out of the lake due to river and/or groundwater flows P [L/T] is the precipitation directly on the lake, and E_v [L/T] is the lake evaporation. Inflows may include surface inflow, subsurface inflow, and water imported into the lake. Outflows may include surface and subsurface outflow from the reservoir and exported water. The change in storage in the lake or reservoir may also include subsurface storage or "bank" storage of water.

In determining the hydrologic balance of a lake, the change in volume and surface inflow and outflow can usually be measured easily. Precipitation can also be measured without difficulty except for large lakes where it must be estimated for the open water. The remaining unknowns include subsurface water movements and the evaporation.

4.1.1 Evaporation

The most important single factor in the evaporative loss of water is the incoming solar radiation (See Chapter 9). The vapor pressure (i.e., the maximum gaseous pressure of water vapor at a given temperature and 100% humidity) is also a primary variable and depends on the water temperature. Wind, air temperature, and water quality also contribute to the evaporative flux.

There are a variety of methods for estimating evaporation; the most simple and direct approach utilizes measurement of water loss in evaporation pans. For lake estimation, the pans may be floated in the water surface and loss of water measured over time or the pan may be established at a ground based station to minimize problems associated with open water measurements (e.g., splashing). A second approach assumes all terms except evaporation in Eq. 4.3 can be measured over a given time period and E_v is then computed by difference, that is,

$$E_v = \frac{1}{A_s}\left[-\frac{\Delta V}{\Delta t} + Q_{in} - Q + PA_s\right] \tag{4.4}$$

The rate of heat loss by evaporation can be estimated by a variety of equations and Eq. 9.6 provides one suggestion. With this heat loss H_e in cal/cm^2 · day, the water loss can be calculated from

$$E_v = \frac{H_e}{\rho(585)}\left[\left(\frac{cal}{cm^2 \cdot day}\right)\left(\frac{cm^3}{g}\right)\left(\frac{g}{cal}\right) = \frac{cm}{day}\right] \tag{4.5}$$

where ρ is the water density in g/cm^3 and the latent heat of vaporization is 585 cal/g.

Table 4.2 shows the water balance computation for the Great Lakes used by O'Connor and Mueller (1970) in their analysis of chlorides in those lakes. (The basic model used by O'Connor and Mueller is discussed below.) Figure 4.3 shows the location of the Great Lakes and the hydraulic profile.

As seen in Table 4.2, the detention times range from 2.6 yr for Lake Erie to 191 yr for Lake Superior. Evaporation varies from 20 to 30 in./yr, a not uncommon range. In the more arid regions of course, evaporation may exceed this range by factors of 2–5.

4.1.2 Temperature Stratification

Lakes in general are not well-mixed. Gradients of temperature may develop along shore as well as with depth. Tropical lakes may stratify only weakly. However, many North Temperate lakes during summer heating develop a warmer layer of water at the surface which overlies a colder deeper layer of water. The prediction of thermal behavior in lakes and reservoirs is important for power plant siting

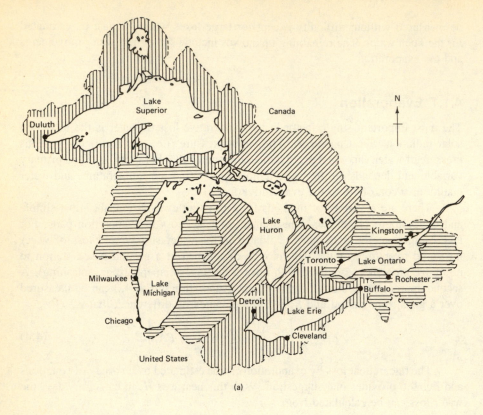

(a)

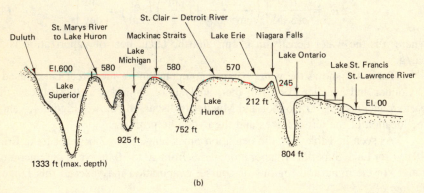

(b)

Figure 4.3 (a) Map of Great Lakes. (b) Hydraulic profile of the Great Lakes. From O'Connor and Mueller (1970). Reprinted by permission of American Society of Civil Engineers.

considerations and thermal effects on sensitive ecosystems. The degree of horizontal or vertical stratification of the lake or reservoir may have a significant effect on other aspects of water quality by "trapping" chemicals in regions of reduced exchange and water interaction. Thus, the extent of thermal stratification also in-

TABLE 4.2 WATER BALANCE COMPUTATIONS FOR GREAT LAKES

Lake	Drainage areas (mi²)		Volume (mi³)	Mean depth (ft)	Mean flow (cfs)	Detention time (yr)
	Water	Land				
Superior	31,800	48,200	2,940	487	71,800	191
Michigan	22,400	45,500	1,170	276	55,000	99.1
Huron	23,000	49,600	850	195	175,400	22.6
Erie	9,930	30,000	113	85	201,000	2.6
Ontario	7,520	27,300	404	283	238,000	7.9

Lake	Runoff (cfs/mi²)	Precipitation on lake surface (in./yr)	Evaporation from lake surface (in./yr) computed
Superior	1.00	29.4	19.4
Michigan	0.86	31.2	21.5
Huron	1.05	31.2	31.4
Erie	0.79	34.0	30.9
Ontario	1.30	34.3	31.4

Source: From O'Connor and Mueller (1970). Reprinted by permission of American Society of Civil Engineers.

fluences the vertical dissolved oxygen profiles where reduced DO often occurs in the lower layer due to minimal exchange with aerated water in the top layer.

The vertical temperature profile at the end of winter for a typical north temperature lake is shown in Fig. 4.4(a) and is often homogeneous from top to bottom. As spring warming begins, the surface layer begins to heat and because of its lower density, begins to stratify and become a distinct layer from the deeper layer beneath it. By midsummer, a strong stratification may have formed and often three distinct vertical regions can be identified [Fig. 4.4(b)]. The epilimnion represents the uppermost layer where the temperature is generally constant with depth. This layer becomes generally well mixed due to surface wind action. The temperature then decreases rapidly with depth in a region called the metalimnion below, which is the hypolimnion which extends to the bottom of the lake. As defined by Hutchinson (1957), the thermocline is the plane of maximum rate of decrease of temperature with respect to depth. For preliminary modeling of stratified lakes, two layers are often considered: the epilimnion and the hypolimnion.

During the fall, surface temperatures begin to cool and subsequently the thermocline penetrates deeper into the lake; fall mixing begins. Isothermal conditions then prevail again during late fall [Fig. 4.4(c)]. As surface temperatures continue to decline and reach levels below the temperature of maximum density, a winter inversion may develop as shown in Fig. 4.4(d). The major reasons for the behavior of vertical temperature in water bodies are the low conductivity of heat (see Chapter 9) and the absorption of heat in the first few meters. As the surface waters begin to heat, transfer to lower layers is reduced and a stability condition develops. As will be seen shortly, these considerations of temperature behavior in lakes and reservoirs will be incorporated into some general modeling frameworks of water quality.

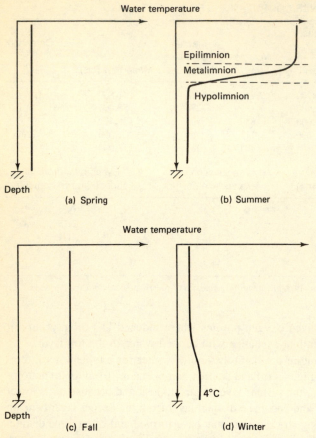

Figure 4.4 Variation of water temperature with depth throughout the year. (a) Spring. (b) Summer. (c) Fall. (d) Winter.

4.2 LAKEWIDE WATER QUALITY RESPONSE TO INPUTS

4.2.1 Lakes as Completely Mixed Systems

It is often useful to describe lakes and reservoirs under the assumption that the body of water is completely mixed horizontally and vertically. This kind of assumption is similar to the one made in the case of rivers and estuaries. The completely mixed assumption is justified on the basis of wind stresses on the water surface resulting in internal mixing. In addition, when the scale of the problem is sufficiently long, as from year to year, then seasonal mixing processes may result in a completely mixed lake over the years. An assumption of this type should be recognized as a gross approximation to the actual lake since variations in concentration of many substances will exist throughout the lake. However, the assumption permits many useful estimates to be made of the behavior of such systems since the relationships between lake inputs due to residual discharges or tributary inputs, flow through the lake, and resulting concentration are made more clear.

Figure 4.5 shows the notation for the completely mixed lake. Consider the discharge of material W [M/T] into the lake from a point input such as a municipality or industry; where $W = Q_e s_e$, for the Q_e = effluent flow and s_e = effluent concentration from the source. Similarly, there may be inputs from a main river ($Q_r s_r$) or tributary ($Q_T s_T$) discharging into the lake. Inputs may also occur from internal sources such as releases of substances from the sediments ($S_D V$). The total mass input, W, is then

$$W = Q_e s_e + Q_r s_r + Q_T s_T + PA_s s_p + S_D V \tag{4.6}$$

for s_p as the concentration in the precipitation. Dry fallout of substances from particle deposition on to the surface of the lake may also have to be included. The lake concentration is given by s [M/L^3], the volume by V [L^3], and the outflow by Q. Note that the outflow is given by

$$Q = Q_e + Q_r + Q_T + PA_s - E_v A_s \tag{4.7}$$

and may have to include other sources or sinks of water. We can also assume that the substance decays with a first-order decay coefficient K [T^{-1}], as we have done before in the rivers and estuaries. A mass balance for the lake is then given by

$$\frac{dVs}{dt} = W(t) - Qs - KVs \tag{4.8}$$

Figure 4.5   Notation for completely mixed lake.

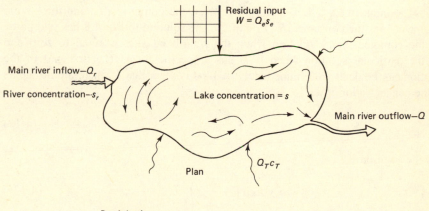

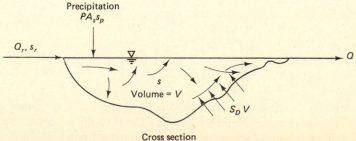

This equation simply states that the time rate of change of the substance s is equal to the amount of input less both the mass discharged with the flow and the mass lost by its decay in the system. Equation 4.8 is written with constant parameters of flow, Q, and decay, K, over time. If the volume of the lake is also assumed temporally constant, then expanding the derivative gives

$$\frac{dVs}{dt} = V\frac{ds}{dt} + s\frac{dV}{dt}$$

$$= V\frac{ds}{dt}$$

since $dV/dt = 0$.

A suitable initial condition is given by

$$s = s_0 \quad \text{at} \quad t = 0 \tag{4.9}$$

Rewriting Eq. 4.8 gives

$$V\frac{ds}{dt} + Qs + KVs = W(t) \tag{4.10}$$

or

$$V\frac{ds}{dt} + K's = W(t) \tag{4.11}$$

where $K' = Q + KV$.

The input in Eqs. 4.10 and 4.11 is $W(t)$, the output is $s(t)$, and the system parameters are the constants, K, V, and Q. The solution to Eq. 4.8 has two parts: (a) the complementary solution when the right-hand side is equal to zero and (b) the particular solution when $W(t)$ has a specific form. The first part of the solution can be dealt with immediately; if $W(t)$ is zero, that is, there are no inputs of the concentration, then

$$\frac{ds}{dt} + \frac{K's}{V} = 0 \tag{4.12}$$

and the solution is

$$s = s_0 \exp\left(-\frac{K'}{V}t\right) \tag{4.13a}$$

$$= s_0 \exp\left[-\left(\frac{Q}{V} + K\right)t\right] \tag{4.13b}$$

$$= s_0 \exp\left[-\left(\frac{1}{t_d} + K\right)t\right] \tag{4.13c}$$

This indicates that, as time increases, the effect of the initial condition s_0 will gradually decay and disappear from the system. The time for the concentration to decrease to a given level depends on two factors: (a) the reciprocal of the detention

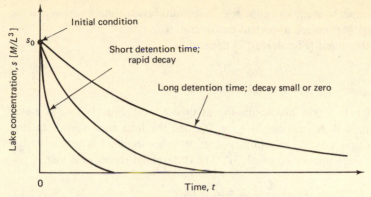

Figure 4.6 "Flushing" of lake initial condition (Eq. 4.13).

time of the lake, that is, the hydraulic flushing of the lake and (b) the decay rate of the substance, K. Figure 4.6 shows this "flushing" of a lake initial condition. As shown, for long detention times and small decay or conservative substances, the time for a substance to be reduced in a lake or reservoir with no input (Eq. 4.13) may be long compared to a short detention time, high decay system.

Of course, some previous load into the lake resulted in the concentration, s_0. Consider now the buildup of concentration due to the load $W(t)$. An important form of $W(t)$ is when the input changes from one constant level (or a zero load) to another load level. This is the so-called "step input" and the response in a lake is of particular importance. This type of input is illustrated in Fig. 4.7 and represents the "stepping-up" at time t_0 to a new constant level of the forcing function.

Figure 4.7 "Step inputs" to lake. (a) Unit step at time t_0. (b) Unit step beginning at $t = 0$. (c) Step input of load to $\overline{W}$ at time $t = 0$.

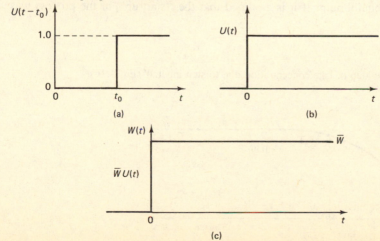

This is a useful input to study since the resulting equilibrium concentration and the time to reach equilibrium are important considerations.

The unit step input [Fig. 4.7(a)] is defined as

$$U(t - t_0) = 0, \qquad t < t_0 \tag{4.14}$$
$$U(t - t_0) = 1, \qquad t > t_0$$

For $t_0 = 0$, $U(t)$, is a unit step beginning at time $t = 0$, Fig. 4.7(b). If a step input of magnitude $\overline{W}$ is imposed at $t_0 = 0$, then the load is as shown in Fig. 4.7(c). Prior to time $t = 0$, an arbitrary starting time, the load is zero. Then at $t = 0$, the load steps up to a constant, $\overline{W}$. The differential equation is then

$$V\frac{ds}{dt} + K's = \overline{W} U(t) \tag{4.15}$$

The solution, which is called the step response, is

$$s_u = \frac{\overline{W}}{K'}\left[1 - \exp\left(\frac{-K'}{V}t\right)\right] \tag{4.16a}$$

or

$$s_u = \frac{\overline{W}}{Q + KV}\left\{1 - \exp\left[-\left(\frac{Q}{V} + K\right)t\right]\right\} \tag{4.16b}$$

$$= \frac{\overline{W}}{Q(1 + Kt_d)}\left\{1 - \exp\left[-\left(\frac{1}{t_d} + K\right)t\right]\right\} \tag{4.16c}$$

The subscript, u, refers to the step response.

This equation indicates that after a sufficient length of time has elapsed the variable s will approach the constant value $\overline{W}/K'$. Figure 4.8 shows this buildup in time. The time to reach the equilibrium condition depends directly on Q/V, the reciprocal of the detention time and K the reaction rate. For $(1/t_d + K)t = 1$, the concentration is at 63% of equilibrium and for $(1/t_d + K)t = 3$, the concentration is at 95% of equilibrium. If it is assumed that the "start-up" of the process was

Figure 4.8   Buildup of lake concentration due to step input $\overline{W}$ (Eq. 4.16).

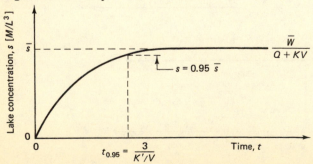

sufficiently far removed in the past then

$$\bar{s} = \frac{W}{Q + KV} = \frac{W}{Q(1 + Kt_d)} \qquad (4.17)$$

For conservative materials $K = 0$ and the equilibrium concentration is

$$\bar{s} = \frac{W}{Q} \qquad (4.18)$$

Note that this water quality response for conservative substances, W/Q, appears in the solutions for riverine, estuarine, and lake systems. The ratio W/Q is then an upper bound first estimate of the steady state response of any system to an input load.

The total response is now given by the sum of Eq. 4.13, the initial condition response, and the response due to the step load, Eq. 4.16. Thus,

$$s = \frac{W}{Q + KV}\left\{1 - \exp\left[-\left(\frac{Q}{V} + K\right)t\right]\right\} + s_0 \exp\left[-\left(\frac{Q}{V} + K\right)t\right] \qquad (4.19)$$

Figure 4.9 shows this total response for two cases: "small" initial condition, that is, initial condition less than the equilibrium concentration due to the step load and "large" initial condition, that is, an initial condition considerably greater than the equilibrium concentration. As shown, for the former case [Figs. 4.9(a) and (c)], the concentration increases to an equilibrium concentration while for the latter case [Figs. 4.9(b) and (d)], the concentration decreases to the new equilibrium concentration from the higher initial concentration.

Figure 4.9   Response due to initial condition and step input. (a) "Small" initial condition. (b) "Large" initial condition. (c) and (d) Sum of components.

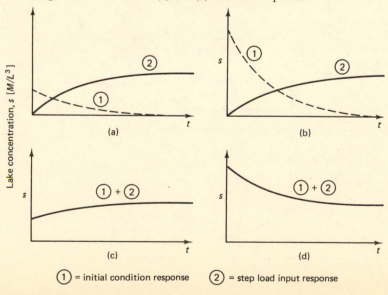

① = initial condition response   ② = step load input response

With Eq. 4.19, the responses to step changes in input load can now be easily computed. Figure 4.10 shows two examples. In Fig. 4.10(a), the load is assumed to begin at time $t = 0$ and step up to $\overline{W}$ to time t_1, at which time the load steps down to zero. The response for $t = 0$ to $t = t_1$ is computed from Eq. 4.19 with s_0 at zero. At time t_1, Eq. 4.19 is used again, beginning now with a new arbitrary starting time, the initial concentration, s_0 given by the solution of Eq. 4.19 at $t = t_1$ and $\overline{W} = 0$. Thus, the "flushing" out of the initial concentration is computed beginning at the new $t = 0$. Figure 4.10(b) shows a similar procedure for a step change in W from $\overline{W}_1$ to $\overline{W}_2$ as in a reduction due to treatment of the input load. Equation 4.19 is used again for each period of time when W is a constant. At a

Figure 4.10 Responses to varying load. (a) Step input and subsequent reduction to zero. (b) Step input with reduction to new load level.

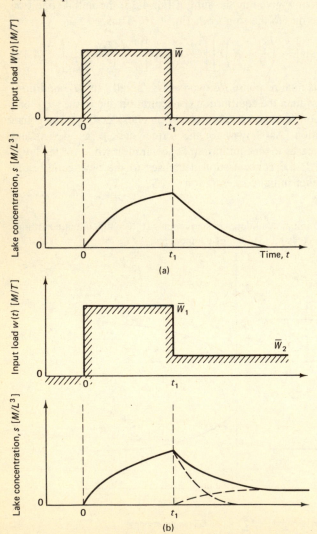

(a)

(b)

SAMPLE PROBLEM 4.1

DATA

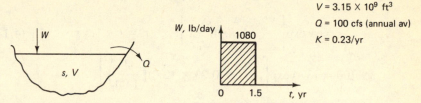

$V = 3.15 \times 10^9 \text{ ft}^3$

$Q = 100$ cfs (annual av)

$K = 0.23/\text{yr}$

Initially, with $s = 0$, a lake receives a loading of a slowly reacting pesticide (triallate) of 1080 lb/day for 1.5 yr and then it is terminated.

PROBLEM

Determine the equilibrium concentration, the maximum concentration, and the time after termination until a level of 100 μg/l is reached.

ANALYSIS

Equilibrium Concentration (Eq. 4.17)

$$\bar{s} = \frac{W}{Q + KV} = \frac{W/Q}{1 + Kt_d}$$

$$t_d = \frac{V}{Q} = \frac{3.15 \times 10^9 \text{ ft}^3}{100 \text{ cfs}} \times \frac{1 \text{ day}}{86{,}400 \text{ sec}} \times \frac{1 \text{ yr}}{365 \text{ days}} = 1.00 \text{ yr}$$

$$\bar{s} = \frac{1080/(100 \times 5.4)}{1 + 0.23 \times 1.00} = 1.63 \text{ mg/l} = \underline{1630 \ \mu\text{g/l}}$$

Maximum Concentration
{Occurs at $t = 1.5$ yr [Eq. (4.16)]}

$$s = \frac{W}{Q + KV}\left\{ 1 - \exp\left[-\left(\frac{Q}{V} + K\right)t \right] \right\}$$

$$s = \bar{s}\left\{ 1 - \exp\left[-(1 + Kt_d)\left(\frac{t}{t_d}\right) \right] \right\}$$

$$s(t = 1.5 \text{ yr}) = 1630\left\{ 1 - \exp\left[-(1 + 0.23 \times 1.00)\left(\frac{1.5}{1.00}\right) \right] \right\} = \underline{1370 \ \mu\text{g/l}}$$

(continued)

Sample Problem 4.1 (continued)

Depuration Time to 100 μg/l

$$s = s_0 \exp\left[-(1 + Kt_d)\left(\frac{t'}{t_d}\right)\right]; \qquad t' = t - 1.5 \text{ yr} \qquad [\text{Eq. (4.13)}]$$

$$100 = 1370 \exp\left[-(1 + 0.23 \times 1.00)\left(\frac{t'}{1.00}\right)\right],$$

$\underline{t' = 2.1 \text{ yr}}$ or $\underline{t = 3.6 \text{ yr}}$

---

change in W, the final condition becomes the initial condition and the calculation is continued. Sample Problem 4.1 illustrates a lake response to a step change in loading.

4.2.1.1 Suspended Solids and Other Particulates The above development applies equally well to water quality constituents that are in the particulate form. Examples include suspended solids and particulate forms of nutrients or toxicants. It can be assumed that the particulates do not decay as a result of any chemical or biological processes. The only mechanism then for the loss of the particulates from the water column is by net settling to the bottom of the lake or reservoir. A more detailed treatment of suspended solids balance in a lake including settling and resuspension is given in Chapter 8 in the context of toxic substances.

Equation 4.8, for settling particulates and constant volume in time, then becomes

$$V\frac{ds}{dt} = W(t) - Qs - v_n A \cdot s \tag{4.20}$$

where v_n is the net loss velocity $[M/T]$ from the water column of the particulates and A is the water area $[L^2]$ over which the settling is occurring, usually approximated by the surface area of the lake. The solution to Eq. 4.20 is the same as Eq. 4.19 except that the loss due to decay is now the loss due to settling of the particulates out of the water column. Thus

$$K_s = \frac{v_n A}{V} = \frac{v_n}{H} \tag{4.21}$$

where H is the mean depth $[L]$, (ratio of the area to volume) and K_s is the loss rate due to settling $[1/T]$.

The flux or mass rate of deposition of the particulates F_s $[M/T]$ out of the water column and into the stationary bed sediment of the lake or reservoir is then

$$F_s = v_n A \cdot s \tag{4.22}$$

$$\left[\frac{M}{T}\right] = \left[\frac{L}{T} \cdot L^2 \cdot \frac{M}{L^3}\right]$$

This flux is of considerable importance in the buildup of nutrients or toxicant in the sediment. Often this mass input to the sediment is estimated from Eq. 4.20 at steady state. Thus,

$$v_n A \cdot s = F_s = \overline{W} - Q \cdot s \tag{4.23}$$

which states that the net mass loss rate out of the water column of particulates, or retention, is the difference of particulates between the incoming load and the mass exiting via the outflow and lake concentration.

The net flux to the bottom can also be measured directly by the deployment of sediment traps which collect the "rain" of particulates over time at a specific location in the water column.

Returning to Eq. 4.20 at steady state, and dividing through by the lake surface area, a useful areal loading equation is obtained as

$$s = \frac{W/A}{Q/A + v_n} \tag{4.24}$$

$$= \frac{W'}{q + v_n} \tag{4.25}$$

where W' is the areal loading of the particulates $[M/L^2 - T;\ g/m^2 \cdot yr]$ and the quantity q $[L/T]$, is given by

$$q = \frac{Q}{A} = \frac{H}{t_d} \tag{4.26}$$

This ratio, q, is identical to the ratio used in the design of sedimentation tanks for water and waste treatment and hence has also been designated the "overflow rate" in the present context. The flux balance is now

$$v_n s = \frac{W}{A} - qs \tag{4.27}$$

where $v_n s$ is now in units of $[M/L^2 \cdot T]$, a convenient areally normalized flux to the bottom of a lake or reservoir.

Table 4.3 shows the solids balance components for the Great Lakes including Saginaw Bay, an embayment of Lake Huron. The net settling velocities of 0.2–1.2 m/day are typical of deeper lakes while the near zero net settling velocity is typical of more shallow wind stirred bays.

4.2.2 Response to an Impulse Input

The lakewide response to an impulse or accidental dump or spill of a substance over a short period of time can also be estimated. For some problems that involve

TABLE 4.3   SOLIDS BALANCE FOR GREAT LAKES

Lake/region	Fine grain solids loading rate (g/m² · yr)[a]	Overflow rate (m/yr)	Solids concentration (mg/l)	Net solids loss rate (m/day)	Sediment deposition rate (g/m² · yr)
Superior	98	0.8	0.5	0.53	98
Michigan	69	0.8	0.5	0.38	69
Huron	109	3.4	0.5	0.58	107
Saginaw Bay	146	10.3[b]	8.0	0.02	64
Erie					
West	3762	101.1[c]	20.0	0.24	1740
Central	1414	66.8[c]	5.0	0.59	1080
East	1658	146.5[c]	5.0	0.51	927
Ontario	229	11.0	0.5	1.22	224

[a]Includes upstream loads and turbulent exchange loads.
[b]Includes turbulent exchange with Lake Huron.
[c]Includes turbulent exchange between Lake Erie basins.
Source: From Thomann and Di Toro (1983). Reprinted by permission of International Association for Great Lakes Research.

an impulse input, the assumption of a completely mixed lake is not appropriate. This would be especially true if the impulse input were discharged in a near-shore region of a large lake. If it can be assumed, however, that the discharge occurs over a short period of time (the impulse) and that the mass input of the waste is mixed quickly throughout the lake then the following analysis is appropriate.

This impulse type of input is represented physically by the discharge of an amount M [e.g., kg] of waste material over a relatively short interval of time. A "short interval of time" is considered relative to the detention time of the lake and the decay coefficient of the substance. Mathematically, this is represented by the Dirac delta function or unit impulse and is represented as

$$\delta(t - t_0) = 0, \qquad t \neq t_0 \tag{4.28}$$

$$\int_{-\infty}^{\infty} \delta(t - t_0) \, dt = 1 \tag{4.29}$$

It should be noted that $\delta(t - t_0)$ has the units of $[1/T]$ as can be seen from Eq. 4.29. Figure 4.11(a) is a plot of the unit impulse input. Equations 4.28 and 4.29 state that the input is zero everywhere except at $t = t_0$, where it approaches infinity such that the area of the discontinuous spike is unity. Strictly speaking, this is a purely hypothetical respresentation since in any practical applications the impulse is discharged over a finite time interval and is therefore, a continuous function. It is introduced, however, to facilitate the description of the lake system response to the impulse input.

The delta function may seem somewhat obscure at first, and Fig. 4.12 shows an approach to understanding its behavior. The left-hand panel shows a load of 10 kg discharged into a completely mixed lake over three time intervals. Note that as Δt is reduced by successive orders of magnitude and the mass is held constant at

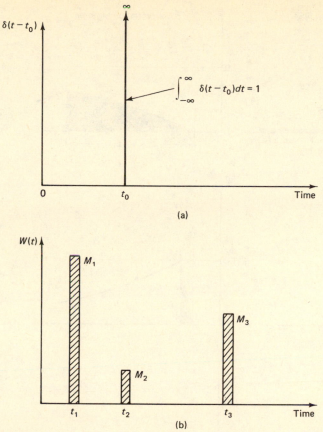

Figure 4.11   (a) The unit impulse, Dirac delta function. (b) Three impulse load inputs.

10 kg, the "spike" grows, that is, the rate of discharge increases by an order of magnitude directly proportional to the reduction in Δt. Since the left panel represents a unit step increase in load followed by a decrease in load at Δt, Eq. 4.19 can be used to calculate the response in the lake under the three conditions. These results are shown in the right panel of Fig. 4.12. As shown, for $\Delta t = 10$ days, the response builds up over the 10-day period in accord with the expected buildup for a unit step input, approaching an equilibrium value of 1.0 $\mu g/l$ up until 10 days and decreasing thereafter. As Δt is reduced however, the concentration of 1.0 $\mu g/l$ is approached sooner because of the increased mass rate of input. At the small Δt of 0.1/day, there is an initial very rapid rise in the concentration in the lake with a subsequent exponential decline following cessation of the input. This can now be shown mathematically using the delta function input.

Operationally, an input of the form

$$W(t) = M \ \delta(t - t_0) \tag{4.30}$$

is interpreted as the discharge of a mass M [M] at time, t_0. The impulse notation indicates that the discharge occurs only at t_0 and is zero everywhere else.

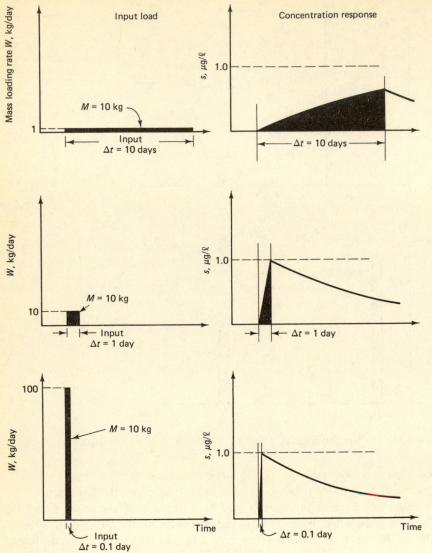

Figure 4.12 Illustration of response due to an input load discharged over various Δt time intervals $(1/t_d + K = 0.10/\text{day}; V = 10^7 \text{ m}^3)$.

If a sequence of n pulses are released at time, t_r, then

$$W(t) = \sum_{r=1}^{n} M_r \, \delta(t - t_r) \tag{4.31}$$

is a mathematical representation of these "spike" releases. Figure 4.11(b) illustrates Eq. 4.31 for three unit impulse inputs.

For a single impulse then, the differential equation is

$$V\frac{ds}{dt} + K's = M \, \delta(t) \tag{4.32}$$

for $t_0 = 0$. The solution to Eq. 4.32 can be shown to be

$$s(t) = \frac{M}{V} \exp \left(\frac{K'}{V} t \right)$$ (4.33)

and as expected from Fig. 4.12, there is an "immediate" rise to the concentration M/V and then an exponential decline.

The delta function solution, Eq. 4.33, can therefore be a convenient and rapid means for estimating the response of a completely mixed lake to a relatively "instantaneous" discharge.

4.2.3 Lakes in Series

In some situations, the output from one lake that has received an input of a water quality variable becomes in turn the input to a second lake further downstream. That lake may also then discharge to a third lake. Reservoir systems often cascade from one reservoir to another as one proceeds down a drainage basin. This sequence of lakes or reservoirs is often called "lakes in series." The Great Lakes is an example (Fig. 4.3). O'Connor and Mueller (1970) and Chapra and Reckhow (1983) have studied the water quality behavior of such lakes in series. Figure 4.13 schematically illustrates the lakes in series. If each lake is considered as a completely mixed body of water, a mass balance equation can be written around each lake as a unit. This approach then is identical to the finite segment development for estuaries discussed in Section 3.6 except that the downstream lake does not interact with the upstream lake.

Thus, following Fig. 4.13 and Eq. 4.8, assuming temporally constant volumes, and subscripting by lake, the equations for the two lakes are

$$V_1 \frac{ds_1}{dt} = W_1(t) - Q_{12}s_1 - V_1 K_1 s_1$$ (4.34)

$$V_2 \frac{ds_2}{dt} = W_2(t) + Q_{12}s_1 - Q_{23}s_2 - V_2 K_2 s_2$$ (4.35)

where the loads W_1 and W_2 include all direct inputs of the variable s as well as any distributed sources to the lake, Eq. 4.6. Temporally constant parameters are

Figure 4.13 Illustration of two lakes in series.

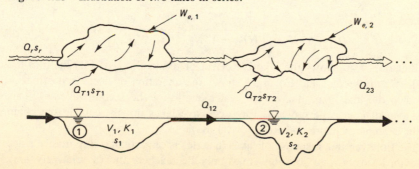

assumed, but the parameters may change from lake to lake. Note that the mass output of the first lake, $Q_{12}s_1$, becomes an input to the second lake in Eq. 4.35. It is assumed in this latter equation that the channel connecting the two lakes is "short" and no change in concentration occurs in the travel time between the lakes. If this distance is significant, any decay process in the connecting river must be taken into account using the techniques of Chapter 2.

At steady state, $ds_i/dt = 0$ and Eqs. 4.34 and 4.35 can be written as

$$(Q_{12} + V_1K_1)s_1 + 0 = W_1 \tag{4.36}$$
$$-Q_{12}s_1 + (Q_{23} + V_2K_2)s_2 = W_2$$

These two equations can be written in the form

$$a_{11}s_1 + 0 = W_1 \tag{4.37}$$
$$a_{21}s_1 + a_{22}s_2 = W_2$$

or in matrix form as

$$\begin{bmatrix} a_{11} & 0 \\ a_{21} & a_{22} \end{bmatrix} \begin{pmatrix} s_1 \\ s_2 \end{pmatrix} = \begin{pmatrix} W_1 \\ W_2 \end{pmatrix} \tag{4.38}$$

where

$$a_{11} = Q_{12} + V_1K_1$$
$$a_{21} = -Q_{12}$$
$$a_{22} = Q_{23} + V_2K_2$$

Equations 4.37 and 4.38 can be recognized as a simple special case of the more general representation of finite segments used in the estuary as in Eqs. 3.36 and 3.37. In this case, however, in contrast to the estuary case, there is no mixing or exchange between the lakes. The steady state solution for the first lake is as in Eq. 4.17, that is,

$$s_1 = \frac{W_1}{Q_{12} + V_1K_1} \tag{4.39}$$

The solution for the second lake at steady state is then

$$s_2 = \frac{W_2 + Q_{12}s_1}{Q_{23} + V_2K_2} \tag{4.40}$$

or

$$s_2 = \frac{W_2}{Q_{23} + V_2K_2} + \left[\frac{Q_{12}}{(Q_{23} + V_2K_2)(Q_{12} + V_1K_1)}\right] W_1 \tag{4.41}$$

Equation 4.40 shows the additive effect of the load, $Q_{12}s_1$ $[M/T]$ from the first lake on the second and Eq. 4.41 shows the alteration or decreased effect of the load to the first lake, W_1, on the water quality of the second lake, s_2. This process can of course be repeated for successive lakes downstream in the chain (see, for example, Chapra and Reckhow, 1983).

The time variable response of lakes in series is of particular importance since the previous steady-state solutions may only be reached after a relatively long period of time has elapsed. O'Connor and Mueller (1970) have explored the time variable responses to the general case of n lakes in series. The solution to the first lake of a constant load entering the lake at time $t = 0$ (the step input) with a zero initial concentration in the lake is given by Eq. 4.19 rewritten here for the first lake

$$s_1 = \frac{W_1}{Q_{12} + K_1 V_1} \left\{ 1 - \exp \left[-\left(\frac{Q_{12}}{V_1} + K_1 \right) t \right] \right\} \tag{4.42}$$

The time variable solution in the second lake due to a step input into the first lake is given by an integrating factor for Eq. 4.35 (O'Connor and Mueller, 1970). The response is (see also Chapra and Reckhow, 1983)

$$s_1 = \frac{Q_{12} W_1}{V_2 V_1 \alpha_1} \left\{ \frac{1}{\alpha_2} [1 - \exp(-\alpha_2 t)] - \frac{1}{\alpha_1 - \alpha_2} [\exp(-\alpha_2 t) - \exp(-\alpha_1 t)] \right\} \tag{4.43}$$

where

$$\alpha_1 = \frac{Q_{12}}{V_1} + K_1 \tag{4.43a}$$

$$\alpha_2 = \frac{Q_{23}}{V_2} + K_2 \tag{4.43b}$$

The relative shift in time and magnitude of the response of the second downstream lake, as seen from Eq. 4.43, depends on the system parameter changes between the two lakes. A summary of the responses of four, initially uncontaminated, sequential lakes to a constant discharge in the first lake is given in Table 4.4. As may be noted, one can easily write the solution for any successive lake. These solutions are a special case of a more general recursion principle (Di Toro, 1972).

If a waste source is suddenly removed from an upstream lake after it has been discharging for some time, the upstream lake will flush out the substance in accordance with Eq. 4.13. This equation is rewritten as

$$s_1 = s_{10} e^{-\alpha_1 t} \tag{4.44}$$

where α_1 is determined from Eq. 4.43a and s_{10} is the concentration in the first lake at the termination of the discharge. This decreasing concentration, coupled with Q_{12}, introduces a load into the next downstream lake, the effect of which is given by the following

$$s_{21} = \alpha_{12} s_{10} \left[\frac{e^{-\alpha_1 t}}{(\alpha_2 - \alpha_1)} + \frac{e^{-\alpha_2 t}}{(\alpha_1 - \alpha_2)} \right] \tag{4.45}$$

where s_{21} is the time variable response in the second lake to the washout of the first lake and

$$\alpha_{12} = \frac{Q_{12}}{V_2} \tag{4.46}$$

as defined in Table 4.4.

TABLE 4.4 CONCENTRATIONS IN DOWNSTREAM LAKES
DUE TO UPSTREAM DISCHARGE

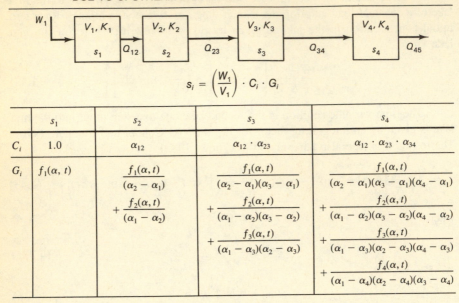

$$s_i = \left(\frac{W_1}{V_1}\right) \cdot C_i \cdot G_i$$

	s_1	s_2	s_3	s_4
C_i	1.0	α_{12}	$\alpha_{12} \cdot \alpha_{23}$	$\alpha_{12} \cdot \alpha_{23} \cdot \alpha_{34}$
G_i	$f_1(\alpha, t)$	$\dfrac{f_1(\alpha, t)}{(\alpha_2 - \alpha_1)}$ $+ \dfrac{f_2(\alpha, t)}{(\alpha_1 - \alpha_2)}$	$\dfrac{f_1(\alpha, t)}{(\alpha_2 - \alpha_1)(\alpha_3 - \alpha_1)}$ $+ \dfrac{f_2(\alpha, t)}{(\alpha_1 - \alpha_2)(\alpha_3 - \alpha_2)}$ $+ \dfrac{f_3(\alpha, t)}{(\alpha_1 - \alpha_3)(\alpha_2 - \alpha_3)}$	$\dfrac{f_1(\alpha, t)}{(\alpha_2 - \alpha_1)(\alpha_3 - \alpha_1)(\alpha_4 - \alpha_1)}$ $+ \dfrac{f_2(\alpha, t)}{(\alpha_1 - \alpha_2)(\alpha_3 - \alpha_2)(\alpha_4 - \alpha_2)}$ $+ \dfrac{f_3(\alpha, t)}{(\alpha_1 - \alpha_3)(\alpha_2 - \alpha_3)(\alpha_4 - \alpha_3)}$ $+ \dfrac{f_4(\alpha, t)}{(\alpha_1 - \alpha_4)(\alpha_2 - \alpha_4)(\alpha_3 - \alpha_4)}$

where

$$f_i(\alpha, t) = \frac{1 - \exp(-\alpha_i t)}{\alpha_i}$$

$$\alpha_i = K_i + \frac{1}{t_{d\,i}}$$

$$t_{d\,i} = \frac{V_i}{Q_{i,i+1}}; \qquad \alpha_{i,i+1} = \frac{Q_{i,i+1}}{V_{i+1}}$$

Assumptions: (1) at $t = 0$, $s_i = 0$ and W_1 begins; (2) V_i, $Q_{i,i+1}$, K_i, W_1 are temporally constant.

Of course, the second lake will have built up its own concentration (s_{20}) at the termination of the discharge in the first lake. This will flush out of the second lake in accordance with

$$s_{22} = s_{20} e^{-\alpha_2 t} \tag{4.47}$$

where s_{22} is the washout of the second lake's initial concentration. The total concentration in the second lake is the sum of the two effects obtained by adding Eq. 4.45 to Eq. 4.47:

$$s_2 = s_{21} + s_{22}$$

$$s_2 = \alpha_{12} s_{10} \left[\frac{e^{-\alpha_1 t}}{\alpha_2 - \alpha_1} + \frac{e^{-\alpha_2 t}}{\alpha_1 - \alpha_2} \right] + s_{20} e^{-\alpha_2 t} \tag{4.48}$$

Since the first term in Eq. 4.48 is a source term, concentrations in the second lake may increase after termination of the load in the first lake depending on the two lakes' characteristics. An example of a washout calculation for the Great Lakes is presented in Sample Problem 4.2.

SAMPLE PROBLEM 4.2

DATA

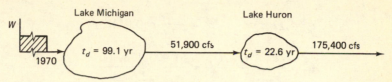

In 1970 the concentration of a conservative substance discharged into Lake Michigan was 270 μg/l and 47 μg/l in Lake Huron. The discharge terminated in 1970, as indicated.

PROBLEM

Using the long-term average outflows and detention times of these Great Lakes, as shown above (O'Connor and Mueller, 1970), determine when the peak concentrations will be halved in each lake.

ANALYSIS

Lake Michigan

Designate Lake Michigan as lake 1 and Lake Huron as lake 2. At termination in 1970 ($t = 0$), the concentration in lake 1 will be at its maximum, and it will decrease thereafter.

$$t_{d1} = \frac{V_1}{Q_{12}} = 99.1 \text{ yr, as given above}$$

$$\alpha_1 = (K_1 = 0) + \frac{1}{t_{d1}} = 1/99.1 = 0.01009/\text{yr}$$

$$\frac{s_1}{s_{10}} = 0.5 = \exp(-0.01009t) \qquad \text{[Eq. (4.44)]}$$

$$t = 68.7 \text{ say } 69 \text{ yr}$$

In year 2039, Lake Michigan will have $s = 135$ μg/l.

(continued)

Sample Problem 4.2 (continued)

Lake Huron

$$t_{d2} = 22.6 \text{ yr}, \qquad \alpha_2 = \frac{1}{22.6} = 0.04425/\text{yr}$$

$$\alpha_{12} = \frac{Q_{12}}{V_2} = \frac{Q_{12}/Q_{23}}{t_{d2}} = \frac{51{,}900/175{,}400}{22.6}$$

$$= 0.01309/\text{yr}$$

$$s_2 = 0.01309 \times 270 \left[\frac{\exp(-0.01009t)}{0.04425 - 0.01009} \right. \qquad \text{[Eq. (4.48)]}$$

$$\left. + \frac{\exp(-0.04425t)}{0.01009 - 0.04425} \right]$$

$$+ 47 \exp(-0.04425t)$$

Solving the above by trial and error, s_2 reaches a maximum value of 62 μg/l at $t = 26$ yr.

Substituting $s_2 = 0.5 \times 62 = 31$ μg/l into the above equation and solving for t by trial and error again, $t = 120$ yr. Thus, the peak concentration in Lake Huron would be halved 120 yr after termination of the discharge in Lake Michigan. The Lake Huron concentration chronology is shown below with its components.

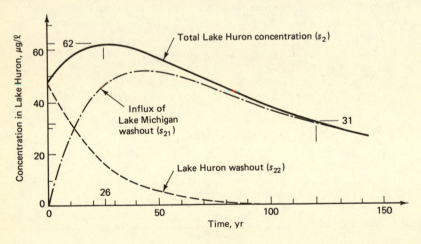

4.3 FINITE SEGMENT (FINITE DIFFERENCE) STEADY STATE LAKE MODELS

An understanding of a lake or reservoir as a completely mixed system can be of significant value in estimating the effects on water quality of man's activities. This is in spite of the rather severe assumption of complete mixing. The completely mixed application can be made for a number of substances including suspended and dissolved substances as well as for heat balance computations. As noted earlier in this chapter, however, the vertical behavior of lakes is of particular importance since, during periods of stratification, both surface and bottom waters exhibit quite different quality. The estimation of vertical gradients is therefore of importance for a number of water quality problems. In addition, horizontal gradients may occur that are of particular significance such as the "trapping" of waste discharges in a near-shore zone or in the interaction between a cove or embayment of the lake and the lake proper.

The approach to the calculation of water quality variables throughout a lake, both horizontally and vertically, is identical to Section 3.6. In that section, the basic structure of the finite segment or finite difference approximation to spatially varying flow, dispersion, and mixing and reaction kinetics is given. The application to lakes is direct. Steady state is assumed throughout. Specific situations are developed here to illustrate the finite segment framework for lakes and reservoirs. The two-layered stratified lake is considered first.

4.3.1 Two-Layer, Stratified Lake

Consider then a lake that has stratified due to temperature differences during the summer months into two layers as shown in Fig. 4.14. At other times, for example, winter, the lake is isothermal and completely mixed vertically and horizontally. The epilimnion receives an incoming load W and flow Q but also mixes and exchanges with the hypolimnion. The lower layer may also be receiving a source of the substance s as, for example, from sediment release of nutrients. The mass balance equations for the epilimnion (seg. 1) and the hypolimnion (seg. 2) are then

Figure 4.14 Schematic of two-layer stratified lake.

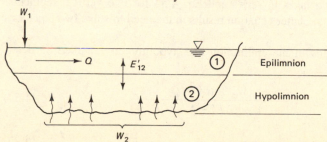

given by

$$V_1 \frac{ds_1}{dt} = W_1 - Qs_1 + E'_{12}(s_2 - s_1) - V_1 K_1 s_1 \qquad (4.49)$$

$$V_2 \frac{ds_2}{dt} = W_2 + E'_{12}(s_1 - s_2) - V_2 K_2 s_2 \qquad (4.50)$$

where the bulk vertical mixing coefficient, E'_{12} $[L^3/T]$, is given by

$$E'_{12} = \frac{E_{12} A_{12}}{\overline{Z}_{12}} \qquad (4.51)$$

for E_{12} as the vertical dispersion coefficient $[L^2/T]$, A_{12} as the interfacial area $[L^2]$ between layers 1 and 2, and $\overline{Z}_{12}$ as the vertical distance from the center of layer 1 to an equal distance into layer 2 when $H_2 \gg H_1$, and equal $t \cdot H/2$ when $H_2 \approx H_1 \cdot H$ is the lake's total depth, and H_1 and H_2 are the depths of the epilimnion and hypolimnion, respectively.

Under a steady state condition, the temporal derivatives in Eqs. 4.49 and 4.50 become zero. The two coupled equations can then be written as

$$(Q + E'_{12} + V_1 K_1)s_1 + (-E'_{12})s_2 = W_1$$

$$(-E'_{12})s_1 + (E'_{12} + V_2 K_2)s_2 = W_2 \qquad (4.52)$$

or in matrix form as

$$\begin{bmatrix} a_{11} & a_{12} \\ a_{21} & a_{22} \end{bmatrix} \begin{pmatrix} s_1 \\ s_2 \end{pmatrix} = \begin{pmatrix} W_1 \\ W_2 \end{pmatrix} \qquad (4.53)$$

where

$$a_{11} = Q + E'_{12} + V_1 K_1$$
$$a_{12} = -E'_{12}$$
$$a_{21} = -E'_{12}$$
$$a_{22} = E'_{12} + V_2 K_2$$

These equations can again be recognized as a special case of the more general matrix equation for water quality in multidimensional estuaries (e.g., Eq. 3.66). In this case the segments are vertical and interactive through the mixing processes across the epilimnion–hypolimnion interface. A comparison can also be made between Eq. 4.38 for two lakes in series and Eq. 4.53 for two vertical segments. Note the interaction in the latter equation results in the need to solve two equations simultaneously.

The concentrations in the epilimnion and hypolimnion can then be determined as

$$s_1 = \frac{(W_1 + \beta W_2)/Q}{1 + (1 - \beta) E'_{12}/Q + V_1 K_1/Q} \qquad (4.54)$$

$$s_2 = \beta \left(s_1 + \frac{W_2}{E'_{12}} \right) \qquad (4.55)$$

where

$$\beta = \frac{E'_{12}}{E'_{12} + V_2 K_2} \qquad (4.55a)$$

Sample Problem 4.3 illustrates use of these equations.

In a stratified system such as this, the concentration, s, of the water quality parameter in the epilimnion and hypolimnion are coupled through the vertical exchange. During the completely mixed period, E_{12} is "high." As a result, gradients in the concentration are not maintained so that the lake concentration is uniform from top to bottom. As E_{12} decreases, the interaction between the epilimnion and hypolimnion decreases and substantial differences in the concentration may result. The estimation of the degree of vertical mixing given by E_{12} is therefore important in estimating water quality during stratification periods.

Before proceeding to the estimation of the vertical dispersion coefficient, it should be noted that if the water quality constituent is of a particulate form, such as suspended solids or particulate nutrient or toxicant, then the basic equations 4.49 and 4.50 are somewhat different. Thus considering $K_1 = K_2 = 0$ and a constant net settling velocity v_n between the two layers, the equations for the particulate in the upper and lower layers are, at steady state

$$0 = W_1 - Qs_1 + E'_{12}(s_2 - s_1) - v_n As_1 \qquad (4.56)$$

$$0 = E'_{12}(s_1 - s_2) - v_n As_2 + v_n As_1 \qquad (4.57)$$

The latter equation assumes no external source of particulates from the bottom to the second layer. Note that the second layer receives the mass flux of s_1 from the upper layer which is a transport term due to settling from the upper to the lower layer. By combining terms with s_1 and s_2, Eq. 4.57 requires that either $E'_{12} + v_n A = 0$, which is impossible, or $s_1 - s_2 = 0$ such that

$$s_2 = s_1$$

thereby indicating that stratification of the particulates is not possible under these steady-state conditions. This is a consequence of assuming the net settling is equal in both layers. There is no mechanism in this simple model for a gradient to develop between layers. The flux of solids from the first layer to the second must equal the flux of solids from the second layer to the bottom or else there would be a buildup or decay of solids. The concentration in the lake is then simply

$$s_1 = s_2 = \frac{W_1}{Q + v_n A} \qquad (4.58)$$

This equation is the same as Eq. 4.24. If an external source of particulate occurs in the hypolimnion or the net settling varies from top to bottom as occasioned by resuspension, then a vertical gradient in particulates may develop.

This simple example of particulate behavior, such as suspended solids in lakes and reservoirs, illustrates the importance of recognizing the settling characteristics of such water quality constituents.

SAMPLE PROBLEM 4.3

DATA

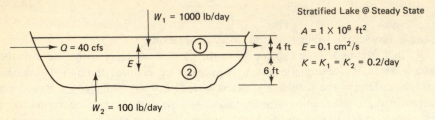

W_1 = 1000 lb/day

Q = 40 cfs

E

W_2 = 100 lb/day

4 ft

6 ft

Stratified Lake @ Steady State
$A = 1 \times 10^6 \text{ ft}^2$
$E = 0.1 \text{ cm}^2/\text{s}$
$K = K_1 = K_2 = 0.2/\text{day}$

PROBLEM

Determine steady state concentrations in layers 1 and 2.

ANALYSIS

Let $\bar{Z}_{12} = (H_1 + H_2)/2$ since $H_1(= 4 \text{ ft}) \approx H_2(= 6 \text{ ft})$.

$$E' = \frac{EA}{\bar{Z}_{12}} = \frac{EA}{(H_1 + H_2)/2} \qquad \text{[Eq. (4.51)]}$$

$$= \frac{0.1 \text{ cm}^2/\text{s} \times (1 \text{ ft}/30.48 \text{ cm})^2 \times 1 \times 10^6 \text{ ft}^2}{(4 + 6)/2 \text{ ft}}$$

$$= 21.5 \text{ cfs}$$

$$V_1 = AH_1 = 1 \times 10^6 \times 4 = 4 \times 10^6 \text{ ft}^3$$

$$V_2 = AH_2 = 1 \times 10^6 \times 6 = 6 \times 10^6 \text{ ft}^3$$

$$\beta = \frac{E'}{E' + V_2 K_2} = \frac{21.5}{21.5 + (6 \times 10^6 \times 0.2)/86,400} \qquad \text{[Eq. (4.55a)]}$$

$$= 0.607$$

Surface Layer

$$s_1 = \frac{(W_1 + \beta W_2)/Q}{1 + (1 - \beta)(E'/Q) + V_1 K_1/Q} \qquad \text{[Eq. (4.54)]}$$

$$s_1 = \frac{(1000 + 0.607 \times 100)/(40 \times 5.4)}{1 + (1 - 0.607)(21.5/40) + (4 \times 10^6 \times 0.2)/(86400 \times 40)}$$

$$= \underline{3.41 \text{ mg/l}}$$

Bottom Layer

$$s_2 = \beta\left(s_1 + \frac{W_2}{E'}\right) \qquad \text{[Eq. (4.55)]}$$

$$s_2 = 0.607\left(3.41 + \frac{100}{21.5 \times 5.4}\right) = \underline{2.59 \text{ mg/l}}$$

4.3.1.1 Estimation of Vertical Dispersion Coefficient   The degree of vertical mixing between two layers in a stratified lake is an important parameter that determines the vertical gradients of water quality as shown in Eqs. 4.54 and 4.55. The vertical dispersion coefficient can be estimated by (1) dye studies or (2) analysis of vertical profiles of tracers where all kinetics and source/sink terms are known.

Water temperature profiles have been used extensively and for many years for estimating vertical dispersion in lakes and reservoirs. Heat balance and temperature models are discussed in Chapter 9. Here a simple procedure is presented.

If it is assumed that the heat entering the lake has been dissipated at the bottom of the first layer, then a simple heat balance equation can be written for the hypolimnion. That is, in Eq. 4.50, consider s to represent the temperature with no sources or sinks or decay of heat; then for the hypolimnion

$$V_2 \frac{\Delta T_2}{\Delta t} = \frac{E_{12} A_{12}}{\bar{Z}_{12}} (T_1 - T_2) \tag{4.59}$$

where the first term on the left is the time rate of change of temperature T_2 in layer 2 in °C, T_1 is the temperature in the upper layer. Equation 4.59 states that the change in temperature in the second layer is due entirely to the heat flux from the exchange between the two layers. If T_1 and T_2 are known from measurements, then Eq. 4.59 can be used to estimate E_{12}. One constraint on E_{12} is that it should always be non-negative; therefore, if $T_2 > T_1$, E_{12} can be arbitrarily set equal to a small number since this condition usually prevails during the period of inverse winter stratification [Fig. 4.4(d)]. Since T_1 may equal T_2, indicating complete mixing to an isothermal condition, an upper bound on E_{12} is also necessary. Therefore, E_{12} is set "large," say 100 cm^2/s under that condition (Kullenberg et al., 1973, 1974). Solving Eq. 4.59 for E_{12} gives

$$E_{12} = \frac{|T_2^{(t-1)} - T_2^{(t)}| V_2 \bar{Z}_{12}}{(T_1 - T_2) \Delta t \cdot A_{12}} \tag{4.60}$$

and

$$E_{12} = 100 \text{ cm}^2/\text{s} \quad \text{for} \quad T_1 = T_2$$
$$E_{12} = 0.1 \text{ cm}^2/\text{s} \quad \text{for} \quad T_2 > T_1$$

The use of Eq. 4.60, as shown in Sample Problem 4.4, provides a first estimate of the vertical exchange between the two layers and, as such, is of particular use in water quality problems requiring a first approximation to the stratified lake case.

Various empirical relationships have also been proposed to estimate the vertical dispersion, E_z from characteristics of the water column, specifically characteristics such as the vertical stability. This characteristic is related to vertical stratification and is expressed through evaluation of the vertical density gradient. For example, Imboden and Gachter (1979) have suggested the following relationship for the vertical dispersion coefficient for depths of 20–40 m,

$$E_z = (1.7 \cdot 10^{-4})(N)^{-0.82} \tag{4.61}$$

SAMPLE PROBLEM 4.4

DATA

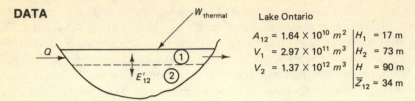

Lake Ontario

$A_{12} = 1.64 \times 10^{10} \ m^2$ | $H_1 = 17 \ m$
$V_1 = 2.97 \times 10^{11} \ m^3$ | $H_2 = 73 \ m$
$V_2 = 1.37 \times 10^{12} \ m^3$ | $H = 90 \ m$
| $\bar{Z}_{12} = 34 \ m$

PROBLEM

Estimate the vertical dispersion coefficients using the L. Ontario temperature data shown in the plot below.

ANALYSIS

For hypolimnion:

$$V_2 \frac{dT_2}{dt} = E'_{12}(T_1 - T_2) \qquad \text{[Eq. (4.59)]}$$

For discrete data:

$$E'_{12} = V_2 \frac{\Delta T_2}{\Delta t}\left(\frac{1}{T_1 - T_2}\right) = \frac{V_2}{\Delta t}\left[\frac{T_2(t + \frac{1}{2}) - T_2(t - \frac{1}{2})}{T_1(t) - T_2(t)}\right]$$

$$E_{12} = \frac{E'_{12}\bar{Z}_{12}}{A_{12}} = \frac{V_2\bar{Z}_{12}}{A_{12}\,\Delta t}\left(\frac{\Delta T_2}{\Delta T} = \tau\right) = \frac{V_2\bar{Z}_{12}}{A_{12}}\left(\frac{\tau}{\Delta t}\right) \qquad \text{[Eq. (4.60)]}$$

Using L. Ontario data:

Day (Julian)	t (days)	Δt (days)	T_1 (°C)	T_2 (°C)	ΔT_2 (°C)	ΔT (°C)	τ	E^a (cm²/s)
138			5.5	3.8				
	152	28	8.75	4.3	1.0	4.45	0.225	1.3
166			12.0	4.8				
	172.5	13	12.55	4.85	0.1	7.7	0.0130	0.16
179			13.1	4.9				
	189	20	14.2	5.05	0.3	9.15	0.0328	0.27
199			15.3	5.2				
	204	10	15.25	5.2	0.0	10.05	0.00	0.00
209			15.2	5.2				
	219.5	21	15.15	5.25	0.1	9.9	0.0101	0.08
230			15.1	5.3				
	240	20	16.1	5.4	0.2	10.7	0.0187	0.15
250			17.1	5.5				
	255	10	17.55	6.3	1.6	11.25	0.142	2.3
260			18.0	7.1				
	265.5	11	16.75	7.05	−0.1	9.70	0.0103	0.15
271			15.5	7.0				
	281	20	13.15	6.9	−0.2	6.25	0.032	0.26
291			10.8	6.8				

$$^a E(\text{cm}^2/\text{s}) = \left(\frac{V_2\bar{Z}_{12}}{A_{12}}\right)\left(\frac{\tau}{\Delta t}\right) = \left(\frac{1.37 \times 10^{12} \times 17}{1.64 \times 10^{10}}\right)\left(\frac{\tau}{\Delta t}\right)\left(\frac{m^2}{\text{day}}\right) \times \left(\frac{10^4 \text{cm}^2}{m^2}\right) \times \left(\frac{\text{day}}{86,400\text{s}}\right) = 164\left(\frac{\tau}{\Delta t(\text{days})}\right)$$

Sample Problem 4.4 (continued)

A plot of the temporal variation of the calculated vertical dispersion coefficients is shown below together with surface and bottom water temperatures.

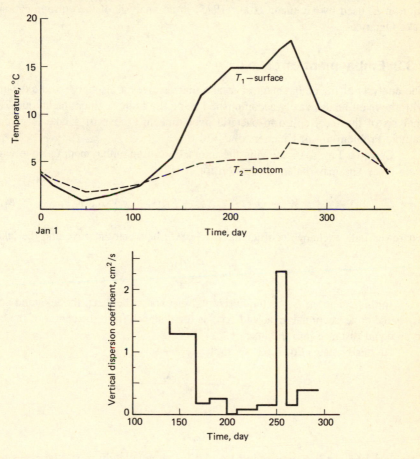

where E_z is in cm^2/s and N is the "stability frequency" (s^{-2}) given by

$$N = \frac{g}{\rho} \frac{\partial \rho}{\partial Z} \qquad (4.62)$$

for g as the acceleration of gravity (9.8 m/s^2), ρ as the density of water in g/cm^3, and Z as the depth in m. As the density gradient ($\partial \rho / \partial Z$) increases, the stability increases and the vertical mixing and dispersion decrease. For example, at $N = 10^{-6}$ (s^{-2}), E_z from Eq. 4.61 is 14 cm^2/s and as N increases to 10^{-5} (s^{-2}), E_z drops to 2 cm^2/s.

Table 4.5 shows some estimates of the vertical dispersion coefficients. The order of magnitude of this mixing is considerably less than the order of horizontal mixing in open bodies of water and estuaries. See, for example, Table 3.3. Generally, the lower values in Table 4.5 apply to the period of stratification and reflect the reduced exchange during that period. For example, Fig. 4.15 shows the variation in E_z used by Thomann et al. (1975) in an analysis of the eutrophication of Lake Ontario.

4.3.2 Embayment of Lake

The analysis of the embayment of a lake that receives a discharge and exchanges with the main lake is easily accomplished using the finite segment techniques. (An analysis of the fecal coliform bacteria over time in a cove of a lake is given in Sample Problem 5.4 of Chapter 5.)

Consider Fig. 4.16 showing the interaction of an embayment or cove with a main lake. The mass balance equation for s_1 is

$$V_1 \frac{ds_1}{dt} = W_e + Q_r s_r - Q_{1L} s_1 + E'_{1L}(s_L - s_1) - V_1 K_1 s_1 \qquad (4.63)$$

where the bulk exchange coefficient, E'_{1L} $[L^3/T]$ between the cove and the lake is

$$E'_{1L} = \frac{E_{1L} A_{1L}}{\Delta x_{1L}} \qquad (4.64)$$

In this equation, E_{1L} is the horizontal dispersion between the lake and cove, A_{1L} is the cross-sectional area, and Δx_{1L} is the distance from the center of the cove to an equal distance into the lake.

At steady state, Eq. 4.63 is simply

$$s_1 = \frac{W + E'_{1L} s_L}{Q_{1L} + E'_{1L} + V_1 K_1} \qquad (4.65)$$

where

$$W = W_e + Q_r s_r$$

Equation 4.65 can be compared to Eq. 4.17, the completely mixed lake. As seen, the cove interacting with the lake receives an additional load, $E'_{1L} s_L$ due to the exchange of the lake concentration into the cove. There is an exchange of the cove concentration into the lake so that E'_{1L} also acts as an outflow transport of mass, as seen in the mass balance equation 4.63, resulting in the additional "dilution" in the denominator of Eq. 4.65. This equation can then be used as a first approximation to estimating the steady state concentration in the embayment of a lake (Sample Problem 4.5).

4.3.3 Multi-Dimensional Lakes

The finite segment (difference) approach given in Section 3.6 can be used for analysis of lakes and reservoirs where water quality variations are to be calculated

TABLE 4.5   SOME VERTICAL DISPERSION COEFFICIENTS

Water body	Length/width (km)	Depth (m)	Vertical dispersion coefficient E (cm²/s)	Reference[a]
Lake Vomb, Sweden	3/5	15	2–20	1
Lake Mein, Sweden	5/5	20	5–10	1
Lake Ontario, United States and Canada	350/75	90	1–10	2
Rochester Bay, Lake Ontario	50/10	50	0.1–2	3
Lake Erie, Central Basin	200/80	24	0.05–1	4
Cayuga Lake, NY	67/3		0.2–1.5	5
Atlantic Ocean, east of Barbados		0–10	7–44	
		30–40	0.25–0.85	6

[a]References: (1) Bengtsson (1973) quoted by Hansen (1978). (2) Kullenberg et al. (1973) and Kullenberg et al. (1974). (3) Kuo and Thomann (1983). (4) Di Toro and Connolly (1980). (5) Sundaram et al. (1969). (6) Young and Silker (1974).

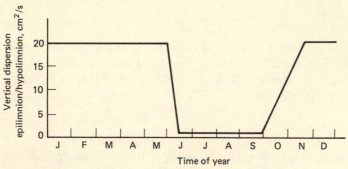

Figure 4.15   Variation in vertical dispersion between epilimnion and hypolimnion of Lake Ontario eutrophication model. From Thomann et al. (1975).

Figure 4.16   Notation for embayment interacting with lake.

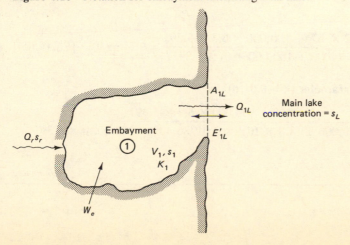

SAMPLE PROBLEM 4.5

DATA

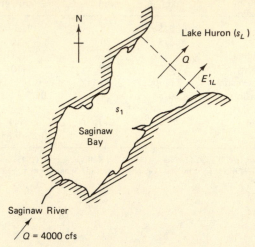

Saginaw Bay is in the southwest quadrant of Lake Huron. It receives the flow of the Saginaw River, which has a high chloride loading of approximately 2×10^6 lb/day. From Thomann and Mueller (1983), $V_1 = 8.8 \times 10^{11}$ ft^3, $s_L = 6.3$ mg/l, $Q = 4000$ cfs, $E'_{1L} = 100,000$ cfs.

PROBLEM

Estimate the baywide average chloride concentration. If the substance were not chlorides, but a nonconservative parameter with $K = 0.01$/day, calculate the concentration (s_1).

ANALYSIS

Chlorides ($K = 0$)

$$s_1 = \frac{2 \times 10^6 + 100,000 \times 6.3 \times 5.391}{(4000 + 100,000 + 0) \times 5.391} = \underline{\underline{9.6 \text{ mg/l}}} \quad \text{[Eq. (4.65)]}$$

Nonconservative Parameter ($K = 0.01$/d)

$$s_1 = \frac{2 \times 10^6 + 100,000 \times 6.3 \times 5.391}{[4000 + 100,000 + (8.8 \times 10^{11} \times 0.01/86,400 = 102,000)] \times 5.391}$$

$$= \underline{\underline{4.9 \text{ mg/l}}}$$

(continued)

Sample Problem 4.5 (continued)

Mass balance (10^6 kg/day)

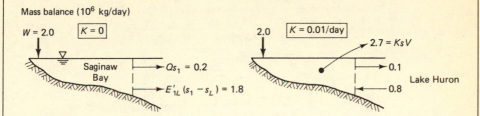

Note that, for $K = 0$, the bay is a source to the lake whereas, when $K = 0.01/\text{day}$, the lake feeds the bay.

in all three directions. The procedure is to write a mass balance around each segment in a manner indentical to Eq. 3.28. An illustration is shown in Fig. 4.17 for a four horizontal segment lake stratified into two layers. The net flow transport is as shown. The mass balance around seg. 4 is then at steady state

$$V_4 \frac{ds_4}{dt} = 0 = Q_{24}s_2 - Q_{43}s_4 + E'_{24}(s_2 - s_4) + E'_{34}(s_3 - s_4) \\ + E'_{84}(s_8 - s_4) - V_4 K_4 s_4 \tag{4.66}$$

The inclusion of the proper sign of mass transport due to flux is to be noted as well as the inclusion of vertical mixing of seg. 4 with seg. 8 below.

As noted in Chapter 3, a lake such as Fig. 4.17 results in eight simultaneous linear equations to be solved for the concentration in each segment. If s is a particulate form of a water quality constituent then the settling flux from the upper layer to the lower layer must be included as discussed previously (see Eqs. 4.56 and 4.57).

Figure 4.17 An eight-segment stratified lake.

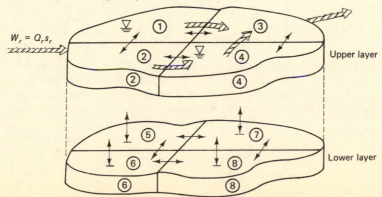

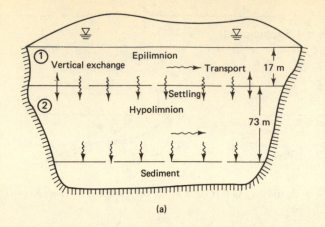

(a)

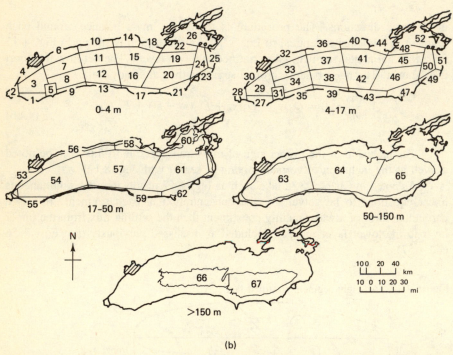

(b)

Figure 4.18 Finite difference segmentation for Lake Ontario. (a) Open lake vertical two-segment model (lake 1) (from Thomann et al., 1975). (b) Three-dimensional grid (from Thomann et al., 1979). (c) Finer scale grid, Rochester Embayment (from Thomann et al., 1979; Kuo & Thomann, 1983). Reprinted by permission of American Society of Civil Engineers.

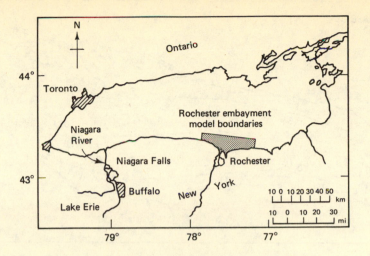

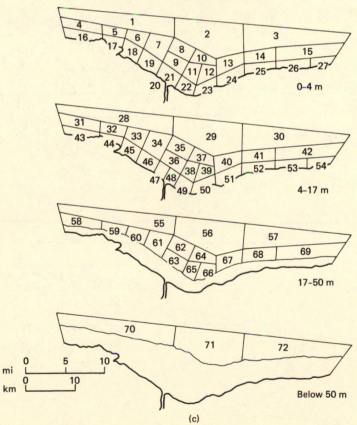

Rochester Embayment Segmentation

(c)

211

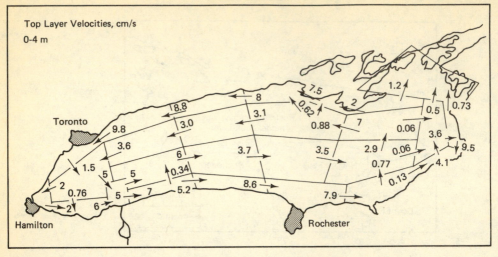

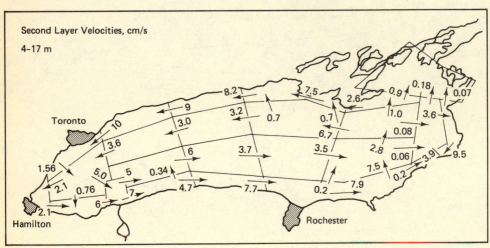

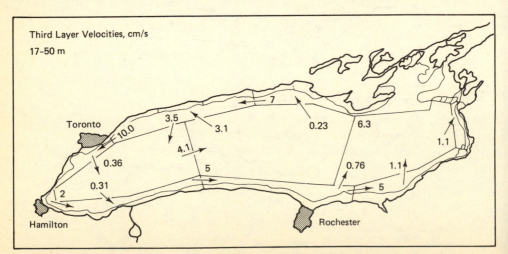

Figure 4.19 Best estimate circulation regime, Lake Ontario three-dimensional model segmentation. From Thomann et al. (1975).

One of the difficulties with multi-dimensional analyses of the water quality of lakes is the specification of the net transport throughout the regions of the lake. Such transport can be either measured (which may be quite difficult and expensive) or computed using lake hydrodynamic models. Such models include transport and mixing due to wind stress at the surface and to internal density gradients. Heat balance models, as discussed in Chapter 9, can also be used to estimate vertical dispersion characteristics.

Figure 4.18 shows the segmentation used by Thomann et al. (1975) and Thomann et al. (1979) for Lake Ontario. Figure 4.18(a) is a simple two-layer model (Lake 1) of Lake Ontario where the emphasis was on open lake vertical variations in water quality. A three-dimensional grid of Lake Ontario (Lake 3) is shown in Fig. 4.18(b) where the net summer flow transport used in the water calculations is shown in Fig. 4.19. The specification of the net velocities in this latter figure was a combination of observed data and hydrodynamic calculation. Figure 4.18(c) shows the segmentation for the Rochester embayment of Lake Ontario. This segmentation is actually embedded in the seg. 17 region of the Lake 3 grid. The boundary conditions of the Rochester embayment were calculated from the Lake 3 model. It should also be noted that the segments are not of equal size. The density of the grid reflects the need to capture increased water quality gradients in the vicinity of inputs.

Lake water quality calculations may therefore range from the simple completely mixed lake, to vertical stratification, and finally to three-dimensional situations. The degree of complexity increases dramatically as the dimensionality of the lake is increased. The principles of lake water quality modeling can now be applied in subsequent chapters to individual water quality problem contexts.

REFERENCES

Bartsch, A. F., and J. H. Gakstatter, 1978. Management Decisions for Lake Systems on a Survey of Trophic Status, Limiting Nutrients, and Nutrient Loadings in American-Soviet Symposium on Use of Mathematical Models to Optimize Water Quality Management, 1975, U.S. Environmental Protection Agency, Office of Research and Development, Environmental Research Laboratory, Gulf Breeze, FL, pp. 372–394. EPA-600/9-78-024.

Bengtsson, L., 1973. Mathematical Models of Wind-Induced Circulations in a Lake, IAHS Publication No. 109, *Proc. Helsinki Symp.: 313–320.*

Brezonik, P. D., and J. J. Messer, 1977. Analysis of Trophic Conditions and Eutrophication Factors in Lake Weir, Florida, in the North American Project—A Study of U.S. Water Bodies, compiled by L. Seyb and K. Randolph, Env. Res. Lab, Corvallis, ORD, USEPA, Corvallis, OR 1–24. EPA-600/3-77-086.

Chapra, S. C., and K. H. Reckhow, 1983. *Engineering Approaches for Lake Management, Volume 2: Mechanistic Modeling,* Butterworth, Boston, MA, 492 pp.

Di Toro, D. M., 1972. Recurrence Relations for First-Order Sequential Reactions in Natural Waters, *Water Resources Research* **8** (1); 50–57.

Di Toro, D. M., and J. P. Connolly, 1980. Mathematical Models of Water Quality in Large Lakes, Part 2: Lake Erie, Env. Res. Lab. ORD, USEPA, Duluth, MN, 231 pp. EPA-600/3-80-065.

Hansen, N-E. O., 1978, Mixing Processes in Lakes, *Nord. Hydro.* **9**:57–74.

Hutchinson, G. E., 1957. *A Treatise on Limnology. Vol. I. Geography, Physics and Chemistry*, Wiley, New York, 1015 pp.

Hutchinson, G. E., 1967. *A Treatise on Limnology. Vol. II. Introduction to Lake Biology and the Limnoplankton*, Wiley, New York, 1115 pp.

Imboden, D. M., and R. Gachter, 1979. *The Impact of Physical Processes on the Trophic State of a Lake in Biological Aspects of Freshwater Pollution*, O. Ravera, Ed., Pergamon, New York, pp. 93–110.

Kochov, M. M., 1962. *The Biology of Lake Baikal*, Academy of Sciences of the USSR, Moscow, 315 pp.

Kullenberg, G., C. R. Murthy, and H. Westerberg, 1973. An Experimental Study of Diffusion Characteristics in the Thermocline and Hypolimnion Regions of Lake Ontario, Proc. 16th Conf. Grt. Lks. Res., IAGLR, Ann Arbor, MI, 774–790.

Kullenberg, G., C. R. Murthy, and H. Westerberg, 1974. Vertical Mixing Characteristics in the Thermocline and Hypolimnion Regions of Lake Ontario (IFYGL), Proc. 17th Conf. Grt. Lks. Res., IAGLR, Ann Arbor, MI, 425–434.

Kuo, J. T., and R. V. Thomann, 1983. Phytoplankton Modeling in the Embayments of Lakes, *J. Environ. Engr., Proc. Am. Soc. Civ. Eng.* **109**(6):1311–1332.

Lerman, A. Ed., 1978. *Lakes, Chemistry, Geology, Physics*, Springer-Verlag, New York, 363 pp.

Linsley, R. K., M. A. Kohler, and J. L. H. Paulhus, 1958. *Hydrology for Engineers* McGraw-Hill, New York, 340 pp.

Mortimer, C. H., 1976. Environmental Status of Lake Michigan Region Vol. 2. Physical Limnology of Lake Michigan, ANL/ES-40, Argonne National Laboratory, Argonne, IL, 121 pp.

O'Connor, D. J., and J. A. Mueller, 1970. Water Quality Model of Chlorides in the Great Lakes, *J. Sanit. Eng. Div., Proc. Am. Soc. Civ. Eng.* **96**(SA4):955–975.

Reckhow, K. H., and S. C. Chapra, 1983. *Engineering Approaches for Lake Management. Vol. 1: Data Analysis and Empirical Modeling*, Butterworth, Boston, MA, 340 pp.

Showers, V., 1979. *World Facts and Figures*, Wiley, New York, 757 pp.

Sundaram, T. R. et al., 1969. An Investigation of the Physical Effects of Thermal Discharges into Cayuga Lake, Cornell Aeronautical Lab., Inc.

Thomann, R. V., D. M. Di Toro, R. P. Winfield, and D. J. O'Connor, 1975. Mathematical Modeling of Phytoplankton in Lake Ontario. 1. Model Development and Verification, NERC, ORD, USEPA, Corvallis, OR, 177 pp. EPA-660/3-75-005.

Thomann, R. V., R. P. Winfield, and J. J. Segna, 1979. Verification Analysis of Lake Ontario and Rochester Embayment Three-Dimensional Eutrophication Models, ERL, ORD, USEPA, Duluth, MN, 135 pp. EPA-600/3-79-094.

Thomann, R. V. and D. M. Di Toro, 1983. Physico Chemical Model of Toxic Substances in the Great Lakes, *J. Grt. Lks. Res.* **9**(4):474–496.

Thomann, R. V., and J. A. Mueller, 1983. Steady-State Modeling of Toxic Chemicals— Theory and Application to PCB's in the Great Lakes and Saginaw Bay, in *Physical Behavior of PCBs in the Great Lakes*, D. Mackay, S. Paterson, S.J. Eisenreich and M.S. Simmons (Eds.), Ann Arbor Science Pub., Ann Arbor, MI, pp. 283–309.

Wetzel, R. G., 1975. *Limnology*. W. B. Suanders Co., Philadelphia, 743 pp.

Young, S. A., and W. B. Silker, 1974. The Determimation of Air-Sea Exchange and Oceanic Mixing Rates Using [37]Be During the BOMEX Experiment, *J. Geophys. Res.* **79**(30):4481–4489.

PROBLEMS

4.1. Lake Weir in Florida has been the subject of some study regarding the effect of nutrients on the lake (Brezonik and Messer, 1977). Characteristics of the lake are as follows:

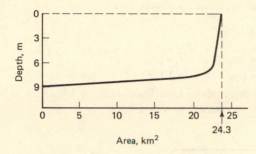

$$\text{max length} = 5.3 \text{ km,} \qquad \text{average depth} = 7 \text{ m}$$
$$\text{max width} = 5.0 \text{ km, } A_s = 24.3 \text{ km}^2$$
$$\text{rainfall} = 1.2 \text{ m/yr.}$$
$$\text{energy loss due to evaporation} = 193 \text{ cal/cm}^2 \cdot \text{day}$$
$$\text{lake outflow} = 1.1 \cdot 10^6 \text{ m}^3/\text{yr}$$
$$\text{estimated seepage in} = 2.9 \cdot 10^6 \text{ m}^3/\text{yr}$$
$$\text{loss due to groundwater recharge} = 4.96 \cdot 10^6 \text{ m}^3/\text{yr}$$
$$\Delta H = -0.1\text{m over the yr}$$

(a) Sketch the cross section of the lake.
(b) Does this water budget balance over the year?
(c) What is the revised contribution from seepage in m^3/yr in order to make the water budget balance over the year?
(d) Determine the detention time of the lake as would be used in water quality calculations.

4.2. Suppose 34 metric tons/day of total phosphorus has been discharged into Lake Ontario "for a long time." The load is then decreased "instantaneously" to 19 metric tons/day. Compute and sketch the concentration (mg/l) response in the lake over time (yr) for:
(a) Assuming total phosphorus is conservative.
(b) Assuming total phosphorus loss at a rate $= 0.001/\text{day}$. Let the volume $= 1.67 \cdot 10^{12}\text{m}^3$ and the outflow $Q = 6555 \text{ m}^3/\text{s}$.

4.3. Consider again Problem 4.2 and assume again that 34 mt/day have been discharged into Lake Ontario "for a long time." The substance decays with a rate of 0.001/day and flow and volume are as before. Now, suppose the concentration of the lake was measured in 1977 at 0.0152 mg/l and in 1979 at 0.0247 mg/l. On the basis of this information, calculate the new (1977) input load to the lake in mt/day.

4.4. Suppose for the conservative case conditions of Lake Ontario given in Problem 4.2 the 34 metric tons/day was dropped instantaneously to zero for 5 yr, but then increased immediately to the 19 metric tons/day as shown:

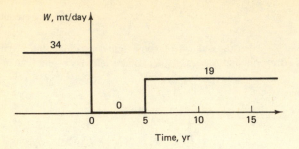

(a) What is the concentration (mg/l) at $t = 5$ yr?
(b) What is the concentration (mg/l) at $t = 10$ yr?

4.5. The input of polychlorinated biphenyls to Lake Ontario can be considered to have begun in 1950. Assume the load history is as shown below.

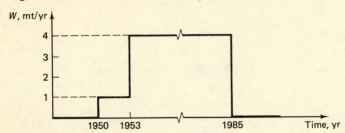

Suppose that in 1985, the load were dropped to zero. How long would we have to wait in years to have the PCB concentration reach 5 ng/l:
(a) If $K = 0$.
(b) If $K = 0.001$/day?

4.6. Given

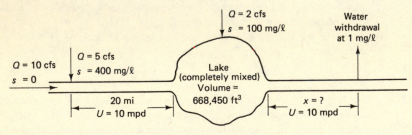

Decay rate throughout = 0.15/day

Assuming the entire system has reached equilibrium:
(a) What is the distance that must be used between the inputs shown and the water withdrawal such that the concentration withdrawn is 1 mg/l?
(b) Suppose that the distance from the lake outlet to the withdrawal cannot be more than 20 mi. What should be the new volume of the lake?

4.7. Two reservoirs discharge to a common stream as shown below. The flow, influent concentrations, and volumes are shown. The substance is conservative. Assume that the flow and influent concentrations have been entering the reservoir for a long enough time to reach steady state in the entire system. At time $t = 0$, the influent concentration to reservoir 1 drops to zero. The flow remains the same. How many days will it take before the concentration at point D is reduced by 50% of the steady state prereduction value?

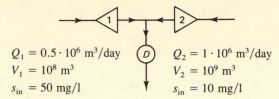

$Q_1 = 0.5 \cdot 10^6$ m³/day D $Q_2 = 1 \cdot 10^6$ m³/day
$V_1 = 10^8$ m³ $V_2 = 10^9$ m³
$s_{in} = 50$ mg/l $s_{in} = 10$ mg/l

4.8. A large number of measurements on a lake gave an average solids concentration in the lake of 0.5 mg/l. The measurements also indicated that there were 10,000 kg solids/day entering the lake, the outflow was 60 m³/s, and the volume was 10^8 m³. What is the estimated loss rate of the solids in the lake in day^{-1} assuming the lake is completely mixed?

4.9. A lake with a volume of 140,000 MG and a flow of 1600 MGD receives an input of 80,000 lb/day beginning at $t = 0$.
  (a) If $K = 0.05$/day, how long will it take in days to reach 90% of the equilibrium value? What is this equilibrium value in mg/l?
  (b) Suppose $K = 0$, how many days will it take to reach 90% of equilibrium? What is the new equilibrium value?
  (c) Suppose the source is stopped at exactly the time of 90% of equilibrium. How long after that will it take to reach a concentration of 0.5 mg/l for the case of $K = 0.05$/day?

4.10. 10,000 lb/day of a slowly reactive chemical ($K = 0.01$/yr) is discharged to Lake Michigan. Using the data of Sample Problem 4.2, calculate and plot the concentrations (μg/l) in Lakes Michigan and Huron until both lakes attain 95% of their equilibrium values. The volumes are 1170 and 850 mi³ for Lakes Michigan and Huron, respectively.

4.11. Repeat Sample Problem 4.3 using $W_1 = 0$ and $W_2 = 1100$ lb/day. In addition, determine the components of the mass balances for each layer.

4.12. A reservoir can be viewed as horizontally well mixed, but divided vertically as shown in the sketch below. The conditions on flow are also shown and represent an input flow to the upper layer with withdrawal through an outlet in the lower level of the reservoir.

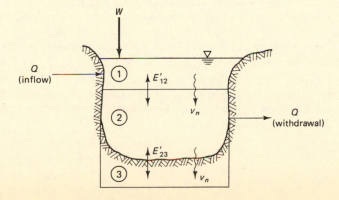

Assume phosphorus as the conservative variable of interest, but phosphorus does have a net effective settling velocity v_n which is constant in 1 and 2.

$$V_1 = 10^6 \text{ m}^3, \qquad V_2 = 10^8 \text{ m}^3, \qquad E'_{12} = 0, \qquad Q = 10^5 \text{ m}^3/\text{day}$$
$$E'_{23} = 10^6 \text{ m}^3/\text{day}$$
$$v_n = \text{settling velocity} = 0.5 \text{ m/day}, \text{ depth } 1 = 10 \text{ m}, \text{ depth } 2 = 100 \text{ m}$$
$$W = 10,000 \text{ kg/day}$$

(a) Set up the steady state equations that express this situation. Express as a set of three equations and in matrix form.
(b) Compute the phosphorus concentration (mg/l in the water column).
(c) Suppose instead of $E'_{12} = 0$ (complete stratification case), $E'_{12} = \infty$, the well-mixed case). Compute the lake concentration and the sediment concentration under this new condition.
(d) Repeat (b) for $E'_{23} \to \infty$.
(e) Repeat (b) for $E'_{23} \to 0$.

4.13. A portion of a lake has the following geometry and characteristics:

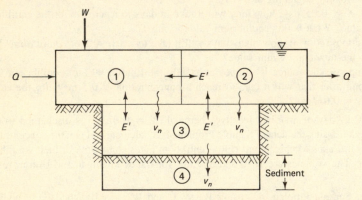

$$E'_{12} = 10^8 \text{ m}^3/\text{day}, \qquad E'_{13} = 10^5 \text{ m}^3/\text{day} \qquad E'_{23} = 2 \cdot 10^5 \text{ m}^3/\text{day},$$
$$A_{13} = 10^5 \text{ m}^2 \qquad A_{23} = 2 \cdot 10^5 \text{ m}^2, \qquad A_{34} = 3 \cdot 10^5 \text{ m}^2,$$
$$V_1 = V_2 = V_3 = 3 \cdot 10^6 \text{ m}^3, \qquad K = 0.1/\text{day everywhere}$$

There is no flow into or out of seg. 3.
(a) If the net settling velocity in this system is 0.2 m/day, and dissolved and detrital phosphorus of 1.0 mg/l has been measured in seg. 1 and 1.5 mg/l in seg. 2, what is the net flux of phosphorus into the sediment in kg/day? Assume backward differences.
(b) If the sediment segment is 10 cm thick, what is the concentration of this phosphorus form in the sediment in mg/l?

4.14. A particulate substance, with a net settling rate of 0.02 m/day, enters Saginaw Bay at a rate of 5000 lb/day. If the concentration in Lake Huron is 5.0 μg/l, determine the concentration in Saginaw Bay and sketch a mass balance. Use the data in Sample Problem 4.5. The mean depth of the bay is 6 m.

Chapter 5

Indicator Bacteria, Pathogens, and Viruses

5.1 INTRODUCTION

It is appropriate to begin the exploration of individual water quality problem contexts with an investigation of the impact of bacteria and other organisms which may cause communicable diseases. This is appropriate because:

1. The kinetics of baterial die-away are usually considered to be first-order so that all of the preceding chapters (2–4) apply directly.
2. It is the oldest of the water pollution problems in the discovery of the links between contaminated water and communicable disease.
3. The problem is still quite relevant today and is manifested in continuing high levels of waterborne diseases in some countries, closing of bathing beaches, and restrictions on water consumption.

The transmission of waterborne diseases (e.g., gastroenteritis, amoebic dysentery, cholera and typhoid fever) has been a matter of concern for many years. The impact of high concentrations of disease-producing organisms on water uses can be significant. Bathing beaches may be closed permanently or intermittently during rainfall conditions when high concentrations of pathogenic bacteria are discharged from urban runoff and combined sewer overflows. Diseases associated with water used for drinking purposes continue to occur. Figure 5.1, from Craun (1978), shows the average annual number of waterborne disease outbreaks in the United States for the period 1920–1975. The outbreaks for 1971–1975 involved a total of 27,829 people and included acute gastrointestinal illness, hepatitis-A, shi-

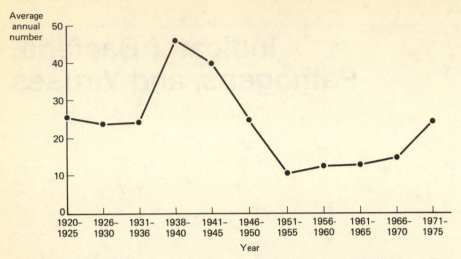

Figure 5.1 Average annual number waterborne outbreaks, 1920–1975, United States. From Craun (1978).

gellosis, giardiasis (caused by the parasite, Giardia), and four cases of typhoid fever (Craun, 1978). Virtually all of the population affected used water from municipal or semi public systems.

The modes of transmission of pathogens are through ingestion of contaminated water and food, and exposure to infected persons or animals. Infections of the skin, eyes, ears, nose and throat may result from immersion in the water while bathing. The water uses impacted by pathogenic bacteria, viruses, and parasites are:

1. Drinking water: municipal, domestic, industrial and individual supplies.
2. Primary contact recreation such as bathing and water skiing.
3. Secondary contact recreation such as boating, fishing.
4. Shellfishing such as harvesting for clams and oysters.

Measurement approaches include analysis for (a) indicator bacterial groups that reflect the potential presence of pathogens, (b) the pathogenic bacteria directly, (c) viruses, and (d) intestinal parasites. The various measures can be summarized as in Table 5.1 (National Academy of Sciences, 1977). Although there is a variety of indicators and direct enumeration of pathogens is possible, primary emphasis in water quality has been placed in the past on the coliform group of bacteria. This is due principally to the fact that the coliform groups meet many of the criteria for a suitable indicator organism, for example easily detected by simple laboratory tests, generally not present in unpolluted waters, and the number of indicator bacteria tends to be correlated with extent of contamination. The National Academy of Sciences (NAS) (1977) indicates, that although other indicators are under study (such as the bacteriophages, Scarpino, 1978), they should not yet be substituted for the coliform group. Also, the World Health Organization (WHO) (1984a, 1984b)

TABLE 5.1 EXAMPLES OF COMMUNICABLE DISEASE INDICATORS AND
ORGANISMS

Indicators	Viruses
Bacteria	Hepatitis A
Total coliform	Enteroviruses
Fecal coliform	Polioviruses
Fecal streptococci	Echoviruses
Obligate anaerobes (*Clostridium*	Coxsackieviruses
perfringins)	
Bacteriophages (Bacterial viruses)	Pathogenic protozoa and helminths
	(intestinal worms)
Pathogenic bacteria	*Giardia lambia*
Vibrio cholerae	*Entamoeba histolytica*
Salmonella species	Facultatively parasitic amoebae
Shigella species	(*Naegloria* and *Hartmanella*)
	Nematodes

recommends guidelines for drinking water based on the coliform group and for virological quality, which is expressed in terms of water turbidity and disinfection.

5.1.1 Indicator Bacteria

The total coliform (TC) bacteria group is a large group of bacteria that has been isolated from both polluted and nonpolluted soil samples as well as the feces of humans and other warm blooded animals. This group was widely used in the past as a measure of health hazard and continues to be used in some areas. More definitively, the coliform group of bacteria comprises all aerobic and facultative anaerobic, gram-negative, nonspore-forming, rod-shaped bacteria that ferment lactose with gas formation within 48 hr at 35°C. *Escherichia coli* is a bacterium of this group.

The fecal coliform (FC) bacteria group are indicative of organisms from the intestinal tract of humans and other animals. The test for FC is at an elevated temperature (44.5°C) where growth of bacteria of nonfecal origin is suppressed.

The fecal streptococci (FS) bacteria group includes several species or varieties of streptococci and the normal habitat of these bacteria is the intestines of humans and animals. Examples include, *Streptococcus faecalis* which represents bacteria of humans and *Streptococcus bovis* and *Streptococcus equinus* which represent bacteria that are indicators of cattle and horses.

The degree to which these tests of indicator organisms represents the presence of pathogens (such as *Salmonella*) has been the subject of continuing investigation (Geldreich, 1970, 1978; Smith and Twedt, 1971; Brezenski and Russomanno, 1969). There is a general correlation between the concentration of FC bacteria and the occurrence of *Salmonella*. Geldreich (1978) has summarized results for estuarine waters and indicated that for FC concentrations less than 200/100 ml, *Salmonella* occurrences ranged from 6.5 to 31%. However, at FC concentrations greater than about 1000/100 ml, the frequency of *Salmonella* occurrence at least doubled. For recreational lakes and streams and FC levels from 1 to 200/100 ml,

Salmonella occurred in 28% of the water samples. When FC were about 1,000/100 ml, *Salmonella* occurrence was 96%. In general, then, at levels of FC less than about 200/100 ml, occurrences of *Salmonella* are less than about 30%. However, at levels of FC at 1,000/100 ml or greater, occurrences of *Salmonella* range to greater than 95%.

Relationships between TC and pathogens are generally not considered to be quantitative. The TC determination has been used for many years in evaluation of the sanitary aspects of water quality. But because of the difficulties associated with the occurrence of nonfecal bacteria in the test and other considerations such as differential resistance to chlorination by nonfecal bacteria (Kabler and Clark, 1960), the use of the total coliform test is being gradually replaced by FC and FS. Attempts are often made to correlate FC to TC levels. As a general rule, the FC levels are about 20% of the TC concentrations (Kenner, 1978) although a wide spread exists in this ratio.

The ratio of FC to FS (FC/FS) has been shown to be a good indicator of whether the sources of bacteria are from humans (FC/FS > 4) or from other warm blooded animals (FC/FS < 1) (Geldreich, 1966). Figure 5.2 shows the behavior of the FC/FS ratio for the Potomac estuary following a transient flow increase (Thomann, 1981). Prior to the increase, the FC/FS ratio is generally greater than 4 indicating some predominance of bacteria of human origin. Following the increase, the ratio declines reflecting the increased runoff and associated bacteria of nonhuman origin. The ratio of FC/FS by itself, however, does not necessarily have any specific sanitary or health significance.

Figure 5.2 Ratio of fecal coliform to fecal streptococci bacteria for transient flow event of 1967, Potomac Estuary. Data from Lear and Jaworski (1969).

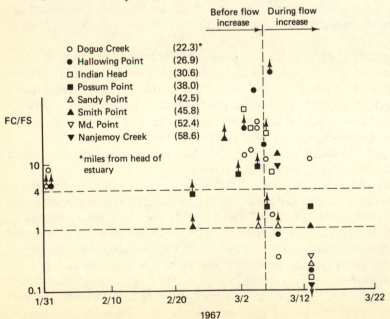

There are some important reservations about the use of this ratio for esti-mating bacterial origin and disagreement with the reliability of the ratio has been registered (Dutka and Kwan, 1980). For example, because of varying decay rates of FC and FS, the ratio may change as a function of distance from an outfall. Thus, if the bacteria at the outfall are primarily of human origin and the rate of decay of FC is greater than FS, then at some point downstream, FS may be greater than FC. On the basis of the FC/FS ratio alone, one might then erroneously conclude that the bacteria at the downstream site are of nonhuman origin. Geldreich (1978) suggests that FC/FS ratios are valid for estimating the origin of the bacteria only during about a 24-hr travel time downstream from the point of pollution discharge. This guideline however was derived from an experiment with stormwater and the distance over which the FC/FS ratio is an indication of bacterial origin may be significantly greater than the 24-hr guideline depending on the die-off rates of FC and FS in various bodies of water. In any event, from a water quality viewpoint, the individual bacterial components (i.e., FC and FS) should be analyzed and modeled and not the ratio.

As a general measure therefore, and within the constraints of such a simplified representation, the FC/FS ratio may only be useful as a broad indicator of the sources of bacterial pollution as suggested by Geldreich (1978) and Kenner (1978).

5.1.2 Bacteriological Standards

Table 5.2 shows some representative bacteriological standards promulgated by regulatory agencies and the associated beneficial water use. For primary water contact recreation, increased emphasis has been placed on the FC group as the bacterial indicator group to be used for regulatory purposes. It should be noted however that the issue of bacteriological standards for bathing waters has been the subject of considerable controversy. Indeed, as Salas (1985) points out in his review, the origins of such standards was sometimes on an aesthetic basis rather than on the basis of impact on public health. The wide range of standards and conditions for primary water contact recreation reflects a range of interpretation of the relative health risk.

The stringent condition in the allowable concentration for shell-fishing (TC < 70/100 ml and FC < 14/100 ml) is aimed at protecting the consumer of clams and oysters from communicable diseases such as hepatitis and gastrointestinal disorders. Since shellfish filter the overlying water and concentrate bacteria as part of the feeding process, the low concentration of bacteria in the water column is intended to result in an acceptable level in the organism itself.

McCabe (1978) summarized information on water supply systems in the United States where the delivered water exceeded the required TC concentration. Table 5.3 adapted from his work shows that the problem of exceeding the bacterial standard tends to be skewed toward communities with smaller populations. The relatively high exceedance percentage for the smaller communities presumably reflects a reduced level of operational concern and equipment maintenance and increased use of groundwaters, such as shallow springs.

TABLE 5.2 SOME BACTERIOLOGICAL STANDARDS AND ASSOCIATED BENEFICIAL USES

Beneficial use	Bacteria group	Concentration (num/100 ml)[a]	Exceedance condition %/t[b]	Exceedance condition Concentration (num/100 ml)	Agency	Reference[c]
Public water supply[d]	Total coliform	0			WHO	1
	Fecal coliform	0			WHO	
Primary water contact recreation, bathing	Total coliform	2400	20/30	1000	California	5
		5000(10,000)[h]	0/2	10000	California	5
		1000			New York	3
					Europe	4, 5
					Japan	5
					Peru	5
	Fecal coliform	200	20/	5000	Washington, D.C.	2
		200	10/30	400	New York	3
		200(May–Sept)			North Carolina	3
		1000(Oct–Apr)[e]	10/60	400	California	5
		100(2000)[h]			North Carolina	5
					Europe	4, 5
					Peru	5
Secondary water contact recreation	Fecal coliform	1000	20/	5000	Washington, D.C.	2
		770–1500	10/30	2000	New Jersey	3
Fishing	Total coliform	5000[f]			Massachusetts	3
		1000[g]			New Hampshire	3
	Fecal coliform	100[e]			North Carolina	3
		1000[e]			South Carolina	3
Shellfishing	Total coliform	70	10/30	230	United States	5
	Fecal coliform	14	10/30	43	Venzuela, Mexico	5

[a]Log mean.

[b]%/t = not to be exceeded percentage over t days (e.g., 10/30, 400 = no more than 10% of samples during 30-day period should exceed 400/100 ml.

[c]References: (1) World Health Organization (1984a, 1984b); (2) Government of District of Columbia (1981); (3) City of New York (1979); (4) Commission of European Communities (1977); (5) Salas (1985).

[d]Treated water entering distribution system; turbidity < 1 NTU, free chlorine residual 0.2–0.5 mg/l following 30 minute contact.

[e]Except during or immediately following rain.

[f]Monthly average in water not subject to urban runoff or CSO.

[g]Average except when occurs naturally or subject to CSO.

[h]First number is a guide, ( ) is mandatory.

TABLE 5.3   WATER SUPPLY SYSTEMS IN THE UNITED STATES
EXCEEDING COLIFORM DENSITY LIMIT

Population range	Number of systems	Percent of systems exceeding coliform limit	Percent of survey population exceeding coliform limit
≤ 500	446	18	15
500–5000	315	10	8
5000–10,000	75	7	7
10,000–25,000	59	9	7
25,000–50,000	36	3	3
50,000–100,000	16	6	6
>100,000	22	0	0
Totals	969	12	2

Source: Adapted from McCabe (1978).

5.1.3 Pathogenic Bacteria

It is clear that in spite of the contemporary concern with chemicals and other substances in water, that the occurrence of diseases associated with pathogenic bacteria continues to be a problem especially on the worldwide scene. Bonde's (1981) summary of the number of notified disease cases for industrialized countries is shown in Table 5.4.

The principal waterborne pathogens of concern include (a) *V. cholerae*, resulting in the disease of cholera; (b) *Salmonellae*, causing typhoid and paratyphoid; and (c) *Shigella* species, causing dysentery.

Determination of specific pathogenic bacteria is severely limited in practice especially for routine water quality surveys. The limitations are due principally to the high degree of expertise needed to perform the specific isolations especially in small numbers. The procedures are accurate only for *Salmonellae* (NAS, 1977, p. 70). Thus, the emphasis continues to be on the coliform group, the indicator organisms of most utility in water quality.

TABLE 5.4   NOTIFIED INFECTIOUS DISEASE CASES
PER 100,000 POPULATION

Cholera	0.0–0.03
Typhoid	0.0–9.1
Paratyphoid	0.0–2.5
Other *Salmonella*	0.3–54.0
Shigella	0.02–22.0

Source: From Bonde (1981). Reprinted by permission of Elsevier Science Publishers.

5.1.4 Viruses

As summarized by the National Academy of Sciences (NAS) (1977, p. 88), viruses are "submicroscopic, inert particles that are unable to replicate or adapt to environmental conditions outside a living host." The viruses are thus parasites, and those that are of most significance in water are the enteric viruses, that is, those that inhabit the intestinal tract. The enteric virus particles, called virions are approximately spherical, acid-stable, and are relatively persistent in the environment. The most important enteric viruses include enteroviruses (e.g., polioviruses, coxsackie viruses) reoviruses, parvoviruses, and adenoviruses and range in size from 20 to 80 nm. The viruses are measured and enumerated from the degree to which a sample infects a cell culture or an experimental animal.

Viruses ingested from water can result in a variety of diseases including hepatitis (a disease of continuing importance in the United States), and diseases of the central nervous system caused by the polioviruses, coxsackieviruses, and echoviruses.

5.1.5 Pathogenic Protozoan—*Giardia lambia*

As noted in Table 5.1, communicable diseases can also be transmitted in parasitic protozoa and helminths (intestinal worms). The eggs and cysts of the organisms may be ingested, protozoa may reproduce, and helminths mature in the human host. Amoebic dysentery (from *Entamoeba histolytica*) is a particularly severe disease in the world. Helminths include the whipworm, hookworm, and dwarf tapeworm. Particular concern is for the protozoa *Entamoeba histolytica* and *Giardia lambia* which may survive public water treatment. In the United Status in 1974, the largest number of cases in a waterborne disease outbreak was due to *Giardia lambia* (NAS, 1977).

The intestinal parasite and protozoan, *Giardia lambia* has been known to exist for several hundred years. Indeed, Leewenhoek was the first to see *Giardia* in 1681 in his own stools, (Levine, 1979). The disease caused by this organism called giardiasis is characterized by diarrhea, nausea, intestinal cramps in the chronic or low level stage, and by vomiting and anorexia at the acute stage. While normally associated with travelers from abroad or from hikers and campers drinking from streams, there is an apparent increase in the incidence of the disease even in protected supplies and systems. Craun (1979) indicates that from 1965 to 1977, a total of 7009 cases of giardiasis were reported in 23 waterborne outbreaks. Although there is uneasiness over the increased incidence of this disease, most of the outbreaks appear to be associated with individual breakdowns of water supply systems.

5.2 INPUTS OF ORGANISMS

The principal sources of organisms associated with waterborne communicable disease are: (a) point sources from domestic, municipal, and some industrial sources, (b) combined sewer overflows, (c) runoff from urban and suburban land through separate sewers, and (d) municipal waste sludges disposed of on land or in water bodies.

5.2.1 Point Source Inputs

Tables 1.2 and 5.5 summarize the magnitudes of the concentrations of some important organisms in raw sewage. Geldreich (1978) in summarizing the mean bacteria data for 14 cities in the United States reported an overall mean of $21.9 \cdot 10^6/100$ ml of TC. Individual city means ranged from 1.6 to $47.4 \cdot 10^6/100$ ml. Davis (1979) reported a mean concentration of $412 \cdot 10^6/100$ ml for 14 days at two Houston, Texas plants with a standard deviation of $1011 \cdot 10^6/100$ ml. These

TABLE 5.5 COMMUNICABLE DISEASE ASSOCIATED ORGANISMS IN RAW SEWAGE

Type	Mean concentration (range of mean)[a]	Remarks	Reference[e]
Total coliform	21.9 (1.6–47.4)	14 U.S. cities	1
	412 (1011)[b]	Two Houston plants—14 days	8
	(70–733)	Mexico	5
	200	Area of Rio de Janeiro, Brazil	6, 7
	180	Lima, Peru	9
Fecal coliform	8.3 (0.3–49)	21 U.S. cities	1
	30(69)[b]	2 Houston plants—14 days	8
	120	Lima, Peru	9
FC/TC	0.38 (0.12–0.48)	14 U.S. cities	1
Fecal streptococci	1.6 (0.064–4.5)	Seven U.S. cities	1
	1.9(20.7)[b]	Two Houston plants—14 days	8
FC/FS	5.1 (4.4–28)	Seven U.S. cities	1
	12.8(9.3)[b]	Two Houston plants—14 days	8
Salmonella Sp.	0.008 (.02)[b]	Houston—10 samples	8
	0.0021	Lima, Peru	9
Virus	(10–40)[c]	Estimated maximum of 10^4 in U.S.	2
	565[c]	Geometric mean	4
Giardia lambia	9.6[d] (9.6–240)	Theoretical calculation	3

[a]All units in 10^6 num/100 ml except where noted by c–d.
[b]Standard deviation.
[c]PFU/100 ml; PFU = plaque-forming unit.
[d]10^3 cysts/l.
[e]References: (1) Geldreich (1978); (2) US EPA (1978); (3) Jakubowski and Ericksen (1979); (4) Haas (1983); (5) Castagnino (1977); (6) Hydroscience (1977a); (7) Coelho et al. (1978); (8) Davis (1979); (9) Yanez (1983).

values exceed the data of Geldreich by about an order of magnitude. Data from Rio de Janeiro indicate TC of $200 \cdot 10^6/100$ ml and generally reflect a lower per capita water usage and hence a more concentrated sewage. The average FC/FS of 5.1 indicates, as noted previously, that the source of the bacteria is primarily of human origin (FC/FS $>$ 4).

Variations in the data of Table 5.5 can be expected on a diurnal, weekly, and seasonal basis depending on the water use patterns of the community. Figure 5.3 shows the diurnal variation in TC as measured in the influent to the Newtown Creek plant in New York City for a weekend and midweek period (Hazen and Sawyer, 1978). The variation within a day of an order of magnitude can be noted.

5.2.2 Combined and Separate Sewer Inputs

Extensive bacteriological data on discharges from both combined and separate sewers have been collected as a result of increased concern about the relative impact of these sources on bacteriological quality. For example, studies were conducted of effects of bypasses during 1964 in the Detroit and Ann Arbor, Michigan areas (Benzie and Courchaine, 1966) and in Syracuse, New York (Drehwing et al., 1979). Some of the results from the combined sewer area in Detroit are shown in Table 5.6 and for Syracuse in Table 5.7. The data appear to be distributed as a lognormal frequency distribution. The ratios of FC to FS indicate pollution of human origin. The order of magnitude of these results bears testimony to the bacteriological problem created by combined sewer overflows. The values of coliforms in some instances equalled the order expected in raw sewage. The maximum observed density of TC in Detroit was $160 \cdot 10^6$ organisms/100 ml. The variation from location to location is also marked as shown in Table 5.7 for two sewers in

Figure 5.3   Variation of total coliforms in influent to Newtown Creek Plant, New York City. Plotted from NYC Dept of Envir. Protect. data in report by Hazen and Sawyer (1978).

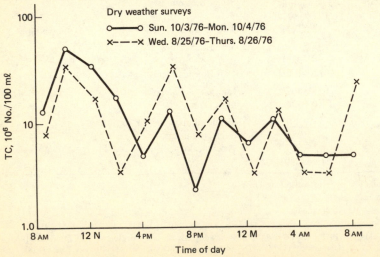

TABLE 5.6   BACTERIOLOGICAL CONCENTRATION FROM COMBINED SEWER AREA IN DETROIT

Variable	10%[a]	50%[a]	90%[a]
Total coliform	$1.4 \cdot 10^6$	$9.4 \cdot 10^6$	$65 \cdot 10^6$
Fecal coliform	$5.5 \cdot 10^5$	$2.7 \cdot 10^6$	$10.3 \cdot 10^6$
Fecal streptococci	$2.6 \cdot 10^5$	$5.8 \cdot 10^5$	$1.2 \cdot 10^6$
FC/FS	2.1	4.7	8.8

[a]Percent of time that concentration (num/100 ml) was equal to or less than value indicated.

Source: From Benzie and Courchaine (1966).

Syracuse, New York. For example, the extreme 95% levels for the W. Newell Street sewer exceed those of the steeper Maltbie Street sewer by about a factor of 10. Across 25 cities, a geometric mean of $6 \cdot 10^6/100$ ml of TC bacteria has been reported (US EPA, 1976), see Table 1.2.

Combined sewer overflows therefore can contribute high waste loads (relative to dry weather loads) but for short periods of time and at randomly distributed intervals. This "impulse" type load can be a point source or the load can be almost uniformly distributed along the length of a stream or estuary where in some large cities bypass locations exist at every block. The short-term nature of this type of load requires particular attention in the mathematical modeling of the effects of combined sewer discharges on water quality.

The increasing urbanization of the environment has also affected the quality of runoff that is physically separated from the sanitary sewerage system. Field sampling and analysis of the water quality characteristics of separate storm water runoff have been conducted in Ann Arbor, Michigan and Cincinnati, Ohio, among

TABLE 5.7   BACTERIOLOGICAL CONCENTRATION (NUM/100 ml) FROM COMBINED SEWERS IN SYRACUSE, NEW YORK

Variable	Geometric mean	95% confidence level
Total coliform	$6.4 \cdot 10^{5a}$	$0.17 \cdot 10^8$
	$13.4 \cdot 10^5$	$9.7 \cdot 10^8$
Fecal coliform	$1.2 \cdot 10^5$	$0.64 \cdot 10^7$
	$4.7 \cdot 10^5$	$11.0 \cdot 10^7$
Fecal streptococci	$3.8 \cdot 10^4$	$3.1 \cdot 10^5$
	$1.4 \cdot 10^4$	$13.4 \cdot 10^5$
FC/FS	3.2	20.6
	33.6	82.1

[a]First number: at Maltbie St., drainage area = 46.5 ha, trib. pop. = 1350, slope = 0.0043. Second number: at W. Newell St., drainage area = 21.9 ha, trib. pop. = 1200, Slope = 0.003.

Source: From Drehwing et al. (1979).

other places. The Ann Arbor studies (Benzie and Courchaine, 1966) were confined primarily to bacteriological quality. The Cincinnati studies (Weibel et al., 1964) covered a wider range of water quality variables.

The bacteriological results from the Cincinnati and Ann Arbor studies are summarized Table 5.8. The substantial difference in values between the two geographical locations is probably due to differing land use patterns and type of area investigated. The Ann Arbor separate sewer drains an area of about 3800 acres in contrast to the Cincinnati area of 27 acres which indicates that for the former area, the sewer system is lengthier and more complex. The ratio of FC to FS is, except for one value, equal to or less than 0.7 which is indicative of bacterial pollution due to nonhuman sources such as rodents, cats, dogs, and other animals (Geldreich, 1966).

A geometric mean for 20 cities of $3 \cdot 10^5/100$ ml of TC has been reported by the US EPA (1976) representing the discharge from storm sewers and unsewered areas (see Table 1.2).

5.2.3 Input Load Calculations

Sample Problem 1.3 shows a typical calculation for continuous steady inputs of bacteria where the resulting load is in units of organisms per day. This calculation of a steady input is straightforward.

However, as noted in this chapter and as sketched in Fig. 1.7, indicator and pathogenic organisms in CSO and runoff from separate sewers are discharged at intermittent intervals. As discussed in Chapter 1, Di Toro et al. (1979) showed that the mean load per overflow event is given by Eq. 1.9 repeated here as

$$W_R = \overline{N}Q_R \tag{5.1}$$

TABLE 5.8 BACTERIOLOGICAL CONCENTRATIONS IN URBAN LAND RUNOFF

Variable	10%[a]	50%[a]	90%[a]
Total coliform			
Cincinnati[b]	2,900	58,000	460,000
Ann Arbor[b]	70,000	1,200,000	20,000,000
Fecal coliform			
Cincinnati	500	10,900	76,000
Ann Arbor	7,000	82,000	1,000,000
Fecal streptococci			
Cincinnati	4,900	20,500	110,000
Ann Arbor	25,000	140,000	770,000
Ratio—FC/FS			
Cincinnati	0.1	0.5	0.7
Ann Arbor	0.3	0.6	1.3

[a]Percent of time number of organisms/100 ml was equal to or less than value indicated.
[b]For Cincinnati, values are for individual counts. For Ann Arbor, values are for median counts over storm.
Source: From Benzie and Courchaine (1966) and Wiebel et al. (1964).

where $\overline{N}$ is the average bacterial concentration [num/L^3; e.g., num/100 ml]. The long-term average loading rate in Eq. 1.10 is

$$W_A = \frac{W_R D}{\Delta} \tag{5.2}$$

for D as the average duration of the events and Δ as the average interval between events. This average loading rate can be used in steady state models of water bodies to compare responses, on the average, between point source inputs and CSO's. In some cases, this average load, W_A may then be used in design load calculations using minimum river flows such as the 7Q10.

To first approximation, Eq. 1.15 and Fig. 1.9 can be used to estimate the frequency of occurrence of various input bacterial loads (see Chapter 1, Section 1.6). These pulse inputs can be used in time variable water quality models to assess the impacts of varying storms on the transient bacterial concentration.

Sample Problem 5.1 (extending Sample Problem 1.1) shows the procedure for calculating the expected exceedance levels of various multiples of the average bacterial load per event. Note that the input load W (10^{14} org/day) ranges over an order of magnitude, but at least the lower value (6.21 · 10^{14} org/day) occurs an estimated 23 times. The upper load of 62.1 · 10^{14} org/day is estimated to be exceeded twice in a year. These load frequencies can then be used to estimate the consequences of various control actions for bacterial reductions.

5.3 ORGANISM DECAY RATE

The survival, fate, and distribution of bacteria and other organisms in natural waters depend on the particular type of water body (i.e. stream, estuary, lake) and associated phenomena that influence the growth, death, and other losses of the organisms. The factors that influence the kinetic behavior of the communicable disease organisms after discharge to a water body are

1. Sunlight
2. Temperature
3. Salinity
4. Predation
5. Nutrient difficiencies
6. Toxic substances
7. Settling of the organism population after discharge
8. Resuspension of particulates with associated sorbed organisms
9. Aftergrowth, that is, the growth of organisms in the body of water

These factors may be present in varying degrees depending on the specific situation. The resultant distribution of the organism concentration will then reflect the net decay (or increase) of the organism as a function of location in the body of water. Figure 5.2 is an example of the changes in bacterial composition that may occur during a transient flow event. Figure 5.4 shows the longitudinal profile

SAMPLE PROBLEM 5.1

DATA

Given the problem conditions of Sample Problem 1.1 and the following additional information:

coefficient of variation of rainfall: $\nu_i = 0.7$
coefficient of variation of runoff: $\nu_q = \nu_i = 0.7$
coefficient of variation of FC: $\nu_N = 2.0$
average concentration of FC: $\overline{N} = 10^5$ org/100 ml

PROBLEM

For various multiples of the mean input FC load, calculate the number of storms per year where the load multiples will be exceeded.

ANALYSIS

Average load:

$$W_R = \overline{N}Q_R \qquad \text{[Eq. (1.9)]}$$

$$= 10^5 \frac{\text{org}}{100 \text{ ml}} \cdot 254 \text{ cfs} \cdot \frac{2.446 \cdot 10^7 \text{ org/day}}{\text{cfs} - \text{org/100 ml}}$$

$$= 6.21 \cdot 10^{14} \text{ org/day}$$

Coefficient of variation of load:

$$\nu_w = \nu_q \nu_N \sqrt{1 + (1/\nu_q)^2 + (1/\nu_N)^2} \qquad \text{[Eq. (1.14)]}$$

$$= 0.7(2.0)\sqrt{1 + (1/0.7)^2 + (1/2)^2}$$

$$= 2.54$$

Number of storm events/year:

$$\overline{S} = \frac{T}{\Delta} \qquad \text{[Eq. (1.16)]}$$

$$= \frac{8760 \text{ hr/yr}}{77 \text{ hr/event}} = 114 \text{ events/yr}$$

W (10^{14} org/day)	W/W$_R$	% of events[a] ≤ W/W$_R$	% of events > W/W$_R$	No. of events per yr > W$_R$
6.21	1	80	20	23
31.05	5	94	6	7
62.1	10	98.2	1.8	2

[a]From Fig. 1.9 and $\nu = 2.54$.

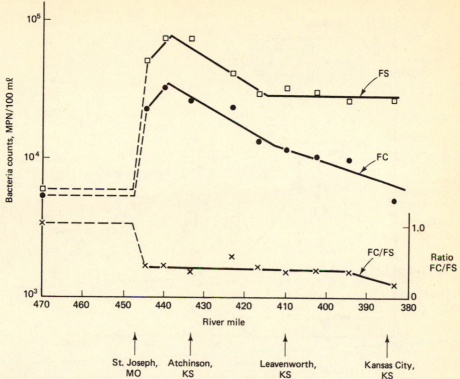

Figure 5.4 Average fecal coliform and fecal streptococci bacteria counts Missouri River, April–May 1959. Plotted from data in Kenner (1978).

for the average FC and FS bacteria concentrations for the Missouri River. The low values of the FC/FS ratio can be noted. Figure 5.5 shows some trends in FC concentrations in the Potomac Estuary in the vicinity of Washington, D.C. The typical shape of the resulting concentration due to a nonconservative variable discharged into an estuary as given by Eq. 3.8 is apparent in the prestorm condition trend. During an area-wide storm, however, with inputs from combined sewers and urban runoff, the trend is markedly different reflecting the large number of discharges over the upper end of the estuary.

   If the overall net first-order decay rate of bacteria and other organisms (i.e., all mortality plus losses minus increases) is designated K_B (1/day), then the principal components of the net decay rate can be written as

$$K_B = K_{B1} + K_{Bl} + K_{Bs} - K_a \qquad (5.3)$$

where K_{B1} = basic death rate as a function of temperature, salinity, predation

K_{Bl} = death rate due to sunlight

K_{Bs} = net loss (gain) due to settling (resuspension)

K_a = aftergrowth rate

   An alternate manner of expressing the overall decay rate finds wide use in describing the decline of bacteria. This alternate expression is the time to obtain

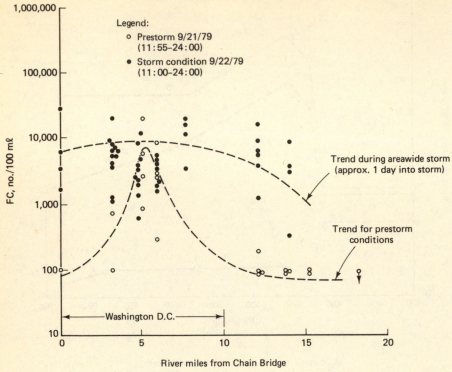

Figure 5.5   Trends in fecal coliform concentrations, Potomac Estuary (data from O'Brien & Gere and Limno-Tech, 1980), plot from Thomann (1981).

90% mortality or loss of the original number of bacteria assuming a first-order loss. Thus, the 90% mortality time, t_{90}, is given by

$$0.10 = \exp(-K_B t_{90}) \tag{5.4}$$

or

$$t_{90} = \frac{2.3}{K_B} \tag{5.5}$$

Table 5.9 is a summary of some reported decay rates for bacteria and viruses. As noted, for the TC group, the freshwater rate is about 1/day while the seawater rate is about 1.4/day but can range up to 84/day in ocean outfall studies. These latter values would include the effects of sunlight as discussed below. The rates for the *S. fecalis* are generally about an order of magnitude lower than the coliform groups, although Fujioka et al. (1981) reported values up to 55/day in sunlighted seawater for FS. Data on *Salmonella* given by Geldreich and Kenner (1979) and Dutka and Kwan (1980) show two phases: a more rapid die-off of about 1/day and then a more resistant phase with a die-off rate of about 0.1/day. Viruses generally also die away at a slower rate than the bacteria by about an order of magnitude.

TABLE 5.9 SOME REPORTED OVERALL DECAY RATES FOR BACTERIA AND VIRUSES

Organism	K_B (day^{-1})	Remarks	Reference[a]
		Coliforms	
Total coliform	1–5.5	Freshwater—summer (or 20°C), seven locations	1
	0.8	Average freshwater, 20°C	2
	1.4 (0.7–3.0)	Seawater, 20°C	2
	48(8–84)	From 14 ocean outfalls (variable temperature)	1
Total or fecal	0–2.4	New York Harbor Salinity: 2–18‰ Dark Samples	3
	2.5–6.1	New York Harbor Salinity: 15‰ Sunlighted Samples	3
Fecal coliform	37–110	Seawater, sunlighted	12
E. coli	0.08–2.0	Seawater, 10–30‰	13
		Fecal streptococci	
S. faecalis	0.4–0.9	Freshwater—20°C	4
	0.1–0.4	Freshwater—4°C	4
	0–0.8	Kanawha River—20°C	5
	0.3	Stormwater—20°C; 0–3 days	6
	0.1	Stormwater—20°C; 3–14 days	6
	1.0–3.0	Hamilton Bay, Lake Ontario, 18°C; 0–10 days	7
	0.05–0.1	Hamilton Bay, Lake Ontario, 10–28 days	7
S. bovis	1.5	Stormwater—20°C	6
Fecal streptococci	18–55	Seawater, sunlighted	12
		Pathogens	
Salmonella typhimurium	1.1	Stormwater—20°C; 0–3 days	6
	0.1	Stormwater—20°C; 3–14 days	6
Salmonella thompson	0.5–3	Hamilton Bay, Lake Ontario; 18°C; 0–10 days	7
	0.1	Hamilton Bay, Lake Ontario; 18°C; 10–28 days	7
		Viruses	
Coxsackie	0.77	Lake Wingra, 21—23°C	8
	0.12	Marine waters—25°C	9
	0.03	Marine waters—4°C	9
Echo 6	0.08	Marine waters—25°C	9
	0.03	Marine waters—4°C	9
Polio type 1	0.16	Marine waters—25°C	9
	0.05	Marine waters—4°C	9
	0.26	Lake Wingra, 21–23°C	8
Enteric (polio, Echo, coxsackie)	0.15	Tanana River Alaska, 0°C ice cover	10
	1.1–2.3	Hawaiian Ocean, 24°C	11

[a]References: (1) Mitchell and Chamberlain (1978); (2) Mancini (1978); (3) Hydroscience (1977b); (4) USEPA (1974); (5) Kenner (1978); (6) Geldreich and Kenner (1969); (7) Dutka and Kwan (1980); (8) Herrmann et al. (1974); (9) Colwell and Hetrick (1975); (10) Dahling and Safferman (1979); (11) Fujioka et al. (1980); (12) Fujioka et al. (1981); (13) Anderson et al. (1979).

Returning to Eq. 5.3, the temperature effect for bacteria is often approximated by

$$(K_B)_T = (K_B)_{20}(1.07)^{T-20} \tag{5.6}$$

and Fig. 5.6 shows that to first approximation this relationship is also appropriate for virus decay rates as a function of temperature.

The effect of sunlight has been evaluated in a variety of studies including the work of Gameson and Gould (1974). Their results for surface data (0–5 cm) in direct sunlight are summarized in Table 5.10. However, it must be recognized that solar radiation varies with depth as a function of the light extinction coefficient. One can obtain from the results of Gameson and Gould (1974) that

$$K_{B0}(t) = \alpha I_0(t) \tag{5.7}$$

where K_{B0} (day^{-1}) is the decay rate at the surface, α is a proportionality constant, and $I_0(t)$ in cal/cm$^2 \cdot$ hr is the surface solar radiation. From the data of Gameson and Gould (1974), α is approximately unity. The depth-averaged effect of sunlight on decay rate can then be shown to be

$$\overline{K}_B = \frac{\alpha I_0(t)}{HK_e}[1 - \exp(-K_e H)] \tag{5.8}$$

where H is the depth in m over which the average is taken and K_e (m^{-1}) is the vertical light extinction coefficient. (This latter quantity is discussed in Chapter 7.)

Figure 5.6   Relationship of decay rates of viruses with temperature.

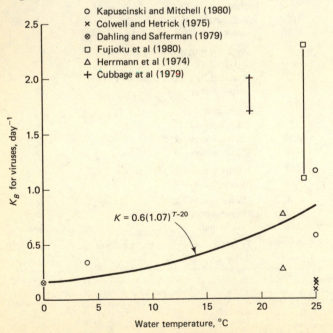

TABLE 5.10 EFFECT OF SUNLIGHT ON BACTERIAL
DECAY RATE

% frequency of occurrence of decay rate ($\geq$)	K_B (day^{-1})	t_{90} (hr)
5	80	0.7
50	18	3.1
95	7	8.0

Source: From Gameson and Gould (1974).

From these considerations, two representations of the loss rate can be utilized. The first most simple model uses the overall net loss rate K_B as the measure of bacterial kinetics and no attempt is made to describe the individual mechanisms or the kinetic structure. At most, K_B is considered as a function of temperature. This simple model recognizes that there may be considerable uncertainty in the input loads in certain problem contexts and that it is really not practical or meaningful to describe the kinetic structure at any significant level of detail.

The second model level incorporates some of the principal mechanisms discussed above. Mancini (1978) has evaluated the available data for incorporation of salinity, temperature, and solar radiation. On the basis of that work and the discussion of sunlight effects above, K_B can be written as

$$K_B = [0.8 + 0.006(\% \text{ seawater})]1.07^{(T-20)}$$

$$+ \frac{\alpha I_0(t)}{K_e H}[1 - \exp(-K_e H)] \qquad (5.9)$$

$$+ \frac{v_s}{H}$$

where v_s is the net loss rate in m/day of the particulate bacterial forms. (Note that v_s can be negative, zero, or positive, depending on the degree of resuspension.) The increased complexity of this formulation of K_B is worthwhile in the situations where input loads are known with some degree of certainty. In Eq. 5.9, the effect of varying sunlight may be significant in attempting to calibrate a short-term transient bacterial situation.

5.3.1 Bottom Sediments as a Reservoir of Organisms

Since the above formulations for K_B include a settling term out of the water column, the sediment may be a significant source of organisms. It has been known for quite some time that organisms in the microscopic range can adhere to particles dispersed in water and wastewater. LaBelle et al. (1980) in referencing Savage (1905) indicated that Savage, "concluded that samples of mud yielded more bacteriological evidence of the degree of fecal pollution of a tidal river than samples of either water or oysters." For example, the data on viruses sorbed to wastewater particles of Hejkal et al. (1981) have been analyzed and plotted in Fig. 5.7. The relationship

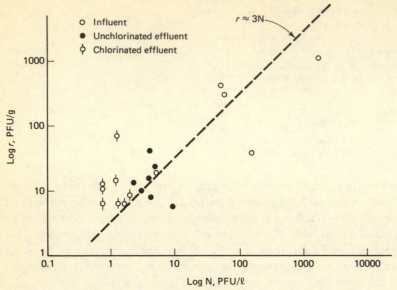

Figure 5.7 Distribution of Enteroviruses on particles in wastewater and in water (centrifuged). Analyzed from data of Hejkal et al. (1981).

between the viruses on the solids and those in the water after centrifugation is approximately

$$r = 3N \qquad (5.10)$$

where r is the viral concentration on the solids (PFU/g) and N is the concentration in the water (PFU/l). Also, LaBelle et al. (1980) noted that viruses from sediments in the Texas Gulf coastal region were found in greater numbers in the sediment than in the overlying water and that some correlation was found between the number of viruses and concentration of FCs in the sediment.

Thus, the discharge of bacteria and viruses to natural waters may result in the sorption of such organisms to particles, settling into bottom sediments and subsequent resuspension of contaminated particles. Microorganisms can apparently survive in the sediments for longer periods of time than in the overlying water column. For example, Roper and Marshall (1979) indicated that the survival of *E. coli* appeared to be increased in the estuarine sediment of the Tamor River in Tasmania, Australia. The decay rate for the E. coli in the sediment (concentration average about 10^6 *E. coli*/g sediment) was 0.14 and 0.21/day in two chamber experiments. This reduced die-off rate is attributed by Roper and Marshall to inhibition of predation and parasitism. Similar increased survival rates for viruses in estuarine sediments have been reported by LaBelle and Gerba (1980) who indicate that at a polluted site the inactivation rate for poliovirus 1 and echovirus 1 decreased from 4.6/day in seawater alone to 1.1/day for virus adsorbed to sediment.

Since the sediment may include large concentrations of microorganisms, the resuspension of such sediment and subsequent desorption may be an important source of contamination in the overlying waters. In the context of the previous models then, the bacteria in the sediments can be considered as a distributed source of organisms.

5.4 FATE OF ORGANISMS

5.4.1 Rivers and Streams

For rivers and streams, following Eqs. 2.22 and 2.23, the downstream distribution of bacteria is

$$N = N_0 \exp(-K_B t^*) \tag{5.11}$$

where N is the concentration of the organism [num/L^3; e.g., num/100 ml] and N_0 is the concentration at the outfall after mixing. Because of the factors listed above that influence the growth and death of organisms, the shape of the bacterial profile downstream from an outfall may take several forms as shown in Fig. 5.8. For a variety of situations, a simple die-away or decay of bacteria as indicated in Fig. 5.8(a) is a good representation of the real data, although as shown in Fig. 5.8(b), often a second region of a reduced decay rate occurs. This region may be representative of the more resistant strains in an indicator group such as the TC bacteria. Figure 5.8(c) shows the case where there may be an initial period of aftergrowth dominating all other loss terms and followed by a first-order decay. Positive exponential growth may then occur over a certain distance and time. Mancini (1978) indicates a coefficient of $K_B = 2.2$/day corresponding to a threefold increase in 12 hr as an estimate of the aftergrowth effect. Finally, Fig. 5.8(d) shows the case where because of a combination of growth just balancing death and losses, there is an initial downstream region where the die-away is zero.

For the distribution of microorganisms in rivers, successive application of Eqs. 2.22 or 2.23 is appropriate as given in Eq. 5.11. Figure 5.9, adapted from Kittrell and Furfari (1963), shows the decrease of coliform bacteria for the Tennessee River below Chattanooga and illustrates the estimation of K_B. The data also indicate that Eq. 5.11 does not apply over the entire reach equivalent to 12 days travel time. A reevaluation of the rate K_B should then be made beginning at about 4–5 days. The rate for the reach from 5 to 12 days is less than that for the first reach. This probably reflects the predominance of more resistant forms of the coliform group. Sample Problem 5.2 shows the considerations that must be included in the modeling analysis of a bacterial distribution in a stream.

5.4.1.1 Source from Sediment If bottom resuspension of organisms is being considered then an appropriate equation is

$$U\frac{dN}{dx} = -K_B N + \frac{v_u}{H} M_s r_N \tag{5.12}$$

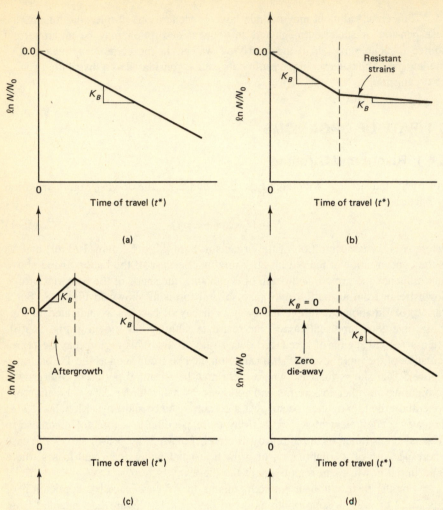

Figure 5.8 Forms of bacterial and other organism distributions in rivers (a) Single die-away. (b) Multiple die-away, resistant strains. (c) Region of aftergrowth. (d) Region of no die-away.

where v_u is the resuspension velocity $[L/T]$, H is the depth of the water column $[L]$, M_s is the bulk density of the sediment solids $[M_s/L_{s+w}^3; M_s = \text{mass of solids}, L_{s+w}^3 = \text{bulk volume} = \text{water plus solids}]$, and r_N is the concentration of bacteria or virus on a solids basis $[M_N/M_s; M_N = \text{number of bacteria}]$. The solution to Eq. 5.12 with the distributed source is as given before (see Eq. 2.28)

$$N = N_0 \exp\left(-K_B \frac{x}{U}\right) + \frac{v_u M_s r_N}{H K_B}\left[1 - \exp\left(-K_B \frac{x}{U}\right)\right] \qquad (5.13)$$

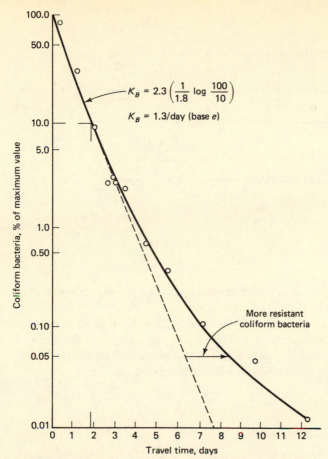

Figure 5.9 Coliform (total) decline in Tennessee River below Chattanooga and overall decay rate. Adapted from Kittrel and Furfari (1963). Reprinted by permisson of Water Pollution Control Federation.

Practical application of Eq. 5.13 would require estimation of the resuspension velocity through analysis of the solids balance of the river. Such estimation is discussed in Chapter 8.

Figure 5.10 (p. 245) (O'Connor, 1968) shows an application of the stream TC model to a reach of the Mohawk River, N.Y. The river was divided into 15 reaches and a decay rate of 0.9/day was used throughout. The increase in flow results in a significant dilution in the downstream direction.

The reduction in coliform concentration after treatment to 99.99% removal at two levels of background coliform is also shown.

SAMPLE PROBLEM 5.2

DATA

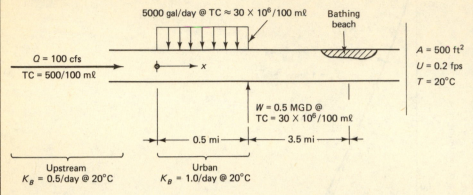

5000 gal/day @ TC ≈ 30 × 10^6/100 mℓ   Bathing beach

Q = 100 cfs
TC = 500/100 mℓ

A = 500 ft²
U = 0.2 fps
T = 20°C

W = 0.5 MGD @ TC = 30 × 10^6/100 mℓ

0.5 mi — 3.5 mi

Upstream
K_B = 0.5/day @ 20°C

Urban
K_B = 1.0/day @ 20°C

PROBLEM

(a) For the conditions above, determine the required percent reduction in the point source to meet New York State standards.

(b) Using the result from (a), will the standard be met under a higher flow of 400 cfs and lower temperature of 15°C? Assume the velocity in the stream is proportional to $Q^{0.6}$ and that the upstream TC count remains 500/100 ml.

Neglect flow inputs from urban distributed and point sources.

ANALYSIS

(a) Lower Flow–Higher Temperature

$$U = 0.2 \text{ fps} \times 16.4 = 3.3 \text{ mpd}$$

Upstream Impact

$$K_B = 0.5/\text{day}$$

$$N = N_0 e^{-K_B x/U} \qquad \text{[Eq. (5.11)]}$$

$$N = 500 e^{-0.5 \times 4.0/3.3} = \underline{273 \text{ organisms}/100 \text{ ml}}$$

Urban Distributed Source Impact

$$K_B = 1.0/\text{day}; \quad 5000 \text{ gal/day} = 0.005 \text{ MGD}$$

$$S_D = \frac{w}{A} \qquad \text{[Eq. (2.26)]}$$

$$S_D = \frac{(0.005 \times 1.548 \text{ cfs}) \times 30 \times 10^6/100 \text{ ml}}{0.5 \text{ mi} \times 500 \text{ ft}^2 \times 5280 \text{ ft/mi}}$$

$$\times \, 86,400 \text{ sec/day} = \underline{15200 \text{ org}/100 \text{ ml·day}}$$

Sample Problem 5.2 (continued)

$$N(x = 0.5 \text{ mi}) = \frac{S_D}{K_B} (1 - e^{-K_B x/U}) \qquad \text{[Eq. (2.28)]}$$

$$N(x = 0.5 \text{ mi}) = \frac{15200 \text{ org}/100 \text{ ml·day}}{1.0/\text{day}} \times (1 - e^{-1.0 \times 0.5/3.3})$$

$$= 2140/100 \text{ ml}$$

$$N(x = 4.0 \text{ mi}) = 2140 \times e^{-1.0 \times 3.5/3.3}$$

$$= \underline{741 \text{ org}/100 \text{ ml}}$$

Urban Point Source Impact

$$K_B = 1.0/\text{day}$$

$$N(x = 0.5 \text{ mi}) = \frac{W}{Q} = \frac{(0.5 \times 1.548 \text{ cfs}) \times 30 \times 10^6/100 \text{ ml}}{100 \text{ cfs}}$$

$$= 232,000 \text{ org}/100 \text{ ml}$$

$$N(x = 4.0 \text{ mi}) = 232,000 e^{-1.0 \times 3.5/3.3} = \underline{80,400 \text{ org}/100 \text{ ml}}$$

Required Point Source (PS) Reduction

Assume upstream and distributed urban sources are not controllable. From Table 5.2, the New York State standard for bathing areas is TC < 2400 org/100 ml.

∴ allowable PS impact = 2400 − (273 + 741) = 1386/100 ml

$$\underline{\text{Percent reduction}} = \left(\frac{80,400 - 1386}{80,400}\right) \times 100 = \underline{98.3\%}$$

(b) Higher Flow–Lower Temperature

$$U \propto Q^{0.6}, \qquad U = 0.2 \text{ fps} \left(\frac{400 \text{ cfs}}{100 \text{ cfs}}\right)^{0.6} = 0.46 \text{ fps} = 7.5 \text{ mpd}$$

$$\text{with } A = \frac{Q}{U}, \qquad A \propto \frac{Q}{Q^{0.6}} = Q^{0.4}$$

$$A = 500 \text{ ft}^2 \left(\frac{400 \text{ cfs}}{100 \text{ cfs}}\right)^{0.4} = 870 \text{ ft}^2$$

$$K_B(\text{upstream}) = 0.5 \times 1.07^{15-20} = 0.36/\text{day} \qquad \text{[Eq. (5.6)]}$$

$$K_B(\text{urban}) = 1.0 \times 1.07^{15-20} = 0.71/\text{day}$$

Upstream Impact

$$N = 500 e^{-0.36 \times 4.0/7.5} = \underline{413 \text{ org}/100 \text{ ml}}$$

(continued)

Sample Problem 5.2 (continued)

Urban Distributed Source Impact

$$N(x = 0.5) = \frac{15,200 \times (500/870)}{0.71}(1 - e^{-0.71 \times 0.5/7.5}) = 569 \text{ org}/100 \text{ ml}$$

$$N(x = 4.0) = 569 \times e^{-0.71 \times 3.5/7.5} = \underline{409 \text{ org}/100 \text{ ml}}$$

Urban Point Source Impact

$$N(x = 4.0) = 232,000 \times \left(\frac{100}{400}\right) \times e^{-0.71 \times 3.5/7.5} = 41,600 \text{ org}/100 \text{ ml}$$

at 98.3% kill, $N = 41,600 \times (1 - 0.983) = \underline{707 \text{ org}/100 \text{ ml}}$

Total Impact

$N = \Sigma N = 413 + 409 + 707 = 1529$ org/100 ml, and since $1529 < 2400$ (std), the water quality standard will be met.

Summary at beach	Beach TC (org/100 ml)	
Source	Low flow–high temperature	Higher flow–lower temperature
Upstream	273	413
Distributed urban	741	409
PS urban	1386	707

Note how the upstream boundary propagates further downstream under the higher flow condition. This is due to decreased travel time and decreased decay rate. For the urban sources, the increased dilution overpowers these effects.

5.4.2 Estuaries and Lakes

For estuaries, Eqs. 3.8, and similarly for lakes, Eq. 4.13, with $K = K_B$ for each case, are the appropriate equations for applications to the fate of microorganisms. Sample Problem 5.3 shows a calculation for an estuary and Sample Problem 5.4 illustrates the technique for constructing a model of the time variable response of a cove of a lake.

Figure 5.11 shows the results of a TC steady-state calibration in Boston Harbor using the two-dimensional grid, flows, and dispersion shown in Fig. 3.26. The calibration as noted is considered satisfactory to order of magnitude which is typical for coliform bacteria analyses. The outer boundary of zero coliform can also be noted.

An application of an analysis of the statistial variation of TC is given by Di Toro et al (1978) for New York Harbor. Figure 6.30 shows the segmentation used

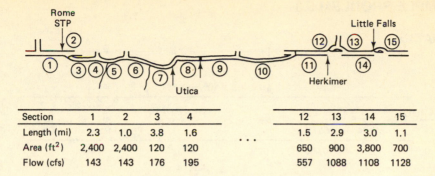

Section	1	2	3	4		12	13	14	15
Length (mi)	2.3	1.0	3.8	1.6	...	1.5	2.9	3.0	1.1
Area (ft^2)	2,400	2,400	120	120		650	900	3,800	700
Flow (cfs)	143	143	176	195		557	1088	1108	1128

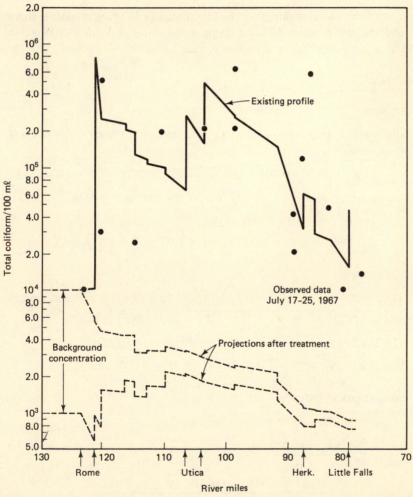

Figure 5.10 Comparison of total coliform calculation with observed data, Mohawk River, New York. 15 river segments used in model (top). Note increase in flow. From O'Connor (1968).

245

SAMPLE PROBLEM 5.3

DATA

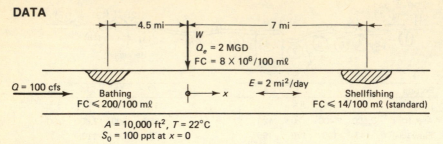

$A = 10,000 \text{ ft}^2$, $T = 22°C$
$S_0 = 100$ ppt at $x = 0$

PROBLEM

What percent reduction is required for the raw discharge in order to meet both the bathing and shellfish harvesting FC standards in the estuary? Assume chlorination is used and assume an effective aftergrowth factor of three.

ANALYSIS

FC Decay Rate

For a daily average decay rate, neglecting solar radiation effects (high K_e) and, assuming no settling,

$$K_B = [0.8 + 0.006 \times (\% \text{ seawater})] \times 1.07^{(T-20)} \qquad [\text{Eq. (5.9)}]$$

With 100% seawater = 35 PPT salinity, the % seawater at $x = 0$ is

$$(10 \text{ PPT}/35 \text{ PPT}) \times 100 = 29\%$$

$$\therefore \underline{K_B(22°C) = [0.8 + 0.006 \times 29] \times 1.07^{22-20} = \underline{1.11/\text{day}}}$$

Estuary Coefficients

$$U = Q/A = 100 \text{ cfs}/10,000 \text{ ft}^2 = 0.010 \text{ fps} \times 16.4 \text{ mpd/fps} = 0.164 \text{ mpd}$$

$$\alpha = \sqrt{1 + (4K_B E/U^2)} = \sqrt{1 + [4 \times 1.11 \times 2/(0.164)^2]} = 18.2$$

$$j_1 = (U/2E)(1 + \alpha) = (0.164/2 \times 2)(1 + 18.2) = +0.787/\text{mi}$$

$$j_2 = (U/2E)(1 - \alpha) = (0.164/2 \times 2)(1 - 18.2) = -0.705/\text{mi}$$

FC Concentrations in Estuary

With no disinfection, at $x = 0$:

$$N_o = \frac{(W = Q_w N_w)}{Q\alpha}$$

$$= \frac{2 \text{ MGD} \times 1.548 \text{ cfs/MGD} \times 8 \times 10^6/100 \text{ ml}}{100 \text{ cfs} \times 18.2}$$

$$= 13,600/100 \text{ ml}$$

Sample Problem 5.3 (continued)

At the bathing area ($x = -4.5$):

$$N(x = -4.5) = N_0^{j_1 x} = 13,600e^{0.787 \times (-4.5)} = 394/100 \text{ ml} (>200)$$

At the shellfishing area ($x = +7$ mi):

$$N(x = +7) = N_0^{j_2 x} = 13,600e^{-0.705 \times 7} = 97.7/100 \text{ ml} (>14)$$

Both FC counts exceed standards and effluent reduction is required.

Effluent FC % Reduction

With disinfection, a threefold aftergrowth is assumed which increases estuary counts.

$$\text{bathing:} \qquad \% \text{ kill} = \frac{3 \times 394 - 200}{3 \times 394} \times 100 = 83\%$$

$$\text{shellfishing:} \qquad \% \text{ kill} = \frac{3 \times 97.7 - 14}{3 \times 97.7} \times 100 = 95.2\%$$

Therefore, the *shellfishing standard* requires the higher removal rate of 95%. Resulting estuary concentrations are:

$$x = 0: \qquad N_0 = \left[\frac{2 \times 1.548 \times 8 \times 10^6 (1 - 0.952)}{100 \times 18.2} \right] \times 3$$

$$= 1960/100 \text{ ml}$$

$$x = -4.5: \qquad N = 1960e^{+0.787 \times (-4.5)} = 57/100 \text{ ml} (\leq 200)$$

$$x = 7: \qquad N = 1960e^{-0.705 \times 7} = 14/100 \text{ ml} (\leq 14)$$

Note: The salinity varies in the estuary segment being considered and, thus, the FC decay rate also varies. With $S/S_0 = e^{Ux/E}$

$$\text{at beach } (x = -4.5), \qquad S = 10 \text{ ppt } e^{0.164 \times (-4.5)/2} = 6.9 \text{ ppt}$$

$$\text{at shellfish area } (x = 7), \qquad S = 10e^{0.164 \times 7/2} = 17.8 \text{ ppt}$$

FC decay rates would be

$$x = -4.5: \qquad K_B = \left(0.8 + 0.006 \times \frac{6.9}{35} \times 100 \right) \times 1.07^{22-20} = 1.05/\text{day}$$

$$x = +7: \qquad K = \left(0.8 + 0.006 \times \frac{17.8}{35} \times 100 \right) \times 1.07^{22-20} = 1.27/\text{day}$$

compared with the value used of 1.11/day.

The impact would be to increase upstream (less saline) FC counts—due to the lower upstream K_B rates—and decrease downstream counts due to the higher decay. A more rigorous analysis would include a spatially varying rate.

SAMPLE PROBLEM 5.4

DATA

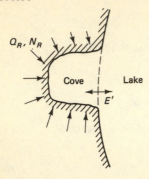

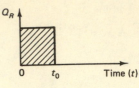

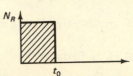

Distributed runoff, Q_R, with a TC concentration assumed temporally constant, enters a cove of a lake for a time t_0, the duration of storm runoff.

PROBLEM

Estimate the maximum TC count in the cove and the time—after the runoff ends—to attain a desired count. Treat the cove as well mixed and exchanging with the lake with a bulk dispersion coefficient E' $[L^3/T]$.

ANALYSIS

Mass Balances

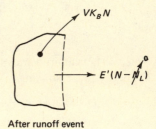

During runoff event

After runoff event

During Runoff Event

$$V \frac{dN}{dt} = W - Q_R N - E'(N - N_L) - V K_B N$$

$$\frac{dN}{dt} = \frac{W}{V} - N\left(\frac{V K_B + Q_R + E'}{V}\right)$$

Letting $\alpha = (V K_B + Q_R + E')/V$, then

$$\frac{dN}{dt} + \alpha N = \frac{W}{V}$$

248

Sample Problem 5.4 (continued)

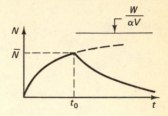

For $N = 0$ at $t = 0$,

$$N = \frac{W}{\alpha V} (1 - e^{-\alpha t})$$

Maximum concentration $\overline{N}$ occurs at t_0, the end of the runoff event.

$$\therefore \overline{N} = \frac{W}{\alpha V} (1 - e^{-\alpha t_0})$$

Note similarity of this expression to Eq. 4.16.

After Runoff Event

$$V \frac{dN}{dt} = -VK_B N - E'(N - N_L)$$

$$\frac{dN}{dt} + \alpha'N = 0 \qquad \text{where} \quad \alpha' = \frac{VK_B + E'}{V}$$

For $N = \overline{N}$ at $t' = 0(t = t_0)$,

$$N = \overline{N}e^{-\alpha't'}$$

or

$$N = \overline{N}e^{-\alpha'(t - t_0)}$$

It is now desired to find the time after storm that is required to reach a desired TC count.

Let desired count $= N_s$

$$N_s = \overline{N}e^{-\alpha't'}$$

$$\frac{N_s}{\overline{N}} = e^{-\alpha't'}$$

$$\ln\left(\frac{N_s}{\overline{N}}\right) = -\alpha't'$$

$$\therefore t' = -\frac{\ln(N_s/\overline{N})}{\alpha'}$$

or

$$t' = \frac{\ln(\overline{N}/N_s)}{\alpha'}$$

(continued)

Sample Problem 5.4 (continued)

Summary

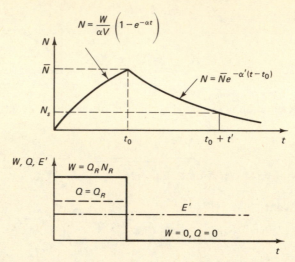

$$N = \frac{W}{\alpha V}\left(1 - e^{-\alpha t}\right)$$

$$N = \bar{N}e^{-\alpha'(t-t_0)}$$

$$W = Q_R N_R$$

$$Q = Q_R$$

$$W = 0, Q = 0$$

Notes: A next level of analysis might include time variable runoff flows and TC counts, varying lake concentration ($N_L \neq 0$) and time variable die-off rate [$K_B = f$(daylight, night)], etc.

for the calculation. Using a statistical method of analysis of estuaries subjected to intermittent loads as developed by Di Toro (1979), and assuming a lognormal distribution of bacteria in the water column, TC bacteria counts were calculated at the beaches in the New York Harbor complex for various percentile occurrences and compared with observed data. At the beaches near Coney Island (Fig. 5.12), the tenth and ninetieth percentile are seen to bound the data and the predicted fiftieth percentile appears to represent the medians of the observed data at various locations. A major portion of the predicted variance in the TCs was ascribed to continuous sources although intermittent sources also contributed significantly.

5.5 ENVIRONMENTAL CONTROLS

There are three broad categories for control of pathogenic bacteria, viruses, and parasites:

1. Control at the input source of the microorganism.
2. Control at the area of water use.
3. Control of the product that is affected by the contamination.

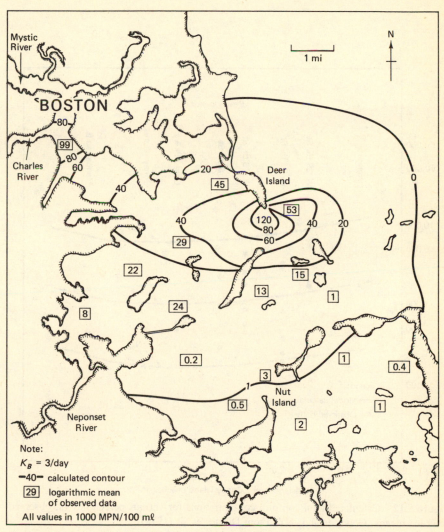

Figure 5.11 Comparison of total coliform calculation with observed data, Boston Harbor, MA. See Fig. 3.25. From Hydroscience (1970).

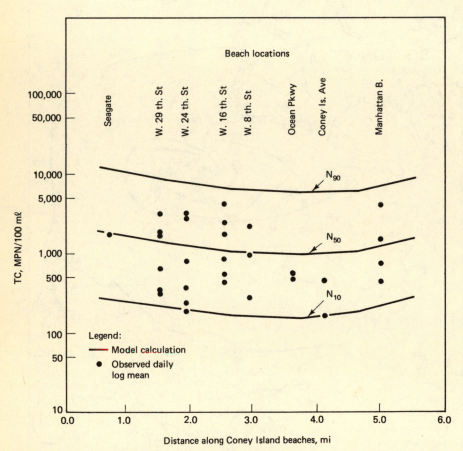

Figure 5.12   Calculated total coliform concentrations for various percentiles vs. observed daily log means, Coney Island, NY. From DiToro et al. (1978).

Point sources of municipal wastes are the usual principal inputs of communicable disease organisms and such inputs can be reduced by:

1. Treatment of wastes without disinfection.
2. Disinfection by (a) chlorination, (b) ultraviolet radiation, (c) ozonation, and (d) chlorine dioxide.

Tables 5.11 and 5.12 summarize some of the information on the efficiency of point source controls in reducing microorganisms discharged to natural waters. Figure 5.10 shows the results of a calculation where a projection is made to test the effectiveness of municipal treatment and chlorination on reducing bacterial levels in the receiving water.

TABLE 5.11 APPROXIMATE REMOVAL EFFICIENCIES OF MICROORGANISMS FROM MUNICIPAL WASTE TREATMENT

Organism type	Treatment	Approximate % removal	
		Without chlorination	With chlorination
Total coliform	Primary	25–75[a]	
	Secondary	90–99	99.9–99.99
Virus	Secondary advanced	90[b]	99–99.99[b]

[a]Fair et al. (1968).
[b]US EPA (1978).

TABLE 5.12 REMOVAL EFFICIENCIES FOR WASTE STABILIZATION PONDS

Organism type	Number of ponds in series	Location	Approximate % remaining of raw influent	Reference[a]
Total coliform	3–4	Logan, Utah	0.01	1
	1	Lima, Peru	5	2
	2	Lima, Peru	4	2
	3	Lima, Peru	0.07	2
Fecal coliform	3–4	Logan, Utah	0.005	1
	1	Lima, Peru	4	2
	2	Lima, Peru	0.4	2
	3	Lima, Peru	0.014	2
Salmonella	1	Lima, Peru	27	2
	2	Lima, Peru	2	2
	3	Lima, Peru	0.2	2

[a]References:
(1) Middlebrooks et al. (1978).
(2) Yanez (1983).

Controls of the area of water use would include:

1. Bathing restrictions on a transient basis, that is, during and after storms resulting in combined sewer overflows.
2. Construction of dikes, and diversion structures to protect a given area.

Controls of the product would include:

1. Chlorination or other forms of disinfection of water used for municipal water supply.
2. Treatment plants for contaminated shellfish to allow depuration of bacteria prior to marketing.
3. Distribution of high quality bottled water during emergencies.

REFERENCES

Anderson, I. C., M. Rhodes, and H. Kator, 1979. Sublethal Stress in Escherichia coli: A Function of Salinity, *Appl. Environ. Micro.* Dec.: 1147–1152.

Benzie, W. J., and R. J. Courchaine, 1966. Discharges from Separate Storm Sewers and Combined Sewers, *J. Water Poll. Contr. Fed.* **38**(3):410–421.

Bonde, G. J., 1981. Salmonella and Other Pathogenic Bacteria, *The Science of the Total Environment*, Elsevier, New York, **18**:1–11.

Brezinski, F. T., and R. Russomanno, 1969. The Detection and Use of Salmonellae in Studying Polluted Tidal Estuaries, *J. Water Poll. Contr. Fed.* **41**:5.

Castagnino, W. A., 1977. Polucion de aqua, modelos y control. centro panonericano de Ingenieria Sanitaria y Ciencias del Ambiente (CEPIS), Pub. CEPIS No. 34, Lima, Peru, 234 pp.

City of New York, 1979. Areawide Waste Treatment Management Planning Program, Section 208. NYC Dept. of Environmental Protection, 8 Chapters.

Coelho, V. M. B., M. R. M. B. Fonseca, J. A. Chipe, A. S. P. Guimaraes, and R. Lozinsky, 1978. Guanabara Bay, Basic Inputs to the Economic Model. Fumdacao Estadual de Engenharia do Meio Ambiente (FEEMA), Rio de Janeiro, Brazil, 64 pp.

Colwell, R., and F. Hetrick, 1975. Survival of Human Pathogens in the Marine Environment, Maryland University, Department of Microbiology, College Park, MD. AD-A012-489, 26 pp.

Commission of the European Communities, 1976. Council Directive of 8 December 1975 Concerning the Quality of Bathing Waters, *Official J. European Communities* **19**:L31.

Craun, G. F., 1978. Impact of the Coliform Standard on the Transmission of Disease, C. W. Hendricks, Ed., *Evaluation of the Microbiology Standards for Drinking Water*, U.S. EPA, Washington, D.C., Office of Drinking Water, pp. 21–35. EPA-570/9-78-00C.

Craun, G. F., 1979. Waterborne Outbreaks of Giardiasis. W. Jakubowski and J. C. Hoff, Eds., *Waterborne Transmission of Giardiasis*, Proceed. of a Symposium, U.S. EPA, ORD, ERC, Cincinnati, OH, pp. 127–149. EPA-600/9-79-001.

Cubbage, C. P., J. J. Gannon, K. W. Coehran, and G. W. Williams, 1979. Loss of

Infectivity of Poliovirus 1 in River Water Under Simulated Field Conditions, *Water Res.* **13**:1091–1099.

Dahling, D. R., and R. S. Safferman, 1979. Survival of Enteric Viruses Under Natural Conditions in a Subarctic River, *Appl. Environ. Micro.* Dec.:1103–1110.

Davis, E. M., 1979. Maximum Utilization of Water Resources in a Planned Community, Bacterial Characteristics of Stormwaters in Developing Rural Areas, MERL, ORD, U.S. EPA, Cincinnati, OH, 83 pp. EPA-600/2-79-050f.

Di Toro, D. M., 1979. Statistics of Receiving Water Response to Runoff, Paper presented at the Urban Stormwater and Combined Sewer Overflow Impact on Receiving Water Bodies, National Conference, Univ. of Central Florida, Orlando, FL, Nov. 1979; 35 pp.

Di Toro, D. M., J. A. Mueller, and M. J. Small, 1978. Rainfall-Runoff and Statistical Receiving Water Models, NYC 208 Task Report 225, by Hydroscience, Inc. for Hazen & Sawyer Engr. and NYC Dept. of Envir. Prot., March 1978, 271 pp.

Di Toro, D. M., E. D. Driscoll, and R. V. Thomann, 1979. A Statistical Method for the Assessment of Urban Stormwater, by Hydroscience, Inc. for the U. S. EPA, Office of Water Planning and Standards, Wash. D. C., May 1979, 7 Chapters & Appendix. EPA 440/3-79-023.

Drehwing, F. J., A. J. Oliver, D. A. MacArthur, and P. E. Moffa. 1979. Disinfection/Treatment of Combined Sewer Overflows, Syracuse, New York, USEPA, MERL, Cincinnati, OH, 244 pp. EPA-600/2-79-134.

Dutka, B. J., and K. K. Kwan, 1980. Bacterial Die-off and Stream Transport Studies, *Water Res.* **14**:909–915.

Fair, G. M., J. C. Geyer, and D. A. Okun, 1968. *Water and Wastewater Engineering*, Vol. 2. *Water Purification and Wastewater Treatment and Disposal*, Wiley, New York, Chapters 19–38.

Fujioka, R. S., P. C. Loh, and L. S. Lau, 1980. Survival of Human Enteroviruses in the Hawaiian Ocean Environment: Evidence for Virus-Inactivating Microorganisms, *Appl. Environ. Micro.* June:1105–1110.

Fujioka, R. S., H. H. Hashimoto, E. B. Siwak, and R. H. F. Young, 1981. Effect of Sunlight on Survival of Indicator Bacteria in Seawater, *Appl. Environ. Micro* March:690–696.

Gameson, A. L. H., and D. J. Gould, 1974. Effects of Solar Radiation on the Mortality of Some Terrestial Bacteria in Sea Water, *Proc. Int. Symp. on Discharge of Sewage from Sea Outfalls, London*, Pergamon Press, Great Britain, Paper No. 22.

Geldreich, E. E., 1966. Sanitary Significance of Fecal Coliforms in the Environment, U.S. Department of the Interior, Publication No. WP-20-3.

Geldreich, E. E., 1970. Applying Bacteriological Parameters to Recreational Water Quality, *J. Am. Water Works Assoc.* **62**:113.

Geldreich, E. E., 1978. *Bacterial Populations and Indicator Concepts in Feces, Sewage, Stormwater and Solid Wastes in Indicators of Viruses in Water and Food*, G. Berg, Ed., Ann Arbor Science Pub., Inc., Ann Arbor, MI, pp. 51–97.

Geldreich, E. E., and B. A. Kenner, 1969. Concepts of Fecal Streptococci in Stream Pollution, *J. Water Poll. Contr. Fed.* **41**(8): R336–R352.

Government of the District of Columbia, 1981. Water Quality Standards for Interstate Waters of the District of Columbia, Dept. of Env. Serv., Env. Health Admin., Bureau of Air and Water Quality, Water Hygiene Division, 16 pp.

Haas, C. N., 1983. Effect of Effluent Disinfection on Risks of Viral Disease Transmission via Recreational Water Exposure, *J. Water Poll. Cont. Fed.* **55**(8):1111–1116.

Hazen and Sawyer, Engineers, 1978. Storm/CSO Laboratory Analysis, NYC 208 Task Report 223, Vol. II. Prepared for NYC Dept. of Environmental Protection, March 10, 1978.

Hejkal, T. W., F. M. Wellings, A. L. Lewis, and P. A. LaRock, 1981. Distribution of Viruses Associated with Particles in Wastewater, *Appl. Environ. Micro.* **41**(3):628–634.

Herrmann, J. E., K. D. Kostenbader, Jr., and D. O. Cliver, 1974. Persistence of Enteroviruses in Lake Water, *Appl. Micro.:* 895–896.

Hydroscience, Inc., 1970. Development of Water Quality Model of Boston Harbor, Interim Report. Comm. of Mass. Water Res. Comm., Boston, Mass., 18 pp. + Figs. + Appendix.

Hydroscience, 1977a. Water Quality Model of Guanabara Bay, Environ. Control Program in State of Rio de Janeiro, Brazil, BRA-73/003, Tech. Report 5, 142 pp. + Appendix.

Hydroscience, 1977b. NYC 208, Task Report, Special Water Quality Studies (PCP Task 317), City of New York, 208 Report, Hazen & Sawyer, Managing Consultants, N.Y., N.Y. 78 pp.

Jakubowski, W., and T. H. Ericksen, 1979. Methods for Detection of *Giardia* Cysts in Water Supplies, W. Jakubowski and J. C. Hoff, Eds., *Waterborne Transmisison of Giardiasis*, Proc. of a Symposium, U. S. EPA, ORD, ERC, Cincinnati, OH, pp. 193–210. EPA-600/9-79-001.

Kabler, P. W., and Clark, H. F., 1960. Coliform Group and Fecal Coliform Organisms as Indicators of Pollution in Drinking Water. *J. Am. Water Works Assoc.* **52**:1577.

Kapuscinski, R. B., and R. Mitchell, 1980. Processes Controlling Virus Inactivation in Coastal Waters, *Water Res.* **14**:363–371.

Kenner, B. A., 1978. Fecal Streptococcal Indicators, G. Berg, Ed., *Indicators of Viruses in Water and Food*, Ann Arbor Science Pub., Ann Arbor, MI, pp. 147–169.

Kittrell, F. W., and S. A. Furfari, 1963. Observations of Coliform Bacteria in Streams, *J. Water Poll. Contr. Fed.* **35**:1361–1385.

LaBelle, R. L., and C. P. Gerba, 1980. Influence of Estuarine Sediment on Virus Survival Under Field Conditions, *Appl. Environ. Micro.* **39**(4):749–755.

LaBelle, R. L., C. P. Gerba, S. M. Goyal, J. L. Melnick, I. Cech, and G. F. Bogdan, 1980. Relationships Between Environmental Factors, Bacterial Indicators, and the Occurrence of Enteric Viruses in Estuarine Sediments, *Appl. Environ. Micro.* **39**(3): 588–596.

Lear, D. W., Jr., and N. A. Jaworski, 1969. Sanitary Bacteriology of the Upper Potomac Estuary, Tech. Report No. 6, CTSL, FWPCA, U.S. Department of the Interior, Washington, D.C., 35 pp.

Levine, N. D., 1979. *Giardia lambia*: Classification, Structure, Identification, W. Sakubowski and J. C. Hoff, Eds., *Waterborne Transmission of Giardiasis*, Proceed. of a Symposium, U.S. EPA, ORD, ERC, Cincinnati, OH, pp. 2–8. EPA-600/9-79-001.

Mancini, J. L., 1978. Numerical Estimates of Coliform Mortality Rates Under Various Conditions, *J. Water Poll. Contr. Fed* **50**(11).

McCabe, L. J., 1978. Chlorine Residual Substitution-Rationale, C. W. Hendricks, Ed., *Evaluation of the Microbiology Standards for Drinking Water*, U.S. EPA, Washington, D.C., Office of Drinking Water, pp. 57–63. EPA-570/9-78-00C.

Middlebrooks, E. J., C. H. Middlebrooks, B. A. Johnson, J. L. Wight, J. H. Reynolds, and A. D. Venosa, 1978. MPN and MF Coliform Concentrations in Lagoon Effluents, *J. Water Poll. Contr. Fed.* **50**(11):2530–2546.

Mitchell, R., and Chamberlin, C., 1978. Survival of Indicator Organisms, G. Berg, Ed., *Indicators of Viruses in Water and Food*, Ann Arbor Sci. Pub., Ann Arbor, MI, pp. 15–37.

National Academy of Sciences, 1977. Drinking Water and Health. Safe Drinking Water Committee, Nat. Res. Council, Washington, D. C., 939 pp.

O'Brien and Gere Eng. Inc., and Limno-Tech., Inc., 1980. Preliminary Review of Potomac and Anacostia River Wet Weather Survey, September 21–28, 1979. Dist. of Columbia, Dept. of Env. Services, Combined Sewer Overflow Study, VI Sections.

O'Connor, D. J., 1968. Water Quality Analysis of the Mohawk River-Barge Canal by Hydroscience, Inc. for NYS Dept. of Health, Div. of Pure Waters, Albany, N.Y., 166 pp. + Appendix.

Roper, M. M., and K. C. Marshall, 1979. Effects of Salinity on Sedimentation and of Particulates on Survival of Bacteria in Estuarine Habitats, *Geomicrobiology Journal* **1**(2):103–116.

Salas, H., 1985. Historia y Applicacion de Normas Microbiologicas de Calidad de Agua en el Medio Marino, En: Hojas de Divulgatión Técnica, CEPIS, OPS/OMS, No. 29, Lima, Peru, 15pp.

Savage, W. G., 1905. Bacteriological Examination of Tidal Mud as an Index of Pollution of the River, *J. Hyg.* **5**:146–174.

Scarpino, P. V., 1978. Bacteriophage Indicators, G. Berg, Ed., *Indicators of Viruses In Water and Food*, Ann Arbor Science Pub. Inc., Ann Arbor, MI, pp. 201–227.

Smith, R. J., and R. M. Twedt, 1971. Natural Relationships of Indicator and Pathogenic Bacteria in Stream Waters, *J. Water Poll. Contr. Fed.* **43**:11.

Thomann, R. V., 1981. Review of Potomac Estuary Bacterial Issues, by HydroQual, Inc. for Government of the District of Columbia, DES, 26 pp.

USEPA, 1974. Analysis and Control of Thermal Pollution, Water Program Operations, p. 8–4. EPA-430/1-74-010.

USEPA, 1976. Areawide Assessment Procedures Manual, Vol. I, MERL, ORD, 6 Chapters. EPA-600/9-76/014.

USEPA, 1978. Human Viruses in the Aquatic Environment: A Status Report with Emphasis on the EPA Research Program, Report to Congress. Office of Drinking Water, Washington, D. C., 37 pp. EPA-570/9-78/006.

Weibel, S. R., R. J. Anderson, and R. L., Woodward, 1964. Urban Land Runoff as a Factor in Stream Pollution, *J. Water Poll. Contr. Fed.* **36**(7):914–924.

World Health Organization, 1984a. Guidelines for Drinking Water Quality, Vol. 1, Recommendations, Geneva, Switzerland, 130pp.

World Health Organization, 1984b. Guidelines for Drinking Water Quality, Vol. 2, Health Criteria and other Supporting Information, Geneva, Switzerland, 335pp.

Yanez, F. A., 1983. Indicators and Pathogenic Organisms Die Away in Ponds Under Tropical Conditions, paper Presented at 56th Annual WPCF Conference, Atlanta, GA, 23 pp.

PROBLEMS

5.1. For a June–September period of 122 days, the following statistics were calculated:

	Mean	Coefficient of variation
Storm intensity	0.055 in./hr	1.55
Duration	3 hr	1.15
Time between storms	80 hr	1.15

Coefficient of variation of TC concentration = 1.0. Use Table 1.3 TC concentrations (urban runoff and CSO).

The drainage area has 1875 acres (1310 separate acres and 565 combined sewer acres). The runoff coefficient is 0.35 for both subareas. Calculate for the *entire area*:
(a) Mean runoff (cfs) during the summer.
(b) Mean loading rate (org/day) of TC bacteria during summer storms.
(c) Long-term summer loading rate (org/day) of TC bacteria including the nonstorm periods.
(d) Average number of storms per summer.
(e) The percent frequency of storms equal to or less than the mean loading rate.
(f) The number of storms per summer that would have a load greater than five times the mean loading rate.

5.2. A river receives an input of FC bacteria as shown.

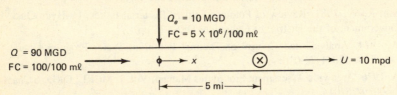

Five miles downstream from the outfall a concentration of 185,000/100 ml is measured.
(a) Determine the decay rate in the stream (day^{-1}).
(b) Determine the effluent percent reduction so that the FC bacterial count at $x = 5$ mi is reduced by a factor of 1000.

5.3. Repeat Sample Problem 5.2 for the lower flow of 100 cfs and a water temperature of 25°C.

5.4. Two equal point sources discharge TC bacteria to the estuary.

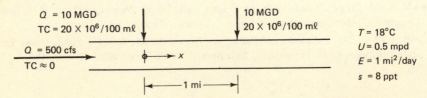

(a) Determine the required percent removal in order that TC $\leq$ 2000/100 ml. Assume equal removals at both plants; neglect wasterwater flows. Hint: Estimate percent removals based on $N(x = 0)$ or $N(x = 1)$.
(b) Confirm (a) by plotting N vs. x for the range of $-3 \leq x \leq 3$ mi.

5.5. If the river in Sample Problem 5.2 is replaced by a tidal river with a dispersion coefficient $E = 1.0$ smpd, can the New York State standard be met at the beach with a PS reduction alone? If so, what reduction is required? Use $Q = 100$ cfs and $T = 25°C$.

5.6. A cove of a large lake receives distributed runoff for 6 hr. The cove volume is 3.6×10^6 ft^3.

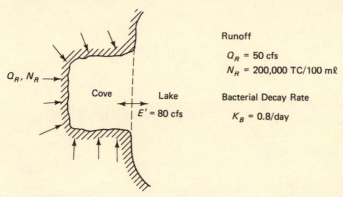

Runoff

$Q_R = 50$ cfs

$N_R = 200,000$ TC/100 mℓ

Bacterial Decay Rate

$K_B = 0.8$/day

(a) Estimate the maximum TC count in the cove.

(b) Calculate the times, after storm, when the cove will have TC counts of 1000, 100, and 10/100 ml.

Chapter 6

Dissolved Oxygen

6.1 INTRODUCTION

The problem of dissolved oxygen (DO) in surface waters has been recognized for over a century. The impact of low DO concentrations or of anaerobic conditions was reflected in an unbalanced ecosystem, fish mortality, odors, and other aesthetic nuisances. While coliform bacteria was the surrogate variable for communicable disease and public health (as discussed in Chapter 5), DO is a surrogate variable for the general health of the aquatic ecosystem. This chapter reviews the basic principles for modeling DO, a continuing important water quality variable.

Early work from circa 1870–1900 on DO and the development of the bio-chemical oxygen demand (BOD) test in the United Kingdom is summarized by Theriault (1927). Stream DO studies and laboratory experiments to determine oxygen demand were well underway by the late nineteenth century. Theriault (1927, p. 2) cites DO investigations in 1870 on the Thames River where it was determined that if ''organic matter be perfectly excluded from the air in a carefully stoppered bottle, the gradual diminution in the amount of dissolved oxygen indicates exactly the progress in the oxidation of the organic matter.'' Further, because of the resulting slow oxidation, ''there is no river in the United Kingdom long enough to effect the destruction of sewerage by oxidation,'' By the turn of the century, following the research reported by the Royal Commission on Sewage Disposal, the science of DO and BOD measurement and interpretation had progressed rapidly.

In the United States, a major effort in water quality evaluations began in 1912 when the U.S. Public Health Service was directed by the Congress to study the ''sanitation and sewage, including the pollution, either directly or indirectly, of navigable streams and lakes of the United States'' (quoted in Crohurst, 1933). Studies on the Ohio River were conducted in 1914–1916 and formed the basis for a seminal work on the mathematical modeling of DO in the work of Streeter and Phelps (1925) (see also Phelps, 1944). That work included the application of some simple mathematical formulations of the major processes associated with a DO balance in a river. As such, this pioneering effort placed the observational past on a more scientific basis including the development of a predictive framework.

Subsequent work included the significant input of Velz (1938, 1939, 1947, 1984), and O'Connor (1967), who continued to develop the mathematical and biochemical bases for DO analyses in streams and most importantly in estuarine systems such as New York Harbor. The massive and detailed effort on the Thames Estuary in the United Kingdom (Department of Scientific & Industrial Research, 1964) continued the development of estuarine DO analyses as did O'Connor (1960, 1962, 1965, 1966), Thomann (1963), and O'Connor and Mueller (1984), among others.

The discharge of municipal and industrial waste, and urban and other non-point source runoff will necessitate a continuing effort in understanding the DO resources of surface waters. The DO problem can thus be summarized as: the discharge of organic and inorganic oxidizable residues into a body of water, which during the processes of ultimate stabilization of the oxidizable material (in the water or sediments), and through interaction of aquatic plant life, results in the decrease of DO to concentrations that interfere with desirable water uses.

6.2 PRINCIPAL COMPONENTS OF DO ANALYSIS

Figure 6.1 (which is a specific representation of the more general problem framework shown in Fig. 1.1) shows the major components of the DO problem. The principal inputs include the BOD of municipal and industrial discharges, the oxidizable nitrogen forms, and nutrients which may stimulate phytoplankton or rooted aquatic plant growth. The nature of the aquatic ecosystem then determines the DO level through such processes as reaeration, photosynthesis, or sediment oxygen demand. For a given desirable water use, or set of uses, a determination of the DO level consistent with that use must be made and compared to the observed or predicted DO level.

The principal modes of engineering control that may have to be employed include primary and secondary (usually biological) waste treatment to reduce BOD and oxidizable nitrogen levels. For the more complex eutrophication–DO problems, nutrient removal may also be required. Each of these components can now be considered, beginning with the DO water quality criteria.

6.2.1 DO Criteria and Standards

The relationships of the level of DO to specific uses has been a continued subject of debate. The principal use affected is fish preservation, including survival and reproduction. This also affects recreational and commercial fishing. Much research has been conducted relating specific levels of DO to fish behavior (see, for example, Doudoroff and Shumway, 1970; Warren, 1971; Warren et al., 1973; Alabaster and Lloyd, 1980). A complete summary is given in USEPA (1985).

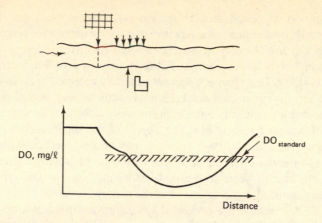

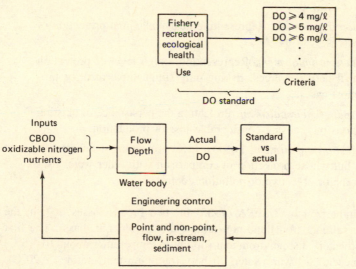

Figure 6.1 Major components of the DO problem.

Warren et al. (1973) summarized the relationship between DO and responses such as survival, development, growth, and swimming performance of chinook and coho salmon, steelhead trout, and largemouth bass. Their results indicated that juveniles survived for "prolonged periods of time at dissolved oxygen concentrations as low as 2 mg/l and less, except at relatively high temperatures." Other conclusions included the result that juvenile largemouth bass experienced some reduction in maximum swimming speeds when DO was reduced to below 5 or 6 mg/l at 25°C. Embryos of the salmonids were impaired at DO levels below 3 mg/l. Warren et al. (1973) recommended that if overall DO criteria were to be applied, then the proposal of Doudoroff and Shumway (1970) should be used. In that work, a relationship was suggested between the natural seasonal minimum DO and the acceptable seasonal minimum as a function of varying degrees of protection.

The approach of Doudoroff and Shumway (1970) was adopted by the National Academy of Sciences (NAS), National Academy of Engineering (NAS, 1973, pp. 131–135). That approach was not generally accepted because of apparent difficulty in identifying natural background conditions and instead the USEPA (1976) suggested a single minimum concentration of 5 mg/l at any time would protect the diversity of aquatic life. The appeal of the single universal minimum is its simplicity. The disadvantage is that it may be cost-inefficient and does not reflect the assignment of varying water uses which could tolerate lower levels of DO. In the ensuing 10 years to the mid-1980s, the issue of DO criteria in the United States continued to be discussed at the national level.

The USEPA, based on a review of laboratory and field data on the impact of varying levels of DO in freshwater ecosystems (USEPA, 1985) found the earlier assignment of risk to the fishery to be valid and useful to decision making. The four levels of protection or risk are:

> *No production impairment.* Representing nearly maximal protection of fishery resources.
>
> *Slight production impairment.* Representing a high level of protection of important fishery resources, risking only slight impairment of production in most cases.
>
> *Moderate production impairment.* Protecting the persistence of existing fish populations but causing considerable loss of production.
>
> *Severe production impairment.* For low level protection of fisheries of some value but whose protection in comparison with other water uses cannot be a major objective of pollution control.

Impacts were summarized by USEPA (1985) for two broad classes of fish; the salmonidae (e.g., salmon, trout) and non-salmonid (e.g. bluegill, bass). The first represents a "coldwater" fish environment (optimum temperature about 10–15°C) and the latter represents a "warmwater" fishery (optimum temperature > 20°C). The DO concentrations (mg/l) judged by the USEPA in the proposed criteria (USEPA, 1985) to be associated for the four risk levels and two classes are

  1. Salmonid waters.
     a. Embryo and larval stages.
        No production impairment   = 11(8)[1]
        Slight production impairment   = 9(6)[1]
        Moderate production impairment = 8(5)[1]
        Severe production impairment   = 7(4)[1]
        Acute mortality limit   = 6(3)[1]

---

[1]Note: These are water column concentrations recommended to achieve the required *intergravel* dissolved oxygen concentrations shown in parentheses. The 3 mg/l difference is discussed in USEPA (1985).

  b. Other life stages.

  No production impairment   = 8

  Slight production impairment   = 6

  Moderate production impairment = 5

  Severe production impairment   = 4

  Acute mortality limit   = 3

2. Non-salmonid waters.

  a. Early life stages.

  No production impairment   = 6.5

  Slight production impairment   = 5.5

  Moderate production impairment = 5

  Severe production impairment   = 4.5

  Acute mortality limit   = 4

  b. Other life stages.

  No production impairment   = 6

  Slight production impairment   = 5

  Modernate production impairment = 4

  Severe production impairment   = 3.5

  Acute mortality limit   = 3

Based on this assessment, the USEPA (1985) suggested national DO criteria at 0.5 mg/l above the "slight production impairment" values. Table 6.1 shows the suggested criteria. Accompanying text discusses the importance of diurnal fluctuations, releases for reservoirs, and recognizes that specific situations may result in minima lower than suggested in Table 6.1. The average however should be met and the issues of the duration and magnitude of the minima must be addressed. National criteria have not been suggested for marine waters.

DO criteria, which are based on scientific investigations and judgment, must

TABLE 6.1 WATER QUALITY CRITERIA FOR AMBIENT DISSOLVED
OXYGEN CONCENTRATION, FRESHWATER

	Cold water criteria		Warm water criteria	
	Early life stages[a][b]	Other life stages	Early life stages[b]	Other life stages
30-day mean	NA[c]	6.5	NA	5.5
7-day mean	9.5(6.5)	NA	6.0	NA
7-day mean minimum	NA	5.0	NA	4.0
1-day minimum[d]	8.0(5.0)	3.0	5.0	3.0

[a]These are water column concentrations recommended to achieve the required *intergravel* DO concentrations shown in parentheses. The 3 mg/l differential is discussed in the criteria document. For species that have early life stages exposed directly to the water column, the figures in parentheses apply.

[b]Includes all embryonic and larval stages and all juvenile forms to 30 days following hatching.

[c]NA (not applicable).

[d]For reservoir or other manipulatable discharges, further restrictions apply.

Source: From USEPA (1985).

be translated into standards that are enforceable as legal objectives. Such assignment of water quality standards in the United States is by the states and other local agencies. Table 6.2 shows a sampling of such standards for states in the United States. (These standards predate the draft national DO criteria suggested by USEPA (1985) and shown in Table 6.1). As noted, the range in the minimum and daily average allowable DO is wide and reflects the application of varying levels of DO to different water uses. The exceptions and frequency of occurrence requirements can also be noted. The standards generally allow for a "mixing zone" within which the level of DO may be less than required. The mixing zone specifications vary widely from state to state. Notably absent from many of the standards is the specification of the depth at which the DO standard is to apply which is important in evaluating deep lakes, reservoirs, and vertically stratified estuaries.

6.3 SOURCES AND SINKS OF DO— KINETIC RELATIONSHIPS

This section summarizes the principal components of the DO problem. The kinetic interactions are discussed without reference to a specific body of water. The purpose of this section is therefore to aid the understanding of the mechanisms of the various sources and sinks of DO. The application of these mechanisms to each water body type, river, estuary, and lake then follows in subsequent sections of this chapter.

The DO problem begins, of course, with the input of oxygen demanding wastes into a water body. In the water body itself, the sources of DO are:

1. Reaeration from the atmosphere.
2. Photosynthetic oxygen production.
3. DO in incoming tributaries or effluents.

Internal sinks of DO are:

1. Oxidation of carbonaceous waste material.
2. Oxidation of nitrogenous waste material.
3. Oxygen demand of sediments of water body.
4. Use of oxygen for respiration by aquatic plants.

With the above inputs and sources and sinks, the following general mass balance equation for DO (designated by c) in a segment volume V, can be written

$$V \frac{dc}{dt} = \text{reaeration} + (\text{photosynthesis} - \text{respiration})$$
$$- \text{oxidation of CBOD, NBOD (from inputs)} \qquad (6.1)$$
$$- \text{sediment oxygen demand} + \text{oxygen inputs}$$
$$\pm \text{oxygen transport (into and out of segment)}$$

This equation is applied to a specific water body where the transport and sources and sinks are unique to that aquatic system. These source–sink components will now be examined in order to develop further the mass balance equation.

TABLE 6.2 SOME U.S. STATE STANDARDS FOR DISSOLVED OXYGEN

State	Surface water use classification				Comments	Reference[a]
	Highest		Lowest			
	Daily average (mg/l)	Minimum (mg/l)	Daily average (mg/l)	Minimum (mg/l)		
Alabama		5		2	Applies @ 5 ft for depth >10 ft and mid for 10 ft	1
Georgia	6	5		3		2
Iowa	7*	5	5*	4	*Not less than value for at least 16/24 hr	3
New Mexico		6		4		4
Ohio		6		3		5
Pennsylvania		7	3.5*		*During 4/1–6/15 and 9/16–12/31, not <6.5 mg/l as seasonal average	6
Texas		6		2	Diurnal decrease of 1 mg/l below standard allowed for no more than 8/24 hr	7
Washington		9.5		4	Natural levels can be decreased by 0.2 mg/l	8

[a]References:

(1) Alabama (1981).
(2) Georgia (1980).
(3) Iowa (1983).
(4) New Mexico (1981).
(5) Ohio (undated).
(6) Pennsylvania (1979).
(7) Texas (1981).
(8) Washington (1982).

6.3.1 Oxygen Demanding Waste Inputs

The principal inputs affecting the DO are municipal and industrial discharges of wastes, combined sewer overflows, and separate sewer discharges. Such discharges include material that exerts a chemcial oxygen demand (COD), carbonaceous biochemical oxygen demand (CBOD), and oxidizable nitrogen which is also represented by the nitrogenous biochemical oxygen demand (NBOD). Tables 6.3 and 6.4 show some data on the levels of these inputs from several sources (Thomann, 1972; Mueller and Anderson, 1983; Castagnino, 1978). Note that although the concentration of municipal waste is higher in a Latin American country such as Uruguay than in the United States, the per capita input of CBOD5 is less due to a reduced per capita flow.

6.3.1.1 Carbonaceous Biochemical Oxygen Demand   A distinction must be made between the oxygen demand of the carbonaceous material in waste effluents and the nitrogenous oxygen demanding component of the effluent. Figure 6.2(a) shows a typical oxygen demand curve of untreated waste that contains nitrogenous material. The carbonaceous demand is usually exerted first, normally as a result of a lag in the growth of the nitrifying bacteria necessary for oxidation of the nitrogen forms. The CBOD is exerted by the presence of heterotrophic organisms that are capable of deriving the energy for oxidation from an organic carbon substrate. Municipal sewage and most rivers, estuaries, and lakes contain large numbers of these heterotrophic organisms. Except for cases where toxic chemicals are present in the water or river, the CBOD is exerted almost immediately. Figure 6.2(b) shows the CBOD and NBOD for an effluent that has received secondary treatment without nitrification. Note the relative significance of the NBOD which has not

TABLE 6.3   INPUTS OF OXYGEN DEMANDING WASTE MATERIAL
BEFORE TREATMENT

Source	CBOD5 (mg/l)	CBODU (mg/l)	NBOD (mg/l)	Organic N (mg/l)	NH$_3$–N (mg/l)
Municipal waste					
U.S. average	180 (100–450)[a]	220 (120–580)	220	20 (5–35)	28 (10–60)
U.S. Rural	220			20	5
Mexico[b]	275			18	20
Uruguay[b]	(250–300)			(10–25)	(12–28)
Argentina[b]	408				
Combined sewer overflow	170 (40–503)	220		5.9 (0.08–25)	2.8 (0–11.5)
Separate urban runoff	19 (2–84)			1.7 (0.2–4.8)	0.6 (0.1–1.9)

[a]Numbers in parentheses indicate approximate range.
[b]Castagnino (1978).

TABLE 6.4 MASS INPUTS OF OXYGEN DEMANDING MATERIAL BEFORE TREATMENT

Source	Population (1000)	Flow (m³/c · day)	Per capita inputs (g/c · day)				
			CBOD5	CBODU	NBOD	Organic N	NH₃-N
U.S. municipal waste							
Average	—	0.55	100 (0.22)[a]	180 (0.27)	180 (0.27)	11	15
Variation with population[b]	Rural	0.27	60			5.3	1.3
	20	0.48	72				
	20–50	0.45	82				
	50–150	0.57	86				
	150	0.49	118				
Cities[c]							
New York	7000	0.61	80			6.3	6.4
N. New Jersey	2800	0.62	226			—	16.7
Los Angeles	3000	0.43	109			6.7	8.3
Uruguay[d]	—	0.19	50				
Japan[e]			44				12 (TN)

[a]lb/c-day @ 145 gcd (0.55 m³/c-day)
[b]From Mueller and Anderson (1983) citing Otis et al. (1977) and Loehr (1968).
[c]Mueller and Anderson (1983).
[d]Castagnino (1978).
[e]Kameda (1980).

269

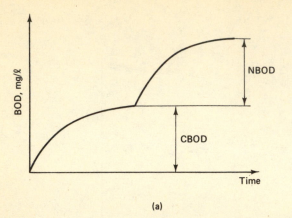

(a)

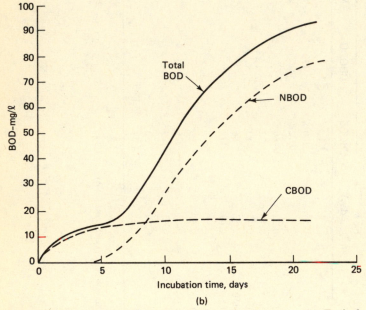

(b)

Figure 6.2 Carbonaceous and nitrogenous BOD curves. (a) Typical two-stage uptake. (b) Westgate Treatment Plant, Potomac Estuary, Va. Secondary treatment, Sept. 1978. From Slayton and Trovato (1979).

been removed by the treatment process. The ultimate BOD at day 20 is therefore due primarily to the nitrogenous oxygen demand.

The CBOD of a waste or river sample is obtained by standard procedures of incubation over a period of time to describe the full course of oxygen utilization as shown in Fig. 6.2(a) (APHA, 1985). The oxidation of organic matter in the BOD bottle involves a complex set of sequential reactions (Gaudy and Gaudy, 1980). A portion of the organic matter oxidized by microbes is used to create new microbes. The organic matter is sequentially oxidized where at each step part of the organic matter is oxidized to CO_2 and H_2O and part is used to create new microbial matter. As a result, the progression of the BOD is almost never a simple mathematical relationship but varies with the seed used as the inoculum in the BOD test. A simplification of the BOD kinetics assumes first-order kinetics and is discussed next. It should be understood, however, that the development is to some extent semiempirical but, as such, has proved to be a useful approximation in practice.

If L = the oxidizable carbonaceous material remaining to be oxidized in the BOD bottle and if first-order kinetics are assumed, then

$$\frac{dL}{dt} = -K_1 L \tag{6.2}$$

where K_1 is the rate of oxidation of the carbonaceous material and t is the incubation time (Fig 6.3). The solution to this equation is

$$L = L_0 \exp(-K_1 t) \tag{6.3}$$

where L_0 is the initial amount of carbonaceous material present in the bottle at the beginning of the test.

If now the oxygen consumed in the stabilization of the organic material is

$$y = L_0 - L$$

then

$$y = L_0[1 - \exp(-K_1 t)] \tag{6.4}$$

where y is the biochemical oxygen demand and L_0 is now seen as the ultimate amount of CBOD or equivalently the amount of oxidizable organic substrate that is available. The organic waste concentration is then given by L while its oxygen uptake is given by y; both are measured in the BOD as oxygen equivalents.

The 5-day CBOD is

$$y_5 = \text{CBOD5} = L_0[1 - \exp(-5K_1)] \tag{6.5}$$

Figure 6.3 Simplified representation of progression of CBOD—first-order kinetics.

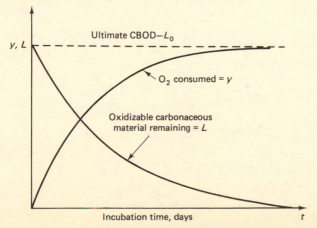

Ultimate CBOD—L_0

y, L

O_2 consumed = y

Oxidizable carbonaceous material remaining = L

Incubation time, days t

and represents the amount of oxygen utilized over a 5-day period, a convenient measurement period. The principal reason for the determination of the long-term BOD time history as given by Eq. 6.5 is to provide a basis for converting 5-day CBOD reported values to ultimate CBOD values.

The ratio, f, of the ultimate BOD to the 5-day BOD can be seen from Eq. 6.5 to be a function only of the laboratory rate of oxidation, K_1, that is,

$$f = \frac{L_0}{y_5} = [1 - \exp(-5K_1)]^{-1} \tag{6.6}$$

As seen in Fig. 6.4 however, care must be taken in distinguishing between the BOD, CBOD, and NBOD. If a BOD test is done without such distinction, then confounded values of f may be obtained. Equation 6.6 is for f defined as the ratio of CBODU to CBOD5.

For municipal waste discharges the BOD rate, K_1, sometimes called a "bottle rate," would be expected to depend on the degree of treatment, that is, as the degree of treatment increases the residual material is increasingly more resistant to treatment; and, hence, the rate decreases. Thus, for an activated sludge treatment plant the BOD rate would decrease, indicating that the residual material after treatment is more refractory, that is, less susceptible to biochemical oxidation. However, HydroQual (1983) indicated no significant trend in the ratio. Their average and standard deviation of f are shown in Table 6.5 together with the ratio of CBODU to BOD5.

Since the BOD reaction is mediated by bacteria, as well as other microorganisms such as protozoa and rotifers, the rate at which the bacteria oxidize the organic material is a function of the temperature of the waste/water mixture. Increased temperature increases the rate of oxidation so that the ultimate BOD is reached sooner than at a lower temperature. In the form shown in Eq. 6.5, the ultimate BOD is independent of the temperature.

The temperature effect is given approximately by

$$(K_1)_T = (K_1)_{20}(1.04)^{T-20} \tag{6.7}$$

where T is in °C, $(K_1)_{20}$ is the BOD rate at 20°C, and $(K_1)_T$ is the rate at T°C.

Figure 6.4   Illustration of relationships between CBOD, NBOD, and total BOD.

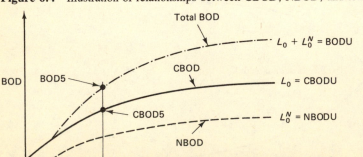

TABLE 6.5 APPROXIMATE CBOD REACTION RATES, K_1, MUNICIPAL WASTE

Treatment level	K_1 (day^{-1}) @ 20°C		$\dfrac{\text{CBODU}}{\text{CBOD5}}$	$\dfrac{\text{CBODU}}{\text{BOD5}}$
	Approximate range	Approximate average		
None	0.3–0.4	0.35	1.2	—
Primary/secondary	0.1–0.3	0.20	1.6	—
Activated sludge	0.05–0.1	0.075	3.2	—
Primary—advanced[a] (35–40 plants; range of effluent BOD = 1–100 mg/l)	—	0.087	2.84 (± 1.17)[b]	2.47 (± 1.52)[b]

[a]From HydroQual (1983).
[b]Mean (± 1 standard deviation).

It should be noted that the "bottle rate," K_1, in general is not equal to the deoxygenation rate for the BOD that occurs in natural waters, although for deep bodies of water, K_1 is a useful first approximation to the deoxygenation rate. This is discussed more completely below in Section 6.3.5.

Another estimate of the ultimate CBOD can be obtained by recognizing that complete stabilization of the organic carbon in an effluent or water body would require about 2.7 mg of oxygen for every mg of carbon that is oxidized. Thus,

$$L_0 \approx 2.7 C_{org} \tag{6.8}$$

where C_{org} = organic carbon (mg/l).

6.3.1.2 Nitrogenous Biochemical Oxygen Demand

The first-stage CBOD is often followed by the second stage representing the oxidation of the nitrogenous compounds in the waste or water body. Some industrial wastes, for example, paper mill wastes, are deficient in any nitrogen forms and therefore do not exhibit the second stage. Nitrogenous matter in waste consists of proteins, urea, ammonia, and, in some cases, nitrate. The intermediate decomposition products of the proteins such as amino acids, amides, and amines are also present in varying degrees. The proteins are broken down by hydrolysis in a series of steps into a variety of amino acids. Both exocellular and endocellular enzymes are involved in the process. The decomposition of the amino acids which can occur in a number of different ways, is endocellular. Ammonia is released in this process of deamination.

The ammonia, which is highly soluble, combines with the hydrogen ion to form the ammonium ion, thus tending to raise the pH. In the neutral pH range, all of the ammonia is present in the ammonia form and at a higher pH is evolved as a gas. The heterotrophic decompostion is the typical reaction in the first stage of the biochemical deoxygenation of natural waters. Ammonia is an end product of both this reaction and the reaction associated with the hydrolytic breakdown of proteins. The ammonia present in natural waters is thus a result of either the direct discharge of the material in wastewaters or of the decomposition of organic matter in various forms.

The ammonia, in turn, is oxidized under aerobic conditions to nitrite by bacteria of the genus *Nitrosomonas* as follows (Hutchinson, 1957; Stratton and McCarty, 1967; Delwiche, 1956; Gaudy and Gaudy, 1980)

$$NH_4 + 1.5O_2 \longrightarrow 2H^+ + H_2O + NO_2^- \tag{6.9}$$

This reaction requires 3.43 g of oxygen utilization for 1 g of nitrogen oxidized to nitrite. The nitrite thus formed is subsequently oxidized to nitrate by bacteria of the genus *Nitrobacter* as follows (Gaudy and Gaudy, 1980):

$$NO_2^- + 0.5O_2 \longrightarrow NO_3^- \tag{6.10}$$

This reaction requires 1.14 g of oxygen utilization for 1 g of nitrite nitrogen oxidized to nitrate. The total oxygen utilization in the entire forward nitrification process is therefore 4.57 g of oxygen per g of ammonia nitrogen oxidized to nitrate. The *Nitrobacter* bacteria use about three times as much substrate as the *Nitrosomonas* bacteria to derive the same amount of energy. Actually, some of the ammonium may be used in cell production so that the oxygen utilization may be less than 4.57 and approach 4.2 g O_2/gNH_3 oxidized (Gaudy and Gaudy, 1980).

Essential factors for nitrification are oxygen, phosphates, and an alkaline environment to neutralize the resulting acids. Nitrifying bacteria are very susceptible to the action of toxic substances. These bacteria are obligate autotrophs, which derive their carbon for cell sythesis from CO_2.

The energy obtained from these reactions is utilized for the assimilation of carbon from either carbon dioxide or bicarbonate, but generally not from organic carbon. The alkaline environment is required to neutralize the acidic end products. Below a pH of 6.0, inhibition occurs.

Given the appropriate conditions, the concentration of the organisms is a significant factor controlling the rate of nitrification, with the concentration of the reactant having a reduced effect. However, if there is an ample supply of organisms, the rate appears to be controlled by the concentration of the reactant. The number of nitrifiers is, of course, determined by the generation time of organisms, which is in the order of one day, by contrast to an order of a few hours for many heterotrophic bacteria.

The common source of the nitrifying organisms is rich soil, where they are usually found in high numbers. In rivers receiving wastewaters, nitrifying bacteria are present in varying degrees depending on the nature and the treatment of wastewaters. Their number in sewage is low (about 100/ml) but increases through biological treatment to approximately 1000/ml. A natural habitat is on biological aggregates or on surfaces where the environment is appropriate for growth. Surfaces such as these are found in trickling filters and activated sludge treatment plants as well as the rocky beds of shallow rivers.

In summary, the nitrogenous BOD (NBOD) results from the oxidation of the ammonia to nitrite and then to nitrate nitrogen when conditions are appropriate. The NBOD is often measured by adding a nitrification suppressant to the BOD bottle and therefore measuring only the CBOD. If the BOD is also measured on an unsuppressed sample, NBOD can be obtained by difference. The suppressant 2-chloro-6-(trichloromethyl) pyridine (TCMP) is recommended (APHA, 1985) and apparently acts as a nitrification suppressant throughout a long-term test without any degradation. Data have indicated that the progression of the NBOD, designated L^N, is approximately first-order given by

$$L^N = L_0^N[1 - \exp(-K_N t)] \tag{6.11}$$

where K_N is the overall NBOD reaction rate. Since organic nitrogen can also be hydrolyzed to ammonia, the NBOD if all reactions go to completion can also be estimated from

$$L_0^N = 4.57(N_0 + N_a) \tag{6.12}$$

where N_0 and N_a are the organic and ammonia nitrogen concentrations, respectively. This estimate is then an upper bound of the NBOD. Tables 6.3 and 6.4 show some typical values of the NBOD for wastes before any treatment.

Figure 6.4 shows the relationships between the CBOD, NBOD, and the total BOD. As seen, if total BOD is measured at 5 days, the BOD5 in such a measurement may include some component of the NBOD. Determinations of the efficiency of removal of CBOD may therefore be confounded (Hall and Foxen, 1983). It is generally suggested that measurements of CBOD and NBOD be obtained separately (as by use of NBOD suppressments) in order to properly assess the individual contributions to the total BOD.

6.3.2 Atmospheric Reaeration

In order to understand the basic mechanism of the transfer of oxygen to any water body from the atmosphere, consider a simple two-phase system shown in Fig. 6.5(a). The vessel of water is exposed to the atmosphere. If the system is allowed to come to equilibrium with the atmosphere over the vessel, a fixed level of oxygen for a given temperature will be reached. This is the oxygen saturation level and is given by Henry's law; that is, "the weight of any gas that dissolves in a given volume of a liquid, at a constant temperature, is directly proportional to the pressure that the gas exerts above the liquid." Oxygen dissolved in water behaves according to Henry's law. Therefore,

$$p = H_e c_s \tag{6.13}$$

where p = partial pressure of O_2 in mm Hg

c_s = saturation conc. of DO in liquid in mg/l

H_e = Henry's constant in mm Hg/mg/l

Henry's constant is the ratio of the partial pressure of the gaseous phase to the solubility of DO in the water phase. A dimensionless form is

$$H_e = 16\frac{M}{T}\frac{p(\text{mmHg})}{c_s(\text{mg/l})} \tag{6.14}$$

where T is temperature in °K and M is the molecular weight in g/g-mole and H_e has units mg/l in the gas phase per mg/l in the water phase. Thus, for oxygen with a molecular weight of 32, a partial pressure of 158 mm Hg, and a saturation

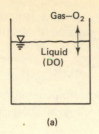

(a)

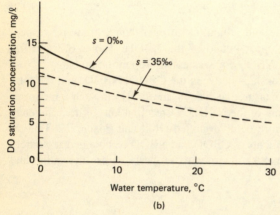

(b)

Figure 6.5 (a) A simple liquid–gas system. (b) DO saturation concentration as a function of temperature and salinity.

value (c_s) of 14.2 mg/l, the value for the Henry's constant is 21. (See also Chapter 8, Eqs. 8.80 to 8.82.)

The saturation value, c_s, of DO in equilibrium with the atmosphere depends on:

1. Temperature.
2. Salinity.
3. Pressure.

An expression for the DO saturation concentration at sea level as a function of temperature (zero salinity) is given in APHA (1985) as

$$\ln c_{sf} = -139.34411 + \frac{1.575701 \times 10^5}{T}$$

$$- \frac{6.642308 \times 10^7}{T^2} + \frac{1.243800 \times 10^{10}}{T^3}$$

$$- \frac{8.621949 \times 10^{11}}{T^4} \qquad (6.15)$$

where c_{sf} = freshwater DO saturation concentration in mg/l at 1 atm
      ln = natural logarithm
      T = temperature in °K

and

$$T(°K) = T(°C) + 273.150 \tag{6.15a}$$

The effect of salinity or chlorides is to reduce the saturation value. APHA (1985) incorporates the effect of salinity, as follows:

$$\ln c_{ss} = \ln c_{sf} - S\left(1.7674 \times 10^{-2} - \frac{1.0754 \times 10^1}{T} + \frac{2.1407 \times 10^3}{T^2}\right) \tag{6.16}$$

where c_{ss} = saline water DO saturation concentration in mg/l
S = salinity in ppt

and

$$S = 1.80655 \times \text{chlorinity} \tag{6.16a}$$

Fig. 6.5b shows the DO as a function of temperature and salinity. Table A.1 in the Appendix contains saturation concentrations calculated from the APHA formulae for a range of temperatures and chlorinities.

For some problems, the effect of pressure is important. Examples include DO analyses in high elevation streams or lakes. The effect of barometric pressure is given by (APHA, 1985)

$$c_{sp} = c_{s0}P\left[\frac{[1 - (P_{wv}/P)](1 - \theta P)}{(1 - P_{wv})(1 - \theta)}\right] \tag{6.17}$$

where c_{sp} = DO saturation (mg/l) at pressure P
c_{s0} = DO saturation at sea level
P = nonstandard pressure in atm (see Eq. 6.17a)
P_{wv} = partial pressure of water vapor, atm, calculated from:
$\ln P_{wv} = 11.8571 - (3840.70/T) = (216961/T^2)$
T = temperature in °K
$\theta = 0.000975 - (1.426 \times 10^{-5}t) + (6.436 \times 10^{-8} t^2)$
t = temperature in °C

The barometric pressure at a given altitude, P in atmospheres, can be estimated from (Trewartha, 1954).

$$P = P_0 - (0.02667)(\Delta H)]/760$$
$$P_0 = \text{barometric pressure at station 0 in mm Hg} \tag{6.17a}$$
$$\Delta H = \text{difference in elevation between stations 0 and } H \text{ in ft}$$

The change in saturation with elevation may also be approximated as

$$\text{Percent of } c_s \text{ @ sea level} = 100 - 0.0035H \tag{6.18}$$

for H (in ft) above sea level (Zison et al., 1978).

For Denver, Colorado at an elevation of 5000 ft, the DO saturation is 82.5% that of the saturation at sea level and for Bogota, Columbia at an elevation of 9000 ft, the DO saturation is only 68% of that at sea level.

Now suppose it is assumed that in the vessel of Fig. 6.5(a), the DO is decreased very rapidly to a value significantly below c_s. This may represent, for example, an instantaneous discharge of a substance that rapidly utilizes the oxygen in the water. The driving force for restoration of the equilibrium DO condition is then given by the difference of the DO in the vessel and the maximum amount that can exist in the vessel, the saturation value. The rate of change of DO with respect to time to the first approximation, can be considered to be proportional to this difference. Since the atmosphere is the reservoir of oxygen, an exchange of oxygen will take place across the interfacial area of the surface of the water.

The derivation of this exchange makes use of the "two-film" theory where a gaseous film is assumed at the atmosphere side of the air–water interface and a liquid film is assumed on the water side of the interface. For oxygen to transfer from the atmosphere to the water, a parcel of water must first travel from the bulk liquid to the interface. Oxygen can then diffuse to the water parcel, through the gaseous film and through the liquid film. Resistances are encountered in each film. The relatively high Henry's constant for oxygen indicates a high "partitioning" of oxygen into the gaseous phase relative to the liquid phase. Therefore, the gas film has relatively low resistance compared to the liquid film and the liquid film resistance is said to control the exchange. (A more general expression for both liquid and gas film control is given in Chapter 8, Eq. 8.79. Oxygen transfer is simply one example of the general exchange of chemicals across the air–water interface.) The flux of oxygen through the controlling liquid film then equals the time rate of change of DO in the vessel assuming complete mixing:

$$V\frac{dc}{dt} = K_L A(c_s - c)$$

$$[L^3]M/(L^3 \cdot T) = [L/T][L^2][M/L^3]$$

$$[M/T] = [M/T] \tag{6.19}$$

where c is the DO in the water $[M/L^3]$; V is the volume of the vessel $[L^3]$; A is the surface area between the water and the atmosphere $[L^2]$; and K_L is the DO interfacial transfer coefficient $[L/T; (m/day)]$.

Note that this equation is analogous to the heat transfer across the interface of a water body as given in Chapter 9. The equation, therefore, indicates that the mass transfer of oxygen into the vessel is proportional (through K_L) to the difference between the saturation value and the DO at any time t. Equation 6.19 can also be written as

$$\frac{dc}{dt} = K_a(c_s - c) \tag{6.20}$$

where K_a = volumetric reaeration coefficient $[T^{-1}]$ given by

$$K_a = \frac{K_L A}{V} \tag{6.20a}$$

or $$K_a = K_L/H \tag{6.20b}$$

where H is the depth or, more properly, the ratio of the surface area to volume. It is also convenient to introduce the DO deficit, that is the difference between c_s and c, as

$$D = c_s - c \tag{6.21}$$

Substitution into Eq. 6.20 gives

$$\frac{dD}{dt} = -K_a D - \frac{dc_s}{dt} \tag{6.22}$$

If it is now assumed that the temperature, salinity, and pressure are constant in time, then c_s = constant and $dc_s/dt = 0$. Therefore,

$$\frac{dD}{dt} = -K_a D \tag{6.23}$$

and for $D = D_0$, the initial deficit $(c_s - c_0)$ at time $t = 0$, the solution to Eq. 6.23 is

$$D = D_0 e^{-K_a t} \tag{6.24}$$

or

$$c = c_s - (c_s - c_0)e^{-K_a t} \tag{6.25}$$

Figure 6.6 shows the relationships of D, the DO deficit, and c, the DO with time. The deficit decreases exponentially to zero or conversely the DO increases to the saturation value. The effect of the reaeration coefficient is therefore to increase or decrease the time it takes for the DO in the vessel to reach the equilibrium value of the DO saturation. ''High'' values of reaeration (representative of a turbulent system) indicate rapid recovery of the DO levels while ''low'' values (representative of quiescent water bodies) indicate slower recovery of the DO concentrations. The oxygen exchange relationship of Eq. 6.19 is generally applicable to all water bodies and requires an evaluation of the reaeration properties of a specific river, estuary, or lake.

6.3.2.1 Reaeration Coefficient

The oxygen transfer coefficient in natural waters depends on:

1. Internal mixing and turbulence due to velocity gradients and fluctuation.
2. Temperature.

Figure 6.6 (a) DO deficit (D) and (b) DO (c) as a function of time and reaeration coefficient (K_a).

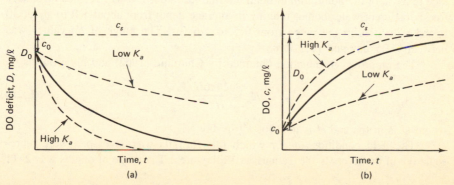

3. Wind mixing.
4. Waterfalls, dams, rapids.
5. Surface films.

Oxygen transfer, as a function of internal mixing and turbulence, has been the subject of much study and investigation in recent years. Holley (1975) has summarized the substantial literature on oxygen transfer, described the theory of oxygen transfer in some depth, and explored the effect of the preceding influences. The studies have ranged from the theoretical investigations of O'Connor and Dobbins (1958) to the empirical field studies of Churchill et al. (1962) and British investigators (Owens et al, 1964; Edwards and Owens, 1965). In these latter field investigations, the general procedure was to measure the reaeration coefficient indirectly by changes in DO under closely controlled prototype conditions. The studies have covered a wide range of river situations from the shallow short-run streams of England to the deep, wide, and slowly moving rivers of the United States. Other work by Tsivoglou et al. (1968) permits the independent and direct measurement of reaeration capacity of natural streams by releasing a gaseous tracer (krypton 85), together with other tracers to measure time of flow and dispersion. In addition, Rathbun (1979) has developed techniques for measurement of field reaeration rates in streams using ethylene gas and a conservative tracer such as dye.

O'Connor's formulation for the reaeration coefficient is

$$K_a = \frac{K_L}{H} = \frac{(D_L U)^{1/2}}{H^{3/2}} \tag{6.26}$$

where D_L is the oxygen diffusivity at 20°C (0.000081 ft^2/hr), U is the average stream velocity, and H is the average depth. For a particular stretch of river, the depth, H, is taken as the ratio of the volume to surface area. For a particular point, H is the ratio of cross-sectional area to width. If U is in fps and H in ft, then for the reaeration coefficient in day^{-1},

$$K_a = \frac{12.9 U^{1/2}}{H^{3/2}} \tag{6.27}$$

This formulation was derived from theoretical considerations regarding surface renewal of the liquid film through internal turbulence. It was verified originally for six different water bodies ranging in average depth from about 1 ft to about 30 ft and velocities in the range from 0.5 to 1.6 fps. The range of K_a values was between 0.05/day and 12.2/day.

The empirical formulation obtained by Churchill (1962) and his group is

$$K_a = \frac{11.6 U}{H^{1.67}} \tag{6.28}$$

where U is in fps and H is in ft for K_a in day^{-1}.

The field conditions under which this formula was developed were representative of the rivers in the Tennessee Valley area. The range of depths was 2–11 ft but the range of velocities was from 1.8 to 5.0 fps, significantly higher than the range used by O'Connor and Dobbins.

Tsivoglou et al. (1968) worked in the Jackson River between Covington and Clifton Forge, Virginia where mean depths are about 2–3 ft and mean velocities about 0.3–0.6 fps. The river is shallow and wide with riffles and natural pools. Flows range up to 200–250 cfs. Observed values of K_a using the gas tracer technique averaged about 3.4/day at 20°C.

Later work by Tsivoglou and Wallace (1972) extended the data base of direct measurement of reaeration and proposed the following expressions:

$$K_a = 0.88US \qquad \text{for} \quad 10 < Q < 300 \text{ cfs} \qquad (6.29a)$$

$$K_a = 1.8US \qquad \text{for} \quad 1 < Q < 10 \text{ cfs} \qquad (6.29b)$$

where K_a is in day^{-1} at 20°C, S is the slope in ft/mi, and U is velocity in fps. More recent comparisons by Grant and Skavroneck (1980) indicated that these expressions were the most accurate for small shallow streams.

British workers measured reaeration rates in several streams in England under controlled conditions where a sulfite dosing of the stream produced a transient well-defined drop in DO. The time rate of recovery is a measure of the reaeration rate. Owens et al. (1964) combined the British data with the collected on the Tennessee Valley streams to produce the equation

$$K_a = \frac{21.6U^{0.67}}{H^{1.85}} \qquad (6.30)$$

for U in fps and H in ft. Owens suggests this formula for the velocity range of 0.1–5.0 fps and depth range of 0.4–11.0 ft. The values of K_a computed from this formulation do not differ significantly from the O'Connor–Dobbins equation. Indeed, the British investigators conclude that the rates are essentially the same for values of K_a less than about 11.0–12.0/day. Since the British work was conducted almost exclusively on shallow (0.4–2.4 ft) fast moving streams (0.1–1.8 fps), Eq. 6.28 may be preferable for these types of streams.

In practice, the O'Connor–Dobbins formulation has formed a reasonable basis for estimating the reaeration coefficients although for small streams it appears to underestimate the coefficient. For small streams (i.e., flows less than about 10 cfs and depths less than about 1–3 ft), the work of Tsivolgou (Eq. 6.29b) appears to compare well to observed reaeration rates. For DO modeling analyses, two courses of action are open: (a) calculation of the sensitivity of DO response to varying K_a values computed from different formulas or (b) if the problem warrants, direct measurement of the reaeration using various field techniques (Tsivoglou et al, 1968; Rathbun, 1979).

Note, however, for some of the equations, K_a approaches zero as the depth increases. This apparently implies that reaeration does not occur or is very low for deep bodies of water. Such is not the case since as discussed earlier, K_L, the oxygen transfer coefficient, as the relevant parameter, reflects oxygen exchange as a surface

process. Hydroscience (1971) then suggests a minimum value of the oxygen transfer coefficient of

$$(K_L)_{min} \approx 2\text{--}3 \text{ ft/day} \qquad (0.6\text{--}1.0 \text{ m/d})$$

so that

$$(K_a)_{min} \approx \frac{(K_L)_{min}}{H} = \frac{(K_L)_{min} A}{V} \tag{6.31}$$

For tidal rivers and estuaries, the reaeration coefficient may be estimated
from the O'Connor–Dobbins equation where the velocity is the average tidal velocity. (A more detailed discussion is in Section 6.5.1). Alternatively, a suggested
range of the transfer coefficient K_L in estuaries is shown in Table 6.6 (Hydroscience, 1971), from which K_a can then be estimated from Eqs. 6.20a or 6.20b.

The reaeration coefficient as a function of temperature is given by

$$(K_a)_T = (K_a)_{20}(\theta)^{T-20} \tag{6.32}$$

Holley (1975) and Zison et al. (1978) point out the need to recognize that the
numerical value of θ depends on the mixing condition of the water body. Values
are generally in the range from 1.005 to 1.030. In practice, a value of 1.024 is
often used.

For lakes and reservoirs, or large open bays, the effect of wind may be
significant in oxygen transfer creating internal turbulence that results in increased
reaeration. Net velocities may be small or random such that the previous equations
which assumed an approximately steady velocity are not appropriate. In water
bodies where wind may be important the mixing may therefore be due entirely to
the winds creating local random turbulent motions. Banks (1975) and Banks and
Herrera (1977) have explored this effect and suggested the following relationship:

$$K_L = 0.728U_w^{1/2} - 0.317U_w + 0.0372U_w^2 \tag{6.33}$$

for U_w = wind speed in m/s at 10 m above the water surface and K_L is the wind
driven oxygen transfer coefficient in m/day. Figure 6.7 shows this relationship and
indicates the rapid rise in the transfer coefficient as the wind speed increases. For

TABLE 6.6   RANGE OF DO TRANSFER COEFFICIENTS
FOR TIDAL RIVERS AND ESTUARIES

Mean tidal depth (ft)	Values of K_L (ft/day) for average tidal velocities (fps) of:		
	1	1–2	2
<10	4	5.5	7
10–20	3	4.5	6
20–30	2.5	3.5	5
>30	2	2.5	4

Source: From Hydroscience (1971).

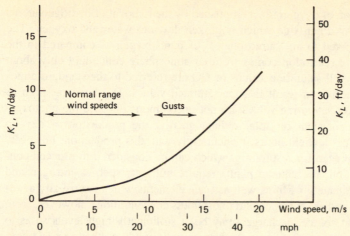

Figure 6.7 Effect of wind speed on oxygen transfer coefficient. After Banks and Herrera (1977).

the normal range of wind speeds from 0 to about 10 m/s, the value of K_L reaches levels of almost 3 m/day or 10 ft/day. This can be compared to the range of K_L for estuaries in Table 6.4 to indicate that the normal range of K_L for lakes due to wind effects is approximately equal to the range for estuaries due to tidal mixing and turbulence. O'Connor (1983) has shown that, for hydraulically smooth surfaces (low wind speeds of less than approximately 5 m/s), K_L varies directly with the wind speed; for hydraulically rough surfaces (high wind speeds of more than approximately 10 m/s), K_L varies as the square root of the wind speed; and in between, a transition function applies. Compared to Fig 6.7, these functions indicate a much smaller increase in K_L at the higher wind speeds.

6.3.3 Photosynthesis and Respiration

The presence of aquatic plants in a water body can have a profound effect on the DO resources and the variability of the DO throughout a day or from day to day. In some cases, emphasis is on the phytoplankton, the microscopic aquatic plants that usually have no motility of their own ("plankton" is from the Greek meaning "wanderer"). Aquatic weeds are also of concern and may be attached (rooted aquatic macrophytes) or free floating. Finally, the attached algae clinging to rocks, plant stems, and other surfaces (part of the periphyton or attached biological community) may also be important.

All of the aquatic plant forms are important because of their ability to photosynthesize. The essence of the photosynthetic process centers about these chlorophyll containing plants which can utilize radiant energy from the sun, convert water and carbon dioxide into glucose, and release oxygen. The photosynthesis reaction can be written as

$$6CO_2 + 6H_2O \xrightarrow{\text{photosynthesis}} C_6H_{12}O_6 + 6O_2 \qquad (6.34)$$

The production of oxygen is accomplished by the removal of hydrogen atoms from water, forming a peroxide which is broken down to water and oxygen. The water is now subjected to an "atmosphere" of pure oxygen as compared to the water surface where reaeration comes from an atmosphere containing only about 21% oxygen. Since all saturation values of DO are referred to the standard atmosphere, photosynthesis can result in supersaturated values. DO levels as high as 150–200% of the air saturation values are not uncommon. Because the photosynthetic process is dependent on solar radiant energy, the production of oxygen proceeds only during daylight hours. Concurrently with this production, however, the algae require oxygen for respiration, which can be considered to proceed continuously. In addition, the aquatic plants require nutrients such as nitrogen and phosphorus for adequate growth as well as trace elements which are usually available in most water bodies. Nitrogen and phosphorus concentrations, however, from point and non-point source discharges may be at sufficiently high levels so as to stimulate excessive growths of plants leading to undesirable aesthetic or DO conditions. This is discussed more fully in Chapter 7, Eutrophication. Figure 6.8 shows a typical diurnal variation in DO due to plant photosynthesis. Minimum values of DO usually occur in the early morning, predawn hours and maximum values occur in the early afternoon. The diurnal range (maximum minus minimum) may be large and if the daily mean level of DO is low, minimum values of DO during a day may approach zero and hence create a potential for a fish kill.

The two principal issues associated with the photosynthesis and respiration components of the DO problem are therefore (a) the degree to which the net effect of photosynthesis and respiration contributes to the average DO resources of the water body and (b) the expected diurnal variability in DO as a result of the presence of aquatic plants.

6.3.3.1 Estimation of Photosynthesis and Respiration Rates There are several methods for estimating the amount of oxygen produced and utilized in the photo-

Figure 6.8 Typical diurnal variation of DO due to photosynthesis by aquatic plants.

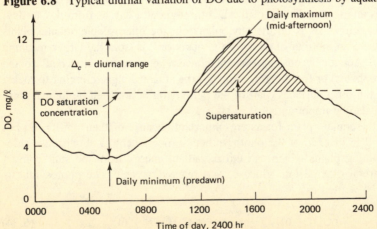

synthesis process. The methods discussed here estimate the average production and respiration of dissolved oxygen for use in the DO balance equation (Eq. 6.1). In that equation, the DO due just to the net production of oxygen by aquatic plants over time and for a segment of volume V is

$$V \frac{dc}{dt} = K_a V(c_s - c) + p_a V - RV \qquad (6.35)$$

where p_a is the average gross photosynthetic production of DO [$(M/L^3 \cdot T)$; mg DO/l · day] and R is the average respiration [$(M/L^3 \cdot T)$]. In general, the total respiration of a segment includes microbial respiration for the CBOD and NBOD. Here R is considered to be plant respiration only, a rate that is sometimes difficult to estimate in the field. Three estimation methods are:

1. "Light and dark" bottle or chamber measurements of DO.
2. Estimation from observed chlorophyll levels.
3. Measurements of diurnal DO range.

The second method estimates production and respiration for water bodies where phytoplankton (i.e., the dispersed microscopic plants) are the principal source of the oxygen. In some water bodies, rooted aquatic plants and attached algae may also contribute substantially to the oxygen budget. Methods 1 and 3 can be used to estimate the DO production due to phytoplankton as well as rooted plants and periphyton.

The "light and dark" method uses the idea that exposing the aquatic plants to the natural light conditions and measuring the change (increase) in DO is an estimate of photosynthetic gross DO production minus the DO utilized in respiration, that is, the net production of DO. The maintenance of a parcel of water in a "dark" environment will provide an estimate of the total DO respiration, since the plants are not photosynthesizing. Actually, the dark bottle includes the uptake of the DO by the bacteria and corrections must be made to estimate the respiration due only to plants. See Eq. 6.37. The addition of the two rates from the light and dark bottles will then be a measure of the gross production of DO.

To determine these rates for phytoplankton in the water column, a series of bottles are suspended in the water body spaced throughout the euphotic zone, that is, the zone from the surface down to 1% of surface light. (This depends on the vertical extinction coefficient as discussed in Chapter 7.) One set of bottles is wrapped in aluminum foil to eliminate any light from entering the bottle. These are the "dark" bottles. The other set is left exposed to the ambient sunlight in the water. These are the "light" bottles. The setup is shown in Fig. 6.9. Light and dark chambers for measurement of the benthic community DO rates are also shown. Probes for measuring the change of DO in the chambers are often used.

The light bottles are filled with river water from a given depth, the DO is measured at time $t = 0$ (DO$_0$), and the bottle is then suspended at depth. A similar procedure is followed for the dark bottle. At some time later, the bottles are retrieved and the DO measured once again (DO_t). This may be done with a DO probe or, if a chemical DO test is carried out, the sample will be destroyed in the

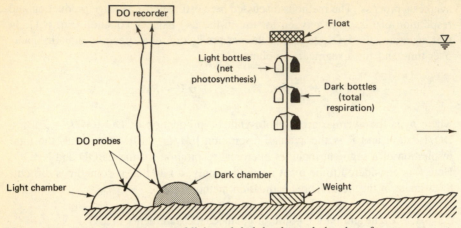

Figure 6.9 Typical arrangement of light and dark bottles and chambers for measurement of photosynthetic oxygen production.

test. (If more data are needed, then additional bottles must be suspended.) The light bottle DO then is an estimate of the net production of oxygen, that is,

$$p_{\text{net}} = \frac{(\text{DO})_t - (\text{DO})_0}{t} = p - R \qquad (6.36)$$

since the resulting change in DO incorporates both the production and the respiration.

The dark bottle measures the total loss of DO due to phytoplankton respiration and the exertion of the BOD. The plant respiration from the dark bottle DO measurement can be estimated by

$$R = \frac{(\text{DO})_0 - (\text{DO})_t}{t} - K_d L_{0f} \qquad (6.37)$$

where $K_d L_{0f}$ is the oxygen uptake correction due to BOD exertion and L_{0f} is the ultimate filtered BOD. The gross production of oxygen during the test is then

$$p = p_{\text{net}} + R \qquad (6.38)$$

If the measurements are carried out during the early afternoon, the gross production will be at approximately its maximum, p_m. Then in Eq. 6.38, $p = p_m$. For a sinusoidal variation in solar intensity, the average daily photosynthetic production for use in DO equations is given by

$$p_a = \frac{2f}{\pi H} (\Sigma p_m \cdot \Delta h) \qquad (6.39)$$

where f is the photoperiod (duration of daylight), Δh is the depth interval over which a given measurement applies, and H is the total depth over which the analysis is being conducted. The summation extends over the number of depth samples. Sample Problem 6.1 illustrates calculation of daily averaged photosynthesis and respiration rates.

The second approach utilizes a direct measurement of the concentration of phytoplankton as represented by the chlorophyll a level in the water. Chapter 7 presents the basic concepts of the growth and death of phytoplankton in some detail. If measurements of chlorophyll are available and if the problem context is such that the chlorophyll itself is not to be predicted, then the phytoplankton chlorophyll can be used to estimate the input of oxygen. However, since photosynthesis is light dependent, the relationship between phytoplankton chlorophyll and photosynthetic production depends on solar radiation, depth, and the extinction coefficient. In addition, the production depends on available nutrients. All of this is discussed in Chapter 7. It is shown there that if nutrients are not limiting, then the average gross production, p_a (mg DO/l · day) is

$$p_a = [a_{op}G_{max}(1.066)^{T-20}P]G(I_a) \tag{6.40}$$

where a_{op} is the ratio of mg DO/μg chl a (range, 0.1–0.3), G_{max} is the maximum growth rate of the phytoplankton (range 1.5–3.0/day), P is the phytoplankton chlorophyll in μg/l, T is water temperature in °C, and

$$G(I_a) = \frac{2.718f}{K_eH}[\exp(-\alpha_1) - \exp(-\alpha_0)] \tag{6.40a}$$

for

$$\alpha_1 = \frac{I_a}{I_s}\exp(-K_ez)$$

$$\alpha_0 = \frac{I_a}{I_s}$$

where $G(I_a)$ is equal to the light attenuation factor over depth and one day (dimensionless), K_e is the extinction coefficient in m^{-1}, I_a is the average solar radiation during the day in ly/day (langleys/day), and I_s is the light at which phytoplankton grow at maximum rate in ly/day (range 250–500). Note that this approach requires estimates of chlorophyll, incoming light and extinction, and the usual temperature and depth measurements. The term in brackets of Eq. 6.40 is the production of oxygen at optimum or "saturated" light conditions (called p_s) which is then modified by the light attenuation factor $G(I_a)$. Thus

$$p_a = p_sG(I_a) \tag{6.41}$$

where

$$p_s = a_{op}G_{max}(1.066)^{T-20}P$$

and for $a_{op} = 0.125$, $G_{max} = 2.0$/day, and $T = 20$°C,

$$p_s = 0.25P \tag{6.42}$$

SAMPLE PROBLEM 6.1

DATA

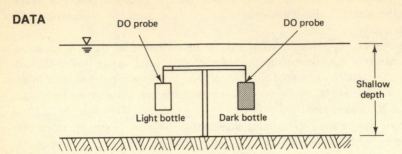

Single light and dark bottles are set up in the shallow, clear stream with DO probes attached to record the DO concentrations during the day. Initial readings are made at 10 A.M. and at 2-hr intervals until 4 P.M. Daylight began at 6 A.M., darkness fell at 7 P.M.

	DO concentrations (mg/l)			
Time	10 A.M.	12 N	2 P.M.	4 P.M.
Light bottle	2.0	4.2	5.7	7.8
Dark bottle	7.7	7.5	7.1	7.0

PROBLEM

Estimate the daily averaged photosynthetic and respiration rates. Assume the bacterial respiration (BOD) is not significant.

ANALYSIS

Obtain Averages for Period of Measurements (p' and R)

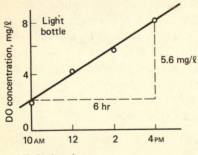 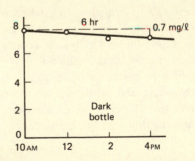

From light bottle,

$$p_{net} = p' - R = \frac{5.6 \text{ mg/l}}{6 \text{ hr}} \times \frac{24 \text{ hr}}{\text{day}} = 22.4 \text{ mg/l} \cdot \text{day} \qquad \text{Eq. (6.36)]}$$

and from dark bottle,

$$R = \frac{0.7 \text{ mg/l}}{6 \text{ hr}} \times \frac{24 \text{ hr}}{\text{day}} = 2.8 \text{ mg/l} \cdot \text{day} \qquad \text{Eq. (6.37)]}$$

Sample Problem 6.1 (continued)

$$\therefore p' = 22.4 + 2.8 = 25.2 \text{ mg/l} \cdot \text{day} \qquad \text{Eq. (6.38)]}$$

Average Daily Rates

R is assumed constant throughout the day.

$$\therefore R = 2.8 \text{ mg/l} \cdot \text{day}$$

Assume photosynthetic oxygen production is sinusoidally distributed over the photoperiod (f) from 6 A.M. to 7 P.M., or $f = 13$ hr.

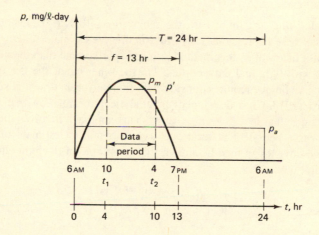

By integrating the area under the half-sine curve between t_1 and t_2 (in terms of f and p_m) and setting this equal to $p'(t_2 - t_1)$, p_m is determined to be

$$p_m = p'(t_2 - t_1) \cdot \frac{\pi}{f(\cos \pi t_1/f - \cos \pi t_2/f)}$$

Setting the area under the half-sine curve ($2\, p_m f/\pi$) equal to the area under the daily averaged rate ($p_a T$), the estimated daily averaged photosynthetic oxygen production rate (p_a) is

$$p_a = p' \cdot \frac{2(t_2 - t_1)/T}{\cos \pi t_1/f - \cos \pi t_2/f}$$

$$p_a = 25.2 \text{ mg/l} \cdot \text{day} \cdot \frac{2 \times 6 \text{ hr}/24 \text{ hr}}{\cos (\pi \times 4/13) - \cos (\pi \times 10/13)}$$

$$p_a = 9.6 \text{ mg/l} \cdot \text{day}$$

In a daily averaged DO model, the net photosynthesis minus respiration rate would then be

$$p_{\text{net}} = 9.6 - 2.8 = 6.8 \text{ mg/l} \cdot \text{day}$$

an oft-cited rule of thumb. The range in the coefficient of 0.25 should be recognized, however, and, in general, Eq. 6.40 is preferred.

Also, as shown in Chapter 7, the phytoplankton respiration, R, in mg/1 · day if given approximately by

$$R = a_{op}(0.1)(1.08)^{T-20}P \tag{6.43}$$

which for $a_{op} = 0.25$ mg $0_2/\mu$g chl a and $T = 20°C$ yields

$$R = 0.025P \tag{6.44}$$

another oft-cited rule of thumb. But, the range in a_{op} (0.1 to 0.3) should be recognized. Since these expressions for p_a and R involve estimates of a_{op} which can vary by about a factor of 3, this method is more approximate than the actual direct measurement with light and dark bottles. On the other hand, the use of the light and dark bottle technique requires considerably more effort in time and survey cost.

Di Toro (1975a) has developed an analytical formulation that relates the diurnal DO range, that is $c_{max} - c_{min}$, to p_a. This approach involves measurement of the DO over a period of at least one day to obtain an estimate of the range throughout the day. It can then be shown that for a time of daylight equal to one-half of the total day,

$$p_a = \left\{ \frac{0.5K_a[1 - \exp(-K_a t)]}{[1 - \exp(-0.5K_a t)]^2} \right\} \Delta_c \tag{6.45}$$

where $\Delta_c = c_{max} - c_{min}$, the diurnal DO range and K_a is the reaeration coefficient in day^{-1} and $t = 1$ day. Respiration must be estimated from one of the preceding techniques. Note that Eq. 6.45 can be used to estimate the diurnal range of DO with an estimate of p_a from the first two methods.

Some approximations can be obtained from Eq. 6.45. As shown by Di Toro (1975a),

$$p_a \approx 2\Delta_c \quad \text{for} \quad K_a(\text{day}^{-1}) < 2$$
$$\approx 3.2\Delta_c \quad \text{for} \quad 2 \leq K_a(\text{day}^{-1}) \leq 10 \tag{6.46}$$

Erdmann (1979), in DO work on the Charles River in Massachusetts, confirmed the former of the two relationships by measurements in the diurnal DO range and gross production estimated from DO surveys.

Conversely, if p_a is estimated from Eq. 6.40 or light and dark bottle measurements, the estimated diurnal range of DO is given by

$$\Delta_c \approx 0.5p_a \quad \text{for} \quad K_a(\text{day}^{-1}) < 2$$
$$\approx 0.3p_a \quad \text{for} \quad 2 \leq K_a(\text{day}^{-1}) \leq 10 \tag{6.47}$$

In order to compare measured photosynthesis and respiration rates, it is useful to eliminate the depth dependence of a particular water body and express the rates in areal units. Thus

$$p_a' = p_a \cdot H$$
$$R' = R \cdot H$$

(6.48)

where p_a' and R' are the areal gross production and respiration, respectively $[(M/L^2 \cdot T)]$. For p_a and R in mg/l $\cdot$ day [g/m^3 $\cdot$ day] and H in m, then p_a' and R' are in g DO/m^2 $\cdot$ day.

Values of p_a' range from 0.3 to 3 g DO/m^2 $\cdot$ day for moderately productive areas to up to 10 g DO/m^2 $\cdot$ day for streams that have a significant biomass of aquatic plants. Respiration levels range over approximately the same values.

6.3.4 Sediment Oxygen Demand

The discharge of settleable waste components may result in the formation of "sludge banks" or deposits of organic material immediately below a waste outfall. These deposits may build up over a period of time if velocities are too low to prevent scouring of the river or estuary bottom. As the depth of the deposited solids increases, anaerobic decomposition of the organic material in the deeper layer begins. The products of this decomposition, CO_2, CH_4, and H_2S proceed up through the sludge layer and into the overlying waters. If gas production is especially high, floating of the bottom sludge may result, leading to a severe aesthetic problem as well as possible transient DO depletion. The surface layer of the bottom deposit in direct contact with the water usually undergoes aerobic decomposition and, in the process, removes oxygen from the supply in the overlying river water, that is, DO diffuses into the surface layer of the sediment for aerobic oxidation.

For some rivers and estuaries, the deposition of solids proceeds only during the low flow summer and fall months when velocities are low. High spring flows the following year may scour the bottom clean and reduce the problem until velocities decrease again. Intermediate cases are common where high flows may scour only a portion of the deposit, oxidize a portion, and then redeposit the material in another location.

The sediment oxygen demand (SOD) may not always be due directly to sewage or industrial sludges. Soluble organic wastes may sometimes result in the growth of attached filamentous bacteria such as *Sphaerotilus* which can withdraw substantial amounts of oxygen. The death of rooted aquatic plants and leaves and detritus in natural runoff may contribute to sediment organic material which will also require oxygen for stabilization. Finally, the settling of phytoplankton in eutrophication systems, particularly lakes, can result in significant SOD levels.

The condition of the bottom may, therefore, vary from deep deposits of sewage or industrial waste origin (depths of several feet have been observed from point sources and combined sewer overflows) to relatively shallow deposits of material of plant origin and, finally, to clean rock and sand. The oxygen utilization by bottom sediments therefore depends on the extent of organic material and the nature of the benthic community.

In situ measurements involve careful submersion of a chamber on the bottom and then measuring the O_2 uptake over time. Laboratory measurements are conducted by removing a sample of the river bottom (preferably undisturbed) placing the sample in a large vessel with oxygenated water. DO reduction over time is a measure of the uptake of the bottom muds. Results are usually reported in g O_2 uptake/$m^2 \cdot$ day.

Some of the earliest work on the magnitude of the SOD was done by Baity (1938) and Fair et al. (1941). A variety of other in situ measurements have been made for numerous water bodies in recent years. Table 6.7 summarizes the range of SOD values.

The demand of g $O_2/m^2 \cdot$ day is exerted from a bottom surface area. Thus, the total flux is

$$S_B' = \frac{S_B A_B}{V} = \frac{S_B}{H} \tag{6.49}$$

or

$$S_B' = \frac{S_B}{A/P_B} = \frac{S_B}{(R_H)_B} \tag{6.49a}$$

where S_B is the SOD [$M/L^2 \cdot T$], A_B is the contributing bottom area [L^2], V is the volume of the overlying water column [L^3], S_B' is the oxygen demand [$M/L^3 \cdot T$], A is the cross-sectional area [L^2], P_B is the length of contact of the water body with bottom deposits [L], and $(R_H)_B$ is the "effective" hydraulic radius [L].

Temperature effects can be approximated in the 10–30°C range by

$$(S_B)_T = (S_B)_{20}(\theta)^{(T-20)} \tag{6.50}$$

where the temperature coefficient, θ, has a reported range of 1.040 to 1.130 (Zison et al., 1978). A value of 1.065 is often used. Below 10°C the SOD probably decreases more rapidly than indicated by Eq. 6.50 and approaches zero in the water temperature range of 0–5°C.

The depth of the deposit is also of some interest since the uptake rate is dependent on this parameter. Oldaker et al. (1966) found a linear relationship

TABLE 6.7   SOME SEDIMENT OXYGEN DEMAND VALUES

	SOD (g $O_2/m^2 \cdot$ day) @ 20°C	
Bottom type and location	Range	Approximate average
Sphaerotilus—(10 g dry wt/m^2)	—	7
Municipal sewage sludge—outfall vicinity	2–10.0	4
Municipal sewage sludge—"aged", downstream of outfall	1–2	1.5
Estuarine mud	1–2	1.5
Sandy bottom	0.2–1.0	0.5
Mineral soils	0.05–0.1	0.07

Source: From Thomann (1972).

between estimates of the ultimate amount of oxygen needed for complete stabilization and depth of sediment (from domestic sewage) over the range 1.5–20 cm. Others (Owens et al., 1964 and Fillos and Molof, 1972) have indicated that uptake was independent of sludge depth for depths greater than about 2 to 8 cm.

It is generally assumed for most calculations that the uptake rate is independent of the oxygen concentration in the overlying waters as concluded by Baity (1938) for the 2–5 mg/l DO range. Edwards and Owens (1965), however, indicated that the uptake rate varies as the DO to about the 0.45 power. The data of Fillos and Molof (1972) result in the following relationship for a 7.6-cm depth of sludge:

$$S_B = 7.2 \left(\frac{c}{0.7 + c} \right) \tag{6.51}$$

for S_B in g/m^2 · day and c, the DO, in mg/l. This relationship indicates independence of S_B for $c > 3$ mg/l and is at one-half its maximum value at 0.7 mg/l DO.

As waste treatment increases to secondary and advanced levels, the occurrence of gross, noxious sludge deposits of raw sewage origin decreases. Nevertheless, transient treatment plant bypasses and combined sewer overflows still can contribute to the benthal demand. In addition, secondary effects of the soluble wastes and materials may produce bacterial and higher order plant growths which can also contribute to the SOD.

6.3.5 Oxidation of CBOD

As noted previously in this chapter (Oxygen Demanding Waste Inputs), the CBOD represents the oxygen demanding equivalent of the complex carbonaceous material present in a waste. Thus CBOD is the concentration of organic material and the oxidation of this organic carbon in a body of water will utilize oxygen at a rate equivalent to the decrease of the CBOD. This discussion summarizes the mechanisms associated with the CBOD oxidation as applied in any body of water.

Consider first a vessel representing a volume segment of a surface water. The vessel is closed to the atmosphere so that reaeration does not occur and at time $t = 0$, an amount of dissolved and particulate CBOD is dispersed throughout the water and the DO is at saturation. (This can be recognized as the condition of the BOD test using a closed BOD bottle.) The organic waste, as measured by the CBOD, will then decline according to biochemical oxidation. Also, the particulate CBOD will settle out of the water column to the bottom of the vessel and it is assumed that once the particulate CBOD reaches the bottom it no longer participates in any interaction with the water. Of course, in actual surface waters, this particulate BOD accumulates as the SOD and its subsequent aerobic and anaerobic stabilization affects the overlying DO. For purposes of this discussion however, the particulate BOD enters a lower compartment and does not subsequently enter the reaction. The SOD would account for this BOD that can settle. Figure 6.10 shows the experiment. Let L_p and L_d be the particulate and dissolved BOD, respectively, of the total BOD, L, given by

$$L = L_p + L_d \tag{6.52}$$

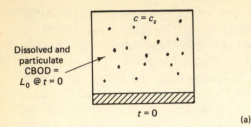
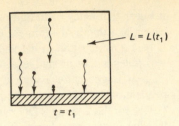

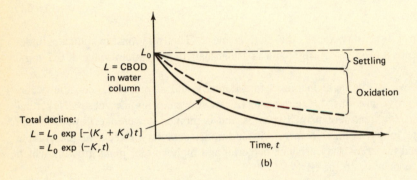

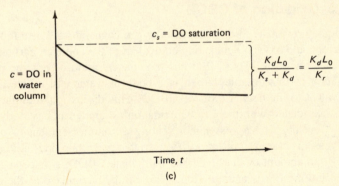

Figure 6.10 (a) Closed vessel with dissolved and particulate CBOD and DO at saturation. (b) Loss of CBOD may be due to settling and oxidation of carbonaceous material. (c) Decline of DO due to oxidation of CBOD.

If L_p only settles without oxidation and L_d only is oxidized, then for a volume V,

$$V \frac{dL_p}{dt} = -v_s A L \qquad (6.53a)$$

and

$$V \frac{dL_d}{dt} = -K_d' V L \qquad (6.53b)$$

where v_s is the settling velocity and K_d' is the CBOD decay rate.

For

$$L_d = f_d L$$
$$L_p = f_p L$$

where f_d and f_p are the fraction dissolved and particulate, respectively, of the total BOD, L, then

$$V \frac{dL}{dt} = -v_s A L_p - K'_d V L_d$$

or

$$\frac{dL}{dt} = -K_r L \tag{6.54}$$

where

$$K_r = v_s \frac{A}{V} f_p + K'_d f_d$$

$$= K_s + K_d \tag{6.55}$$

where K_s is the effective loss rate due to settling, K_d is the effective deoxygenation rate, and K_r is the overall loss rate $[T^{-1}]$ of CBOD from the water column due both to settling and oxidation of soluble BOD.

The solution to Eq. 6.54 is

$$L = L_0 \exp(-K_r t)$$
$$= L_0 \exp[-(K_s + K_d)t]$$
$$= L_0[\exp(-K_s t)\exp(-K_d t)] \tag{6.56}$$

and is seen to be composed of two parts; the loss due to settling and that due to biochemical oxidation. Figure 6.10(b) shows the decline of BOD. Now consider the DO. The appropriate equation for zero reaeration (recall that the vessel is closed) recognizes that a decline of CBOD due to oxidation is equivalent to a decline in DO, by the definition of a BOD. Therefore,

$$\frac{dc}{dt} = -K_d L \tag{6.57}$$

Note that in Eqs. 6.54 and 6.56, L has units mg CBOD/l. In Eq. 6.57, 1 mg CBOD is equivalent to 1 mg DO. For L given by Eq. 6.56 the time variation of the DO in the vessel is

$$c = c_s - \frac{K_d L_0}{K_r}[1 - \exp(-K_r t)] \tag{6.58}$$

and is plotted in Fig. 6.10(c). The DO drops at a rate of K_r, that is, settling plus oxidation because that loss rate represents the total loss of BOD from the water. The final value that the DO reaches after a long time c_∞ is

$$c_\infty = c_s - \frac{K_d L_0}{K_r} \tag{6.59}$$

and indicates that not all of the initial CBOD, L_0, results in an equivalent decline in DO because some portion settled out of the system. This simple example illustrates one aspect of the complexity of estimating the deoxygenation rate of the CBOD for use in DO balance equations. In raw sewage or effluents with a high concentration of solids (greater than about 100 mg/l solids), the percentage of particulate BOD may be high and must be taken into account. Of course, as noted earlier, any loss of BOD from the water column in a natural system is not really a loss at all but contributes to the SOD discussed previously. For effluents at the secondary treatment level with suspended solids of less than 30 mg/l, K_s may not be important due to the absence of any significant particulate BOD. Good practice would indicate that measurements be made of filtered effluent BOD and total BOD to estimate the fraction of the total load that may be subject to settling. This determination might be particularly important for non-point source or CSO problems.

In addition to the possible settling of BOD, the estimation of K_d cannot usually be made from incubation of effluent to determine the BOD bottle rate, (see Eq. 6.4) because the oxidation of BOD in a natural body of water includes phenomena that are not part of the BOD bottle rate. Such phenomena include biosorption by biological slimes on river bottoms. Stream turbulence and roughness, and the density of attached organisms also affect the degree of this type of BOD removal. It is for this reason that attempts have been made to correlate K_d to channel characteristics such as depth, flow, and wetted perimeter.

Finally, for increasing levels of treatment, the residual waste that is discharged will represent the less easily oxidizable fraction and subsequently will be oxidized in the natural water body at a slower rate than untreated waste (see Table 6.5).

As an approximate guide, values of K_d range from about 0.1 to 0.5/day for deeper bodies of water (depths greater than about 5 ft) to 0.5 to 3.0/day for shallow streams (depths less than about 5 ft).

In Hydroscience (1971), the relationship between K_d at 20°C and depth is suggested as

$$K_d = 0.3 \left(\frac{H}{8} \right)^{-0.434} \qquad \text{for} \quad 0 \le H \le 8 \qquad (6.60a)$$

$$= 0.3 \qquad \qquad \text{for} \quad H > 8 \qquad (6.60b)$$

where H is the depth in ft and K_d is in day^{-1}. Wright and McDonnell (1979) from a review of 23 river systems and one laboratory flume resulting in 45 coefficient estimates from field data (including the earlier Hydroscience data) obtained a range of K_d from 0.08 to 4.24/day, for flows from 4.6 to 8760 cfs, wetted perimeters from 11.8 to 686 ft, and depths from 0.9 to 32 ft. They suggest the following relationship (at 20°C) which gave a good fit to the observed data:

$$K_d = 10.3 Q^{-0.49} \qquad (6.61)$$

Wright and McDonnell also observed that for flow conditions greater than about 800 cfs, K_d rates were consistent with bottle rates for the effluent. They concluded that conventional laboratory procedures for estimating BOD decay rate could be used for the larger streams.

Since the oxidation of the CBOD is a bacterially mediated process, the rate K_d is a function of the water temperature. The effect of temperature K_d may be approximated by

$$(K_d)_T = (K_d)_{20} 1.047^{T-20} \tag{6.62}$$

where $(K_d)_T$ and $(K_d)_{20}$ are the decay coefficients at water temperature $T(°C)$ and 20°C respectively. The base 1.047 is reported to range from 1.02 to 1.09 (Zison et al. 1978).

6.3.6 Nitrification and NBOD Rates

As discussed in 6.3.1.2, the oxidizable nitrogen (organic and ammonia nitrogen) may be oxidized by nitrifying bacteria to nitrite and nitrate and utilizing DO as part of the process. The conditions for nitrification include:

1. Nitrifying bacteria.
2. Optimum pH in the alkaline (pH $\cong$ 8) range.
3. DO levels greater than about 1–2 mg/l.

The discharge of the oxidizable nitrogen may therefore deplete the oxygen resources of the water body if the conditions are favorable for nitrification. Just as with the CBOD oxidation, it is necessary to estimate the rate at which such an uptake of DO may occur. If it is assumed that conditions are favorable for nitrification, then there are two analysis routes. The first approach is to analyze the individual components of the nitrification process—the organic, ammonia, nitrite, and nitrate nitrogen. This is a relatively complex analysis and requires some data on each of the nitrogen forms to estimate the various reaction rates. The second approach simply combines the organic and ammonia nitrogen (the total Kjeldahl nitrogen, TKN) and assumes that all of the TKN will be oxidized. This is the NBOD. Measurements of NBOD are made as discussed previously and the loss rate of NBOD representing an overall nitrification rate is estimated. Each approach is now considered.

6.3.6.1 Nitrification Kinetics
In order to describe the basic kinetic framework for nitrification, consider a completely mixed vessel and define the various nitrogen forms as follows:

$$N_1 = \text{organic nitrogen}$$
$$N_2 = \text{ammonia nitrogen}$$
$$N_3 = \text{nitrite and nitrate nitrogen}$$

Nitrite and nitrate are combined since nitrite is generally rapidly converted to nitrate. If nitrite is considered separately, an additional equation must be added. A simplified approach to the progression of nitrification in natural waters can be derived by assuming that all reactions are first-order reactions. Such an assumption has been shown by Di Toro (1975b) to be reasonable with a well-seeded nitrifier bacteria population and "dilute" nitrogen concentrations on the order of less than several mg/l.

Under the first-order assumption, the differential equations expressing the transformation of each of the nitrogen forms are given by

$$\frac{dN_1}{dt} = -K_{11}N_1 \qquad (6.63a)$$

$$\frac{dN_2}{dt} = K_{12}N_1 - K_{22}N_2 \qquad (6.63b)$$

$$\frac{dN_3}{dt} = K_{23}N_2 - K_{33}N_3 \qquad (6.63c)$$

In these equations, the various rate coefficients are:

K_{11} = overall loss coefficient of organic nitrogen due to settling of particulate forms and hydrolysis and bacterial decomposition to ammonia

K_{12} = rate of formation of ammonia due to decomposition of organic nitrogen

K_{22} = overall loss of ammonia due to uptake by aquatic plant forms and oxidation to nitrite/nitrate

K_{23} = rate of formation of nitrite/nitrate due to oxidation of ammonia

K_{33} = overall loss of nitrite/nitrate due to uptake by aquatic plant forms or through denitrification

Note that not all of N_1 goes forward to N_2 (the ammonia) because of settling of particulate organic forms. Thus, in general $K_{11} \geq K_{12}$.

Under conditions of low concentrations of DO, the bacterial reduction of nitrate can occur. This reaction is to be distinguished from the utilization of nitrate and subsequent reduction by aquatic plants. The mechanism of denitrification may be important as a sink of nitrate in the water column. The process is as follows: under anaerobic conditions in the sediment, nitrate is reduced to nitrogen gas which is liberated. The low level of nitrate in the sediment sets up a gradient with the higher nitrate concentrations in the water column. Diffusion of the nitrate from the water column into the interstitial water of the sediment may occur leading to a loss of nitrate from the water column.

All of the rate coefficients that involve nitrification assume that a sufficient population of nitrifying bacteria is present and that conditions for nitrification are suitable.

Equations 6.63 represent a series of first-order coupled equations. The solution of Eq. 6.63a is

$$N_1 = N_{01} \exp(-K_{11}t) \qquad (6.64)$$

where N_{01} is the initial value of the organic nitrogen at time $t = 0$.

Not all of the organic nitrogen may hydrolyze to ammonia in the time scale of a given problem due to the refractory nature of some organic nitrogen forms.

The refractory fraction, if known from experiments can be subtracted from the more labile organic nitrogen.

Substitution of this solution into Eq. 6.63b and integration gives the solution for ammonia, N_2, as

$$N_2 = \frac{K_{12}N_{01}}{K_{22} - K_{11}} \left[\exp(-K_{11}t) - \exp(-K_{22}t) \right] + N_{02} \exp(-K_{22}t) \quad (6.65)$$

where N_{02} is the initial value of ammonia. Sequential substitution and integration gives the concentration of nitrite and nitrate, that is,

$$N_3 = \frac{K_{12}K_{23}}{K_{22} - K_{11}} \left[\frac{\exp(-K_{11}t) - \exp(-K_{33}t)}{K_{33} - K_{11}} - \frac{\exp(-K_{22}t) - \exp(-K_{33}t)}{K_{33} - K_{22}} \right] N_{01}$$

$$+ \frac{K_{23}}{K_{33} - K_{22}} \left[\exp(-K_{22}t) - \exp(-K_{33}t) \right] N_{02} + N_{03} \exp(-K_{33}t) \quad (6.66)$$

In terms of the uptake of oxygen to complete the nitrification step from ammonia to nitrite/nitrate, the DO equation is

$$\frac{dc}{dt} = -4.57 K_{23} N_2 \quad (6.67)$$

which expresses the uptake of DO due to the complete oxidation of NH_3 via nitrification where N_2 is given by Eq. 6.65.

The loss of oxygen due to these reactions is therefore

$$c = c_s - 4.57 \left(\frac{K_{12}K_{23}}{K_{22} - K_{11}} \left\{ \left[\frac{1 - \exp(-K_{11}t)}{K_{11}} \right] - \left[\frac{1 - \exp(-K_{22}t)}{K_{22}} \right] \right\} N_{01} \right.$$

$$\left. + \frac{K_{23}}{K_{22}} \left[1 - \exp(-K_{22}t) \right] N_{02} \right) \quad (6.68)$$

The form of these solutions is shown schematically in Fig. 6.11. As shown, the decline in DO due to the nitrification process is a complex interaction between the various pathways that a given nitrogen form might take. This approach therefore requires estimates of the various kinetic coefficients and data on each of the nitrogen forms.

6.3.6.2 NBOD Rates The second approach makes no attempt to describe the above kinetic reactions but utilizes an overall oxidation rate of the organic plus ammonia nitrogen (the TKN). Measurements of the equivalent oxygen demand of the TKN are made to determine the NBOD in the water body. The rate of change of the NBOD is given by

$$\frac{dL^N}{dt} = -K_N L^N \quad (6.69)$$

where K_N is the overall oxidation rate of the NBOD given by Eq. 6.11. This equation cannot be derived from the preceding kinetic equations and therefore

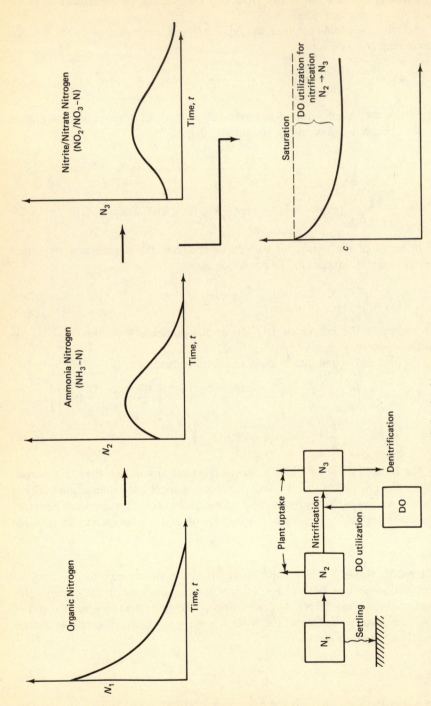

Figure 6.11 Steps in nitrification and utilization of dissolved oxygen.

represents a semiempirical approach. The uptake of DO is then

$$\frac{dc}{dt} = -K_N L^N$$

and under no reaeration, the decline in DO due to the NBOD is

$$c = c_s - L_0^N[1 - \exp(-K_N t)] \tag{6.70}$$

where L_0^N is the initial concentration of NBOD.

The range of values for K_N is approximately the same as for the deoxygenation coefficient of the CBOD. Therefore, for the deeper larger bodies of water, values of K_N of 0.1–0.5/day at 20°C are typical. For smaller streams, values greater than 1/day are not uncommon.

The effect of temperature on the nitrification rate is given by

$$(K_N)_T = (K_N)_{20} 1.08^{T-20} \quad \text{for} \quad 10° \leq T \leq 30°C \tag{6.71}$$

where a range of 1.0548 to 1.0997 is reported for the temperature coefficient, 1.08 (Zison et al, 1978).

If this relationship were used for all temperatures, values of the nitrification for temperatures below 10°C would be too high since below this temperature, the nitrifying bacteria apparently do not multiply in any significant amount. Therefore, the rate K_N is usually set equal to zero at about 5–10°C.

6.4. DISSOLVED OXYGEN ANALYSIS— STREAMS AND RIVERS

6.4.1 Single Point Source

The principal equations that govern the spatial distribution of DO in a river follow the development of the distribution of residuals into rivers as discussed in Chapter 2. Steady-state and plug flow, that is, no longitudinal mixing are assumed. Here, however, the situation becomes more complex because of the interaction between oxidation of the BOD and the resulting uptake of DO and subsequent reaeration of DO from the atmosphere. First, consider the longitudinal distribution of the carbonaceous BOD drawing on Eq. 2.21 for the equation representative of any substance decaying according to first-order decay and Eq. 6.56 which represents the kinetics for the carbonaceous BOD. The full CBOD equation under a steady-state input of a point source is then

$$0 = -U\frac{dL}{dx} - K_r L \tag{6.72a}$$

with boundary condition at the outfall of

$$L = L_0 \quad \text{at} \quad x = 0 \tag{6.72b}$$

The concentration of CBOD at the outfall is given as before (see Eq. 2.18) by

$$L_0 = \frac{W + L_u Q_u}{Q_u + Q_e} = \frac{W + L_u Q_u}{Q} \qquad (6.73)$$

where W is the mass rate of discharge of CBOD from a waste source tributary, Q_u and Q_e are the upstream river and waste flow, respectively, and L_u is the upstream CBOD concentration. The solution of Eq. 6.72a with condition 6.72b is

$$L = L_0 \exp\left(-K_r \frac{x}{U}\right) = L_0 \exp(-K_r t^*) \qquad (6.74)$$

The distinction between the "bottle" BOD rate (Section 6.3.1.1) and the loss rate of BOD in surface waters, K_r, (Section 6.3.3) should be reviewed again at this point. Figure 6.12 (Eckenfelder and O'Connor, 1961) further explains this difference for the case of rivers. Consider a river (at temperature $T°C$) with an input at the location $x = 0$ as shown in Fig. 6.12(a) and that all the BOD is CBOD. If a sample is taken at location A, brought back to the laboratory, incubated at a standard temperature of 20°C, and analyzed for CBOD over a period of say 20 days, the curve labeled A in Fig. 6.12(b) would result. This is the normal progression of BOD in the laboratory as given by Eq. 6.4. The rate of BOD is the "bottle" rate, K_1, and the ultimate BOD is the CBOD, L_0 at location A. This BOD represents the ultimate CBOD that could be oxidized in the river at the input location. This measured BOD should agree with the BOD that would be calculated from the mass balance equation of Eq. 6.73. If the measured and calculated BOD do not agree, then either incomplete mixing exists in the cross section or the loads are inaccurate or both.

If the sampling crew now proceeds downstream to point B, and collects another sample from the river, returns to the laboratory, and incubates the sample, curve B will result. The reduction in the ultimate CBOD from point A reflects the oxidation of the BOD by the microorganisms of the stream and any effects of settling to the river bottom. The decline in the CBOD from A to B reflects the rate of oxidation at the river temperature. The CBOD plotted at point B in Fig. 6.12(a) therefore represents the CBOD remaining in the river. A similar procedure yields curve C and point C in Figs. 6.12(b) and (a). The CBOD loss as a function of river distance or travel time is then given by the curve described by the points A through C. The sketch clearly indicates that the loss of BOD in the river itself may be quite different from the progression of BOD in the laboratory. The preceding discussion on oxidation of CBOD has addressed this issue. For shallow streams, the difference between total loss of BOD in the river, the deoxygenation rate, and the laboratory uptake rate is particularly significant because of the influence of the benthic community on oxidation of the BOD. Note also that if there are no changes in the characteristics of the CBOD downstream then the river decay rate, K_r, will be the same for the CBOD5 as well as for the ultimate CBOD profile.

Figure 6.13 shows the effect of different river water temperatures on the progression of the BOD. At a low river temperature of, for example, 5°C, the decline in CBOD with distance downstream is greatly reduced. The river decay

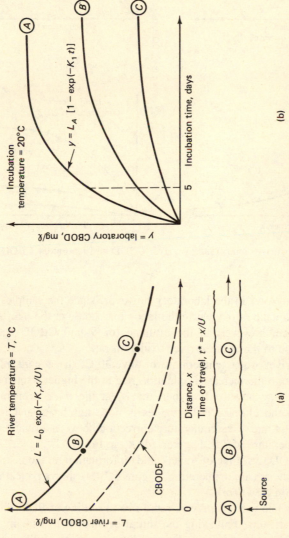

Figure 6.12 Illustration of river and laboratory BOD. (a) Typical decay of BOD in a river. (b) BOD in long-term laboratory determinations. Adapted from Eckenfelder and O'Connor (1961). Reprinted with permission from ''Biological Waste Treatment,'' copyright 1961, Pergamon Press.

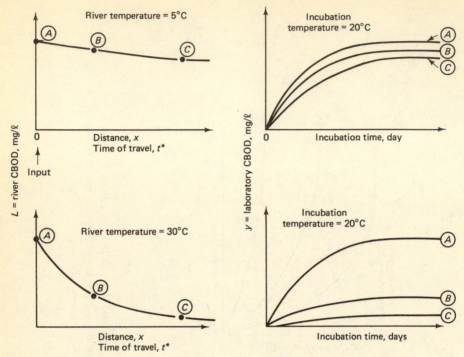

Figure 6.13   Effect of river temperature on river CBOD and laboratory CBOD.

rate, K_r, is therefore low. In the laboratory however, since the samples are incubated at 20°C, the uptake of oxygen in the laboratory bottle at the rate, K_1, is still high. The difference, however, in the ultimate (or 5-day) CBOD between the samples is small due to the reduced river oxidation rate.

    At a higher river water temperature of say, 30°C, the decline in the river BOD is marked due to the increased oxidation rate at the higher temperature. The laboratory K_1 rate may now be less than the K_r at the river temperature. The difference between the ultimate BOD values is now significant and reflects the higher river oxidation rate. The temperature correction between a river temperature of 20°C and a temperature of T°C is given for K_d in Eq. 6.62.

    A plot of CBOD5 vs. time of travel yields information on the instream BOD removal rate K_r at the stream temperature. Figure 6.14(a) shows typical results for the Mohawk River in New York.

    The initial decline in BOD may be rapid due to settling of BOD and/or rapid uptake in shallow streams. Following the initial decline, the instream BOD rate may be less and reflect only oxidation. This is illustrated in Fig. 6.14b which shows two reaches where the BOD rate is different due to these effects. The procedure is simply to follow the basic techniques of calculating the distribution of BOD from beginning of one reach to the beginning of the next. A mass balance is then performed (Eq. 6.73) at the next junction point and the calculation using Eq. 6.74 is repeated for the next reach.

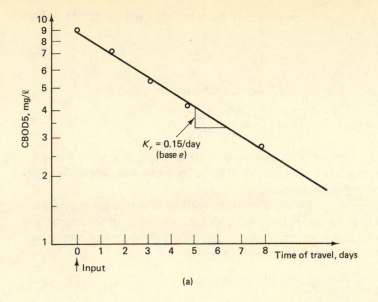

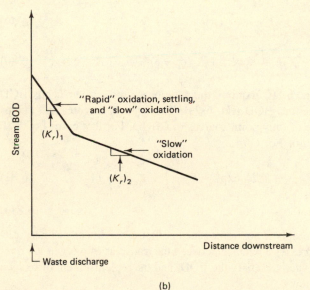

(b)

Figure 6.14 Variation of BOD downstream of input in a river. (a) Mohawk River, NY (1966). (b) Variable BOD decay rates. From Thomann (1972).

If river inflows vary along the length of the stream, the effect of dilution must be taken into account. This can be easily accomplished by recognizing that Eq. 6.72a can be written as

$$0 = -\frac{1}{A}\frac{d(QL)}{dx} - K_r L \qquad (6.75)$$

If $L' = QL$ $[M/T]$ is introduced as a new variable, this equation can be written as

$$0 = -U \frac{dL'}{dx} - K_r L' \tag{6.76}$$

which is the same form as Eq. 6.72a so that the solution is also

$$L' = L_0' \exp\left(-\frac{K_r x}{U}\right) \tag{6.77}$$

A plot of the mass of BOD moving past a station per unit time, therefore, provides a proper estimate of the overall BOD decay coefficient, K_r, for spatially varying river flow with constant velocity.

The model equations for the distribution of DO in a river incorporate the mechanisms described earlier. Following Eq. 6.1 the sum of all of the sources and sinks of DO, and treating the nitrogenous components as equivalent nitrogenous BOD, the BOD and DO equations are given by

$$0 = -U \frac{dL}{dx} - K_r L \tag{6.78}$$

$$0 = -U \frac{dL^N}{dx} - K_N L^N \tag{6.79}$$

$$0 = -U \frac{dc}{dx} - K_d L - K_{dN} L^N + K_a(c_s - c) + p_a - R - S_B' \tag{6.80}$$

The DO balance equation, 6.80, requires the solution of the carbonaceous BOD (Eq. 6.78) and the nitrogenous BOD (Eq. 6.79). The work of Streeter and Phelps (1925) dealt with the case of only point sources of CBOD. For that case, Eq. 6.80 becomes after substitution of Eq. 6.74.

$$0 = -U \frac{dc}{dx} - K_d L_0 \exp\left(-K_r \frac{x}{U}\right) + K_a(c_s - c) \tag{6.81}$$

with boundary condition of $c = c_0$ @ $x = 0$. $\tag{6.81a}$

Equation 6.81 represents a nonhomogeneous first-order linear ordinary differential equation and expresses a balance between the transport of oxygen via river velocity, the uptake of oxygen to satisfy the BOD, and the replenishment of oxygen from atmospheric reaeration.

Using the DO deficit equation (6.21) and assuming no change of the saturation value with distance (i.e., $dc_s/dx = 0$), then the DO deficit equation to be solved is

$$U \frac{dD}{dx} = K_d L_0 \exp\left(-K_r \frac{x}{U}\right) - K_a D \tag{6.82}$$

The solution to this equation with $D = D_0$ at $x = 0$ is given by

$$D = \left\{ \frac{K_d}{K_a - K_r} \left[\exp\left(-K_r \frac{x}{U}\right) - \exp\left(-K_a \frac{x}{U}\right) \right] \right\} L_0 + D_0 \exp\left[-\left(\frac{K_a}{U}\right)x \right] \tag{6.83}$$

In terms of DO, this equation is the famous DO sag equation of Streeter and Phelps:

$$c = c_s - \left\{ \frac{K_d}{K_a - K_r} \left[\exp\left(-K_r \frac{x}{U} \right) - \exp\left(-K_a \frac{x}{U} \right) \right] \right\} L_0$$

$$- (c_s - c_0) \exp\left(-K_a \frac{x}{U} \right) \tag{6.84}$$

Figure 6.15 shows the distribution of DO deficit and DO with distance downstream as given by these equations. The DO deficit reaches a maximum at the location given by t_c^*, the critical location. At that point, the uptake of oxygen by the BOD is just balanced by the input of oxygen from the atmosphere. Equation 6.82 then becomes for the location at which $dD/dx = 0$:

$$K_a D_c = K_d L_0 \exp(-K_r t_c^*)$$

where D_c is the critical maximum deficit of DO given by

$$D_c = \frac{K_d}{K_a} L_0 \exp(-K_r t_c^*) \tag{6.85a}$$

or $$c_{\min} = c_s - D_c \tag{6.85b}$$

where $c_{\min}$ is the minimum concentration of DO downstream of the input.

Figure 6.15   (a) DO deficit variation with distance. (b) Typical DO sag, in rivers.

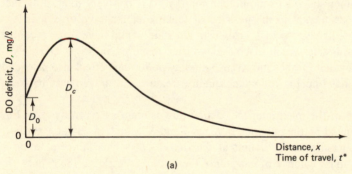

(a)

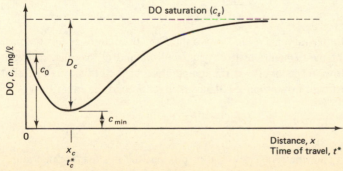

(b)

t^* is given by differentiating Eq. 6.83 and setting $dD/dx = 0$. Therefore

$$t_c^* = \frac{1}{K_a - K_r} \ln\left\{\frac{K_a}{K_r}\left[1 - \frac{D_0(K_a - K_r)}{K_d L_0}\right]\right\} \tag{6.86}$$

If t_c^* is calculated to be less than zero, then D_c occurs at the outfall.

Equations 6.83 to 6.86 represent the DO response in a river due to a single point source discharge of CBOD. It can be noted (Eq. 6.85) that the critical deficit (D_c) or minimum level of DO (c_{min}) is linear to the CBOD concentration at the outfall only for the case of zero initial DO deficit. A reduction in this concentration (as for example through waste treatment) results in a reduction of D_c or an increase of the minimum level of DO. This method of improving DO is explored later in this chapter. Sample Problem 6.2 illustrates application of the above equations for a constant parameter stream and Sample Problem 6.3 examines a stream having two reaches.

The simple expression of the DO sag given by Eq. 6.84 and illustrated in Fig. 6.15 shows that the maximum impact of the waste discharge is downstream of the input location. Thus the water use interference, as, for example, the impact of low DO on a fishery, is not in general at the location of the discharge but may be displaced some distance downstream. Depending on the parameters of Eq. 6.86, the time of travel to the minimum value of DO may be a day or longer or equivalent to tens of miles. The impact may then be outside the limits of the municipal or industrial boundary of the discharge and indeed the minimum concentration of DO may occur in another state or even another country. It is this very interesting fact that was partly instrumental in the passage of federal interstate enforcement legislation in the United States so that a downstream state would have some recourse against an upstream state which discharged wastes resulting in a low DO downstream in the other state. Further, this transboundary pollutional effect also provided some of the original impetus for water quality treaties and agreements between countries.

The location of the maximum DO deficit downstream as given by t_c^* in Eq. 6.86 is seen to depend on the parameters K_a, K_r, and K_d, the CBOD at the outfall, and the initial concentration of deficit at the outfall, D_0. This latter quantity is calculated from a mass balance of oxygen at the outfall (see Eq. 2.18). Thus

$$c_0 = \frac{c_u Q_u + c_e Q_e}{Q_u + Q_e} \tag{6.87}$$

where c_0 is the DO after mixing, c_u and c_e are the DO concentrations upstream of the discharge and of the effluent, respectively, and Q_u and Q_e are the upstream flow and effluent flow, respectively. If the temperature of the effluent is different from that of the upstream flow, then a balance of heat around the outfall gives

$$T_0 = \frac{T_u Q_u + T_e Q_e}{Q_u + Q_e} \tag{6.88}$$

where T_0, T_u, and T_e are the temperature after mixing, the upstream temperature,

SAMPLE PROBLEM 6.2

DATA

Q_e = 4 MGD
CBOD5 = 30 mg/ℓ, f = 2.0
NH_3–N = 10 mg/ℓ

Q_u = 20 cfs

$c_u = c_s$

L_u = 0

N_u = 0

T = 28°C, U = 0.5 fps

Altitude of
stream = 6000 ft

K_a (meas.) = 0.80/day @ 20°C (base e)

From calibration
of BOD data
$\left\{ \begin{array}{l} K_r = K_d = 0.40/\text{day @ 28°C} \\ K_n = 0.40/\text{day @ 28°C} \end{array} \right.$

PROBLEM

Determine the maximum allowable ultimate oxygen demand (UOD) in the effluent entering the stream if the DO concentration is to equal or exceed 5 mg/l. Assume the effluent DO is equal to the stream's DO saturation concentration.

ANALYSIS

Loading

CBOD and NBOD may be treated as one UOD load since the decay rates are the same in the stream. Assume only the ammonia is significant in the NBOD.

$$\text{UOD(mg/l)} = f \times \text{CBOD5} + 4.57 \times NH_3\text{–N}$$

$$= 2.0 \times 30 + 4.57 \times 10 = 105.7 \text{ mg/l}$$

$$W(\text{UOD}) = 4 \text{ MGD} \times 105.7 \text{ mg/l} \times 8.34 = 3530 \text{ lb/day}$$

$$L_0 = \frac{W(\text{UOD})}{Q_u + Q_e} = \frac{3530 \text{ lb/day}}{(20 \text{ cfs} + 4 \times 1.548 = 26.2 \text{ cfs}) \times 5.4}$$

$$= 25.0 \text{ mg/l UOD}$$

Reaction Rates (Stream temperature = 28°C)

$$K_r = K_d = K_N = 0.40/\text{day @ 28°C}$$

$$K_a = K_a(20°) \times \theta^{T-20} = 0.8 \times (1.024)^{(28-20)} = 0.97/\text{day @ 28°C}$$

Maximum DO Deficit

With no initial deficit (D_0 = 0), maximum deficit will occur at a travel time of (Eq. 6.86)

$$t_c^* = \frac{1}{K_a - K_r} \ln\left(\frac{K_a}{K_r}\right) = \frac{1}{0.97 - 0.40} \ln\left(\frac{0.97}{0.40}\right) = 1.55 \text{ days}$$

(continued)

Sample Problem 6.2 (continued)

or

$$x_c = U \times t_c^* = (0.5 \text{ fps} \times 16.4 \text{ mpd/fps} = 8.2 \text{ mpd}) \times 1.55 \text{ days} = 12.7 \text{ mi}$$

From Eq. 6.85a

$$D_c = \frac{K_d}{K_a} L_0 \exp(-K_r t_c^*) = \frac{0.40}{0.97} \times 25.0 \times e^{-0.40 \times 1.55} = \underline{5.55 \text{ mg/l}}$$

DO Saturation Concentration

$$T(^\circ K) = 28 + 273.150 = 301.15$$

Using Eq. 6.15, $\ln c_s = 2.0577$ and $c_s = 7.83$ mg/l (see also Appendix C).
Correcting for altitude of 6000 ft (Eq. 6.18),

$$c_s = 7.83 \left(\frac{100 - 0.0035 \times 6000}{100} = 0.79 \right) = \underline{6.19 \text{ mg/l}}$$

Minimum DO Concentration

With existing load, DO(min) $= c_s - D_c = 6.19 - 5.55 = 0.64$ mg/l, well
below desired concentration of 5 mg/l. $\therefore$ Must reduce UOD load.

Since allowable DO deficit $= 6.19 - 5 = 1.19$ mg/l, and since D_c is
directly proportional to L_0, thus $W(\text{UOD})$, and finally to the effluent UOD concentration, $(L_0)_e$

$$\frac{D_c}{(L_0)_e} = \frac{D_{c(\text{allowable})}}{(L_0)_e^{(\text{allowable})}} \quad \text{and} \quad \frac{5.55}{105.7} = \frac{1.19}{(L_0)_e^{(\text{allowable})}}$$

and $(L_0)_e(\text{allowable}) = (1.19/5.55) \times 105.7 = \underline{23 \text{ mg/l UOD}}$

Note: Reviewing Table 6.11, it is seen that a secondary plant with nitrification,
with an effluent UOD concentration range of approximately 10–30 mg/l, would
prove satisfactory.

SAMPLE PROBLEM 6.3

DATA

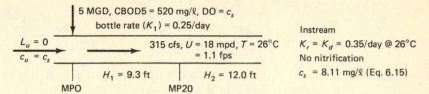

5 MGD, CBOD5 = 520 mg/ℓ, DO = c_s
bottle rate (K_1) = 0.25/day

Instream
$K_r = K_d$ = 0.35/day @ 26°C
No nitrification
c_s = 8.11 mg/ℓ (Eq. 6.15)

$L_u = 0$
$c_u = c_s$

315 cfs, U = 18 mpd, T = 26°C
= 1.1 fps

H_1 = 9.3 ft H_2 = 12.0 ft

MPO   MP20

PROBLEM

Calculate the minimum DO concentration in the stream due to the discharge at mile point zero (MP0). Note the changing depth.

ANALYSIS

Loading

None from upstream; no initial deficit at MP0.

$$W(CBOD5) = 5\ \text{MGD} \times 520\ \text{mg/l} \times 8.34 = 21{,}700\ \text{lb/day}$$

With bottle rate of 0.25/day, ultimate CBOD:CBOD5 is (Eq. 6.6)

$$f = \frac{L_0}{y_5} = \{1 - \exp[-5 \times (K_1 = 0.25)]\}^{-1} = 1.40$$

$$W(CBODU) = 1.4 \times 21700 = 30400\ \text{lb/day}$$

At MP0, then, the instream UOD concentration (L_0) is

$$L_0 = \frac{W(CBODU)}{Q} = \frac{30{,}400\ \text{lb/day}}{315\ \text{cfs} \times 5.4} = 17.9\ \text{mg/l}$$

Reaeration Coefficients (Eqs. 6.27 and 6.32)

$$K_a(26°C) = K_a(20°)\theta^{T-20} = K_a(20°) \times 1.024^{(26-20)} = 1.16K_a(20°)$$

In reach (1) (MP0–20),

$$K_a = 1.16 \times \left(12.9\ \frac{U^{1/2}}{H^{3/2}} = 12.9 \times \frac{1.1^{1/2}}{9.3^{3/2}} \right) = 0.553/\text{day}$$

and in (2) (MP20→),

$$K_a = 1.16\left(12.9 \times \frac{1.1^{1/2}}{12.0^{3/2}} \right) = 0.337/\text{day}$$

(continued)

Sample Problem 6.3 (continued)

Check Reach (1) for Maximum Deficit

Let $x = 0$ @ MP0; $L_0 = 17.9$ mg/l, $D_0 = 0$.

$$x_c = \left(\frac{U}{K_a - K_r}\right) \ln \left[\frac{K_a}{K_r}\left(1 - \frac{K_a - K_r}{K_d}\frac{D_0}{L_0}\right)\right]$$ [Eq. (6.86)]

$$x_c = \frac{18 \text{ mpd}}{(0.553-0.35)/\text{day}} \ln \left(\frac{0.553}{0.35}\right) = 40.5 \text{ mi}$$

Since $x_c > 20$ mi, the end of reach (1), calculate L_{20} and D_{20} at $x = 20$ mi. Then compare D_{20} with maximum deficit in reach (2). At $x = 20$ mi, $t^* = x/U = 20/18 = 1.111$ days.

$$L_{20} = L_0 e^{-K_r t^*} = 17.9 \times e^{-0.35 \times 1.111} = 12.1 \text{ mg CBODU/l}$$

$$D_{20} = \frac{K_d L_0}{K_a - K_r}(e^{-K_r t^*} - e^{-K_q t^*}) = \frac{0.35 \times 17.9}{0.553 - 0.35}(e^{-0.35 \times 1.111} - e^{-0.553 \times 1.111})$$

$$\underline{D_{20} = 4.22 \text{ mg/l}} \text{ [maximum deficit in reach (1)]}.$$

Maximum Deficit in Reach (2)

Let $\bar{x} = 0$ at MP20, where $L_0 = 12.1$ mg/l, $D_0 = 4.22$ mg/l, and $K_a = 0.377$/day.

$$\bar{x}_c = \frac{18.0}{(0.377 - 0.35)} \ln \left[\frac{0.377}{0.35}\left(1 - \frac{0.377 - 0.35}{0.35}\cdot\frac{4.22}{12.1}\right)\right] = 31.4 \text{ mi}$$

$\therefore$ Maximum DO deficit occurs in reach (2) at MP51.4 where

$$\bar{t}_c^* = \frac{\bar{x}_c}{U} = \frac{31.4}{18} = 1.744 \text{ days from MP20}$$

$$D_c = \frac{K_d L_0}{K_a - K_r}(e^{-K_r \bar{t}_c^*} - e^{-K_a \bar{t}_c^*}) + D_0 e^{-K_a \bar{t}_c^*}$$ [Eq. (6.83)]

$$D_c = \frac{0.35 \times 12.1}{0.377 - 0.35}(e^{-0.35 \times 1.744} - e^{-0.377 \times 1.744})$$

$$+ 4.22 e^{-0.377 \times 1.744} = \underline{6.11 \text{ mg/l}}$$

Since D_c @ MP51.4 $> D_{20}$, $\underline{\underline{D(\max) = 6.11 \text{ mg/l}}}$, and $\underline{\underline{\text{minimum DO}}}$ is $8.11 - 6.11 = \underline{\underline{2.00 \text{ mg/l}}}$.

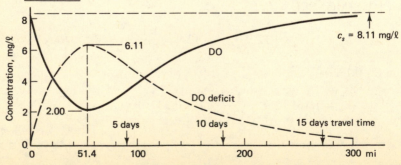

and the effluent temperature, respectively. The DO deficit at the outfall is then given by

$$D_0 = c_s(T_0) - c_0 \qquad (6.89)$$

where $c_s(T_0)$ is the saturation value at temperature T_0, (Strictly speaking if the temperature is changing downstream, the saturation level of DO is also varying and the simple solutions must be modified. For practical purposes, however, the saturation concentration is usually taken as a constant over a given reach of the river as determined from temperature measurements.)

6.4.1.1 Anaerobic Conditions

Some rivers and streams may be so heavily loaded with CBOD, as for example where there is no treatment of the discharge, that the DO would approach complete depletion and anaerobic conditions would result. Gundelach and Castillo (1976) have analyzed this situation in some detail. Assuming no other sources and sinks of BOD or DO, a single point source and $K_r = K_d$, then the rate of change of the CBOD downstream will be satisfied by the rate at which oxygen can be transferred across the surface of the stream and mixed into the river. Thus

$$\frac{dL}{dt} = -K_a c_s \qquad \text{for} \quad x_i \leq x \leq x_f \qquad (6.90)$$

for x_i and x_f as the beginning and end of the anaerobic reach (see Fig. 6.16).

The beginning of the anaerobic reach can be obtained from Eq. 6.84 when $c = 0$. At that location and downstream, Eq. 6.90 gives as a solution for the CBOD profile,

$$L = L_i - K_a c_s \left(\frac{x - x_i}{U} \right) \qquad (6.91)$$

where L_i is the CBOD at the beginning of anaerobic conditions, and x_i is the location where the DO reaches zero. The location at which the DO will begin to be aerobic is given by

$$K_a c_s = K_d L_f \qquad (6.92)$$

for L_f as the BOD at the end of the anaerobic reach. The downstream point of aerobic recovery is then given by Eqs. 6.91 and 6.92 as

$$x_f = x_i + \frac{U}{K_d} \left(\frac{K_d L_i - K_a c_s}{K_a c_s} \right) \qquad (6.93)$$

Equation 6.93 can be used to estimate the length of the anaerobic reach of the river and then Eq. 6.91 can be used to estimate the CBOD at the end of the anaerobic reach. Once aerobic conditions are reached again, the preceding equations (6.83 and 6.84) will apply again. Sample Problem 6.4 incorporates an anaerobic reach.

The above formulation for both the aerobic and anaerobic cases were developed for the response of DO due to the discharge of CBOD. If the waste source also discharges oxidizable nitrogen, that is, ammonia plus organic nitrogen and

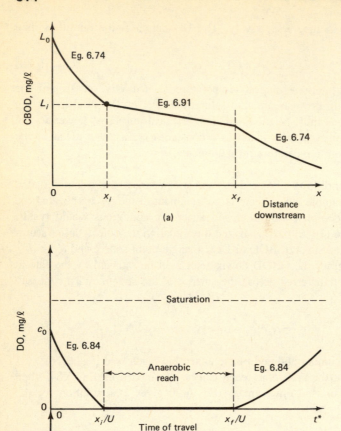

Figure 6.16 Variation of CBOD and DO for complete oxygen depletion in rivers.

conditions are judged suitable for nitrification, then this additional source of oxygen demand must be included. Since the nitrifying bacteria are aerobic organisms, nitrification does not proceed under conditions of DO less than about 1 mg/l. The anaerobic case therefore need not include the oxidation of nitrogen until the DO begins to exceed about 1 mg/l. Considering the oxidizable nitrogen in terms of the NBOD, the appropriate equation for the distribution of the NBOD in the downstream direction is

$$L^N = L_0^N \exp\left(-\frac{K_N X}{U}\right) \tag{6.94}$$

where it can be recalled (Eq. 6.11) that L^N is given approximately by 4.57 times the organic plus ammonia nitrogen. The DO deficit due to the oxidation of the NBOD designated D^N is

$$D^N = \frac{K_n L_0^N}{K_a - K_{dN}}\left[\exp\left(-K_N \frac{x}{U}\right) - \exp\left(-K_a \frac{x}{U}\right)\right] \tag{6.95}$$

SAMPLE PROBLEM 6.4

DATA

Q_e = 10 cfs
L_e = CBODU = 100 mg/ℓ
c_e = 4 mg/ℓ

$K_r = K_d$ = 0.2/day
K_a = 0.6/day

Q = 10 cfs
L = 0
c = 7 mg/ℓ

$\phi \longrightarrow x$ U = 8 mpd, c_s = 8.00 mg/ℓ

PROBLEM

Determine the dissolved oxygen concentration (c) profile in the stream. Assume all reaction rates are at stream temperature.

ANALYSIS

Initial Deficit

DO balance at $x = 0$

$$10 \text{ cfs} \times 7 \text{ mg/1} + 10 \text{ cfs} \times 4 \text{ mg/1} = 20 \text{ cfs} \times c_0 \qquad \text{[Eq. (6.87)]}$$

$$c_0 = 5.5 \text{ mg/1}$$

With c_s = 8.00 mg/1, D_0 = 8.00 − 5.5 = 2.50 mg/1. Also initial CBODU = L_0 = $(10 \times 0 + 10 \times 100)/20$ = 50 mg/1.

Maximum Deficit

Assume an aerobic stream, $c > 0$:

$$t_c^* = \frac{1}{0.6 - 0.2} \ln \left[\frac{0.6}{0.2} \left(1 - \frac{2.50}{50} \times \frac{0.6 - 0.2}{0.2} \right) \right]$$

$$= 2.48 \text{ days} \qquad \text{[Eq. (6.86)]}$$

$$D_c = \frac{0.2 \times 50}{0.6 - 0.2} (e^{-0.2 \times 2.48} - e^{-0.6 \times 2.48})$$

$$+ 2.50 e^{-0.6 \times 2.48} = 10.14 \text{ mg/1} \qquad \text{[Eq. (6.83)]}$$

Since $c_{\min} = c - D_c$ = 8.00 − 10.14 = −2.14 mg/1, the stream will go anaerobic and an analysis according to Section 6.4.1.1 must be performed.

(continued)

Sample Problem 6.4 (continued)

Anaerobic Analysis

Locate x_i (Upstream End of
Anaerobic Reach)

Set deficit = 8.00 mg/l (where DO = 0), and, using Eq. 6.83, solve for x_i by
trial and error.

$$D_1 = \frac{0.2 \times 50}{0.6 - 0.2} (e^{-0.2x/8}) + 2.50e^{-0.6x/8} \qquad \text{[Eq. (6.83)]}$$

Trial x (mi)	2	4	6	7	7.5	7.72
Deficit (mg/l)	4.41	5.95	7.17	7.67	7.91	8.00

Therefore DO is first zero at $x = x_i = 7.72$ mi.

Calculate x_f (Downstream End of
Anaerobic Reach)

$$x_f = x_i + \frac{U}{K_d} \left(\frac{K_d L_i - K_a c_s}{K_a c_s} \right) \qquad \text{[Eq. (6.93)]}$$

where L_i = CBODU concentration at $x = x_i$

$L_i = L_0 e^{-K_r x_i / U} = 50e^{-0.2 \times 7.72/8} = 41.22$ mg/l

$$x_f = 7.72 + \frac{8}{0.2} \left(\frac{0.2 \times 41.22 - 0.6 \times 8.00}{0.6 \times 8.00} \right) = \underline{36.42 \text{ mi}}$$

Deficit for $x \geq x_f$

At x_f, the CBODU remaining is given by Eq. 6.91:

$$L_f = L_i - K_a c_s \frac{(x_f - x_i)}{U} = 41.22 - 0.6 \times 8.00 \left(\frac{36.42 - 7.72}{8} \right)$$

$L_f = 24.00$ mg/l   and   the deficit = 8.00 mg/l

$$\therefore D_3 = \frac{0.2 \times 24.00}{0.6 - 0.2} (e^{-0.2(x - x_f)/8} - e^{-0.6(x - x_f)/8}) + 8.00e^{-0.6(x - x_f)/8}$$

Dissolved Oxygen Concentrations

$$c_1 = 8.00 - D_1 \qquad 0 \leq x \leq 7.72 \text{ mi}$$
$$c_2 = 0.00 \qquad 7.72 \leq x \leq 36.42 \text{ mi}$$
$$c_3 = 8.00 - D_3 \qquad x \geq 36.42 \text{ mi}$$

Sample Problem 6.4 (continued)

The resulting dissolved oxygen concentrations are shown below.

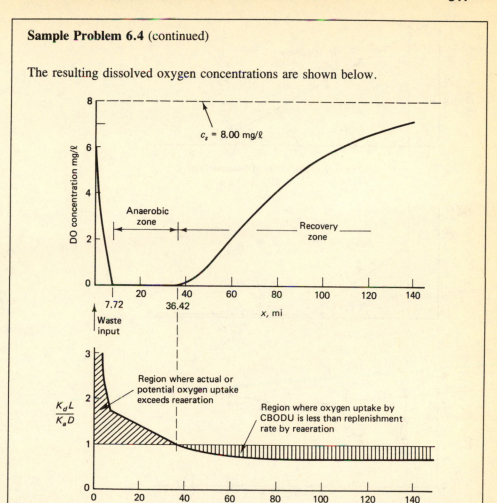

The plot above of the rate of oxygen uptake by CBODU (K_dL) divided by the DO replenishment rate (K_aD) indicates that recovery begins when $K_dL/K_aD \leq 1$.

Figure 6.17 shows an example of the decay of CBOD, the TKN, and the resulting sag and recovery of DO for Flat Creek, Virginia (Bathala et al., 1979).

6.4.2 Multiple Point Sources

When there is more than one point source along the length of the stream, the effect on the DO is cumulative. That is, each source contributes to the DO deficit depending on its BOD loading and the individual contributions are additive. Mathematically this results from the fact that the basic BOD–DO equations are linear so that the principle of superposition applies. Figure 6.18 illustrates two individual DO sag curves and their sum. Computationally, the most direct way to compute

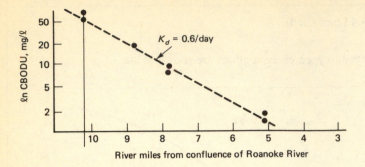

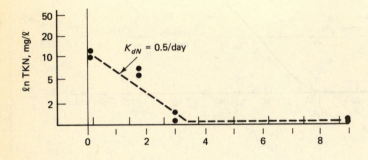

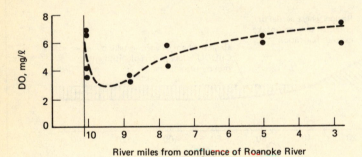

Figure 6.17 Example of BOD, TKN, and DO sag, Flat Creek, a tributary of Roanoke River, VA. Redrawn from Bathala et al. (1979). Reprinted by permission of American Society of Civil Engineers.

the resulting total DO sag is to calculate the BOD and DO at the downstream location just before the next input. A mass balance can then be made to calculate the new concentration at the outfall after mixing with the next downstream point source. The preceding equations can then be applied to the next reach. Thus let $L_1(x_1)$ and $c_1(x_1)$ be the BOD and DO at the end of the first reach given by x_1 but just before the next input W_2 is discharged. Then the new BOD concentration at the second outfall after mixing are

$$L_{02} = \frac{L_1(x_1)Q_1 + L_{e2}Q_{e2}}{Q_1 + Q_{e2}} \qquad (6.96)$$

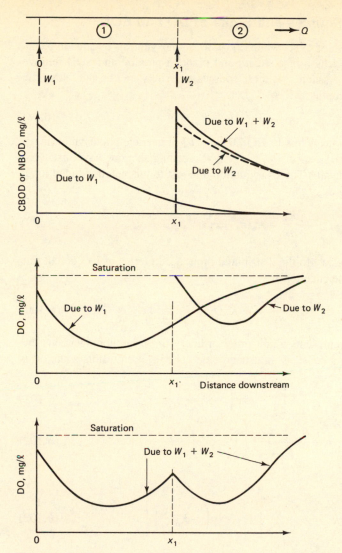

Figure 6.18 Superposition of BOD and DO, two waste inputs in a stream.

where L_{02} is the new initial concentration for the second reach and L_{e2} and Q_{e2} are the effluent BOD concentration and flow, respectively, for the second input. Similarly, the new DO concentration at the beginning of the second reach, c_{02}, is

$$c_{02} = \frac{c_1(x_1)Q_1 + c_{e2}Q_{e2}}{Q_1 + Q_{e2}} \tag{6.97}$$

for c_{e2} as the effluent DO for the second source. With these new initial concentrations, Eqs. 6.83 and 6.84 can be used for the second reach to obtain the resulting DO profile.

6.4.3 Distributed Sources and Sinks of DO and BOD

As discussed previously and as shown in Eq. 6.80, photosynthesis by aquatic plants adds oxygen along the length of a stream and plant respiration and SOD remove oxygen from the river. Assume that these sources and sinks are constant along the length of the river beginning at $x = 0$. Let

$$S_d = p_a - R - S_B' \tag{6.98}$$

be the net distributed source/sink of DO $[M/L^3 \cdot T]$ of average photosynthesis p_a, respiration R, and SOD S_B' all in $[M/L^3 \cdot T]$. (Preceding sections have discussed each of these in some detail.) The DO equation due just to this net distributed input is given from Eq. 6.80 as

$$U \frac{dc_d}{dx} = K_a(c_s - c_d) + S_d \tag{6.99}$$

where c_d is the DO due just to the distributed input S_d. In terms of DO deficit due to the net input, D_d, the equation is

$$U \frac{dD_d}{dx} = -K_a D_d - S_d \tag{6.100}$$

Equation 6.100 represents a first-order ordinary linear nonhomogeneous differential equation with a constant right-hand side. The solution to this equation is

$$D_d = -\frac{S_d}{K_a}\left[1 - \exp\left(-K_a \frac{x}{U}\right)\right] \tag{6.101}$$

$$= \frac{R + S_B' - p_a}{K_a}\left[1 - \exp\left(-K_a \frac{x}{U}\right)\right] \tag{6.102}$$

In terms of DO,

$$c_d = c_s + \frac{S_d}{K_a}\left[1 - \exp\left(-K_a \frac{x}{U}\right)\right] \tag{6.103}$$

Figure 6.19 illustrates these solutions. The important point to notice about this response is that the DO does not respond to its maximum at the location of the beginning of the distributed source but some distance must elapse before the equilibrium or maximum level is reached. The distance to reach that maximum response is a function of the reaeration rate. Thus, at $x = 64$ mi, if $U = 1$ fps and $K_a = $ day^{-1}, 95% of the full effect of the distributed source would be exerted. After this "spatial transient" has been reached, the deficit response is

$$D_d = \frac{-S_d}{K_a} \tag{6.104}$$

For streams with low reaeration (slow moving, sluggish, and deep), the equilibrium DO deficit may be large but the time of travel to reach that maximum

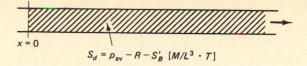

$$S_d = p_{av} - R - S'_B \quad [M/L^3 \cdot T]$$

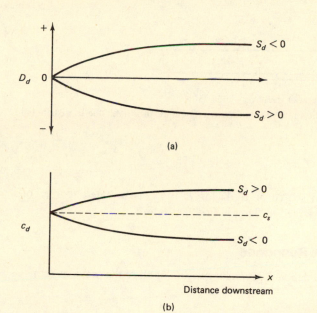

(a)

(b)

Figure 6.19 Response in river due to net distributed source/sink of DO. (a) DO deficit. (b) DO.

response is also great so that the full DO deficit response given by Eq. 6.104 may never be reached. Conversely, for streams with high reaeration (rapidly moving, shallow), the maximum DO deficit response is reduced but so is the time of travel to reach the maximum. Figure 6.20 shows these effects for a range of K_a values.

In addition to the above sources and sinks of DO, there may be a distributed source of BOD to a given reach of river. Such a source of BOD may result from degradation products from sediment decomposition and subsequently diffusing from the sediment into the overlying water where products are exerted as a BOD. Also, BOD may be leaching into a river reach from a landfill along the length of the river. In both cases, there is no flow added to the river by the distributed BOD source. The DO deficit response due to this source of BOD, D_r, is

$$D_r = \left(\frac{K_d}{K_a K_r}\left[1 - \exp\left(-K_a \frac{x}{U}\right)\right]\right.$$
$$\left. - \left\{\frac{K_d}{(K_a - K_r)K_r}\left[\exp\left(-K_r \frac{x}{U}\right) - \exp\left(-K_a \frac{x}{U}\right)\right]\right\}\right)L_{rd} \quad (6.105)$$

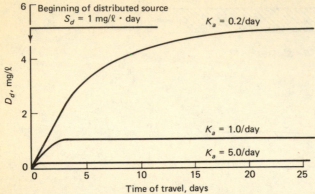

Figure 6.20 Effect of reaeration rate on maximum DO deficit in a river due to distributed source/sink.

where L_{rd} is the distributed source of BOD $[M/L^3 \cdot T]$. Distributed sources of BOD and DO deficit are analyzed in Sample Problem 6.5.

6.4.4 Total DO Deficit Response

The full solution for the DO deficit is given by the sum of all of the previous terms. Thus

$$D = D_0 \exp\left(-K_a \frac{x}{U}\right) \tag{a}$$

$$+ \left\{ \frac{K_d}{K_a - K_r} \left[\exp\left(-K_r \frac{x}{U}\right) - \exp\left(-K_a \frac{x}{U}\right) \right] \right\} L_0 \tag{b}$$

$$+ \left\{ \frac{K_N}{K_a - K_N} \left[\exp\left(-K_N \frac{x}{U}\right) - \exp\left(-K_a \frac{x}{U}\right) \right] \right\} L_0^N \tag{c}$$

$$+ \left(\frac{K_d}{K_a K_r} \left[1 - \exp\left(-K_a \frac{x}{U}\right) \right] \right.$$

$$\left. - \left\{ \frac{K_d}{(K_a - K_r) K_r} \left[\exp\left(-K_r \frac{x}{U}\right) - \exp\left(-K_a \frac{x}{U}\right) \right] \right\} \right) L_{rd} \tag{d}$$

$$- \left[1 - \exp\left(-K_a \frac{x}{U}\right) \right] \frac{P_a}{K_a} \tag{e}$$

$$+ \left[1 - \exp\left(-K_a \frac{x}{U}\right) \right] \left(\frac{R}{K_a} \right) \tag{f}$$

$$+ \left[1 - \exp\left(-K_a \frac{x}{U}\right) \right] \left(\frac{S_B'}{K_a} \right) \tag{g} \quad (6.106)$$

SAMPLE PROBLEM 6.5

DATA

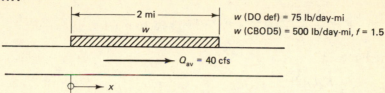

$K_r = 0.4/\text{day}$, $K_d = 0.3/\text{day}$, $K_a = 0.5/\text{day}$, $A = 200\,\text{ft}^2$

All reaction rates are at stream temperature.

PROBLEM

Determine the DO deficit in the stream due to the distributed sources of CBOD and DO deficit.

ANALYSIS

Effect of $w(\text{CBOD5})$

$w(\text{CBODU}) = w(\text{CBOD5}) \times f = 500\,\text{lb/day-mi} \times 1.5 = 750\,\text{lb/day} \cdot \text{mi}$

$$L_{rd} = \frac{w(\text{CBODU})}{A} = \frac{750\,\text{lb/day} \cdot \text{mi}}{200\,\text{ft}^2}\left(\frac{1\,\text{mi}}{5280\,\text{ft}} \times \frac{1\,\text{ft}^3}{62.4\,\text{lb}} \times \frac{10^6\,\text{mg/l}}{\text{lb/lb}}\right)$$

$$= 11.38\,\text{mg/l} \cdot \text{day}$$

For $0 \le x \le 2$ mi

$$D_r = \frac{K_d L_{rd}}{K_a K_r}(1 - e^{-K_a x/U}) - \frac{K_d L_{rd}}{(K_a - K_r)K_r}(e^{-K_r x/U} - e^{-K_a x/U}) \quad \text{[Eq. (6.105)]}$$

With $U = Q/A = (40\,\text{cfs}/200\,\text{ft}^2 = 0.2\,\text{fps}) \times 16.4\,\text{mpd/fps} = 3.28\,\text{mpd}$,

$$D_r = \frac{0.3 \times 11.38}{0.5 \times 0.4}(1 - e^{-0.5x/3.28})$$

$$- \frac{0.3 \times 11.38}{(0.5 - 0.4)0.4}(e^{-0.4x/3.28} - e^{-0.5x/3.28})$$

$$D_r = 17.07(1 - e^{-0.152x}) - 85.35(e^{-0.122x} - e^{-0.152x}) \quad\quad (A)$$

For $x \ge 2$ mi

$$D_r(2) = 17.07(1 - e^{-0.304}) - 85.35(e^{-0.244} - e^{-0.304})$$

$$= 0.58\,\text{mg/l}$$

(continued)

Sample Problem 6.5 (continued)

$$L_u = \frac{L_{rd}}{K_r} (1 - e^{-K_r x/U}) \qquad [\text{Eq. (2.24)}]$$

$$L_u(2) = \frac{11.38}{0.4} (1 - e^{-0.4 \times 2/3.28}) = 6.16 \text{ mg/l}$$

∴ Using Eq. 6.83

$$D = \frac{K_d L_u(2)}{K_a - K_r} (e^{-K_r(x-2)/U} - e^{-K_a(x-2)/U}) + D_r(2)e^{-K_a(x-2)/U}$$

$$D = \frac{0.3 \times 6.16}{0.5 - 0.4} (e^{-0.4(x-2)/3.28} - e^{-0.5(x-2)/3.28})$$

$$+ 0.58e^{-0.5(x-2)/3.28}$$

$$D = 18.48(e^{-0.122(x-2)} - e^{-0.152(x-2)}) + 0.58e^{-0.152(x-2)} \qquad (B)$$

Effect of w(DO deficit)

For $0 \leq x \leq 2$ mi

Treat as distributed source as per Eq. 2.22, where

$$S_D = \frac{w(D_0 \text{ deficit})}{A} = \frac{75 \times 10^6}{200 \times 5280 \times 62.4} = 1.138 \text{ mg/l} \cdot \text{day}$$

$$D_d = \frac{S_D}{K_a} (1 - e^{-K_a x/U}) \qquad [\text{Eq. (2.24)}]$$

$$D_d = \frac{1.138}{0.5} (1 - e^{-0.5x/3.28})$$

$$D_d = 2.276(1 - e^{-0.152x}) \qquad (C)$$

For $x \geq 2$ mi

$$D_d(2) = 2.276(1 - e^{-0.152 \times 2}) = 0.60 \text{ mg/l}$$

$$D = D_d(2)e^{-K_a(x-2)/U} = 0.60 \times e^{-0.5(x-2)/3.28}$$

$$D = 0.60e^{-0.152(x-2)} \qquad (D)$$

Composite Effect of w(CBOD5) and w(DO deficit)

Using the principle of superposition, the total DO deficit equals the sum of its two components in each reach:

$$\underline{0 \leq x \leq 2} \qquad D = \text{Eq. (A)} + \text{Eq. (C)}$$

$$\underline{x \geq 2} \qquad D = \text{Eq. (B)} + \text{Eq. (D)}$$

Sample Problem 6.5 (continued)

In the reach $x \geq 2$, the initial conditions are $L_0 = 6.16$ mg/l and $D_0 = D_r(2) + D_d(2) = 0.58 + 0.60 = 1.18$ mg/l. The maximum deficit may be located using Eq. 6.86.

$$\bar{t}_c^* = \frac{1}{K_a - K_r} \ln \left[\frac{K_a}{K_r} \left(1 - \frac{K_a - K_r}{K_d} \frac{D_0}{L_0} \right) \right]$$

$$\bar{t}_c^* = \frac{1}{0.5 - 0.4} \ln \left[\frac{0.5}{0.4} \left(1 - \frac{0.5 - 0.4}{0.3} \times \frac{1.18}{6.16} \right) \right] = 1.57 \text{ days}$$

and

$$\bar{x}_c = U \times \bar{t}_c^* = 3.28 \text{ mpd} \times 1.57 \text{ days} = 5.16 \text{ mi}$$

$$\therefore D_c = \frac{K_d}{K_a} L_0 e^{-K_r \bar{t}_c^*} = \frac{0.3}{0.5} \times 6.16 \times e^{-0.4 \times 1.57}$$

$$= 1.97 \text{ mg/l} \qquad \text{[Eq. (6.85a)]}$$

Summary Plots

The composite DO deficit, in addition to the component effects, are shown below.

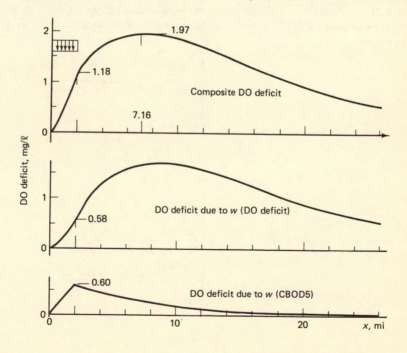

where the various parts of the solution are, respectively,

(a) the initial value of DO deficit;
(b) the deficit due to point source of CBOD;
(c) the point source of NBOD contribution to DO deficit;
(d) the distributed source of BOD input with no significant addition to river flow;
(e) the deficit due to distributed gross photosynthesis;
(f) the distributed plant respiration effect; and
(g) the distributed SOD effect.

The component parts of Eq. 6.106 are illustrated in Fig. 6.21. In this equation, additions of deficit along the stream have not been listed although they can be included where necessary. Since all systems are linear, the independent sources and sinks can be summed for the total steady-state deficit response. The essence of the steady-state DO analysis is to choose a suitable model from the available component parts. In some instances, for example, the distributed source of CBOD may be quite small relative to point loads. In other cases, field data may indicate that plant respiration balances photosynthesis so that this input does not affect the overall deficit.

Figure 6.21 DO deficits in a stream due to (a) initial value, (b) point CBOD or NBOD source, (c) distributed BOD source, (d) algal respiration or benthal demand, (e) gross photosynthesis.

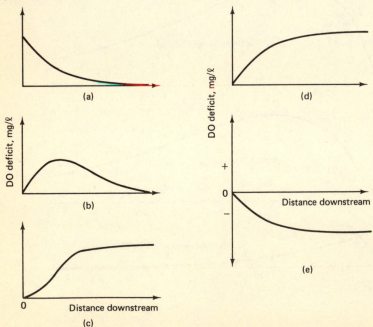

The procedure for application of a DO river model of any complexity is identical to the case with multiple point sources. A suitable form of Eq. 6.106 is applied to a specific reach of the river with appropriate constant system parameters. Values of D as a function of distance x are computed at some convenient spatial interval. When any of the system parameters change significantly or a new waste source is introduced, new initial values are computed. These values include the new initial deficit D_0, and the new initial carbonaceous BOD, L_0, and nitrogenous BOD, L_0^N. These values are computed using simple mass balances. Any new point sources of CBOD or NBOD are incorporated in the mass balance for L_0 and L_0^N.

The Black River, New York (Hydroscience, 1974) is an example of a riverine analysis with multiple sources. In Fig. 6.22, it is seen that a 59 reach model was used for the 92-mi-long river and its two main tributaries. The large number of reaches was required due to the eight municipal and eight paper industry discharges, the tributaries, the 16 dams on the river (see Section 6.8.4.1), and the variable flow and geometry—all of which require new reaches in order to reinitialize BOD and DO concentrations or reset kinetic coefficients. Verifications of observed DO concentrations were performed under three conditions, two of which are shown in Fig. 6.22 for relatively low flows of 1700 and 1400 cfs at Watertown and under a higher temperature condition in Aug. 1973 (18.5–26.5°C) and a lower temperature in Nov. 1973 (7.7–11.0 °C). In Figs. 6.22(b) and (c), note the increase in the saturation concentration as the temperature decreases and the decrease in DO deficit as kinetic coefficients (K_r) decrease with temperature. The sharp increases in DO are the result of dilution by the tributaries or oxygenation as the river falls from 10 to 35 ft over instream dams. A projection made for a very low flow condition (800 cfs) and high temperature everywhere (25 °C), for reduced municipal and industrial sources (secondary treatment for municipalities, best practicable treatment for industries) without taking credit for reaeration over the dams, showed that water quality standards could be met [Fig. 6.22(d)].

6.5 DISSOLVED OXYGEN ANALYSIS— TIDAL RIVERS AND ESTUARIES

6.5.1 Single Point Source

The variation of DO in estuarine systems must include the effects of tidal mixing and dispersion. The basic concepts of this effect are discussed in Chapter 3 where it is shown that the influence of a point waste discharge extends both upstream and downstream from the outfall, The BOD–DO equations that must be solved are

$$0 = -U\frac{dL}{dx} + E\frac{d^2L}{dx^2} - K_rL \qquad (6.107a)$$

$$0 = -U\frac{dL^N}{dx} + E\frac{d^2L^N}{dx^2} - K_NL \qquad (6.107b)$$

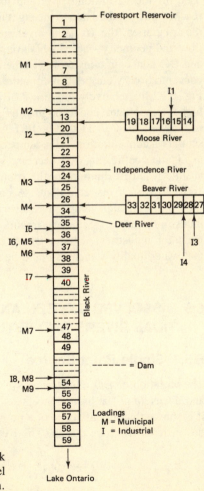

Figure 6.22 Dissolved oxygen model of Black River, NY. From Hydroscience (1974). (a) Model segmentation. (b) and (c) Calibration/verification. (d) Projection.

328

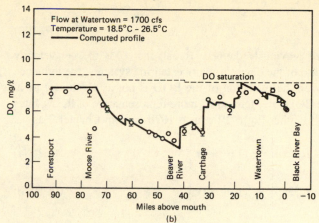

(b)

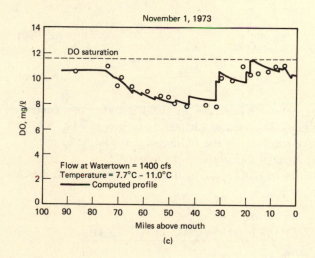

(c)

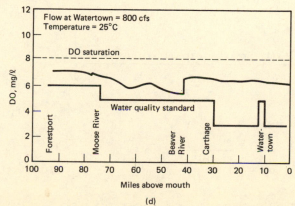

(d)

$$0 = -U\frac{dc}{dx} + E\frac{d^2c}{dx^2} - K_dL(x) - K_NL^N(x)$$

$$+ K_a(c_s - c) - S_B' + p_a - R \qquad (6.108)$$

where all terms have been previously defined. Recall that U is the net velocity through the estuary. Because of the depth of many estuarine systems, the influence of the benthic community on the oxidation of the BOD is not as significant as in streams. Therefore, bottle BOD rates tend to be more representative of the K_d rates in estuaries. The solution to either Eq. 6.107a or 6.107b is from Chapter 3 as

$$L = L_0 \exp\left[\frac{U}{2E}(1 + \alpha_r)x\right] \qquad x \le 0 \qquad (6.109a)$$

$$= L_0 \exp\left[\frac{U}{2E}(1 + \alpha_r)x\right] \qquad x \ge 0 \qquad (6.109b)$$

where

$$L_0 = \frac{W}{Q\alpha_r} \qquad (6.109c)$$

$$\alpha_r = \sqrt{1 + \frac{4K_rE}{U^2}} \qquad (6.109d)$$

The equations are identical to the equations for the distribution of a nonconservative substance in an estuary discussed in Chapter 3 and given in Eq. 3.8. In this case, the nonconservative substance is the BOD and it should be recalled in Eq. 6.109c that Q is the net nontidal flow through the estuary.

For the estuary distribution of BOD given by these equations, the DO deficit solution can be obtained as

$$D = \frac{K_d}{K_a - K_r}\frac{W}{Q}\left\{\frac{\exp[(U/2E)(1 + \alpha_r)x]}{\alpha_r} - \frac{\exp[(U/2E)(1 + \alpha_a)x]}{\alpha_a}\right\}$$

$$x \le 0 \qquad (6.110a)$$

$$= \frac{K_d}{K_a - K_r}\frac{W}{Q}\left\{\frac{\exp[(U/2E)(1 - \alpha_r)x]}{\alpha_r} - \frac{\exp[(U/2E)(1 - \alpha_a)x]}{\alpha_a}\right\}$$

$$x \ge 0 \qquad (6.110b)$$

where

$$\alpha_a = \sqrt{1 + \frac{4K_aE}{U^2}} \qquad (6.110c)$$

Figure 6.23 shows the form of the distribution of the BOD, DO deficit, and DO in an estuary. In general, because of the tidal mixing upstream, the magnitude of

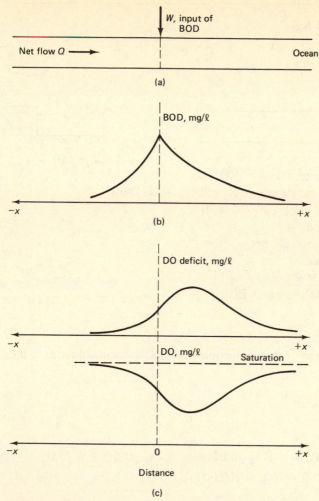

Figure 6.23 Distribution of (a) BOD, (b) DO deficit, and (c) DO in an estuary due to BOD load W.

the peak DO deficit (or, conversely, minimum DO concentration) decreases in an estuary in comparison to a stream for the same BOD input. Also, the location of the minimum value of the DO tends to move closer to the outfall. An estuarine dissolved oxygen calculation due to a single point source is found in Sample Problem 6.6. This sample problem also shows the equations and calculations for the maximum DO deficit and its location.

In the computation of the reaeration coefficient for an estuary using Eq. 6.27, the velocity may be selected according to the following guideline:

1. When the net nontidal velocity, U_0, exceeds the average tidal velocity (U_T), the net nontidal velocity is used.

SAMPLE PROBLEM 6.6

DATA

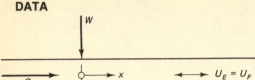

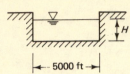

$$Q = \text{freshwater flow} = 1000 \text{ cfs}$$
$$U_E = U_F = \text{maximum tidal velocity} = 2 \text{ knots}$$
$$H(\text{MLW}) = 18 \text{ ft; tidal range} = 4 \text{ ft}$$
$$E = 5 \text{ smpd}$$
$$S = \text{salinity} = 25 \text{ ppt}$$
$$T = 20°C$$
$$K_r = K_d = 0.25/\text{day} \quad @ \ 20°C$$
$$W = 668{,}000 \text{ lb/day CBODU}$$

PROBLEM

For the constant area estuary, calculate the minimum DO concentration due to the discharge.

ANALYSIS

Reaeration Coefficient

$$H(\text{MW}) = H(\text{MLW}) + 0.5 \times \text{tidal range} = 18 + 0.5 \times 4 = 20 \text{ ft}$$
$$A = 5000 \text{ ft} \times 20 \text{ ft} = 100{,}000 \text{ ft}^2$$
$$U = \frac{Q}{A} = \frac{1000 \text{ cfs}}{100{,}000 \text{ ft}^2} = 0.01 \text{ fps} = 0.164 \text{ mpd}$$
$$\text{av } U(\text{tidal}) = \frac{U_E + U_F}{\pi}$$
$$= \frac{2 + 2}{\pi} \text{ knots} \times 1.688 \text{ fps/knots} = 2.15 \text{ fps}$$

Since average tidal velocity > freshwater velocity (2.15 > 0.01), use average tidal velocity to calculate reaeration coefficient (see Section 6.5.1).

$$K_a = \frac{12.9 U^{1/2}}{H^{3/2}} \qquad \text{where} \quad H = H(\text{MW}) \qquad \text{[Eq. (6.27)]}$$

$$\underline{K_a} = 12.9 \times \frac{2.15^{1/2}}{20^{3/2}} = \underline{0.211/\text{day}} \ @ \ 20°C$$

Sample Problem 6.6 (continued)

Numerical Values of Basic Coefficients

$$\alpha_r = \sqrt{1 + \frac{4K_rE}{U^2}} = \sqrt{1 + \frac{4 \times 0.25 \times 5}{0.164^2}} = 13.67 \qquad \text{[Eq. (6.109c)]}$$

$$\alpha_a = \sqrt{1 + \frac{4K_aE}{U^2}} = \sqrt{1 + \frac{4 \times 0.211 \times 5}{0.164^2}} = 12.57$$

$$j_{1r} = \frac{U}{2E}(1 + \alpha_r) = \frac{0.164}{2 \times 5}(1 + 13.67) = 0.241/\text{mi}$$

$$j_{1a} = \frac{U}{2E}(1 + \alpha_a) = \frac{0.164}{2 \times 5}(1 + 12.57) = 0.223/\text{mi}$$

$$j_{2r} = \frac{U}{2E}(1 - \alpha_r) = \frac{0.164}{2 \times 5}(1 - 13.67) = -0.208/\text{mi}$$

$$j_{2a} = \frac{U}{2E}(1 - \alpha_a) = \frac{0.164}{2 \times 5}(1 - 12.57) = -0.190/\text{mi}$$

Substituting $T = 20 + 273.15 = 293.15°K$ and $S = 25$ ppt, Eq. 6.16 yields a DO saturation concentration of $c_s = 7.85$ mg/l

Locate Maximum Deficit

For no freshwater flow, the maximum DO deficit occurs at the point source ($x = 0$) and a deficit profile symmetric about $x = 0$ results. If $U > 0$, the maximum deficit is translated downstream to $x = x_c$, where $x_c > 0$. By differentiating Eq. 6.110b, setting the derivative equal to zero, and solving for x_c, the following results:

$$x_c = \frac{\ln[(\alpha_r/\alpha_a) \cdot (1 - \alpha_a)/(1 - \alpha_r)]}{j_{2r} - j_{2a}}$$

$$\therefore x_c = \frac{\ln[(13.67/12.57) \cdot (1 - 12.57)/(1 - 13.67)]}{-0.208 - (-0.190)} = 0.4 \text{ mi}$$

Substitute x_c into Eq. 6.110b to obtain D_c.

$$D_c = \frac{K_d}{K_a - K_r} \frac{W}{Q} \left(\frac{e^{j_{2r}x_c}}{\alpha_r} - \frac{e^{j_{2a}x_c}}{\alpha_a}\right)$$

$$D_c = \left(\frac{0.25}{0.211 - 0.25}\right)\left(\frac{668,000}{1000 \times 5.391} \text{ mg/l}\right)\left(\frac{e^{-0.208 \times 0.4}}{13.67} - \frac{e^{-0.190 \times 0.4}}{12.57}\right)$$

$$D_c = 5.10 \text{ mg/l}$$

$$\therefore \underline{c(\text{min})} = c_s - D_c = 7.85 - 5.10 = \underline{2.75 \text{ mg/l}}$$

(continued)

Sample Problem 6.6 (continued)

A DO profile is obtained by plotting $c_s - D$, where

$$D = \left(\frac{0.25}{0.211 - 0.25}\right)\left(\frac{668{,}000}{1000 \times 5.391}\right)\left(\frac{e^{0.241x}}{13.67} - \frac{e^{0.223x}}{12.57}\right), \quad x \leq 0$$

$$D = \left(\frac{0.25}{0.211 - 0.25}\right)\left(\frac{668{,}000}{1000 \times 5.391}\right)\left(\frac{e^{-0.208x}}{13.67} - \frac{e^{-0.190x}}{12.57}\right), \quad x \geq 0$$

The solid line shows the resulting DO profile. For comparison, using the procedure above, the profile for a freshwater flow of 10,000 cfs is shown by the dashed line. Since U changes, α_r, α_a, j_{1r}, j_{1a}, j_{2r}, and j_{2a} must be recalculated. Note the increased minimum DO and its downstream displacement.

2. When the average tidal velocity exceeds U_0, then the average tidal velocity is used.

Note that the averge tidal velocity (U_T) is given by

$$U_T = \frac{2}{T} \int_0^{T/2} u_1 \sin\left(\frac{2\pi t}{T}\right) dt = \frac{2}{\pi} u_1 \tag{6.111}$$

for a sinusoidally varying tidal velocity.

The guideline above is an approximation and, for steady-state conditions, one way of examining the effect of using the guideline would be to obtain a time averaged velocity ($\overline{U}$) across the tidal cycle and compare resulting reaeration coefficients using U_0 or U_T and $\overline{U}$. This averaged velocity is obtained by integrating over the tidal period the *absolute* values of the instantaneous velocities (u) and dividing by the tidal period (T):

$$\overline{U} = \frac{1}{T} \int_0^T |u| \, dt \tag{6.112}$$

Assuming that the tidal velocity is sinusoidal and that the vector sum of the freshwater velocity and the instantaneous tidal velocity adequately represents the true velocity,

$$u = \left| U_0 + u_1 \sin\left(\frac{2\pi t}{T}\right) \right| \tag{6.113}$$

where U_0 is the freshwater velocity and u_1 is the amplitude of the tidal velocity. This may also be expressed as

$$u = U_0 |1 + \alpha \sin \theta|$$

where

$$\alpha = \frac{u_1}{U_0} \quad \text{and} \quad \theta = \frac{2\pi t}{T} \tag{6.114}$$

Substituting Eq. 6.113 into Eq. 6.111,

$$\overline{U} = U_0, \quad \text{for} \quad \alpha \leq 1 \tag{6.115a}$$

and

$$\overline{U} = \int_0^{T_1} |u| \, dt + \int_{T_1}^{T_2} |u| \, dt + \int_{T_2}^{T} |u| \, dt, \quad \text{for} \quad \alpha > 1 \tag{6.115b}$$

where T_1 and T_2 are the times in the tidal cycle when $U = 0$ and the velocity changes direction. In the latter case $(\alpha > 1)$, $\overline{U} > U_0$. Note that at $\alpha = \pi/2$, $U_0 = U_T$ and, thus, for $\alpha > \pi/2$, $U_T > U_0$ and U_T would be used to calculate the reaeration coefficient.

Assuming the O'Connor–Dobbins reaeration coefficient formulation is applicable, Eq. 6.27, and the mean water depth is relatively constant, the ratio of the reaeration coefficients is given by β; where

$$\beta = \frac{K_a(U_0)}{K_a(\overline{U})} \quad \text{or} \quad \frac{K_a(U_T)}{K_a(\overline{U})}$$

and

$$\beta = \sqrt{\frac{U_0}{\overline{U} = U_0}} = 1, \quad \text{for} \quad \alpha < 1 \tag{6.116a}$$

$$\beta = \sqrt{\frac{U_0}{\overline{U}}} < 1, \quad \text{for} \quad 1 \leq \alpha \leq \frac{\pi}{2} \tag{6.116b}$$

$$\beta = \sqrt{\frac{U_T}{\overline{U}}} < 1, \quad \text{for} \quad \alpha > \frac{\pi}{2} \tag{6.116c}$$

The result of the above comparison, shown on Fig. 6.24, indicates the following:

1. For $0 \leq \alpha \leq 1$, $\overline{U} = U_0$ and no error results in using the guideline.

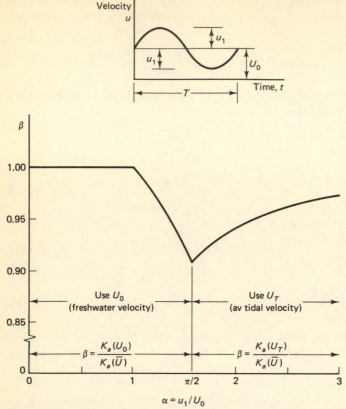

Figure 6.24 Estimate of error introduced in reaeration coefficient using freshwater (u_o) or average tidal (u_T) velocity in lieu of temporally averaged absolute velocity ($\overline{u}$) (see Section 6.5.1). From Mueller (1976). Reprinted by permission of HydroQual, Inc.

2. For $1 \leq \alpha \leq \pi/2$, $\overline{U} > U_0$ and a maximum underestimation of the reaeration coefficient by approximately 9% results.
3. For $\alpha > \pi/2$, $\overline{U} > U_T$ and the maximum difference of 9% decreases as the tidal velocity begins to dominate the system.

This simplified analysis indicates that the reaeration coefficient may be calculated from freshwater *or* tidal velocities selected according to the guideline. The maximum deviation from the value calculated from tidally averaged absolute velocities amounts to no more than 9% of the latter, a difference well within the limits imposed by all other assumptions used in the calculation of the reaeration coefficient.

6.5.2 DO Response due to Sediment Oxygen Demand

The response of an estuary to a distributed source or sink of a substance has been given in Chapter 3. Here, that response is applied to the DO response resulting from SOD distributed over a given reach of the estuary. As with streams, the SOD of an estuary may result from the discharge of settleable organic solids, urban runoff, and upstream non-point sources of organic material.

For a given SOD of S_B' (mg $0_2/l \cdot$ day) given by Eq. 6.49, the DO deficit response is given from solution of Eq. 6.108 for DOD input only, that is,

$$0 = -U\frac{dD}{dx} + E\frac{d^2D}{dx^2} - K_aD + S_B' \qquad (6.117)$$

Equations 3.15 showed the solution for this type of equation where it can be noted now that the reaction rate K of the solution given is the reaeration rate and, thus, α is calculated from Eq. 6.110c. The DO deficit response is then, for $x \leq 0$,

$$D_1 = D_0 \exp(j_1 x) \qquad (6.118a)$$

where

$$D_0 = \frac{S_B'}{K_a}\left(\frac{\alpha - 1}{2\alpha}\right)[1 - \exp(-j_1 a)] \qquad (6.118b)$$

For $0 \leq x \leq a$,

$$D_2 = \frac{S_B'}{K_a}\left(1 - \left(\frac{\alpha - 1}{2\alpha}\right)\{\exp[j_1(x - a)]\} - \left(\frac{\alpha + 1}{2\alpha}\right)[\exp(j_2 x)]\right) \qquad (6.119)$$

where a is the estuary distance over which the SOD is applied. For $x \geq a$,

$$D_3 = D_a \exp[j_2(x - a)] \qquad (6.120)$$

where

$$D_a = \frac{S_B'}{K_a}\left(\frac{\alpha + 1}{2\alpha}\right)[1 - \exp(j_2 a)] \qquad (6.120a)$$

It is seen then that the magnitude of the DO deficit and its upstream and downstream extent are due to a complex interaction between net river flow, tidal dispersion, and estuarine reaeration rate.

6.5.3 Multiple Sources and Superposition

Since all of the estuarine BOD and DO equations are linear, the same principles of superposition apply as in the cases discussed under Chapter 3 and for the stream DO situation discussed in this chapter (see Fig. 6.18). The individual DO deficit responses due to each point source and/or SOD response can be summed to obtain the total response (see Sample Problem 6.7). If, however, the estuary geometry, flow, or dispersion characteristics change, then the simple DO equations discussed in this section do not apply. A more complicated finite difference must then be used (see Section 6.7).

SAMPLE PROBLEM 6.7

DATA

The industry discharging 668,000 lb/day CBODU untreated into the estuary of Sample Problem 6.6 is required to reduce its discharge so that the minimum DO is 5 mg/l for low flow high temperature conditions. An SOD estimated at 2 g/m²·day exists 5 mi upstream and downstream of the discharge. A reserve of 0.25 mg/l is mandated by a regional water resource planning agency and an allowance of 0.5 mg/l is required for random variations.

Low Flow Conditions

$$Q = 800 \text{ cfs} \quad \text{(lower than 1000 cfs in Sample Problem 6.6)}$$
$$E = 4.5 \text{ smpd} \quad \text{(lower than 5 smpd in Sample Problem 6.6,} \\ \text{since lower } Q)$$
$$\text{salinity } (S) = 28 \text{ ppt (higher than 25 ppt in Sample Problem 6.6,} \\ \text{since more landward intrusion of ocean salts)}$$
$$T = 25°C \text{ (higher than 20°C in Sample Problem 6.6)}$$
$$K_r = K_d = 0.15/\text{day @ 20°C (assumes a high degree of treatment)}$$

PROBLEM

Determine the degree of treatment required of the industry.

ANALYSIS

Values of Basic Coefficients

$$U = \frac{Q}{A} = \frac{800}{100,000} = 0.008 \text{ fps} = 0.13 \text{ mpd}$$

Since tidal velocity > freshwater velocity, K_a @ 20°C is the same as in Sample Problem 6.6. Correct for new water temperature

$$K_a(25°C) = 0.21 \times (1.024)^{25-20} = 0.19/\text{day}$$
$$K_r = K_d(25°C) = 0.15 \times (1.047)^{25-20} = 0.24/\text{day}$$
$$\alpha_r = \sqrt{1 + (4 \times 0.19 \times 4.5/0.13^2)} = 14.1$$
$$\alpha_a = \sqrt{1 + (4 \times 0.24 \times 4.5/0.13^2)} = 15.9$$

Sample Problem 6.7 (continued)

From Eq. 6.16, with $T = 298.150°K$ and $S = 28$ ppt, $c_s = 7.05$ mg/l. Based on the plot of DO for $Q = 1000$ cfs, the minimum DO occurs very close to $x = 0$ for $Q = 800$ cfs. Consider this the critical location. At $x = 0$, the deficit due to the sediment demand is also near its maximum (Fig. 3.15).

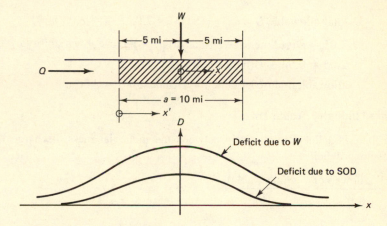

Deficit due to CBODU (Eq. 6.110)

$$D_c = D(x = 0) = \frac{K_d}{K_a - K_r}\left(\frac{1}{\alpha_r} - \frac{1}{\alpha_a}\right), \qquad \text{where} \quad e^0 = 1$$

$$D_c = \left(\frac{0.19}{0.24 - 0.19}\right)\left(\frac{668,000}{800 \times 5.391}\right)\left(\frac{1}{14.1} - \frac{1}{15.9}\right) = 4.72 \text{ mg/l}$$

Deficit due to SOD
(Fig. 3.15, dispersive)

$$D_c = s_{\max} = \frac{S_B/H}{K_a}(1 - e^{-\sqrt{K_a \cdot a^2/4E}}), \qquad H = \frac{20 \text{ ft}}{3.28 \text{ ft/m}} = 6.10 \text{ m}$$

$$D_c = \frac{(2 \text{ g/m}^2\cdot\text{day}/6.10 \text{ m} = 0.328 \text{ mg/l}\cdot\text{day})}{0.24/\text{day}}(1 - e^{-\sqrt{(0.24 \times 10^2)/(4 \times 4.5)}})$$

$$D_c = 0.94 \text{ mg/l}$$

(continued)

Sample Problem 6.7 (continued)

Maximum Allowable Industrial Deficit

Assuming the sediment deposits are not immediately affected by the industrial reduction of load, it will be considered a "background" deficit or uncontrollable deficit.

$$\text{total allowable } D = c_s - c(\text{STD}) = 7.05 - 5 = 2.05 \text{ mg/l}$$

$$\therefore 2.05 = D(\text{sediment}) + D(\text{industrial}) + \text{reserve} + \text{variability}$$

$$2.05 = 0.94 + D(\text{industrial}) + 0.25 + 0.5$$

$$\text{allowable industrial deficit} = D(\text{industrial}) = 0.36 \text{ mg/l}$$

Required Industry Treatment

Since there is a linear relationship between the industrial mass discharge W and the resulting deficit,

$$\frac{W(\text{allowable})}{W} = \frac{D(\text{allowable})}{D},$$

and

$$W(\text{allowable}) = W \times \frac{D(\text{allowable})}{D} = 668,000 \times \frac{0.36}{4.72} = \underline{50,900 \text{ lb/day}}$$

which requires

$$\left(\frac{668,000 - 50,900}{668,000}\right) \times 100 = \underline{92\% \text{ removal}}$$

Notes: A constant parameter estuary was used in the analysis even though the industrial flow of 100 MGD (155 cfs) augmented the freshwater flow of 800 cfs by 19%. In the above analysis this was immaterial since the flow had little effect on the DO deficit distribution.

For cases where flow is more important, the maximum deficits due to the point and distributed sources may not occur at the same location and trial and error solutions may be required.

6.6 DISSOLVED OXYGEN ANALYSIS— LAKES AND RESERVOIRS

One of the principal mechanisms of importance in the variation of DO in lakes and reservoirs is the vertical stratification of temperate lakes and the long residence times of organic material in the lake (see Chapter 4). Figure 6.25 shows the typical vertical variation of DO during summer stratification conditions. Low values of

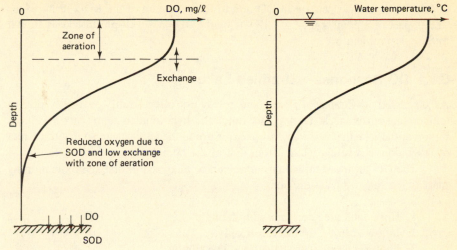

Figure 6.25 Vertical variation of DO during lake stratification.

DO in the hypolimnion result from the flux of oxygen into the sediments to satisfy the SOD and from the poor aeration of the hypolimnion due to the stratified conditions.

First, consider the lake to be completely mixed. The basic equations then for the BOD–DO system are, at steady state:

$$V\frac{dL}{dt} = 0 = W - QL - VK_rL \tag{6.121}$$

$$V\frac{dc}{dt} = 0 = Qc_{in} - Qc + K_LA(c_s - c) - VK_dL \pm W_c \tag{6.122}$$

where all terms are as previously defined, c_{in} is the DO in the incoming flow to the lake, and $\pm W_c$ are all other sources and sinks of DO (photosynthesis, respiration, SOD). For a steady-state condition, the BOD for the completely mixed lake is then given by

$$L = \frac{W}{Q + K_rV} \tag{6.123}$$

which is the same as Eq. 4.17 where the reaction rate K_r is the loss rate of the BOD from the lake. Note that like the stream case, K_r is the sum of the net settling of BOD and the deoxygenation rate.

Equation 6.122 gives for the DO

$$c = \left(\frac{Q}{Q + K_LA}\right)c_{in} + \left(\frac{K_LA}{Q + K_LA}\right)c_s - \left(\frac{VK_d}{Q + K_LA}\right)L \pm \frac{W_c}{Q + K_LA} \tag{6.124}$$

The effect of the BOD discharge W on the DO is then from Eq. 6.123

$$(c)_{BOD} = -\left(\frac{VK_d}{Q + K_LA}\right)\left(\frac{W}{Q + K_rV}\right) \tag{6.125}$$

where $(c)_{BOD}$ is the decrease in DO due to the oxidation of the BOD in the lake. The impact of a point source of CBODU and SOD on a well-mixed lake is shown in Sample Problem 6.8.

6.6.1 DO Response—Stratified Lake

As discussed in Chapter 4, lakes and reservoirs often stratify vertically due to temperature differences. During this time, the hypolimnion represents a volume region of the lake that is isolated from exposure to the atmosphere so that the effect of reaeration is severely altered from that of the completely mixed lake case.

The DO for the stratified lake case can be analyzed to first approximation by making the following assumptions:

1. Horizontal area constant with depth.
2. Inflow is to the surface layer only.
3. Photosynthesis is in the surface layer only.
4. Respiration occurs throughout the lake at an equal rate.
5. Lake is at steady state.

The lake is then divided into two layers with a vertical exchange due to mixing occurring between the epilimnion and the hypolimnion. Chapter 4 discusses this vertical exchange and the use of temperature data for its calculation.

The DO equation for the top layer (1) is

$$0 = Qc_0 + K_L A(c_s - c_1) - Qc_1 + pV_1 - RV_1 - E'(c_1 - c_2) - K_{d1}V_1L_1 \quad (6.126)$$

and for the bottom layer (2) the DO equation is

$$0 = E'(c_1 - c_2) - S_B A - RV_2 - K_{d2}V_2L_2 \quad (6.127)$$

where all terms have been previously defined. Adding the two equations dividing by A and letting

$$q = \frac{Q}{A}$$

be a hydraulic overflow rate $[L/T]$ gives the solution for the two layers. For the top layer,

$$c_1 = \left(\frac{q}{K_L + q}\right)c_0 + \left(\frac{K_L}{K_L + q}\right)c_s$$
$$+ \frac{pH_1 - RH - S_B}{K_L + q} - \frac{K_{d1}H_1L_1 + K_{d2}H_2L_2}{K_L + q} \quad (6.128)$$

and for the hypolimnion,

$$c_2 = c_1 - \left(\frac{S_B + RH_2 - K_{d2}H_2L_2}{E/H_i}\right) \quad (6.129)$$

where $H_i = H/2$ when $H_1 \approx H_2$ and $H_i = H_1$ when $H_2 \gg H_1$. In Eq. 6.129, the bulk dispersion across the epilimnion–hypolimnion boundary is given by Eq. 3.27.

SAMPLE PROBLEM 6.8

DATA

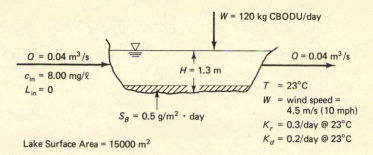

W = 120 kg CBODU/day

Q = 0.04 m³/s

c_{in} = 8.00 mg/ℓ

L_{in} = 0

H = 1.3 m

Q = 0.04 m³/s

T = 23°C

W = wind speed = 4.5 m/s (10 mph)

K_r = 0.3/day @ 23°C

K_d = 0.2/day @ 23°C

S_B = 0.5 g/m² · day

Lake Surface Area = 15000 m²

PROBLEM

Determine the DO concentration in the completely mixed lake and show mass balances for CBODU and DO. Assume effluent is at lake's DO saturation.

ANALYSIS

Basic Parameters

volume:

$$V = AH = 15,000 \text{ m}^2 \times 1.3 \text{ m}$$
$$= 19,500 \text{ m}^3$$

inflow/outflow:

$$Q = 0.04 \text{ m}^3/\text{s} \times 86400 \text{ s/day}$$
$$= 3460 \text{ m}^3/\text{day}$$

hydraulic detention time (Eq. 4.1):

$$t_d = \frac{V}{Q} = \frac{19,500 \text{ m}^3}{3460 \text{ m}^3/\text{day}} = 5.6 \text{ days}$$

DO transfer rate (Eq. 6.33):

$$K_L = 0.728 \times U_w^{1/2} - 0.317 \times U_w$$
$$+ 0.0372 \times U_w^2, \; U_w(\text{m/s})$$

$$K_L = 0.728\sqrt{4.5} - 0.317 \times 4.5$$
$$+ 0.0372 \times 4.5^2$$

$$K_L = 0.87 \text{ m/day} \; (= 2.9 \text{ ft/day; ok}$$
$$\text{compared to } K_L(\text{min})$$
$$= 2\text{--}3 \text{ ft/day})$$

DO saturation concentration (Appendix C):

$$c_s = 8.58 \text{ mg/l} \quad \text{for} \quad T = 23°C$$

Lakewide CBODU Concentration (Eq. 6.123)

$$L = \frac{W}{Q + K_r V} = \frac{120 \text{ kg/day} \times 1000 \text{ g/kg}}{(3460 + 0.3 \times 19500)\text{m}^3/\text{day}} = 12.89 \text{ mg/l}$$

(continued)

Sample Problem 6.8 (continued)

Lakewide DO Concentration (Eq. 6.124)

$$c = \left(\frac{Q}{Q + K_L A}\right) c_{in} + \left(\frac{K_L A}{Q + K_L A}\right) c_s - \left(\frac{VK_d}{Q + K_L A}\right) L - \frac{S_B A}{Q + K_L A}$$

$$Q + K_L A = 3460 \text{ m}^3/\text{day} + 0.87 \text{ m/day} \times 15,000 \text{ m}^2 = 16,500 \text{ m}^3/\text{day}$$

$$c = \frac{3460}{16,500} \times 8.00 + \frac{0.87 \times 15,000}{16,500}$$

$$\times 8.58 - \frac{19,500 \times 0.2}{16,500} \times 12.89$$

$$- \frac{0.5 \text{ g/m}^2\cdot\text{day} \times 15,000 \text{ m}^2}{16,500 \text{ m}^3/\text{day}}$$

$$\underline{c} = 1.68 + 6.79 - 3.05 - 0.45 = \underline{\underline{4.97 \text{ mg/l}}}$$

Mass Balances (all in kg/day)

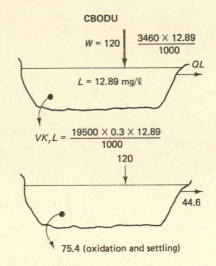

CBODU   in: 120 kg/day

                   out: 44.6 + 75.4 = 120.0 kg/day

Note: Since $VK_d L = 50.3$ kg/day, a net of $75.4 - 50.3 = 25.1$ kg/day of CBODU settles.

Sample Problem 6.8 (continued)

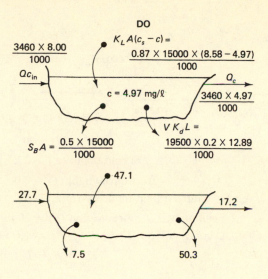

DO   in: $27.7 + 47.1$ $= 74.8$ kg/day

               out: $7.5 + 50.3 + 17.2 = 75.0$ kg/day

Note: Some fraction of the settled CBODU (25.1 kg/day) is included in the SOD (7.5 kg/day).

These equations can then be used for estimating the DO response in a stratified lake (Sample Problem 6.9). Note that the DO in the hypolimnion is lower than that of the epilimnion by the source strength of the SOD and respiration and the degree of mixing across the layers (given by E). It is also seen that c_1 is a constant for different levels of exchange across the epilimnion. This is explored in Sample Problem 6.9.

6.7 FINITE SEGMENT (DIFFERENCE) DO MODELS

As noted in Chapters 3 and 4, in many cases, the geometry of an estuary or lake is not regular and may vary considerably in depth or width. Furthermore, the reaction kinetics may change along the length of the esuary. Then, a finite difference approach is necessary as discussed in Sections 3.6 and 4.3. For DO, similar situations may arise where the preceding solutions may excessively oversimplify the actual water body.

SAMPLE PROBLEM 6.9

DATA

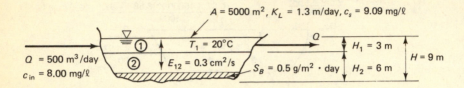

$A = 5000\ m^2$, $K_L = 1.3\ m/day$, $c_s = 9.09\ mg/\ell$

$T_1 = 20°C$

$Q = 500\ m^3/day$

$c_{in} = 8.00\ mg/\ell$

$E_{12} = 0.3\ cm^2/s$

$S_B = 0.5\ g/m^2 \cdot day$

$H_1 = 3\ m$

$H_2 = 6\ m$

$H = 9\ m$

$$\text{Secchi depth } (z_s) = 2.25\ m$$
$$\text{fraction daylight } (f) = 0.5$$
$$\text{average: saturated light intensity } (I_a/I_s) = 3$$
$$\text{chlorophyll } a \text{ concentration } (P) = 20\ \mu g/l$$
$$\text{CBODU concentration } (L)\ L_1 = L_2 = L = 1.0\ mg/l$$
$$K_{d1} = K_{d2} = K_d = 0.1/day$$

PROBLEM

Determine the steady state DO concentrations in the stratified lake which has surface flow alone.

ANALYSIS

Photosynthesis

For layer 2, the hypolimnion, assume that photosynthetic oxygen production $(p_a) \approx 0$.

For Epilimnion

$$G(I_a) = \frac{2.718f}{K_e H_1}\ (e^{-\alpha_1} - e^{-\alpha_0}) = \frac{2.718 \times 0.5}{0.8 \times 3}$$
$$\times (e^{-0.272} - e^{-3}) = 0.40 \qquad \text{[Eq. (6.40a)]}$$

where:

$$\text{extinction coefficient } (K_e) = \frac{1.8}{z_s} = \frac{1.8}{2.25} = 0.8/m \qquad \text{[Eq. (7.28)]}$$

$$\alpha_1 = \left(\frac{I_a}{I_s}\right) e^{-K_e H_1} = 3 \times e^{-0.8 \times 3}$$

$$= 0.272$$

$$\alpha_0 = \frac{I_a}{I_s} = 3$$

Sample Problem 6.9 (continued)

With $p_s = 0.25P$(Eq. 6.42), $p_s = 0.25 \times 20 = 5$ mg/l·day,

$$p_a = p_s G(I_a) = 5 \times 0.40 = \underline{2\ \text{mg/l·day}} \qquad [\text{Eq. (6.41)}]$$

Respiration

Assume same for layers (1) and (2)

$$\underline{R} = 0.025P = 0.025 \times 20\ \mu g/l = \underline{0.5\ \text{mg/l·day}}$$

DO Concentration

$$q = \frac{Q}{A} = \frac{500\ \text{m}^3/\text{day}}{5000\ \text{m}^2} = 0.1\ \text{m/day}$$

$$K_L = 1.3\ \text{m/day}, \qquad K_L + q = 1.4\ \text{m/day}$$

$$\frac{E}{H_i} = \frac{0.3\ \text{cm}^2/\text{s} \times 10^{-4}\ \text{m}^2/\text{cm}^2 \times 86{,}400\ \text{s/day}}{(H = H_1 + H_2 = 3 + 6 = 9\ \text{m})/2} = 0.576\ \text{m/day}$$

where $H_i = H/2$ since H_1 and H_2 are approximately same.

Surface Layer

(Eq. 6.128, as modified for constant values of L, R, and, K_d)

$$c_1 = \left(\frac{q}{K_L + q}\right) c_{\text{in}} + \left(\frac{K_L}{K_L + q}\right) c_s + \frac{p_a H_1 - RH - S_B}{K_L + q} - \frac{K_d L H}{K_L + q}$$

$$c_1 = \frac{0.1}{1.4} \times 8.00 + \frac{1.3}{1.4} \times 9.09 + \frac{2 \times 3 - 0.5 \times 9 - 0.5}{1.4} - \frac{0.1 \times 1.0 \times 9}{1.4}$$

$$\underline{c_1} = 0.57 + 8.44 + 0.71 - 0.64 = \underline{9.08\ \text{mg/l}}$$

Bottom Layer (Eq. 6.129)

$$c_2 = c_1 - \frac{S_B + RH_2}{E/H_i} - \frac{K_d L_2 H_2}{E/H_i}$$

$$\underline{c_2} = 9.08 - \frac{0.5 + 0.5 \times 6}{0.576} - \frac{0.1 \times 1.0 \times 6}{0.576} = 9.08 - 6.08 - 1.04 = \underline{1.96\ \text{mg/l}}$$

The plot shows the constant value of the surface DO for increasing degrees of mixing (E/H_i) and the increasing hypolimnetic DO as the vertical mixing increases.

(continued)

Sample Problem 6.9 (continued)

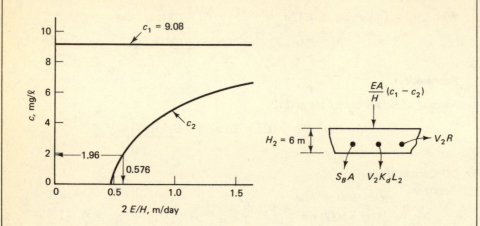

From a mass balance of the hypolimnion it is seen that the mass flux out of the epilimnion is constant and equal to the sum of the hypolimnetic uptake rates (assuming L_2 is independent of the degree of mixing between layers 1 and 2). From Eq. 6.128, it is seen that c_1 is, therefore, constant. Thus, to maintain the dispersive flux constant, c_2 must increase as E increases.

6.7.1 DO Matrix Equation

A finite difference DO mass balance around finite sections can be applied in a manner similar to that used in deriving Eq. 3.28. A steady-state equation comparable to Eq. 3.28 must first be written for the CBOD around segment i as

$$V_i \frac{dL_i}{dt} = 0 = Q_{i-1,i}L_{i-1} - Q_{i,i+1}L_i + E'_{i-1,i}(L_{i-1} - L_i)$$

$$- E'_{i,i+1}(L_i - L_{i+1}) - K_{ri}V_iL_i + W_{Li} \qquad i = 1 \ldots, n \quad (6.130)$$

A similar equation set can be written for the nitrogenous BOD but, for the sake of simplicity, only the CBOD will be considered. Equation 6.130 is identical to Eq. 3.28 except the context is now CBOD. Thus, the general decay coefficient in Eq. 3.28 is now interpreted as the overall BOD loss rate, K_r. The set of equations 6.130 can therefore be expressed as

$$[A](L) = (W_L) \qquad (6.131)$$

The solution in all segments then for the CBOD is

$$(L) = [A]^{-1}(W_L) \qquad (6.132)$$

If W is in lb/day, L in mg/l, then the units of $[\mathbf{A}]^{-1}$ are

$$\frac{\text{mg/l CBOD response in } i}{\text{lb/day CBOD input into } j}$$

The DO equation for the ith segment is

$$V_i \frac{dc_i}{dt} = 0 = Q_{i-1,i}c_{i-1} - Q_{i,i+1}c_i + E'_{i-1,i}(c_{i-1} - c_i)$$

$$- E'_{i,i+1}(c_i - c_{i+1}) + K_{ai}V_i(c_{si} - c_i) + W_{ci} - K_{di}V_iL_i$$

$$+ p_i - R_i - S_{Bi} \qquad i = 1, \ldots, n \qquad (6.133)$$

It can be noted that this equation is similar to Eq. 6.130 but includes the utilization of DO by the oxidation of the BOD given by $-K_{di}V_iL_i$. For the DO deficit, D, the equation is

$$0 = Q_{i-1,i}D_{i-1} - Q_{i,i+1}D_i + E'_{i-1,i}(D_{i-1} - D_i) - E'_{i,i+1}(D_i - D_{i+1})$$

$$- K_{ai}D_i + W_{Di} + K_{di}V_iL_i - p_i + R_i + S_{Bi} \qquad i = 1, 2, \ldots, n \qquad (6.134)$$

Equation 6.134 is formally the same as Eq. 3.28 with the exception that K_a appears instead of the K of the single system. A development similar to that followed for Eqs. 3.28 to 3.38 gives a matrix equation for the DO deficit as

$$[\mathbf{B}](D) = (VK_dL + W'_D) \qquad (6.135)$$

where the $\mathbf{B}$ matrix is identical to the $\mathbf{A}$ matrix of Eq. 3.37 except that the diagonal terms contain V_iK_{ai} instead of V_iK_i for the single system and where W'_D includes all sources and sinks of DO deficit (W_D, p, R, and S_B). Equation 6.135 can be written as

$$[\mathbf{B}](D) = [\mathbf{VK}_d](L) + (W'_D) \qquad (6.136)$$

where $[\mathbf{VK}_d]$ is an $n \times n$ diagonal matrix. The solution for D is then given by

$$(D) = [\mathbf{B}]^{-1}[\mathbf{VK}_d](L) + [\mathbf{B}]^{-1}(W'_D)$$

But Eq. 6.132 provides the solution for (L) which, upon substitution in the above equation gives

$$(D) = [\mathbf{B}]^{-1}[\mathbf{VK}_d][\mathbf{A}]^{-1}(W_L) + [\mathbf{B}]^{-1}(W'_D) \qquad (6.137)$$

The meaning of the BOD–DO deficit system, or a "coupled system," now becomes clear when the first term on the right-hand side is examined. The input vector of the CBOD system is operated on by the steady-state transfer function matrix of the first system, $[\mathbf{A}]^{-1}$, and then further operated on by the transfer function matrix of the second system, $[\mathbf{B}]^{-1}$. The input vector (W'_D) which represents inputs of DO deficit is operated on only by $[\mathbf{B}]^{-1}$. Equation 6.137 can be further "synthesized" by writing

$$(D) = [\mathbf{C}]^{-1}(W_L) + [\mathbf{B}]^{-1}(W'_D) \qquad (6.138)$$

where

$$[\mathbf{C}]^{-1} = [\mathbf{B}]^{-1}[\mathbf{VK}_d][\mathbf{A}]^{-1}$$

Therefore, the solution to the DO deficit finite difference problem that includes spatially variable coefficients involves inversion of two matrices and their subsequent multiplication. The matrix $[\mathbf{C}]^{-1}$ represents the DO deficit response matrix and relates the response in deficit due to discharges of CBOD(L). The columns of $[\mathbf{C}]$ then indicate the DO deficit response in all segments of the estuary (or other body of water) due to a unit load input of CBOD into the given column. The rows of $[\mathbf{C}]^{-1}$ represent the response of DO deficit at a given location due to loads of CBOD in every other location. The units of $[\mathbf{C}]^{-1}$ are

$$\frac{\text{mg/l DO deficit response in segment } i}{\text{lb/day CBOD input into segment } j}$$

The units of $[\mathbf{B}]^{-1}$ are

$$\frac{\text{mg/l DO deficit } (+ \text{ or } -) \text{ response in segment } i}{\text{lb/day DO deficit input } (-p, +R, +S_B)}$$

It can be noted here that the matrix equation (6.138), in general, represents a multidimensional system as discussed in Section 3.6. Thus, the DO deficit steady-state response for multidimensional systems is readily computed from the finite difference approach as a set of simultaneous linear algebraic equations. The inverse of the matrix of coefficients, the steady-state response matrix, has a particular meaning since each element represents the DO response per unit load input. This approach forms the basis for several computer programs such as HARO3 (Chapra and Nosser, 1974) and SPAM (HydroQual, 1984). These programs solve the general multidimensional DO system using the basic set of equations given by 6.132.

6.7.2 Application of Finite Segment Model for DO

A typical example of the application of the finite segment approach is given by the DO deficit in the Delaware Estuary (Pence et al., 1968, Thomann, 1972). The technique described above was applied to a 30-section representation of the Delaware Estuary from Trenton to Delaware Bay (Fig. 6.26). Table 6.8 gives the spatial values of the system parameters required for the finite elements. Note that for area, flow, and dispersion coefficients, values are specified for the boundary sections at the upper and lower ends of the estuary. The spatial changes in the system parameters are readily noted as one proceeds down the estuary.

The values given in Table 6.8 are used to form the elements of the steady-state response matrix as given by Eq. 3.32. Table 6.9 shows a portion of the steady-state response matrix $[\mathbf{B}]^{-1}$ where the reaction coefficient on the main diagonal of the matrix before inversion is taken as the reaeration coefficient. The matrix $[\mathbf{B}]^{-1}$ is interpreted, therefore, as the DO response in the estuary due to benthal demand, photosynthesis, respiration, or other sources and sinks of DO. Thus, if 100,000 lb/day of oxygen deficit is exerted in section 16, the response in all other sections is given by column 16 in Table 6.9. A maximum DO deficit due to this steady-state source of deficit is 1.55 mg/l in section 16.

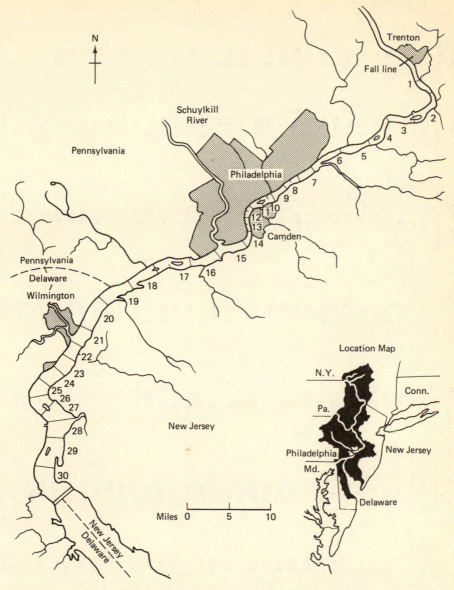

Figure 6.26 Finite segment model of Delaware Estuary. From Thomann (1972).

It should be noted that for the finite segment approach, there is no distinction between point and distributed sources or sinks. Thus for DO deficit inputs such as SOD (see Eq. 6.49) that are expressed in $g/m^2 \cdot day$, the input for the finite section model is

$$S_b(\text{lb/day}) = \frac{S_b(\text{g/m}^2 \cdot \text{day})}{H(m)}\, V(\text{MG})\, (8.34)\, (1\text{b/MG-mg/l})$$

TABLE 6.8 SYSTEM PARAMETERS FOR DELAWARE ESTUARY (inflow at Trenton = 3000 cfs)

Section	Length (ft)	Volume (10^6 ft^3)	Average tidal velocity (ft/hr)	Cross-sectional area, $i, i + 1$ (10^3 ft^2)	Freshwater flow, $i, i + 1$ (cfs)	Dispersion coefficient $i, i + 1$ (smpd)	Decay coefficient (day^{-1})	Reaeration coefficient (day^{-1})
0	—	—	—	6.5	3000	4.0	—	—
1	21,000	242	1500	15.8	3040	4.0	0.40	0.31
2	20,000	364	1500	21.4	3140	4.0	0.40	0.23
3	20,000	460	1500	24.6	3150	4.0	0.40	0.18
4	20,000	532	1500	28.5	3180	4.0	0.40	0.20
5	20,000	636	2000	34.1	3220	4.0	0.40	0.20
6	20,000	756	2000	41.4	3420	4.0	0.40	0.25
7	10,000	455	2000	49.6	3150	4.0	0.40	0.20
8	10,000	504	2000	51.2	3170	5.0	0.40	0.19
9	10,000	533	2000	55.4	3180	5.0	0.40	0.23
10	10,000	582	2000	60.9	3390	5.0	0.40	0.16
11	10,000	630	2500	65.0	3420	5.0	0.40	0.18
12	10,000	655	2500	66.0	3420	5.0	0.40	0.10
13	10,000	694	2500	72.7	3470	5.0	0.40	0.09
14	10,000	805	2500	88.3	3660	5.0	0.40	0.11
15	20,000	1860	2500	98.0	4060	5.0	0.40	0.11
16	20,000	2030	3000	104.9	4250	5.0	0.40	0.13
17	20,000	2184	3000	113.4	4380	5.0	0.40	0.13
18	20,000	2396	3000	126.3	4440	5.0	0.40	0.13
19	20,000	2692	3000	142.8	4460	5.0	0.40	0.13
20	20,000	2932	3000	150.4	4460	5.0	0.40	0.13
21	10,000	1512	3500	151.9	4650	5.0	0.40	0.13
22	10,000	1574	3500	162.9	4650	5.0	0.40	0.13
23	10,000	1698	3500	176.8	4660	5.0	0.40	0.18
24	10,000	1792	3500	181.7	4660	7.0	0.40	0.18
25	10,000	1850	3500	188.4	4660	7.0	0.40	0.28
26	10,000	1924	4000	196.4	4670	7.0	0.40	0.38
27	10,000	2054	4000	214.5	4670	7.0	0.40	0.35
28	10,000	2248	4000	235.0	4720	7.0	0.40	0.35
29	20,000	4896	4000	254.5	4750	7.0	0.40	0.25
30	20,000	5620	4000	307.4	4780	7.0	0.40	0.24

Source: From Thomann (1972).

TABLE 6.9 PORTION OF STEADY-STATE TRANSFER MATRIX $[B]^{-1}$ DELAWARE ESTUARY

DO response (mg/l) in section	Input (bottom demand, photosynthesis, respiration) of 100,000 lb/day of oxygen deficit into section					
	14	15	16	17	18	19
8	0.04	0.02	—	—	—	—
9	0.10	0.04	0.01	—	—	—
10	0.22	0.10	0.03	0.01	—	—
11	0.40	0.17	0.06	0.02	0.01	—
12	0.71	0.31	0.11	0.04	0.01	—
13	1.25	0.55	0.18	0.06	0.02	0.01
14	2.06	0.90	0.31	0.10	0.04	0.01
15	1.47	1.73	0.59	0.20	0.07	0.03
16	1.00	1.17	1.55	0.53	0.19	0.07
17	0.66	0.79	1.04	1.44	0.51	0.18
18	0.44	0.52	0.69	0.95	1.33	0.49
19	0.28	0.33	0.44	0.61	0.86	1.21
20	0.18	0.21	0.28	0.39	0.55	0.77
21	0.13	0.15	0.20	0.28	0.39	0.55
22	0.10	0.11	0.15	0.21	0.29	0.41
23	0.07	0.08	0.11	0.15	0.21	0.30
24	0.05	0.06	0.07	0.10	0.14	0.20
25	0.03	0.04	0.05	0.07	0.10	0.14
26	0.02	0.03	0.04	0.05	0.07	0.10
27	0.02	0.02	0.02	0.03	0.05	0.06
28	0.01	0.01	0.01	0.02	0.03	0.04
29	0.01	0.01	0.01	0.01	0.02	0.03
30	—	—	—	0.01	0.01	0.01

Source: From Thomann (1972).

where H is the average depth in meters. The ratio of the volume of the section to the average depth is the average surface area of the section. Similar computations are necessary for the inclusion of the effects of photosynthesis and respiration.

Table 6.10 shows a portion of the steady-state matrix, $[C]^{-1}$, for the values shown in Table 6.8. The columns of Table 6.10 indicate the DO deficit response for each 100,000 lb/day of CBOD (or NBOD with an appropriate change of K_r to K_n, the NBOD decay coefficient) discharged into segments 14–19. Thus, if 100,000 lb/day is discharged into segment 14, the maximum DO deficit is estimated to be 1.18 mg/l in segment 15. The order of magnitude of the response is interesting and reflects the volume and dispersive characteristics of the Delaware, a typical East Coast estuary subject to large waste discharges. A discharge of 100,000 lb/day equivalent to a population of about 500,000 produces a DO deficit of less than 1 mg/l from section 16 to the mouth of the estuary. This has important consequences for waste removal programs since it indicates that relatively large amounts of waste must be removed in order to obtain substantial changes in DO. The coupled system DO deficit response matrix can then be used to produce the responses due to the vector of CBOD waste load inputs. The resulting DO deficit vector can then be added to the DO deficit vector due to nitrogenous BOD.

TABLE 6.10   PORTION OF STEADY-STATE MATRIX $[C]^{-1} = [A]^{-1}[VK_d][B]^{-1}$
DELAWARE ESTUARY

DO response (mg/l) in section	Waste input of 100,000 lb/day of BOD into section					
	14	15	16	17	18	19
8	0.05	0.02	0.01	—	—	—
9	0.11	0.05	0.02	0.01	—	—
10	0.22	0.11	0.04	0.01	0.01	—
11	0.37	0.19	0.07	0.03	0.01	—
12	0.59	0.31	0.12	0.05	0.02	0.01
13	0.88	0.50	0.20	0.08	0.03	0.01
14	1.13	0.73	0.31	0.12	0.05	0.02
15	1.18	1.10	0.53	0.22	0.09	0.03
16	1.00	1.05	0.98	0.48	0.20	0.08
17	0.76	0.84	0.94	0.91	0.46	0.21
18	0.54	0.62	0.75	0.86	0.85	0.45
19	0.37	0.43	0.54	0.67	0.79	0.79
20	0.24	0.28	0.37	0.48	0.60	0.71
21	0.17	0.21	0.27	0.36	0.46	0.58
22	0.13	0.16	0.21	0.28	0.36	0.47
23	0.10	0.11	0.15	0.20	0.27	0.35
24	0.06	0.08	0.10	0.14	0.19	0.25
25	0.05	0.06	0.07	0.10	0.14	0.18
26	0.03	0.04	0.05	0.07	0.09	0.13
27	0.02	0.03	0.03	0.05	0.06	0.09
28	0.02	0.02	0.02	0.03	0.05	0.06
29	0.01	0.01	0.01	0.02	0.03	0.04
30	—	—	0.01	0.01	0.01	0.02

Source: From Thomann (1972).

6.8 ENGINEERING CONTROL FOR DISSOLVED OXYGEN

As can be seen from the equations that express the resulting DO in streams, estuaries, or lakes (i.e., Eqs. 6.84, 6.110, and 6.124) there are several points at which engineering control can be utilized to improve the DO. These points can be grouped as follows:

1. Point and non-point reduction source of CBOD and NBOD through reduction of effluent concentration and/or effluent flow.
2. Aeration of the effluent of a point source to improve initial value of DO.
3. Increase in river flow through low flow augmentation to increase dilution.
4. Instream reaeration by turbines and aerators.
5. Control of SOD through dredging or other means of inactivation.
6. Control of nutrients to reduce aquatic plants and resulting DO variations.

The reduction of nutrients to control DO and the control of SOD through dredging are discussed in Chapter 7, Eutrophication. The first four DO controls are discussed here.

6.8.1 Point Source CBOD and NBOD Reduction

Table 6.11 summarizes the typical average effluent concentrations from different levels of municipal waste treatment. The differences between the CBOD5 and the BOD5 concentrations for secondary plants reflects the effect of nitrification in the 5-day BOD test (Hall and Foxen, 1983) (see also Fig. 6.4). Also, the close relationship between the total suspended solids and the effluent BOD concentration reflects the fact that the residual BOD at higher levels of treatment is associated with the finely dispersed suspended solids. The effect of the NBOD is quite marked as can be seen in the comparison of the ultimate BOD for the activated sludge (AS) level without nitrification to secondary treatment with nitrification. The CBODU is about 30 mg/l for AS without nitrification but the total ultimate oxygen demand (CBODU + NBODU) is about twice that much or about 60 mg/l. After nitrification, the total UOD drops to about 15 mg/l. The importance of the NBOD to the total oxygen demand of the effluent is therefore quite dramatic.

Sample Problem 6.10 reviews a complete analysis of DO to determine required treatment to meet a DO standard. This sample problem also reviews the calculation to meet a required un-ionized ammonia standard. This latter water quality variable is an important consideration in toxic effects in the aquatic ecosystem. (See also Chapter 8.)

It should also be noted that the effluent concentrations from any plant can be quite variable reflecting incoming load variations, process fluctuations, and changing environmental conditions at the plant. Roper et al. (1979) have obtained relationships from statistical analyses of secondary and advanced treatment plant data. Such analyses relate the 95th percentile concentration to the long-term arithmetic

TABLE 6.11 TYPICAL RANGE OF AVERAGE EFFLUENT CONCENTRATIONS OF CBOD AND NITROGEN FORMS, MUNICIPAL POINT SOURCE (ALL UNITS—mg/l)

Treatment level	BOD5	CBOD5	CBODU	Ammonia nitrogen	NBODU	Total suspended solids
Raw		150–450	120–580	10–60	100–400	100–400
Primary	100–200		150–300		95	40–160
Trickling filter	20–50		50–125	5–20	25–100	10–40
Activated sludge (AS)	10–50	5–15	15–45	5–15	25–75	10–30
Secondary and nitrification[a]	5–20	2–7	6–20	0.5–2.0	2.5–10	5–20
Secondary, chemical, and filtration	1–8		2–20			1–4

[a]Trickling filter and AS plants, summertime conditions.

Source: Adapted and approximated from: Hall and Foxen (1983), HydroQual (1983), Roper et al. (1979), Thomann (1972).

SAMPLE PROBLEM 6.10 (Adapted from Driscoll et al., 1981)

DATA

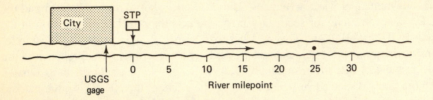

A city of approximately 60,000 people discharges its wastewater into a relatively small river with an average annual flow of about 250 cfs. The city's wastewater is presently treated by a trickling filter plant which provides about 85% BOD removal and has reached its design capacity of 7.5 MGD. The population is projected to increase by more than 50% to 92,000 people (with a range of 75,000 to 120,000 people) by the year 2000. Expansion of the treatment plant to a capacity of 11.5 MGD and provision of an activated sludge system for more efficient secondary treatment are proposed. The STP effluent characteristics are as follows:

		Present	Future[a]
Flow	MGD	7.5	11.5
CBOD5	mg/l	40	30
	lb/day	2502	2877
CBODU[b]	mg/l	80	60
	lb/day	5004	5754
NH_3–N	mg/l	15	15
	lb/day	938	1439
NBOD[c]	mg/l	68	68
	lb/day	4221	6475

[a]Proposed standard secondary treatment.

[b]Long-term BOD tests indicate ratio of CBODU/CBOD5 = 2.0.

[c]NBOD = stoichiometric oxygen requirements for oxidation of reduced nitrogen forms = 4.57 × NH_3–N (effluent oxidizable organic nitrogen is negligible).

PROBLEM

Perform a water quality analysis to determine whether the proposed treatment will meet DO and un-ionized ammonia standards.

Sample Problem 6.10 (continued)

ANALYSIS

River Characteristics

A. Classification and use.
State classifiction—B1.
Designated use—Fish and wildlife propagation.
Water quality standards (partial)
    a. DO concentration—greater than 5.0 mg/l.[1]
    b. Un-ionized ammonia—less than 0.02 mg/l.[1]

B. Geometry.
Relatively constant spatially over reach of interest. From three field surveys at different flows, the following area flow and depth flow relationships were derived:

$$A = 19.5Q^{0.6}; \ A \ (\text{ft}^2), \ Q \ (\text{cfs})$$
$$H = 0.312Q^{0.5}; \ H \ (\text{ft}), \ Q \ (\text{cfs})$$

C. Flow.
USGS gaging station upstream of STP
long-term average summer flow = 100 cfs
7Q10 = 30 cfs.

D. Velocity.
From time of travel studies at various flows, $U = 0.0513Q^{0.4}$; U (fps), Q (cfs).

E. Water Temperature.
25°C during low flow summer survey; 27°C maximum monthly average temperature (August) to be used with 7Q10 flow to check conformance with standards.

Water Quality Analyses

A calibration analysis at $Q_U = 100$ cfs is done first to establish kinetic rates based on observed in-stream data. Then, after temperature correcting these rates, projections of DO and NH_3–N are made for the 7Q10 flow, followed by a unit response analysis since water quality violations occur. Final treatment is then selected. Equations for generating all concentrations shown are found at the end of this Sample Problem.

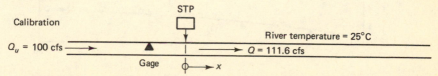

[1]For minimum seven-consecutive-day flow with a return period of once in 10 years (7Q10).

(continued)

Sample Problem 6.10 (continued)

Upstream conditions	Plant effluent	Stream @ $x = 0$
$Q_u = 100$ cfs	$Q_e = 7.5$ MGD (11.6 cfs)	$Q = 111.6$ cfs
CBOD5 = 1.0 mg/l (540 lb/day)	CBOD5 = 40 mg/l (2502 lb/day)	CBOD5 = 5.05 mg/l
NH_3–N = 0.2 mg/l (108 lb/day)	NH_3–N = 15 mg/l (938 lb/day)	NH_3–N = 1.74 mg/l
DO = 8.3 mg/l (sat.)	DO = 8.3 mg/l (sat.)	DO = 8.3 mg/l
		CBODU = $2.0 \times 5.05 = 10.1$ mg/l
		NBOD = $4.57 \times 1.74 = 7.8$ mg/l

$$H = 0.312 \times (111.6)^{0.5} = 3.3 \text{ ft} \qquad U = 0.0513 \times (111.6)^{0.4}$$
$$= 0.34 \text{ fps} = 5.6 \text{ mpd}$$
$$K_a = [12.9 \times 0.34^{0.5}/(3.3)^{1.5}] \times (1.024)^{25-20} = 1.42/\text{day @ } 25°C$$

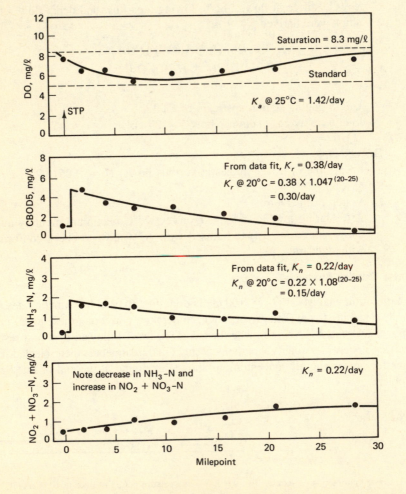

Sample Problem 6.10 (continued)

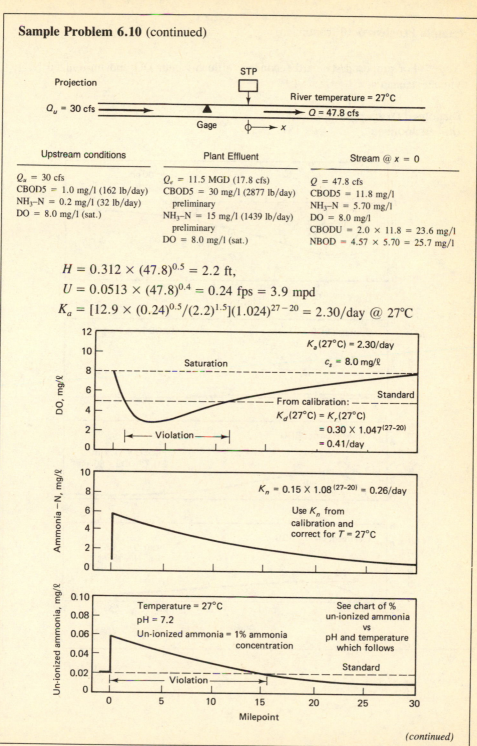

$$H = 0.312 \times (47.8)^{0.5} = 2.2 \text{ ft},$$

$$U = 0.0513 \times (47.8)^{0.4} = 0.24 \text{ fps} = 3.9 \text{ mpd}$$

$$K_a = [12.9 \times (0.24)^{0.5}/(2.2)^{1.5}](1.024)^{27-20} = 2.30/\text{day} @ 27°C$$

(continued)

Sample Problem 6.10 (continued)

Note: For proposed standard secondary effluent, both DO and un-ionized NH_3 violate standards.

Dissolved Oxygen Component
Unit Responses

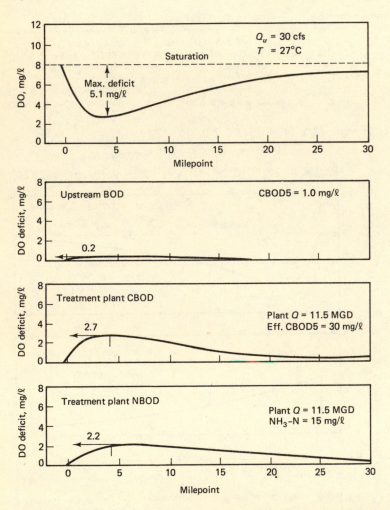

Sample Problem 6.10 (continued)

Final Treatment

Assuming upstream sources are uncontrollable and noting that the proposed effluent BOD5 is under high removal efficiency already ($\geq 85\%$), *add in-plant nitrification*. This will reduce NH_3–N concentration by about 90%, thus reducing the in-stream deficit due to NBOD from 2.2 to 0.2 mg/l. In addition, in-stream un-ionized NH_3 will decrease from a maximum of 0.06 to 0.006 mg/l and, thus, conform with the standard.

Although the minimum DO is 4.9 mg/l ($8.0 - 0.2 - 2.7 - 0.2$) which is below the standard of 5.0 mg/l, this slight deviation should not cause major revisions to the treatment level. Before that is done, additional surveys, better population estimates, distibuted source evaluations, refined upstream concentration estimates, requirements for any reserve, etc, should be considered.

Equations for Concentrations

$$l = CBOD5, \quad L = CBODU, \quad N = NBOD, \quad N_1 = NH_3\text{–}N,$$

$$N_2 = NO_2 + NO_3 - N \quad D = DO \text{ deficit}, \quad c = DO$$

$$l = l_0 \exp\left(-K_r \frac{x}{U}\right), \quad N = N_0 \exp\left(-K_n \frac{x}{U}\right)$$

$$N_1 = N_{10} \exp\left(-K_n \frac{x}{U}\right), \quad N_2 = N_{20} + N_{10}\left[1 - \exp\left(-K_n \frac{x}{U}\right)\right]$$

$$D = \frac{K_d}{K_a - K_r} L_0 \left[\exp\left(-K_r \frac{x}{U}\right) - \exp\left(-K_a \frac{x}{U}\right)\right]$$

$$+ \frac{K_n}{K_a - K_n} N_0 \left[\exp\left(-K_n \frac{x}{U}\right) - \exp\left(-K_a \frac{x}{U}\right)\right]$$

where $c = c_s - D$, $L_0 = 2.0 \times l_0$, $N_0 = 4.57 \times N_{10}$,

$$c_s = DO \text{ saturation}.$$

(continued)

Sample Problem 6.10 (continued)

Percent Un-ionized Ammonia in
Ammonia–Water Solution for Various
pH and Temperature Values (from
Willingham, 1976).

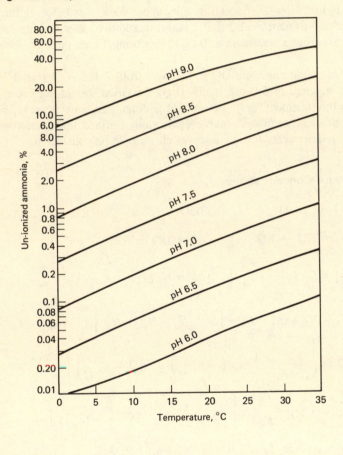

mean concentration. For $\overline{L}$ as the average BOD5 concentration and L_{95} as the concentration that is expected to be exceeded 5% of the time,

$$L_{95} = 2.14\overline{L} + 1.15 \quad \text{(daily average)}$$
$$L_{95} = 1.75\overline{L} + 1.74 \quad \text{(weekly average)} \quad (6.139)$$
$$L_{95} = 1.64\overline{L} + 0.65 \quad \text{(30-day running average)}$$

Thus if the arithmetic average BOD5 is 10 mg/l, we might expect that 95% of the effluent concentrations would be equal to or less than 22.5, 19 and 17 mg/l on a daily, weekly, and 30-day average basis, respectively. Five percent of the time, the daily average could be greater than twice the long-term average BOD5.

Of course, as one proceeds to higher degrees of treatment to reduce the effluent oxygen demanding load, the total cost of treatment increases. For example, Dames and Moore (1978a,b) reported the capital construction and operation costs for municipal plants. Table 6.12 shows a summary of their work in 1977 dollars and as seen a 10 MGD secondary plant capital cost is about $16 million (1977). For a plant beyond secondary treatment, the cost rises to about $28 million (1977), indicating that increased removals of BOD require progressively greater increases in cost. This is illustrated in Fig. 6.27 which shows that as the % removal increases, the treatment cost increases at first slowly between primary and secondary and then very rapidly beyond secondary. Therefore, the last remaining BOD to be removed becomes more and more costly. Since all of the preceding BOD–DO equations for any body of water are linear to the input load, the cost of DO improvement is directly proportional to the cost of BOD load reduction. Increasing DO improvements therefore cost progressively more and more to achieve.

Figure 6.28 shows the effect of point source BOD reduction on the DO of rivers and streams. As noted from Eq. 6.83, the DO deficit due to the BOD load is directly proportional to that load. This is shown in Fig. 6.28(a) which indicates that the DO deficit profile decreases directly as treatment increases. Figure 6.28(b) shows the effect of the initial DO deficit assumed here to be independent of the treatment level. Figure 6.28(c) shows the resulting DO profile for the levels of treatment. While the DO deficit due to the discharged load decreases proportionately to the decreased BOD load, the resulting DO profile depends on the other components of the DO balance such as the initial DO deficit (as well as SOD and photosynthesis and respiration). The same basic principles apply for the DO response due to BOD reduction in estuaries and lakes.

The impact of reducing point source discharges on stream and river DO concentrations is shown in Fig. 6.29 (HydroQual, 1983). In the left column, Wilson

TABLE 6.12 APPROXIMATE COSTS OF MUNICIPAL TREATMENT
FOR DO CONTROL (1977 COSTS)

Design flow (MGD)	Treatment level	Capital construction cost (million $)	Operation and maintenance cost ($100,000/yr)
1	Secondary AS	2.1	0.68
	> Secondary[a]	2.9	0.82
10	Secondary AS	16.1	0.75
	> Secondary	28.1	1.9
100	Secondary AS	122.0	6.9
	> Secondary	275.0	52.0

[a] > Secondary for capital costs, BOD/SS < 30 mg/l; for OMC, includes some advanced waste treatment facilities for other than DO control (e.g., chemical treatment for phosphorus removal).
Source: Calculated from Dames and Moore (1978a, 1978b)

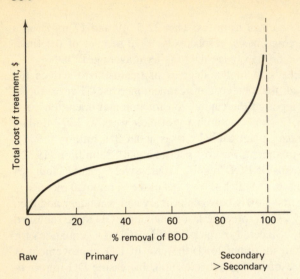

Figure 6.27 Generalized relationship between municipal treatment plant cost and % BOD removal.

Creek, Missouri, is seen to receive the effluent of the Springfield Southwest sewage treatment plant (STP). In 1968, the 9.1-MGD flow generated an ultimate oxygen demand (UOD) loading of approximately 10,000 lb/day from the activated sludge secondary plant, causing the low DO levels observed in Fig. 6.29(a) when the upstream river flow was 30 cfs. Using a river model with coefficients calibrated from July 1968 data, and under a low flow of 5.4 cfs upstream of the STP, a waste load allocation (WLA) was performed which indicated that a flow of 19 MGD generating an STP UOD loading of approximately 3000 lb/day would allow the stream to meet a DO standard of 5 mg/l [Fig. 6.29(b)]. After the plant was upgraded to secondary treatment with nitrification, plus ozone disinfection and final filters, a stream survey was conducted in 1979 when the STP was discharging 17 MGD with approximately 1700 lb/day of UOD and DO concentrations were markedly improved even with the upstream flow of zero. Using the same model coefficients from the calibration, the mathematical model is seen to reproduce the observed DO very well [Fig. 6.29(c)]. The increase in the DO in the stream just downstream of the STP is due to oxygenation of the effluent to a level of approximately 14 mg/l from the ozone disinfection and pure-ox treatment process.

In contrast to Wilson Creek, an effluent dominated stream, the upper main stem of the Patuxent River in Maryland received STP flows that even at the stream 7Q10 flow of 16.5 cfs amount to no more than one-half of the combined flow. In 1968, under an upstream flow of 30 cfs, the Laurel Parkway and Maryland City STPs discharged approximately 1900 lb/day UOD with a flow of 2.4 MGD, resulting in DO concentrations approaching the standard of 4 mg/l [Fig. 6.29(a)]. A WLA resulted in an allowable discharge of approximately 1500 lb/day UOD for an upstream flow of 16.5 cfs and a combined STP outflow of 9.1 MGD [Fig. 6.29(b)]. A postaudit of the river was conducted in 1978, when the river flow of

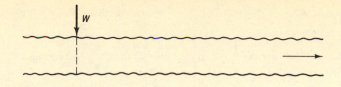

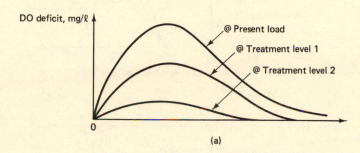

(a)

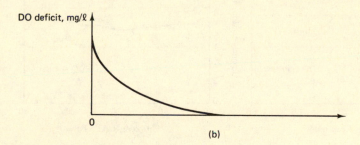

(b)

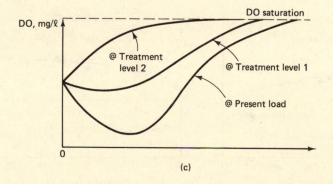

(c)

Figure 6.28 Effect of point source BOD reduction on DO for a given initial deficit—rivers and streams.

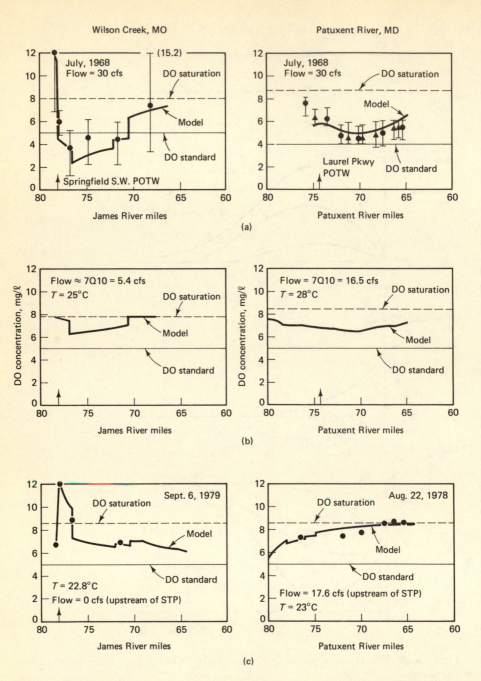

Figure 6.29 Examples of effects of point source reductions on riverine dissolved oxygen concentrations. From HydroQual (1983). (a) Model calibration. (b) Low flow DO projection. (c) Postimprovement conditions.

17.6 cfs was close to the 7Q10 low flow and the STPs were discharging 394 lb/day UOD with a flow of 5.0 MGD. Resulting DO concentrations are seen to approach saturation in Fig. 6.29(c), and the previously calibrated model was able to reasonably reproduce observed DO values.

The postaudit analysis of HydroQual (1983) indicated that root mean square (RMS) errors of DO between model and data during calibration and verification analyses averaged abut 0.7 mg/l but for data following treatment improvement, the RMS increased to 0.9 mg/l. Simplified desk-top calculations performed poorly with RMS errors of 50 to 200% higher than the errors from modeling analyses of a more intensive nature. However, up to nitrification, the simplified analyses resulted in similar decisions to increase treatment, but beyond nitrification the simplified models gave different treatment strategies.

The effect of point source reductions on the DO in an estuary is shown in Fig. 6.30 (O'Connor and Mueller, 1984) for the Hudson River, New York. A 425-segment finite difference model incorporated a vertically stratified, two-layer section in the Hudson River. Twenty-six major STPs discharge to the New York Harbor complex [Fig. 6.30(a)], as well as a host of smaller municipal and industrial sources and over 700 combined sewer overflows and storm drains. Model coefficients were calibrated for six different flow and temperature conditions and an example of the DO calibration is shown in Fig. 6.30(b)—in which the calculated bottom layer concentrations are seen to be lower than those in the aerated upper layer. During the summer 1965 period, approximately 3000 MGD of treated and untreated sewage carried 2,800,0000 lb/day of 5-day BOD into the harbor. Dissolved oxygen deficits due to various classes of sources [Fig. 6.30(c)] indicated that the treated and untreated sewage caused approximately 2.5 mg/l of deficit, approximately 55% of the total, indicating that point source reductions would be effective. Treatment of the untreated sewage (baseline) is seen to increase the river DO so that it almost meets New York State standards [Fig. 6.30(d)] and adoption of secondary treatment (a total of 790,000 lb/day of 5-day BOD) by all dischargers would raise the DO above the 4 mg/l standard. The above technique of determining the impact of various classes of sources is most useful and can also be used to distinguish effects of both carbonaceous and nitrogenous BOD.

6.8.2 Effluent Aeration

As noted in Eqs. 6.83 and 6.84 and Fig. 6.15, the distribution of the DO downstream from a discharge in a river depends on the initial value of the DO or DO deficit at the outfall after mixing. To a lesser degree this is also true in estuaries although for that water body, the degree of the importance of the initial DO is reduced by the usually sizable volume of the estuary. For lakes, the increasing DO mass from an input (c_{in} or W_c in Eq. 6.118) can have an important effect on the DO resources of the lake, again depending on the volume and flow of the lake. In all these cases, the interest centers on the amount of DO that may be inputted by a point source in contrast to the amount of BOD in the effluent.

For rivers, the initial DO and DO deficit after mixing is given by Eqs. 6.87 to 6.89 and is seen to depend on the relative magnitude of the upstream flow and

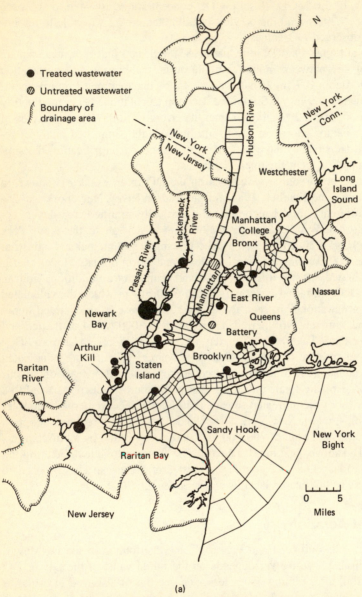

(a)

Figure 6.30 Example of effect of point source reductions on estuarine dissolved oxygen concentrations—Hudson River, NY. From O'Connor and Mueller (1984) from data in Leo et al. (1978) and Higgins et al. (1978). (a) Model segmentation. (b) Calibration. (c) Unit responses. (d) Projections. Reprinted by permission of American Society of Civil Engineers.

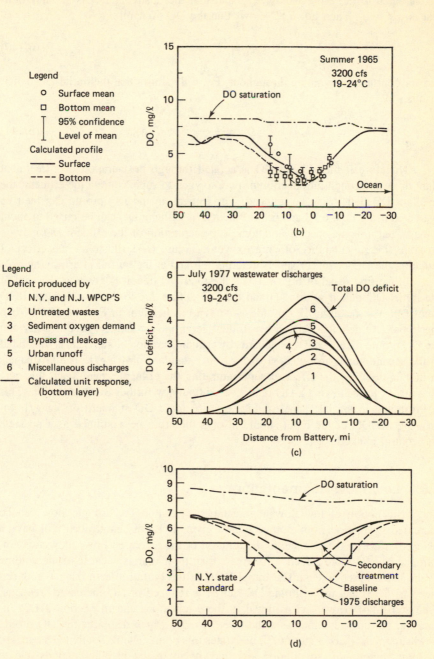

Legend

○ Surface mean
□ Bottom mean
⊤ 95% confidence
⊥ Level of mean

Calculated profile
—— Surface
‐‐‐‐ Bottom

Summer 1965
3200 cfs
19–24°C

DO saturation

Ocean →

(b)

Legend

Deficit produced by

1 N.Y. and N.J. WPCP'S
2 Untreated wastes
3 Sediment oxygen demand
4 Bypass and leakage
5 Urban runoff
6 Miscellaneous discharges
—— Calculated unit response, (bottom layer)

July 1977 wastewater discharges
3200 cfs
19–24°C

Total DO deficit

Distance from Battery, mi

(c)

DO saturation

N.Y. state standard

Secondary treatment

Baseline

1975 discharges

(d)

DO and the effluent flow and DO. To illustrate, consider that the upstream flow, Q_u, is equal to the effluent flow, Q_e, and that the upstream DO is at saturation, that is, $c_u = c_s$. Then Eq. 6.87 shows that the DO after mixing is

$$c_0 = \frac{c_s + c_e}{2} \tag{6.140}$$

and for constant temperature throughout, Eq. 6.89 shows that the initial DO deficit is given by

$$D_0 = \frac{c_s - c_e}{2} \tag{6.141}$$

Therefore if the effluent DO is aerated through postaeration in a treatment plant or even through introduction of pure oxygen to supersaturate the effluent, the resultant DO deficit can be reduced. The impact on the DO profile can then be evaluated using Eqs. 6.83 and 6.84. Practically, effluent DO can be raised to about 7 mg/l, but in some special situations, supersaturation of the effluent DO may be possible (i.e., availability of pure oxygen). Figure 6.31 illustrates the effect of reducing the DO deficit or supersaturating the DO at the outfall after mixing.

In the first panel of Fig. 6.31(a), the total DO deficit is given as the sum of the initial DO deficit profile (1) and the DO deficit due to the point source BOD input (2). The resulting DO profile is shown on the right in Fig. 6.31(b). When the initial DO deficit is halved by effluent aeration, the DO deficit decreases and the DO profile may meet DO standards by effluent aeration alone. This is illustrated in the second panel of Fig. 6.31(b) for a DO standard of 4 mg/l. The last panels in this figure show the effect of supersaturating the effluent where again this may be a means for achieving a DO objective without any further BOD reduction. This is illustrated in the last panel of Fig. 6.31(b) for a DO standard of 5 mg/l. Increasing the DO of the input should always therefore be examined as a possible DO control method.

6.8.3 Low Flow Augmentation

The basic notion of low flow augmentation is that if the drought flow can be increased by release from storage then a given BOD discharge load will have a reduced effect on the DO due to the dilution of the increased flow. Flow augmentation is usually more effective for rivers than for estuaries because of the volume of the latter water bodies. The initial BOD at the outfall after mixing is reduced (see Eq. 6.73) and the resulting DO deficit is reduced or DO increased (see Eqs. 6.83 and 6.84). However, as noted in these equations and in Section 2.1.3, increasing the flow also increases the river velocity which displaces the DO profile downstream in rivers. Figure 6.32 illustrates this point. The DO deficit is decreased in some regions due to dilution and may be increased in other regions due to increased transport. Normally, the overall effect is positive (i.e., DO is increased) in the region of lowest DO.

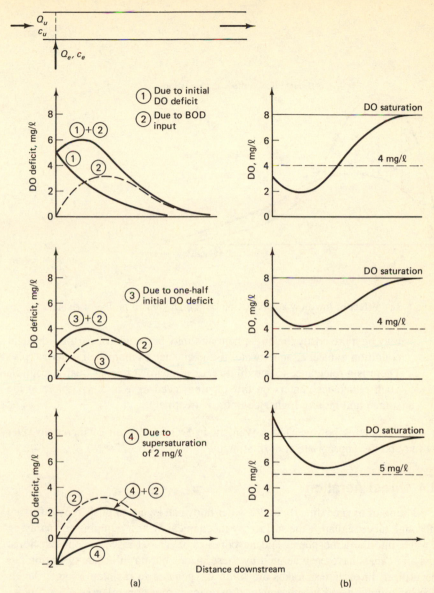

Figure 6.31 Effect of changes in initial DO. (a) DO deficit profiles. (b) Resulting DO profile in rivers.

There are two difficulties with low flow augmentation as a means for DO control:

1. The flow must be available during a drought condition and therefore storage must be provided in reservoirs for water quality control. Such storage

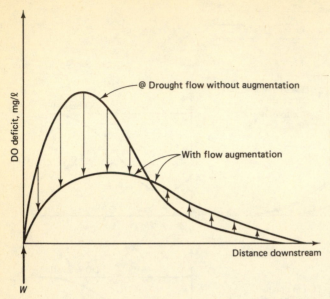

Figure 6.32   Effect of low flow augmentation on the DO deficit profile of rivers.

may be more costly than treatment and may be "raided" during a drought condition to meet a higher water use need such as municipal water supply.

2. There is a reluctance to see dilution as a means for water quality control. Such reluctance centers on the apparent need to treat the problem at the source and require higher degrees of treatment.

For some problems however, improvement in DO as a result of increased river flow may be the appropriate control strategy.

6.8.4 Direct Aeration

The addition of oxygen directly to the water body can be accomplished by diffusing air through tubes, agitating the surface by mechanical means, pumping and spraying the water into the atmosphere, and introducing oxygen at reservoir outlets. Some rivers and canals have been aerated by these means but no full scale operation has been utilized in estuaries. Lakes are often oxygenated to maintain oxygen levels in the hypolimnion and in subsequent downstream reservoir releases.

For rivers, the amount and spacing of the aerators can be determined from solution of the multiple source stream equation (see Chapter 2) coupled with the DO equations (6.83 and 6.84). Figure 6.33 schematically illustrates the effect of instream DO aeration to meet a specific DO standard. The oxygen input requirement (lb DO/hr) and spacing is determined for a given BOD input condition to meet the standard. The oxygen transfer rate is given by (Susag et al., 1966, O'Connor, 1968)

$$R_s = \alpha R_a \frac{c_s - c_u}{9.17} (1.024)^{T-20} \qquad (6.142)$$

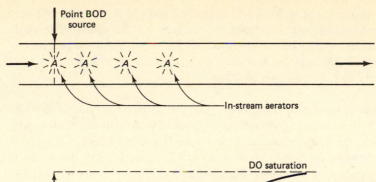

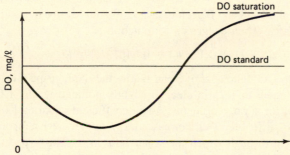

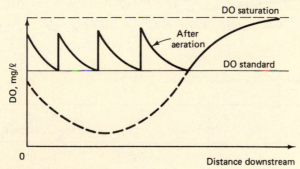

Figure 6.33 Schematic of effect of in-stream DO aeration.

where R_s is the oxygen transfer rate in lb DO/hp-hr, α is the ratio of oxygen transfer of river water to that of pure water (approximately 0.85), and R_a is the oxygen transfer efficiency of the aerator in lb/hp-hr. This value under standard conditions (temperature $= 20°C$, DO $= 0$ mg/l, pressure $= 1$ atm) is about 3 lb/hp-hr although values determined from field runs show that the actual transfer was about 1–2 lb/hp-hr at DO concentrations of 0–3 mg/l (O'Connor, 1968). The required aerator horsepower A(hp) is then

$$A = \frac{c_r}{R_s} \tag{6.143}$$

where c_r is the oxygen requirement to meet the DO standard in lb/hr. The obvious advantage of instream aerators is cost since such aerators need be operated only

during critical conditions of high water temperatures and low flows. As such, there
is a considerable cost savings of aerators over treatment at the source. The disadvantages are (1) the impact on the aesthetics of the stream from the presence of
the aerators and (2) the aerators deal only with DO whereas treatment has other
beneficial effects such as reduction of solids.

6.8.4.1 Aeration at Hydraulic Structures

Oxygen as noted previously can also
be transferred into the stream by construction of cascades or small dams. Of course,
in some systems, the dams or cascades exist already and the effect of the fall of
water on DO must be estimated. Hydroscience (1972) summarizes the works of
the British (Gameson et al., 1958 and Barrett et al., 1960) and Mastropietro (1968)
on the Mohawk River dams. Their suggested formulation is

$$D_a - D_b = \{1 - [1 + 0.11ab(1 + 0.046T)H]^{-1}\}D_a \tag{6.144}$$

where D_a and D_b are the DO deficits above and below the dam, respectively, in
mg/l; a is a coefficient equal to 1.25 in clear to slightly polluted water, 1.00 in
polluted water, and 0.80 for sewage effluents; b is a coefficient equal to 1.00 for
a weir with free fall and 1.3 for step weirs or cascades; T is the water temperature
in °C; and H is the height through which the water falls in ft. This type of formulation was used in the Black River, New York for 16 dams with heights of fall
from 10 to 35 ft (Fig. 6.22).

Butts and Evans (1978) explored a further refinement of Eq. 6.144 given by

$$r = 1 + 0.38abH(1 - 0.11H)(1 + 0.046T) \tag{6.145}$$

TABLE 6.13 WATER QUALITY COEFFICIENTS AND DAM
AERATION COEFFICIENTS

Polluted state	Water quality coefficient a^a	Dam type	Dam aeration coefficient $b^{a,b}$
Gross	0.65	Flat broad crested regular step	0.70
Moderate	1.0	Flat broad crested irregular step	0.80
Slight	1.6	Flat broad crested vertical face	0.60
Clean	1.8	Flat broad crested straight slope face	0.75
		Flat broad crested curved face	0.45
		Round broad crested curved face	0.75
		Sharp crested straight slope face	1.00
		Sharp crested vertical face	0.80
		Sluice gates	0.05

[a] In Eq. 6.145.
[b] "Best estimates" given by Butts and Evans (1978).
Source: From Butts and Evans (1978).

where

$$r = \frac{D_a}{D_b}$$ (6.145a)

and H is the difference in water elevations in meters.

In the work of Butts and Evans, the coefficient *a* was assigned as shown in Table 6.13. Forty-eight dams were evaluated in the field and the best estimates of the dam aeration coefficients, *b*, were determined as shown in Table 6.13.

REFERENCES

Alabama, State of, 1981. State of Alabama, Water Improvement Commission, Water Quality Criteria, Amended Feb. 4, 1981, 28 pp.

Alabaster, J. S., and R. Lloyd, 1980. *Water Quality Criteria for Freshwater Fish*, Butterworths, London, 297 pp.

APHA (American Public Health Association), 1985. *Standard Methods for the Examination of Water and Waste Water*, 16th ed., Washington, D.C., 874 pp.

Baity, H. G., 1938. Some Factors Affecting the Aerobic Decomposition of Sewage Sludge Deposits, *Sewage Works J.* **10**:539–568.

Banks, R. B., 1975. Some Features of Wind Action on Shallow Lakes, *Am. Soc. Civ. Eng., J. Environ. Eng. Div.* **101**(EE5):813–827.

Banks, R. B., and F. F. Herrera, 1977. Effect of Wind and Rain on Surface Reaeration, *Am. Soc. Civ. Eng., J. Environ. Eng. Div.* **103**(EE3):489–504.

Barrett, M. J., A. L. Gameson, and C. G. Dyden, 1960. Aeration Studies of Four Weir Systems, *Water and Water Eng.* **64**:407–413.

Bathala, C. T., K. C. Das, and W. D. Jones, 1979. Assimilative Capacity of Small Streams, *Am. Soc. Civ. Eng., J. Environ. Eng. Div.* Vol. 105(EE6):1049–1061.

Butts, T. A., and R. L. Evans, 1978. Effects of Channel Dams on Dissolved Oxygen Concentrations in Northeastern Illinois Streams, Circular 132, State of Illinois, Dept. of Reg. & Educ., Illinois Water Survey, Urbana, IL, 153 pp.

Castagnino, W. A., 1978. Polucion do aqua, modelos y control, CEPIS, Organ. Panamericana de la Salud, Lima, Peru, 234 pp.

Chapra, S. C., and G. A. Nosser, 1974. Documentation for HAR03: A Computer Program for the Modeling of Water Quality Parameters in Steady State Multi-Dimensional Natural Aquatic Systems, for USEPA, Region II, New York, NY, 2nd ed, Oct. 1974, 4 Chapters + Appendix.

Churchill, M. A., H. L. Elmore, and R. A. Buckingham, 1962. Prediction of Stream Reaeration Rates, *J. Sanit. Eng. Div., Proc. Am. Soc. Civ. Eng.*, **SA4**:1, Proc. Paper 3199.

Crohurst, H. R., 1933. A Study of the Pollution and Natural Purification of the Ohio River, IV. A Resurvey of the Ohio River Between Cincinnati, Ohio and Louisville, Ky, Including a Discussion of the Effects of Canalization and Changes in Sanitary Conditions Since 1914–1916. U.S. Treasury Dept., Public Health Service, Pub. Health Bull. No. 204, Washington, D.C., 111 pp.

Dames and Moore, Inc., 1978a. Construction costs for Municipal Waste Treatment Plants: 1973–77, USEPA, Office of Water Prog. Oper., Washington D. C., 7 Chapters. EPA 430/9-77-013.

Dames and Moore, Inc., 1978b. Analysis of Operation and Maintenance Cost for Municipal Waste Treatment Systems, USEPA, Office of Water Prog. Oper., Washington, D.C., 5 Chapters + 7 Appendixes. EPA 430/9-77-015.

Delwiche, C. C., 1956. Biological Transformations of Nitrogen Compounds, *Ind. Eng. Chem.* **48**(9):1421–1427.

Department of Scientific & Industrial Research, 1964. Effects of Polluting Discharges on the Thames Estuary, Water Pollution Research Tech. Paper No. 11, The Reports of the Thomas Survey Committee and of the Water Pollution Research Laboratory, London: Her Majesty's Stationary Office, 609 pp.

Di Toro, D. M., 1975a. Algae and Dissolved Oxygen, Summer Institute in Water Pollution Notes, Manhattan College, Bronx, NY.

Di Toro, D. M., 1975b. Nitrification, Manhattan College Summer Institute Notes, Bronx, NY.

Doudoroff, P. and D. L. Shumway, 1970. Dissolved Oxygen Requirements of Freshwater Fishes, Food and Agricultural Organization of the United Nations. FAO Fisheries Tech. Paper No. 86, 291 pp, pp. 225–275 reprinted in Warren et al. (1973).

Driscoll, E. D., J. L. Mancini, and P. A. Mangarella, 1981. *Technical Guidance Manual for Performing Waste Load Allocations, Book II, Streams and Rivers*, Chapter 1, Biochemical Oxygen Demand, Dissolved Oxygen and Ammonia Toxicity; for US EPA, Office of Water Regulations and Standards, Monitoring Branch, Washington, D. C., Jan. 1981. 4 Chapters & Appendix.

Eckenfelder, W. W., Jr., and D. J. O'Connor, 1961. *Biological Waste Treatment*, Pergamon, New York, 299 pp.

Edwards, R. W., and M. Owens, 1965. The Oxygen Balance of Streams, in *Ecology and The Industrial Society*, Fifth Symposium of the British Ecological Society, Blackwell, Oxford, England, pp. 149–172.

Erdmann, J. B., 1979. Systematic Diurnal Curve analysis, *J. Water Poll. Contr. Fed.* **51**(1):78–86.

Fair, G. M., E. W. Moore, and H. A. Thomas, 1941. The Natural Purification of River Muds and Pollutional Sediments, *Sewage Works J.,* **13**:270–307, 756–799, 1209–1228.

Fillos, J., and A. H. Molof, 1972. Effect of Benthal Deposits on Oxygen and Nutrient Economy of Flowing Waters, *J. Water Poll. Contr. Fed.* **44**(4):644–662.

Gameson, A. L., K. G. Vandyke, and C. G. Ogden, 1958. The Effect of Temperature on Aeration at Weirs, *Water and Water Eng.,* **62**:489–492.

Gaudy, A. F., and E. T. Gaudy, 1980. *Microbiology for Environmental Scientists and Engineers*, McGraw-Hill, New York, 736 pp.

Georgia, State of, 1980. Rules and Regulations for Water Quality Control, Chapter 391–396, Revised Oct. 1980, Dept. of Natural Resources, Env. Protection Div., Atlanta, GA, pp. 701–765.

Grant, R. S., and S. Skavroneck, 1980. Comparison of Tracer Methods and Predictive Equations for Determination of Stream Reaeration Coefficients on Three Small Streams in Wisconsin, USGS, Water Resource Investigation, 80–19, Madison, Wisc., 36 pp.

Gundelach, J. M., and J. E. Castillo, 1976. Natural Stream Purification Under Anaerobic Conditions, *J. Water Poll. Contr. Fed.* **48**(7):1753–1758.

Hall, J. C., and R. J. Foxen, 1983. Nitrification in BOD5 Test Increases POTW Noncompliance, *J. Water Poll. Contr. Fed.* **55**(12):1461–1469.

Higgins, J. J., J. A. Mueller, and J. P. St. John, 1978. Baseline and Alternatives: Modeling, NYC 208 Task Report 512/522, by Hydroscience, Inc., for Hazen and Sawyer, Engineers, Mar. 1978, 126 pp.

Holley, E. R., 1975. Oxygen Transfer at the Air-Water Interface, in Interfacial Transfer Processes in Water Resources, State University of New York at Buffalo, Report 75-1.

Hutchinson, G. E., 1957. *A Treatise on Limnology, Vol. I*, Wiley, New York, 1015 pp. + xiv.

HydroQual, Inc., 1983. Before and After Comparisons of Water Quality Following Municipal Treatment Plant Improvements, USEPA, Office of Water Prog. Operations, Facility Requirements Div., Washington, D.C.

HydroQual, Inc., 1984. SPAM, HydroQual User's Manual, Version 4.0, 40 pp.

Hydroscience, Inc., 1971. Simplified Mathematical Modeling of Water Quality, prepared for the Mitre Corporation and the USEPA, Water Programs, Washington, D.C., Mar. 1971, 127 pp., 4 Appendixes, 5 Monographs.

Hydroscience, Inc., 1972. Addendum to Simplified Mathematical Modeling of Water Quality, USEPA, Washington, D.C., 43 pp.

Hydroscience, Inc., 1974. Water Pollution Investigation: Black River, New York, Prepared for USEPA, Regions II and V. pp. 1–95. EPA-905/9-74-009.

Iowa, State of, 1983. Water, Air and Waste Management [900], Chapter 61, Water Quality Standards, IAC, 12/21/83, Des Moines, IA, pp. 1–65.

Kameda, Y., 1980. Comprehensive Basin-Wide Planning of Sewer Systems, *J. Water Poll. Contr. Fed.* **52**(5):884–889.

Leo, W. M., J. P. St. John, and D. J. O'Connor, 1978. Seasonal Steady State Model, NYC 208 Task Report 314, Hydroscience, Inc. for Hazen & Sawyer Engineers, Mar. 1978, 659 pp.

Loehr, R. C., 1968. Variation of Wastewater Quality Parameters, Public Works, May, pp. 81–83.

Mastropietro, M. A., 1968. Effects of Dam Reaeration on Waste Assimilation Capacities of the Mohawk River, Proc. 23rd Industrial Waste Conf., Purdue University.

Mueller, J. A., 1976. Review of Basis for Selecting Appropriate Velocity for Computing Reaeration Coefficients in Estuarine Systems with Freshwater Inflows, Technical Memorandum, Hydroscience, Inc., 4 pp.

Mueller, J. A., and A. R. Anderson, 1983. Municipal Sewage Systems, Chapter 2 in *Ocean Disposal of Municipal Wastewater: Impacts on the Coastal Environment*, Vol. 1. Ed. by E. P. Myers and E. T. Harding, Marine Resources Information Center, Mass. Instit. of Tech., Cambridge, MA, MITSG 83-33: pp. 39–92

National Academy of Sciences/National Academy of Engineering, 1973. Water Quality Criteria, 594 pp. EPA R/73/033.

New Mexico, State of, 1981. Water Quality Standards for Interstate and Intrastate Streams in New Mexico, Water Quality Control Commission, WQCC 81-1, Sante Fe, NM, pp. 1–65.

O'Connor, D. J., 1960. Oxygen Balance of an Estuary, *Am. Soc. Civ. Eng., J. San. Eng. Div.* **86**(SA3): 35–55.

O'Connor, D. J., 1962. Organic Pollution of New York Harbor—Theoretical Considerations, J. Water Poll. Contr. Fed. **34**(9):905–919.

O'Connor, D. J., 1965. Estuarine Distribution of Non-conservative Substances, *Am. Soc. Civ. Eng., J. Sanit. Eng. Div.* **91**(SA1):23–42.

O'Connor, D. J., 1966. An Analysis of the Dissolved Oxygen Distribution in the East River, *J. Water Poll. Contr. Fed.* **38**(11):1813–1830.

O'Connor, D. J., 1967. The Temporal and Spatial Distribution of Dissolved Oxygen in Streams, *Water Resources Research* **3**(1):65–79.

O'Connor, D. J., 1968. Reactions, Mechanical Aeration, Summer Institute in Water Pollution Notes, Manhattan College, Bronx, NY.

O'Connor, D. J., 1983. Wind Effects on Gas-Liquid Transfer Coefficients, *Am. Soc. Civ. Eng., J. Environ. Eng.* **109**(3):731–752.

O'Connor, D. J., and W. E. Dobbins, 1958. Mechanism of Reaeration in Natural Streams, *Trans. Am. Soc. Civil Eng.* **123**:655.

O'Connor, D. J., and J. L. Mancini, 1972. Water Quality Analysis of the New York Harbor Complex, *J. Water Poll. Contr. Fed* **44**(11):2129–2139.

O'Connor, D. J., and J. A. Mueller, 1984. Water Quality Analysis of New York Harbor Complex, *Am. Soc. Civ. Eng., J. Environ. Eng. Div.* **110**(6): 1027–1047.

Ohio, State of, Undated. Water Quality Standards, Chapter 3745, #1 of the Administrative Code, Ohio EPA, pp. 1–117.

Oldaker, W. H., A. A. Burgum, and H. R. Pahren, 1966. Report on Pollution of the Merrimack River and Certain Tributaries, Part IV—Pilot Plant Study of Benthic Oxygen Demand, U.S. Dept. of the Int., FWPCA, Northeast Region, Lawrence, MA, August 1966, 14 pp.

Otis, R. J., W. C. Boyle, J. C. Converse, and E. J. Tyler, 1977. On-site Disposal of Small Wastewater Flows, Prepared for EPA Tech. Trans. Seminar Program on Small Wastewater Treatment Systems.

Owens, M., R. Edwards, and J. Gibbs, 1964. Some Reaeration Studies in Streams, *Int. J. Air Water Poll.* **8**:469–486.

Pence, G. D., J. M. Jeglic, and R. V. Thomann, 1968. Time-Varying Dissolved Oxygen Model *Am. Soc. Civ. Eng., J. Sanit. Eng. Div.* **94**(SA2):381–402.

Pennsylvania, State of, 1979. Chapter 93 Water Quality Standards, Title 25, Rules and Regulations, Part I, Dept. of Env. Resources, Subpact C, Prot. of Natural Resources, Article II, Water Resources, Harrisburg, PA, pp. 93.1–93.15b.

Phelps, E. B., 1944. *Stream Sanitation*, Wiley, New York, 276 pp.

Rathbun, R. E., 1979. Estimating the Gas and Dye Quantities for Modified Tracer Technique Measurements of Stream and Reaeration Coefficients, US GS, Water Res. Invest., 79–27, Gulf Coast Hydro. Center, NSTL Station, MI, 42 pp.

Roper, R. E., Jr., R. O. Dickey, S. Marman, S. W. Kim, and R. W. Yandt, 1979. Design Effluent Quality, *Am. Soc. Civ. Eng., J. Environ. Eng. Div.* Vol. 105(EE2):309–321.

Slayton, J. L., and E. R. Trovato, 1979. Biochemical Studies of the Potomac Estuary— Summer 1978. Annapolis Field Office, Annapolis, MD, US EPA Region III, 35 pp. EPA 903/9-79-005.

Stratton, F. E., and P. L. McCarty, 1967. Prediction of Nitrification Effects on the Dissolved Oxygen Balance of Streams, *Env. Sci. Tech.* **1**(5):405–410.

Streeter, H. W., and E. B. Phelps, 1925. A Study of the Pollution and Natural Purification of the Ohio River, III, Factors Concerned in the Phenomena of Oxidation and Reaeration, U.S. Pub. Health Serv., Pub. Health Bulletin No. 146, February 1925, 75 pp. Reprinted by U.S., DHEW, PHA, 1958.

Susag, R. H., R. C. Pelta, and G. J. Schroepfer, 1966. Mechanical Aeration of Receiving Waters, *J. Water Poll. Contr. Fed.* **38**(1):53–68.

Texas, State of, 1981. Texas Surface Water Quality Standards, Texas Dept. of Water Res. [P-7]m, Austin, TX, 108 pp.

Theriault, E. J., 1927. The Oxygen Demand of Polluted Waters, I. Bibliographical, A Critical Review, II. Experimental, The Rate of Deoxygenation, Treasury Dept. USPHS, Pub. Health Bull. No. 173, Washington, D.C., 185 pp.

Thomann, R. V., 1963. Mathematical Model for Dissolved Oxygen, *Am. Soc. Civ. Eng., J. Sanit. Eng. Div.* **89**(SA5):1–30.

Thomann, R. V., 1972. *Systems Analysis and Water Quality Management*, McGraw-Hill, New York, Reprinted by J. Williams Book Co., Oklahoma City, OK, 286 pp.

Trewartha, G. T., 1954. An Introduction to Climate, McGraw-Hill, New York, 402 pp.

Tsivoglou, E., J. Cohen, S. Shearer, and P. Godsil, 1968. Tracer Measurement of Stream Reaeration, II. Field Studies, *J. Water Poll. Contr. Fed.* **40**(2):285–305.

Tsivoglou, E. C., and S. R. Wallace, 1972. Characterization of Stream Reaeration Capacity, USEPA, Report No. EPA-R3-72-012.

USEPA 1976, *Quality Criteria for Water* Washington, D.C., 256 pp.

USEPA, 1985, Ambient Water Quality Criterion for Dissolved oxygen, freshwater aquatic life draft submitted for comments in *Federal Register* **50**(76): 15634, April 11, 1985, 34 pp.

Velz, C. J., 1938. Deoxygenation and Reoxygenation, *Proc. Am. Soc. Civ. Eng.,* **64**(4):767–779.

Velz, C. J., 1939. *Discussion, Proc. Am. Soc. Civ. Eng.,* **65**(4):677–680.

Velz, C. J., 1947. Factors Influencing Self-Purification and Their Relation to Pollution Abatement, *Sewage Works J.* **19**(4):629–644.

Velz, C. J., 1984. Applied Stream Sanitation, 2nd ed., Wiley, New York, 800 pp.

Warren, C. E., 1971. *Biology and Water Pollution Control*, Saunders, Philadelphia, 434 pp.

Warren, C. E., P. Doudoroff, and D. L. Shumway, 1973. Development of Dissolved Oxygen Criteria for Freshwater Fish, USEPA, ORD, Washington, D.C., 121 pp. EPA-R3-73-019.

Washington, State of, 1982. Water Quality Standards for the State of Washington, Chapter 173–201, WAC, pp. 1–11.

Willingham, W. T., 1976. Ammonia Toxicity, USEPA, Report No. 908/3-76-001, Feb. 1976, 19 pp. + 2 Appendixes.

Wright, R. M., and A. J. McDonnell, 1979. In-stream Deoxygenation Rate Prediction, *Am. Soc. Civ. Eng., J. Env. Eng. Div.* **105**(EE2):323–335.

Zison, S. W., W. B. Mills, D. Diemer, and C. W. Chen, 1978. Rates, Constants and Kinetic Formulations in Surface Water Quality Modeling, by Tetra Tech, Inc. for USEPA, ORD, Athens, GA, ERL, EPA 600-3-78-105, 317 pp.

PROBLEMS

6.1. The city of Bogota, Colombia is located on the Bogota River at an elevation of 2600 m. Average temperature is 13°C. Ten kilometers downstream from the city, the river drops to 700 m over a distance of 60 km. The temperature at that point is 28°C. Compute and plot the variation of DO saturation (mg/l) as a function of distance.

6.2. Suppose in a large vessel, with a reaeration rate of 0.8/day and a temperature of 20°C, the initial DO is 1 mg/l. The DO then increases due to reaeration to a level of 5.6 mg/l. At exactly that moment, the DO is dropped again to 3.6 mg/l. How many days will it take, counting from the time the DO started at 1 mg/l, to reach a level of 7.0 mg/l?

6.3. Suppose the temperature profile of a pristine river is:

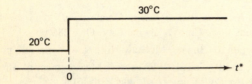

If the reaeration rate is 0.5/day @ 30°C, plot the DO profile in mg/l from $t^* = 0$ to $t^* = 5$ days.

6.4. The following data were collected in Guanabara Bay, Rio de Janeiro using the light and dark bottle method of estimating photosynthesis DO inputs. The bottles were set at a depth of 0.3m. The photoperiod is 11 hours with sunrise at 0630.

	Time	DO(mg/l) in light bottle	DO(mg/l) in dark bottle
June 9	0930	8.0	8.0
	1230	8.4	7.5
	1530	9.6	7.4
	1830	10.0	7.1
	2130	9.8	6.4
	2430	9.8	6.8
	0330	8.9	6.6
	0630	9.6	6.8
	0930	10.4	6.1

(a) Plot the data vs. time of day.

(b) Estimate the average photosynthetic production of oxygen during a day in mg/l · day.

(c) Estimate the respiration due to photosynthesis during the day in mg/l · day.

(d) Suppose an ultimate filtered CBOD was obtained several times from the dark bottle and the average value was 4 mg/l. If the deoxygenation rate is assumed at 0.15/day, what is the new estimated value of the respiration rate in mg/l · day?

6.5. As shown in the sketch, a river receives an input of 5 cfs with an ultimate CBOD of 100 mg/l and saturated DO. Upstream of the input, the flow is 45 cfs, the CBOD ultimate is 5 mg/l, and the DO is also saturated. The depth is 5 ft and the velocity is 0.5 fps. $K_r = K_d = 0.2$/day. Suppose right at the point where the DO deficit is the worst, another 50 cfs is added with zero CBOD and saturated DO. If the velocity does not change because of this flow, what is the dissolved oxygen (mg/l) 10 mi below the 50 cfs addition? The temperature throughout is 20°C.

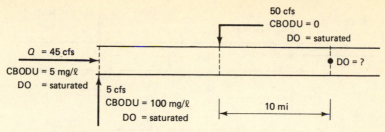

6.6. A river has an upstream flow of 80 cfs, an ultimate first-stage CBOD of 5 mg/l, and a DO of 6 mg/l. An effluent enters the river with a flow of 10 cfs, a 5-day BOD of 20 mg/l, and a DO of zero. A long-term CBOD test on the effluent has indicated a $K_1 = 0.25$/day. DO saturation in the river is 8.26 mg/l at a temperature of 25°C. The CBOD decay rate in the river is 0.6/day at 25°C, the velocity is 0.8 fps, and the depth is 5 ft.

  (a) What is the ultimate BOD (mg/l) 15 mi downstream from the outfall?

  (b) What is the DO (mg/l) at this point, due *only* to the initial DO deficit at the outfall after mixing of the effluent DO and the river DO?

6.7. What is the *total* DO deficit (mg/l) at mile 15 due to the BOD input, background BOD, and initial DO deficit for Problem 6.6? Use Eq. 6.60 for K_d.

6.8. At mile 15, how much oxygen (lb/day) would have to be added to have the DO at 7.0 mg/l at that point in Problem 6.6?

6.9. In problem 6.6, if the initial deficit were reduced to zero, what should the 5-day CBOD of the effluent be to have the DO above 7 mg/l at all places in the river? Assume no change in the "bottle" BOD rate K_1.

6.10. Given the following conditions for a river receiving a waste input from a single point source:

Waste input	Upstream condition before input	River conditions
Q = 5 MGD	Q = 50 MGD	c_s = 8 mg/l
c = 2 mg/l	CBODU = 6 mg/l	reaeration rate = 1.0/day
CBOD5 = 60 mg/l	c = 6 mg/l	
K_1 = 0.2/day		

A river survey indicated that the CBODU at a point 2 days travel time was 6 mg/l. Assuming $K_r = K_d$ and a river temperature of 26.8°C, what is the minimum value of DO resulting from the waste discharge in mg/l?

6.11. **(a)** Continuing with Problem 6.10, suppose that at the point of minimum DO, it is desired to reaerate the river to the DO saturation level. How many pounds of oxygen/day must be pumped into the river to achieve this objective?

  (b) What will the DO (mg/l) be at 1-day travel time below this aeration input?

6.12. A river receives two inputs of BOD as shown in sketch on p. 382. The second input discharges at the foot of a dam. At the foot of the dam reaerated water from the flow going over the dam results in saturated DO conditions after the dam and at the location of load 2. What is the maximum DO deficit downstream of the dam and where does that maximum deficit occur? (All BOD input concentrations are first-stage CBODU.)

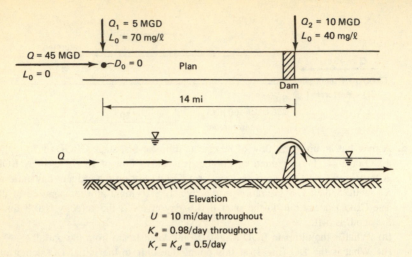

$$U = 10 \text{ mi/day throughout}$$
$$K_a = 0.98/\text{day throughout}$$
$$K_r = K_d = 0.5/\text{day}$$

6.13. The chlorophyll concentration in a river over an 8-day travel time has been observed to have an exponential increase as shown.

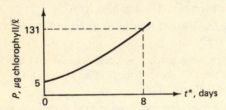

The temperature of the river is 25°C, $G_{max} = 1.8/\text{day}$, and $G(I_a) = 0.264$.

(a) Using $P = P_0 \exp(G_n t^*)$ (where G_n is the net growth rate of the phytoplankton in the river) calculate the average phytoplankton chlorophyll concentration for $t^* = 0$ to $t^* = 8$ days

(b) For $a_{op} = 0.133 \text{ mg DO}/\mu\text{g chlor}$ and using the average chlorophyll from (a), what is the average photosynthetic DO production, p_a, in mg/l · day?

(c) What is the DO deficit response (mg/l) at $t^* = 8$ days due to the average photosynthetic production of DO in this reach? Reaeration rate $= 0.6/\text{day}$.

6.14. A river has the following characteristics for the stretch from $x = 0$ to $x = 66$ mi.

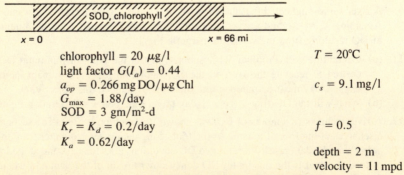

chlorophyll $= 20 \ \mu\text{g/l}$ $T = 20°C$
light factor $G(I_a) = 0.44$
$a_{op} = 0.266 \text{ mg DO}/\mu\text{g Chl}$ $c_s = 9.1 \text{ mg/l}$
$G_{max} = 1.88/\text{day}$
SOD $= 3 \text{ gm/m}^2\text{-d}$
$K_r = K_d = 0.2/\text{day}$ $f = 0.5$
$K_a = 0.62/\text{day}$

depth $= 2 \text{ m}$
velocity $= 11 \text{ mpd}$

at $x = 0$, $D_0 = 3 \text{ mg/l}$ and $L_0 = 29 \text{ mg/l (CBODU)}$

What is the *minimum* diurnal DO expected at $x = 66$ mi?

6.15. Using the projection conditions for the stream in Sample Problem 6.10, plot the ammonia–nitrogen, nitrite + nitrate–nitrogen, and DO deficit concentrations due to the ammonia–nitrogen in the STP effluent (1439 lb/day). Perform the analysis for an elevated stream temperature of 30°C. Assume instream nitrification occurs at the rate determined in the sample problem, that is, $K_N = 0.15$/day at 20°C.

6.16. For the Delaware Estuary, a load of 200,000 lb BODU/day is proposed to be discharged. Assume all parameters are constant with distance. The flow is 3000 cfs, the cross-sectional area is 100,000 ft², the reaeration coefficient is 0.11/day, and $K_r = K_d = 0.4$/day. The tidal dispersion coefficient is 5 smpd. Plot the DO deficit profile and the DO profile for $c_s = 7.5$ mg/l.

6.17. A reservoir has a volume of 100 MG. The average flow through the reservoir is 20 MGD and the low flow is 2 MGD. An ultimate BOD load of 200 lb/day is discharged into the reservoir. $K_r = K_d = 0.2$/day and $K_a = 0.05$/day at 20°C, the temperature of the well-mixed reservoir. Assuming the reservoir volume does not change with the flow and that $c_{in} = c_s$:
(a) Calculate the DO deficit and DO for the average flow condition.
(b) Repeat (a) for the low flow condition.

6.18. For the Delaware Estuary steady-state response matrices given in the text, Tables 6.9, 6.10, compute the DO (mg/l) in seg. 16 for the following configuration.

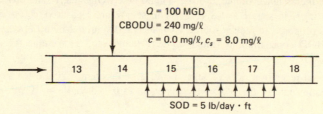

$$Q = 100 \text{ MGD}$$
$$\text{CBODU} = 240 \text{ mg/}\ell$$
$$c = 0.0 \text{ mg/}\ell, \; c_s = 8.0 \text{ mg/}\ell$$

| 13 | 14 | 15 | 16 | 17 | 18 |

SOD = 5 lb/day · ft

Assume $c_s = 7.0$ mg/l for the estuary.

6.19. A three-segment estuary is given as follows:

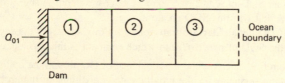

Q_{01} → ① ② ③ Ocean boundary

Dam

The geometry, flow, and dispersion are:

geometry: $V_1 = 10^7 \text{ m}^3$, $V_2 = 10^6 \text{ m}^3$, $V_3 = 10^8 \text{ m}^3$
flows: $Q_{01} = 50 \cdot 10^5 \text{ m}^3/\text{day}$
   $Q_{12} = 55 \cdot 10^5 \text{ m}^3/\text{day}$
   $Q_{23} = 60 \cdot 10^5 \text{ m}^3/\text{day}$
   $Q_{30} = 65 \cdot 10^5 \text{ m}^3/\text{day}$
dispersion: $E' = 10^7 \text{ m}^3/\text{day}$ for all interfaces

The concentration of CBOD of the incoming flow is 10 mg/l and the boundary concentration is 2 mg/l. Toxic chemicals in the incoming flow create a toxic condition in seg. 1 and 2 such that the decay of the CBOD is zero in those segments. The CBOD begins to be oxidized in seg. 3 at a rate of $K_r = 0.1$/day. Using backward differences:
(a) What is the net flux of CBOD from upstream into seg. 3 in kg/day?

(b) What is the CBOD concentration in seg. 3 in mg/l?

(c) What is the response in seg. 3 due just to the CBOD load coming over the dam?

(d) If the load coming over the dam were diverted to seg. 2, what is the CBOD in seg. 3?

6.20. An estuary flows into an enclosed marsh area as follows:

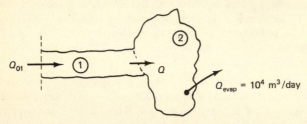

The flow entering the estuary is 10^4 m³/day. It flows through seg. 1 and into seg. 2, but then because of evaporation, there is no water outflow from seg. 2. The exchange between 1 and 2 is $2 \cdot 10^4$ m³/day. Initially a BOD source of 200 kg/day enters seg. 1 and steady-state BOD concentrations of 7.18 mg/l in seg. 1 and 6.16 mg/l in seg. 2 are measured. Later, however, another source of 50 kg/day is added into seg. 2 (together with the assigned source into seg. 1) and new BOD steady-state concentrations of 8.18 mg/l in seg. 1 and 8.46 mg/l in seg. 2 are measured. $K_r = K_d = 0.1$/day. It is proposed to reduce source 1 by 80% and source 2 by 50%: what will be the new BOD concentrations (mg/l) in seg. 1 and seg. 2? Assume the ocean boundary $L_0 = 0$.

6.21. Suppose that the reaeration rates of segs. 1 and 2 of Problem 6.20 are the same as the deoxygenation rate of the BOD which is equal to 0.1/day. $V_1 = 10^4$ m³, $V_2 = 15 \times 10^4$ m³.

(a) How many mg/l of DO deficit would be generated in the first segment for every 100 kg BOD/day into the second segment?

(b) Suppose there is an SOD in seg. 2 of 6 g/m² · day for a depth in seg. 2 of 3m. How much DO deficit will this SOD create in seg. 1?

(c) How many kg of DO deficit/day are transported by mixing across the seg. 1–2 interface due to the SOD of part (b)? Into which segment is this flux directed?

6.22. (a) In Sample Problem 6.9, calculate the mass balance of DO around the epilimnion and the hypolimnion segments for the intially stated conditions of the problem.

(b) Verify that the dispersive flux across the epilimnion–hypolimnion interface is a constant for an E of 1 cm²/s.

Chapter 7

Eutrophication

7.1 INTRODUCTION

Even the most casual observer of water quality has probably had the dubious opportunity of walking along the shores of a lake that has turned into a sickly green "pea soup." Or perhaps, one has walked the shores of a slow moving estuary or bay and had to step gingerly to avoid rows of rotting, matted, stringy aquatic plants. These problems have been grouped under a general term called eutrophication. The unraveling of the causes of eutrophication, the analysis of the impact of human activities on the problem, and the potential engineering controls that can be exercised to alleviate the condition have been a matter of special interest for the past several decades.

Aquatic plants can be thought of in two very broad categories: (a) those that move freely with the water (planktonic aquatic plants) and (b) those that remain fixed—attached or rooted in place. The first category includes the microscopic phytoplankton as well as free floating water weeds or certain types of plants such as the blue-green algae which may float to the surface and move with the surface current. The second category includes rooted aquatic plants of various sizes and attached microscopic plants (the benthic algae). Algae therefore is an all-inclusive designation of simple plants, mostly microscopic, which includes both the free moving plants, the phytoplankton, and the attached benthic algae. In all cases, the plants obtain the primary energy source from sunlight through the photosynthesis process.

Eutrophication is the excessive growth of aquatic plants, both attached and planktonic to levels that are considered to be an interference with desirable water uses. The growth of aquatic plants results from many causes which will be explored in this chapter. One of the principal stimulants, however, is an excess level of nutrients such as nitrogen and phosphorus. In recent years, this problem has been increasingly acute due to the discharge of such nutrients by municipal and industrial sources, as well as agricultural and urban runoff. It has often been observed that there is an increasing tendency for some water bodies to exhibit increases in the severity and frequency of phytoplankton blooms and growth of aquatic weeds apparently as a result of elevated levels of nutrients. Figure 7.1 shows the general problem framework.

This increased production of aquatic plants has several consequences regarding water uses:

1. Aesthetic and recreational interferences—algal mats, decaying algal clumps, odors, and discoloration may occur.
2. Large diurnal variations in dissolved oxygen (DO) can result in low levels of DO at night; which, in turn, can result in the death of desirable fish species.
3. Phytoplankton and weeds settle to the bottom of the water system and create a sediment oxygen demand (SOD), which, in turn, results in low values of DO in the hypolimnion of lakes and reservoirs, and the bottom waters of deeper estuaries.
4. Large diatoms (phytoplankton that require silica) and filamentous algae can clog water treatment plant filters and result in reduced time between backwashing.
5. Extensive growth of rooted aquatic macrophytes (larger plant forms) interfere with navigation, aeration, and channel carrying capacity.
6. Toxic algae have sometimes been associated with eutrophication in coastal regions and have been implicated in the occurrence of "red tide," which may result in paralytic shellfish poisoning.

The eutrophication problem can then be further defined as the input of organic and inorganic nutrients into a body of water which stimulates the growth of algae

Figure 7.1 Sources of nutrients in eutrophication problems.

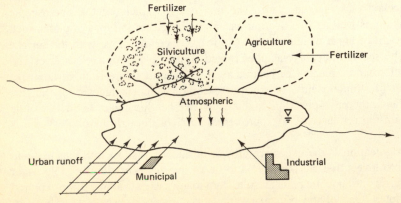

or rooted aquatic plants resulting in the interference with desirable water uses of aesthetics, recreation, fish maintanance, and water supply.

The condition of a water body is then described in terms of its trophic state, that is, its degree of eutrophication or lack thereof. For lakes, three designations have been used:

1. Oligotrophic—clear, low productivity lakes.
2. Mesotrophic—intermediate productivity lakes.
3. Eutrophic—high productivity lakes relative to a basic natural level.

Table 7.1 shows the range of water quality variables for these trophic classifications.

The level of eutrophication due to excessive amounts of phytoplankton can be measured using several criteria:

1. Counts (number/ml) of specific phytoplankton species, for example, *Asterionella formosa*. This is the most direct count of species trends but requires a considerable effort by trained specialists in phytoplankton identification. Also, conversion to biomass measures is difficult because of variations in cell sizes for given species.
2. Cell volume ($\mu m^3/\mu l$) of species. An excellent measure permitting ready conversions to biomass as dry weight or carbon and grouping of data into different categories (e.g., "diatoms" and "others"). However, this measure requires an extensive analytical and data reduction effort.
3. Chlorophyll *a* concentrations ($\mu g/l$). A measure of the gross level of phytoplankton, easily obtained without extensive effort in laboratory. However, chlorophyll does not provide information on species levels nor does it permit grouping into classes of phytoplankton. Chlorophyll *a* is the most common measure used in eutrophication studies.

"Undesirable" levels of phytoplankton vary considerably depending on water body. For example, the following levels represent levels which, in some way, are considered undesirable:

TABLE 7.1 TROPHIC STATUS OF LAKES

Water quality variable	Oligotrophic	Mesotrophic	Eutrophic	Reference[a]
Tp ($\mu g/l$)	<10	10–20	>20	1
Chlorophyll ($\mu g/l$)	<4	4–10	>10	2
Secchi depth (m)	>4	2–4	<2	1
Hypolimnetic oxygen (% saturation)	>80	10–80	<10	1

[a]References:
(1) USEPA (1974).
(2) NAS, NAE (1972).

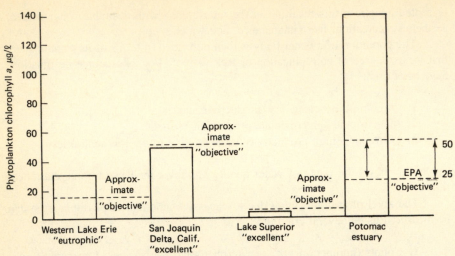

Figure 7.2   Range of chlorophyll *a* average concentrations and "objectives."

Open Lake Michigan—2–5 μg chl/l.

Open Lake Ontario—5–10 μg chl/l.

Western Lake Erie—30 μg chl/l.

Sacramento, San Joaquin Delta—50–100 μg chl/l.

Potomac Estuary—> 100 μg chl/l.

Objectives for "desirable" levels of chlorophyll vary widely depending on the type of problem and the nature of the water body. Figure 7.2 illustrates the range in present observed chlorophyll concentrations and that level considered desirable in some sense. In general, lakes and reservoirs tend to have lower desirable levels of phytoplankton. Chapra and Tarapchak (1976) have summarized several objectives for North Temperate Lakes as follows:

Eutrophic—> 5 to 10 μg chl *a*/l.

Mesotrophic—1 to 15 μg chl *a*/l.

Oligotrophic—< 1 to 4 μg chl *a*/l.

Some other levels that have been discussed include the Potomac Estuary at 25–50 μg chl *a*/l and the Sacramento–San Joaquin Delta at 25–50 μg/l.

7.2 BASIC MECHANISMS OF EUTROPHICATION

The growth and proliferation of aquatic plants is a result of the utilization and conversion of inorganic nutrients into organic plant material through the photosynthesis mechanism. Section 6.3.3 discussed this process from the point of view of dissolved oxygen. Here the initial emphasis is on the role of nutrients in both stimulating and controlling plant biomass.

The fundamental driving force for the process is the incoming solar radiation. Therefore, the eutrophication of a given water body may vary depending on the geographical location of the surface water, the degree of penetration of the solar radiation to different depths, the magnitude and type of nutrient inputs, and the particulars of the water movement through flow transport and dispersion. Since the aquatic plants may vary widely in species composition, the impact of each of the aforementioned factors may differ widely. For example, some species may require significantly less light and nutrients for growth than others. Some forms may remain rooted (as noted earlier), others may be buoyant, while still other forms may sink under different physiological conditions.

In spite of the complexities, an engineering-scientific predictive modeling approach for eutrophication control can be constructed. The principles of the approach are now discussed with the emphasis initially on the phytoplankton.

The basic phenomena underlying the process of phytoplankton growth in the North Temperate regions are summarized in Fig. 7.3. Increasing solar radiation provides the energy source for the photosynthesis reaction. Phytoplankton biomass then begins to increase as water temperature increases and as a result, nutrients in dissolved form are utilized by the plankton. This mechanism continues until nutrients reach levels that will no longer support growth at which the increase in phytoplankton biomass ceases. A decline is then observed, due often to zooplankton

Figure 7.3 Basic process of phytoplankton–nutrient interaction.

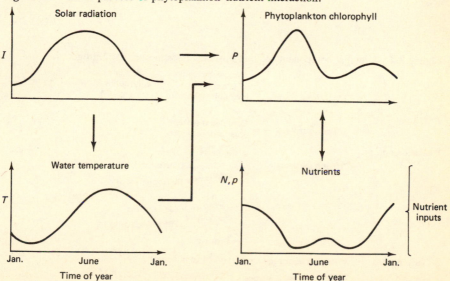

predation and often a late summer–early fall bloom may be observed again due to nutrient recycling. Biomass then declines as solar radiation and temperature decrease to the lower levels of late fall and early winter.

The principal variables of importance then in the analysis of eutrophication are:

1. Solar radiation at the surface and with depth.
2. Geometry of water body; surface area, bottom area, depth, volume.
3. Flow, velocity, dispersion.
4. Water temperature.
5. Nutrients.
    a. Phosphorus.
    b. Nitrogen.
    c. Silica.
6. Phytoplankton—chl *a*.

It should be recognized that the nutrients above are present in several forms in a body of water, and not all forms are readily available for uptake by the phytoplankton. The more important categories are shown in Figs. 7.4(a) and 7.4(b). Total phosphorus is composed of two principal components: the total dissolved form and the total particulate form. The dissolved form, in turn, is composed of several forms, one of which is the dissolved reactive phosphorus. This form of phosphorus is available for phytoplankton growth. The particulate phosphorus forms include inorganic soil runoff phosphorus particulates and the organic particulate phosphorus which include detritus and the phytoplankton phosphorus.

Total nitrogen is composed of four major components, the organic, ammonia, and nitrite plus nitrate forms. The latter three forms make up the total inorganic

Figure 7.4 Principal components of nutrients. (a) Phosphorus. (b) Nitrogen.

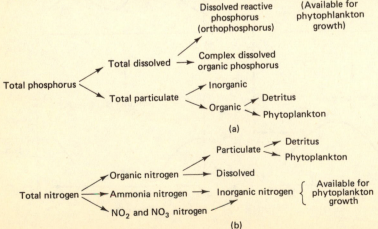

nitrogen which is the form utilized by the phytoplankton for growth. The organic form of nitrogen represents both a dissolved and particulate component. The particulate form, in turn, is composed of organic detritus particles and the phytoplankton.

Figure 7.5 shows the phytoplankton dynamic behavior for the epilimnion of Lake Ontario. The schematic of the lake shown in Fig. 7.5(a) shows the principal inputs and physical mechanisms that influence the phytoplankton growth. As shown in Fig. 7.5(b), the chl a level for the upper layer of Lake Ontario is low until the spring when solar radiation and water temperature increases. A peak value of about 5–10 μg chl a/l is reached in the May–June period. Concurrently, available phosphorus levels [Fig. 7.5(d)] were reduced to levels that caused a cessation of phytoplankton growth. Figure 7.5(c) shows the growth of zooplankton in the summer through predation of the phytoplankton.

Figure 7.6 shows similar dynamic interactions between chlorophyll and phosphorus for Saginaw Bay and southern Lake Huron. The concentrations of chlorophyll in Saginaw Bay of between 20 and 40 μg chl a/l are indicative of the eutrophic status of that body of water at the time of the study.

7.3 EXTERNAL SOURCES OF NUTRIENTS

As shown in Fig. 7.1, the principal external sources of nutrient inputs are:

1. Municipal wastes.
2. Industrial wastes.
3. Agricultural runoff.
4. Forest runoff.
5. Urban and suburban runoff.
6. Atmospheric fallout.

Tables 7.2 to 7.4 and Fig. 7.7 summarize some of the data on these inputs from several sources (Eisenreich et al., 1977; Gakstatter et al., 1978; Hydroscience, 1977a, 1977b; Rast and Lee, 1978; Thomann, 1972; IJC, 1978).

TABLE 7.2 MEAN NUTRIENT INPUTS FROM MUNICIPAL WASTEWATERS

Nutrient form	Influent (mg/l)	Conventional secondary treatment effluent (mg/l)	After phosphorus removal processes (mg/l)
Phosphorus (as p)			
Tp with detergent	5–10	7	1–3
Tp without detergent	2–5	4	
Ortho p with detergent	2–5	5	1–2
Nitrogen (as N)			
TN	50	18	14
IN	30	8	7

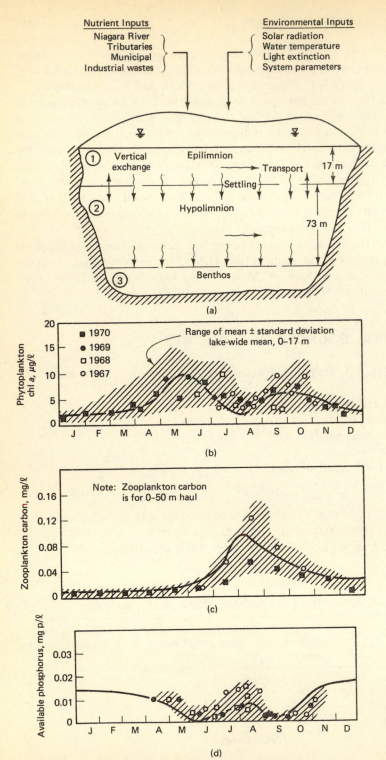

Figure 7.5 Phytoplankton dynamics in Lake Ontario. (a) Schematic of vertical regions. (b) Phytoplankton chlorophyll *a*, epilimnion. (c) Zooplankton carbon. (d) Available phosphorus. From Thomann et al. (1974). Reprinted by permission of International Association for Great Lakes Research.

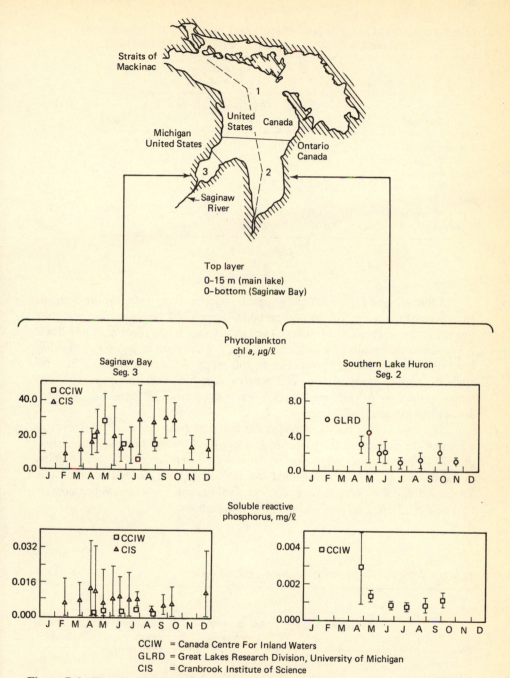

Figure 7.6 Phytoplankton dynamics, Saginaw Bay, Michigan and Southern Lake Huron. After DiToro and Matystik (1980).

393

TABLE 7.3   GREAT LAKES BASIN—RURAL LAND PHOSPHORUS LOADS
DEVELOPED BY INTERNATIONAL JOINT COMMISSION (1978)

Land use	kg Tp/ha · yr, type of soil			
	Sand	Coarse loam	Fine loam	Clay
Row cropping (> 50% row crops)	0.25	0.65	1.05	1.25
Mixed farming (25–50% row crops)	0.10	0.20	0.55	0.85
Forage (< 25% row crops)	0.05	0.05	0.40	0.60
Grassland	0.05	0.05	0.05	0.25

Table 7.2 and Figure 7.7 show typical concentration values for nitrogen and phosphorus; the former from municipal point sources and the latter from non-point sources in the eastern United States. The Great Lakes Basin loads shown in Table 7.3 represent the results of a wide-spread sampling program of phosphorus loading from differing agricultural practices and soil types. As shown in Table 7.4, the range of loading from all non-point sources is rather wide and reflects various ranges in the flow or precipitation rates. The values given in Table 7.4, however, are useful as general guidelines.

Omernik (1977) has summarized nutrient data from non-point sources from a nationwide network of 928 watersheds under the National Eutrophication Study of 1972–1974. A simple regression model was used to relate runoff concentrations to the percent of land in agricultural use and the percent of land in urban use. Tables 7.5 and 7.6 show these results for total phosphorus and nitrogen concentrations for different regions and land use distributions.

TABLE 7.4   OTHER NON-POINT SOURCE LOADINGS[a]

Type	Tp		TN	
	Approximate mean	Approximate range	Approximate mean	Approximate range
Forest, natural	0.4	(0.01–0.9)	3.0	(1.3–10.2)
Atmospheric				
rainfall	0.2	(0.08–1.0)	8.0	
dry fallout	0.8		16.0	
Urban	1.0	(0.1–10)	5.0	(1–20)
Agricultural, general	0.5	(0.1–5)	5.0	(0.5–50)

[a]All values in kg/ha · yr; values from IJC (1978) and Rast and Lee, (1978).

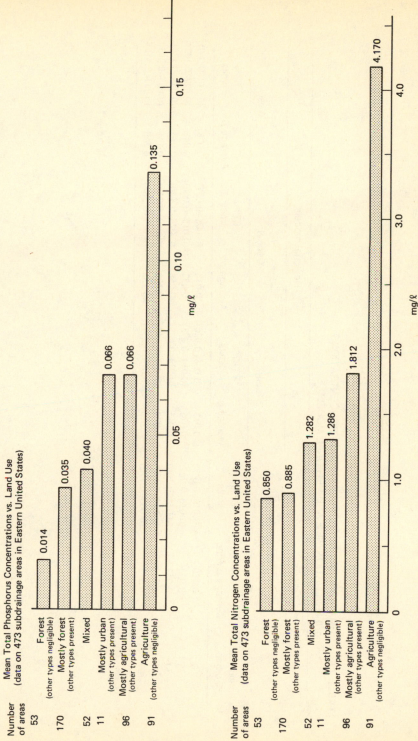

Figure 7.7 Mean total phosphorus and total nitrogen concentrations in streams draining different land use categories. From Bartsch and Gakstatter (1978).

TABLE 7.5 ESTIMATED MEAN TOTAL PHOSPHORUS CONCENTRATIONS (mg/l) FOR NATIONWIDE AND REGIONAL MODELS

% agricultural + % urban	Nationwide model		Regional models					
			East		Central		West	
	Average p conc	67% limits	Average p	67% limits	Average p	67% limits	Average p	67% limits
0	0.020	(0.010–0.042)	0.015	(0.008–0.026)	0.018	(0.008–0.037)	0.028	(0.013–0.060)
25	0.033	(0.016–0.067)	0.026	(0.014–0.048)	0.032	(0.015–0.066)	0.048	(0.022–0.103)
50	0.052	(0.025–0.106)	0.045	(0.024–0.083)	0.058	(0.028–0.120)	0.083	(0.038–0.178)
75	0.083	(0.041–0.170)	0.078	(0.042–0.144)	0.103	(0.049–0.213)	0.141	(0.065–0.303)
100	0.133	(0.065–0.272)	0.136	(0.073–0.251)	0.185	(0.089–0.382)	0.240	(0.111–0.516)

Source: From Omernik (1977).

TABLE 7.6 ESTIMATED MEAN TOTAL NITROGEN CONCENTRATIONS (mg/l) FOR NATIONWIDE AND REGIONAL MODELS

% agricultural + % urban	Nationwide model		Regional models					
			East		Central		West	
	Average N	67% limits	Average N	67% limits	Average N	67% limits	Average N	67% limits
0	0.57	(0.35–0.92)	0.51	(0.33–0.77)	0.57	(0.37–0.88)	0.63	(0.34–1.16)
25	0.90	(0.56–1.46)	0.87	(0.57–1.32)	0.83	(0.54–1.28)	0.95	(0.52–1.75)
50	1.45	(0.89–2.34)	1.49	(0.98–2.26)	1.22	(0.79–1.88)	1.42	(0.77–2.61)
75	2.31	(1.43–3.74)	2.55	(1.68–3.88)	1.78	(1.15–2.74)	2.13	(1.16–3.92)
100	3.69	(2.28–5.97)	4.37	(2.87–6.64)	2.61	(1.69–4.02)	3.19	(1.73–5.87)

Source: From Omernik (1977).

7.4 SIGNIFICANCE OF THE *N/p* RATIO

As discussed previously, the plant biomass increases through the uptake of available phosphorus and nitrogen in the water. Such nutrients result from discharges from point and non-point sources. If the nutrients discharged to the body of water are reduced, the available pool of nutrients that can be utilized by the plants will be reduced and, in general, the total biomass will be reduced. Two important questions that naturally arise then are:

1. Should inputs of phosphorus or nitrogen or both nutrients be controlled?
2. What is the allowable amount of nutrients that can be discharged to the body of water so that some desired level of plant biomass is maintained?

One simple approach to the first question is to examine the relative nutrient requirements by plants for nitrogen and phosphorus (the *N/p* ratio). The second question can be approached using the basic concepts of the modeling frameworks developed earlier.

The nutrient that will control the maximum amount of plant biomass is the nutrient that is in the lowest concentration, or stated another way, the nutrient that "runs out" or reaches a minimum before other nutrients. Figure 7.8 illustrates the principle. Panel (a) of this figure shows that under certain conditions nitrogen may reach a minimum value before phosphorus and as a result controls the maximum amount of biomass (chlorophyll) that might be attained. At the time (or location) where nitrogen has run out, phosphorus is still present at sufficient concentrations. On the other hand, Fig. 7.8(b) shows that if the initial phosphorus concentration were reduced, then phosphorus may run out first, the plant biomass is then controlled by the phosphorus, and the maximum amount may be reduced. Which situation will prevail depends on two considerations:

1. The relative amount of nitrogen and phosphorus required by aquatic plants.
2. The relative amount of nitrogen and phosphorus available for growth initially in the body of water.

The first consideration involves the cell stoichiometry of aquatic plants. For phytoplankton, as an example, cells contain approximately 0.5–2.0 μg phosphorus/μg chl and 7–10 μg nitrogen/μg chl.

Thus in Fig. 7.9, if the available nitrogen initially were 5 mg N/l, and a stoichiometry of 10 μg N/μg chl is assumed, then the nitrogen would potentially result in 500 μg/l chl, that is,

$$5000 \ \mu\text{g N/l} \cdot \mu\text{g chl}/10 \ \mu\text{g N} = 500 \ \mu\text{g chl/l}$$

If 1 mg p/l were available initially, and a stoichiometry of 1 μg p/μg chl is assumed, then twice as much biomass is calculated, that is,

$$1000 \ \mu\text{g p/l} \cdot 1 \ \mu\text{g chl}/\mu\text{g p} = 1000 \ \mu\text{g chl/l}$$

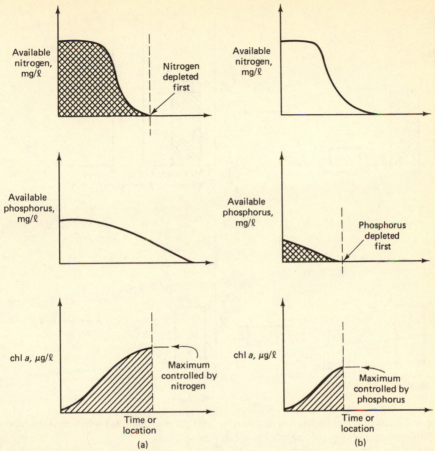

Figure 7.8 Principle of minimum nutrient controlling plant biomass. (a) Nitrogen controlling. (b) Phosphorus controlling.

But the latter biomass level would not be attained since nitrogen would control. A very simplified analysis would then suggest that if phosphorus were the nutrient to be controlled (because of the lower cost to treat phosphorus) and a desirable level of 50 μg chl/l was sought, then the initial phosphorus level must be reduced to 50 μg p/l (0.05 mg p/l).

Letting N = available nitrogen $[M/L^3]$, a_N = the nitrogen/chlorophyll ratio, p = available phosphorus $[M/L^3]$, a_p = the phosphorus/chlorophyll ratio, and P be the resulting chlorophyll $[M/L^3]$, then from the preceding considerations,

$$P = \min\{N/a_N; \ p/a_p\} \tag{7.1}$$

and if the nitrogen concentration is plotted against the phosphorus concentration, Fig. 7.10 results. A value of a_N/a_p of 10 is used and as seen when the ratio of N/p is less than 10, nitrogen controls and for $N/p > 10$, phosphorus controls. At $N/p = 10$, neither nutrient controls.

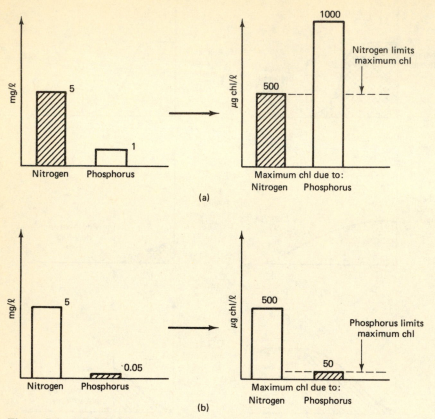

Figure 7.9 Numerical illustration of nitrogen, phosphorus, and chlorophyll relationships. (a) Nitrogen controlling. (b) Phosphorus controlling.

Figure 7.10 Regions of nitrogen and phosphorus control as related to N/p ratio.

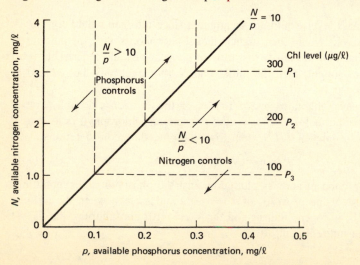

Of course, one should recognize that this simple analysis is subject to some variation due, for example, to the variability in plant stoichiometry. As a result, N/p ratios of 20 or greater generally reflect phosphorus limited systems while N/p ratios of 5 or less reflect nitrogen limited systems.

7.4.1 Controlling Nutrients for Different Water Bodies

Table 7.7 shows N/p ratios for point sources, non-point sources (from Tables 7.5 and 7.6), and marine waters and as indicated the nutrient that limits growth depends on the nature of the source (point and non-point). The N/p ratio for marine waters is low and therefore such waters tend to be nitrogen limited. Table 7.8 shows the

TABLE 7.7 *N/p* RATIOS FOR POINT SOURCE, NON-POINT SOURCE, AND MARINE WATERS[a]

Source type	TN/Tp	IN/Ip	Limiting nutrient	Reference
Point source input				
Activated sludge without phosphorus removal	3	4	Nitrogen	
Activated sludge with phosophorus removal	27	22	Phosphorus	
Non-point source input				
50% agricultural & urban	28	25[b]	Phosphorus	Omernik (1977)
Marine waters		2[c]	Nitrogen	Goldman et al. (1973)

[a]TN/Tp = total nitrogen/total phosphorus; IN/Ip = inorganic nitrogen/inorganic phosphorus, both on mg/l/mg/l basis.
[b]For IN = 0.53 mg N/l and Ip = 0.021 mg p/l in non-point surface runoff.
[c]For IN = 33 µg N/l and Ip = 15 µg p/l.

TABLE 7.8 *N/p* RATIO AND LIMITING NUTRIENT FOR DIFFERENT TREATMENT LEVELS

Treatment level[a]	Effective concentration (mg/l)[b]				TN/Tp	IN/Ip	Limiting nutrient[c]
	TN	IN	Tp	Ip			
Raw	40	25	10	7	4	3.6	Nitrogen
HRAS	27	22	8	5	3.4	4.4	Nitrogen
HRAS + nitrification	26	22	7	5	3.7	4.4	Nitrogen
HRAS + phosphorus removal	27	22	1	1	27.0	22.0	Phosphorus
HRAS + nitrogen removal	3	2	8	5	0.4	0.4	Nitrogen
HRAS + phosphorus removal & nitrogen removal	3	2	1	1	3.0	2.0	Nitrogen

[a]HRAS—high rate activated sludge.
[b]TN, IN—total and inorganic nitrogen, respectively; Tp, Ip—total and inorganic phosphorus, respectively.
[c]Assuming effluent flow dominates the inputs into the water body.

effect of treatment on the N/p ratio and the limiting nutrient for effluent dominated systems.

As a result of the combination of point and non-point sources and marine waters some general guidelines can be offered as shown in Table 7.9. Rivers and streams that are dominated by point sources such as small upland streams will tend to be nitrogen limited but can be made to be controlled by phosphorus if phosphorus is removed at the point source. Larger rivers that may be dominated by non-point source runoff will tend to be controlled by phosphorus.

Estuaries are a more complex situation. In the freshwater tidal river region, either phosphorus or nitrogen may limit plant growth depending on the relative amounts of point and non-point inputs. In the saline regions influenced by signficant intrusion of fully saline marine water, N/p ratios are low and nitrogen tends to control growth. In the brackish area the situation is mixed and represents a transition zone.

For large lakes that have significant inputs of non-point source nutrients, phosphorus controls growth. Smaller lakes that may have significant point source

TABLE 7.9 LIMITING NUTRIENTS FOR VARIOUS WATER BODIES

Rivers and streams

Point source dominated

Without phosphorus removal: $\dfrac{N}{p} \ll 10$: nitrogen limited

With phosphorus removal: $\dfrac{N}{p} \gg 10$: phosphorus limited

Non-point source dominated: $\dfrac{N}{p} \gg 10$: phosphorus limited

Estuaries

Freshwater region

Non-point source dominated: $\dfrac{N}{p} \gg 10$: phosphorus limited

Point source dominated: $\dfrac{N}{p} \ll 10$: nitrogen limited

Saline region

Marine water intrusion: $\dfrac{N}{p} \ll 10$: nitrogen limited

Transition, brackish region: $\dfrac{N}{p} \approx 10$: nitrogen or phosphorous limited

Lakes

Large

Non-point source dominated: $\dfrac{N}{p} \gg 10$: phosphorous limited

Small

Point source dominated: $\dfrac{N}{p} \ll 10$: nitrogen limited inputs may be nitrogen limited, but can be made to be phosphorus limited by phosphorus removal at the treatment plant.

The N/p ratio is therefore a useful measure for understanding at a first level the relationship between nitrogen, phosphorus, and plant biomass. As with all such measures, however, care must be taken in interpretation and use. For example, removal of nitrogen from freshwater nitrogen controlled systems to control phytoplankton will result in low N/p ratios for such waters. It has been observed that low N/p ratios (< 4) in freshwaters may stimulate the growth of noxious blue-green algae which can obtain nitrogen for growth by fixing atmospheric nitrogen. Such algae tend to form unsightly surface mats which can result in serious impairment of water uses. N/p is therefore not the whole story of eutrophication. Much work has been done in this area in lakes because of the significance of the problem in those bodies of water. We will continue this exploration of aquatic plant growth by examining phosphorus mass balances in lakes.

7.5 SIMPLIFIED LAKE PHYTOPLANKTON MODELS

A considerable effort has been expended in the past 10–20 years to develop empirical and theoretical analyses of lake eutrophication that could be easily applied. Since many of the problems of lakes have been with phytoplankton, most of the efforts to date have focused on lake phytoplankton eutrophication and incorporate several key assumptions in the analysis. The simplified phytoplankton models have proven to be of significant value in first estimates of the probable effects of reduction of nutrient inputs.

The basic approach underlying several of the models is the mass balance of the assumed limiting nutrient, that is, phosphorus. Total phosphorus is used as the indicator variable of trophic status. Vollenweider (1968) in his first paper, of a largely empirical nature, related external nutrient loading (in g/m^2 of lake surface area per year) to lake depth and noticed that the population of lakes divided into two broad areas related to the state of eutrophication. From that work, other investigators (Dillon and Rigler, 1974, 1975; Rast and Lee, 1978) have continued to incorporate more data and additional mechanisms and have necessarily relaxed some of the earlier levels of judgment that were required in the analysis. These later efforts have also attempted in semiempirical ways to estimate not just the phosphorus level but the more relevant variable of phytoplankton chlorophyll. Chapra and Tarapchak (1976) have summarized the simplified scheme as the following series of steps:

1. Estimate loading of total phosphorus to the lake.
2. Determine mean annual concentration of total phosphorus in the lake.
3. Estimate the spring concentration of total phosphorus in the lake from mean annual concentration.
4. Compute mean summer concentrations of chlorophyll *a* from spring concentrations of total phosphorus.

7.5.1 Phosphorus Mass Balance

The assumptions used in this analysis are:

1. Completely mixed lake.
2. Steady state conditions, representing a seasonal/annual average.
3. Phosphorus limited.
4. Total phosphorus used as measure of trophic status.

Considering the severity of the assumptions, it is at first surprising that the development to follow works at all. The first assumption ignores lake stratification and subsequent intensification of phytoplankton in the epilimnion. The second assumption ignores the dynamic behavior as shown in Figs. 7.3, 7.5, and 7.6. The third assumption indicates that only one nutrient need be considered, that is, that all other nutritional requirements for phytoplankton are met. As noted above for lakes, phosphorus is often, but not always, the nutrient that limits growth and, therefore, is often the cause of increased phytoplankton resulting from the uptake of phosphorus by the phytoplankton. The fourth assumption indicates that total phosphorus is the measure of concern, not phytoplankton and implies that a relationship exists between total phosphorus and biomass. In spite of these assumptions, the following analysis has been shown to produce very useful results from a water quality control point of view.

The basic mass balance equation for total phosphorus is obtained by following the principles of Chapter 4 for a completely mixed lake. Figure 7.11 shows the schematic of the system. The equation is

$$V \frac{dp}{dt} = W - v_s A_s p - Qp \qquad (7.2)$$

or

$$V \frac{dp}{dt} = W - K_s p V - Qp \qquad (7.2a)$$

Figure 7.11   Schematic of phosphorus mass balance—completely mixed lake.

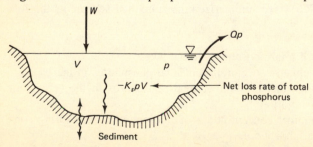

for

$$K_s = \frac{v_s}{H} \qquad (7.2b)$$

where V = volume of lake $[L^3]$.
p = phosphorus (total) in lake, $[M/L^3]$; for example, $\mu g/l$.
Q = outflow.
A_s = lake surface area $[L^2]$.
W = external sources of phosphorus, $[M/T]$; for example, g/s.
K_s = overall loss rate of total phosphorus $[1/T]$.
H = depth of lake $[L]$.

At steady state, Eq. 7.2 reduces to

$$p = \frac{W}{Q + v_s A_s} \qquad (7.3)$$

This equation can also be expressed using an areal loading rate

$$W' = \frac{W}{A_s} \qquad [M/L^2 \cdot T, \text{ e.g., g/m}^2 \cdot \text{yr}] \qquad (7.4)$$

Equation 7.3 is then

$$p = \frac{W'}{q + v_s} \qquad (7.5)$$

where q is the hydraulic overflow rate (Q/A_s) $[L/T]$. An alternative expression using Eq. 7.2a can be derived as

$$p = \frac{W'}{H(\rho + K_s)} \qquad (7.6)$$

where

$$\rho = \frac{Q}{V} = \frac{1}{t_d} \qquad (7.7)$$

for t_d as the detention time of the lake.

The difficulty of using Eq. 7.5 or 7.6 is that v_s, the net settling velocity or K_s, the net loss rate, are not readily known nor can such parameters be measured in a direct experimental way. However, if information is available on the input and output from the lake, an estimate can be made of K_s. Suppose, therefore, that a number of lakes are examined where p, W', H, and ρ can be estimated, then Eq. 7.6 permits estimation of K_s. Vollenweider (1975) deduced that

$$K_s = \frac{10}{H} \qquad (7.8)$$

for H in meters and K_s in yr^{-1}.

Since $K_s = v_s/H$ where v_s is the settling rate $[L/T]$, then Eq. 7.8 is equivalent to $v_s = 10$ m/yr (0.0274 m/day). Other estimates of 16 m/yr and 12.4 m/yr are discussed by Chapra and Tarapchak (1976). Given the empirical relationship of Eq. 7.8, Eq. 7.5 can be written as

$$p = \frac{W'}{10 + H\rho} \tag{7.9}$$

Note that $H\rho = V/A_s \cdot Q/V = Q/A_s = q$.

Now the question of what level of total phosphorus is "acceptable" and what level is "excessive" must be addressed so that estimates of the allowable phosphorus loading W' can be obtained. One objective set that has been suggested is:

acceptable total phosphorus level = 0.010 mg p/l(g/m³) = 10 μg p/l(mg/m³)
excessive total phosphorus level = 0.020 mg p/l = 20 μg p/l

Then the loading rates can be computed from Eq. 7.9 using these levels as

$$W'_{10} = 0.01(10 + H\rho); \quad (g/m^2 \cdot yr) = (g/m^3)(m/yr) \tag{7.10}$$
$$W'_{20} = 0.02(10 + H\rho) \tag{7.11}$$

Figure 7.12 shows these relationships from Vollenweider (1975). With the estimated $H\rho$, and the annual loading rate, the trophic status of the lake can be estimated. If the level is eutrophic, the required load reduction can be read directly from the plot.

7.5.2 Other Semiempirical Approaches

Since the net loss rate depends on lake characteristics, such as resuspension and sediment interactions, modifications of earlier work (Vollenweider, 1976) have also

Figure 7.12 Nutrient loading/lake trophic condition. After Vollenweider (1975).

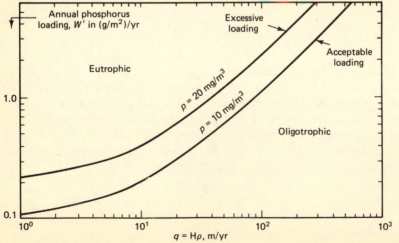

indicated that the following expression may be appropriate:

$$K_s = \sqrt{\rho}$$

Then from Eq. 7.6,

$$p = \frac{W'}{H\rho(1 + \sqrt{1/\rho})} \tag{7.12}$$

The approach of Chapra and Tarapchak (1976) follows the steps outlined in the introduction above and begins with Eq. 7.3 which relates external loading to total annual phosphorus in the lake. The spring concentration of total phosphorus is related empirically to the summer Tp concentration by

$$p = 0.9 p_s \tag{7.13}$$

where p_s is the spring total phosphorus. The final step of computing mean summer chlorophyll is given by Dillon and Rigler (1974) as

$$\text{chl}\, a = 0.0731\, (p_s)^{1.449} \tag{7.14}$$

Equations 7.13, 7.14, and 7.5 were then combined to give

$$\text{chl}\, a = 1866[W'/(q + 12.4)]^{1.449} \tag{7.15}$$

where chl a is in μg/l, W' is in g/m$^2 \cdot$ yr, and q is in m/yr and a net sedimentation velocity of 12.4 m/yr was used. This equation permits estimation of the mean summer chlorophyll directly from the external loading, a very desirable trait.

Equation 7.15 can also be rearranged to evaluate the loading required for a given chlorophyll level. Thus,

$$W' = 0.0055(\text{chl}\, a)^{0.69}(q + 12.4) \tag{7.16}$$

Chapra and Tarapchak (1976) chose the following mean summer chlorophyll levels as objectives:

acceptable: chl a = 2.7 μg/l
excessive: chl a = 9.0 μg/l

These levels are equivalent to total annual phosphorus levels of 13.4 and 30.7 μg p/l for the acceptable and excessive levels, respectively. With these chlorophyll objectives, the loadings are

$$W'_{\text{acceptable}} = 0.011(q + 12.4) \tag{7.17a}$$
$$W'_{\text{excessive}} = 0.025(q + 12.4) \tag{7.17b}$$

These expressions are essentially identical to Eqs. 7.10 and 7.11. As such, the loading rates of Fig. 7.12 separate a trophic status of lakes from about 3 μg chl/l at the acceptable level to about 9 μg chl/l at the excessive level. It should be stressed again that all these results apply as a general criterion for lakes only and as a rule, only for lakes in the North Temperate regions of the world. The simplified lake model is used in Sample Problem 7.1.

SAMPLE PROBLEM 7.1

DATA

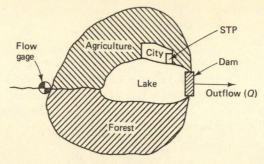

Lake Geometry

$$\text{volume } (V) = 6.22 \times 10^8 \text{ m}^3$$
$$\text{surface area } (A) = 7.77 \times 10^7 \text{ m}^2$$
$$\text{depth } (H) = 8 \text{ m}$$

Sewage Treatment Plant (STP)

$$\text{population served} = 50,000$$
$$\text{water use} = 150 \text{ gcd}$$
$$\text{influent Tp} = 6 \text{ mg/l}$$
$$\text{plant removal} = 20\%$$

Combined Sewers (CSO)

$$\text{runoff coefficient } (C) = 0.45$$
$$\text{service area} = 6 \text{ mi}^2$$
$$\text{capture by STP} = 5\%$$
$$\text{overflow Tp concentration} = 4 \text{ mg/l}$$

Storm Drains (SW)

$$\text{service area} = 4 \text{ mi}^2$$
$$\text{runoff coefficient} = 0.27$$
$$\text{Tp concentration} = 0.7 \text{ mg/l}$$

Upstream Gage

$$\text{annual average flow} = 500 \text{ cfs}$$
$$\text{Tp (virgin land)} = 0.02 \text{ mg/l}$$

Sample Problem 7.1 (continued)

Agricultural Land

$$\text{draingage area} = 60 \text{ mi}^2$$
$$\text{Tp loading} = 0.5 \text{ lb/mi}^2\text{-day}$$
$$\text{runoff} = 30\% \text{ rainfall}$$

Forest

$$\text{draingage area} = 80 \text{ mi}^2$$
$$\text{Tp loading} = 0.15 \text{ lb/mi}^2\text{-day}$$
$$\text{runoff} = 30\% \text{ rainfall}$$

PROBLEM

The lake basin receives 30 in./yr of rainfall. On an annual average basis estimate the total phosphorus (Tp) concentration in the lake.

ANALYSIS

Lake Outflow

Assume precipitation onto the lake surface and evaporation from the surface are equal and do not affect the hydrologic balance. Thus, the upstream flow plus the sum of the incremental flows from the areas draining to the lake equals the outflow.

$$Q(\text{STP}) = 50,000 \text{ cap} \times 150 \text{ gcd} \times \text{MGD}/10^6 \text{ gal}$$
$$\times 1.548 \text{ cfs/MGD} = 11.6 \text{ cfs}$$
$$Q(\text{CSO}) = CIA(1 - \text{capture}) = 0.45 \times (30 \text{ in/yr} \times 1 \text{ yr}/365 \text{ days}$$
$$\times 1 \text{ day}/24 \text{ hr} = 0.00342 \text{ in./hr})$$
$$\times (6 \text{ mi}^2 \times 640 \text{ acre/mi}^2 = 3840 \text{ acres}) \times (1 - 0.05)$$
$$= 5.61 \text{ cfs}$$
$$Q(\text{SW}) = 0.27 \times \left(\frac{30}{365 \times 24}\right) \times (4 \times 640) = 2.36 \text{ cfs}$$
$$Q(\text{agricultural}) = (30 \times 0.3) \text{ in./yr} \times \frac{0.07367 \text{ cfs/mi}^2}{\text{in./yr}} \times 60 \text{ mi}^2 = 39.8 \text{ cfs}$$
$$Q(\text{forest}) = 30 \times 0.3 \times 0.07367 \times 80 = 53.0 \text{ cfs}$$
$$\therefore Q = 500 + 11.6 + 5.61 + 2.36 + 39.8 + 53.0 = 612 \text{ cfs}$$
$$\underline{Q = 612 \text{ cfs} \times \frac{1 \text{ m}^3/\text{s}}{35.4 \text{ cfs}} = 17.3 \text{ m}^3/\text{s}}$$

(continued)

Sample Problem 7.1 (continued)

Lake Tp Loading

$$W(\text{STP}) = 11.6 \text{ cfs} \times [6(1 - 0.20)\text{mg/l}] \times 5.4 \frac{\text{lb/day}}{\text{mg/l-cfs}}$$

$$= 301 \text{ lb/day}$$

$$W(\text{CSO}) = 5.61 \times 4 \times 5.4 = 121 \text{ lb/day}$$

$$W(\text{SW}) = 2.36 \times 0.7 \times 5.4 = 9 \text{ lb/day}$$

$$W(\text{upstream}) = 500 \times 0.02 \times 5.4 = 54 \text{ lb/day}$$

$$W(\text{agricultural}) = 0.5 \text{ lb/mi}^2\text{-day} \times 60 \text{ mi}^2 = 30 \text{ lb/day}$$

$$W(\text{forest}) = 0.15 \times 80 = 12 \text{ lb/day}$$

$$\therefore W = (301 + 121 + 9 = 431) + (54 + 30 + 12 = 96)$$

$$= \underline{527 \text{ lb/day}}$$

Areal Loading

$$W = 527 \text{ lb/day} \times 365 \text{ days/yr} \times 454 \text{ g/lb} = 8.73 \times 10^7 \text{ g/yr}$$

$$W' = W/A_s = 8.73 \times 10^7 / 7.77 \times 10^7 = 1.12 \text{ g/m}^2\text{·yr}$$

Lake Tp Concentration

$$t_d = \text{hydraulic detention time} = \frac{V}{Q}$$

$$t_d = \frac{6.22 \times 10^8 \text{ m}^3}{17.3 \text{ m}^3/\text{s}} \times \frac{1 \text{ yr}}{3.154 \times 10^7 \text{s}} = 1.14 \text{ yr}$$

$$q = \text{overflow rate} = \frac{Q}{A_s} = \frac{1}{V/H \cdot 1/Q} = \frac{H}{t_d} = \frac{8 \text{ m}}{1.14 \text{ yr}} = 7.02 \text{ m/yr}$$

Assume a net loss rate of $\text{Tp}(v_s) = 12.4 \text{ m/yr}$.

$\therefore$ Total phosphorus concentration (p) is

$$p = \frac{W'}{q + v_s} = \frac{1.12 \text{ g/m}^2\text{·yr}}{7.02 \text{ m/yr} + 12.4 \text{ m/yr}} \qquad \text{[Eq. (7.5)]}$$

$$p = 0.058 \text{ g/m}^3 = 0.058 \text{ mg/l} = \underline{\underline{58 \ \mu\text{g/l}}}$$

Notes: The concentration of 58 μg/l would place the lake in the eutrophic status (see Table 7.1 and Fig. 7.12).

Since the range of net loss rates is quite large ($\approx 1 \text{ m/yr}$ to $\approx 20 \text{ m/yr}$), Fig. 7.13, lake specific data should be used to estimate this parameter.

Reductions in the urban sources would be more effective than reductions in the rural sources since the urban sources amount to (431 lb/day $\div$ 527 lb/day) or 82% of the total lake loading.

7.5.3 Variability of Net Effective Settling Velocity

The completely mixed nutrient mass balance equation which forms the basis for the simplified Vollenweider analysis of Fig. 7.12 rests heavily on the assumption that the net effective settling velocity is 10 m/yr (or for Chapra and Tarapchak, 12.4 m/yr). When only external loads are considered, this parameter in the mass balance incorporates phenomena such as

1. Fraction of incoming phosphorus in particulate form.
2. Concentration of suspended solids in the lake.
3. Resuspension of particulate phosphorus from bottom sediments.
4. Diffusive flux of dissolved phosphorus from the sediments.

The net effective settling velocity may, therefore, be expected to vary as a function of these mechanisms. There is no simple direct way to represent all these phenomena in a single parameter reflective of all lake and impoundment situations. The range in v_s is large as shown in Fig. 7.13. The data for this figure are from U.S. lakes and impoundments and were compiled from Rast and Lee (1978), Bachmann and Canfield (1979), Schreiber and Rausch (1979), and Higgins and

Figure 7.13 Calculated net sedimentation velocities.

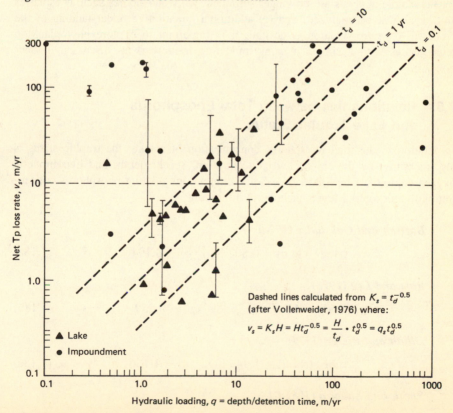

Dashed lines calculated from $K_s = t_d^{-0.5}$ (after Vollenweider, 1976) where:

$$v_s = K_s H = H t_d^{-0.5} = \frac{H}{t_d} \cdot t_d^{0.5} = q_s t_d^{0.5}$$

▲ Lake

● Impoundment

Net Tp loss rate, v_s, m/yr

Hydraulic loading, q = depth/detention time, m/yr

Kim (1981). Only positive values for v_s are plotted; for some reservoirs (e.g., see Higgins and Kim, 1981), the computed net settling velocity is negative which indicates that the mass output of the impoundment exceeds the input mass. This could result from net positive resuspension in the reservoir.

It is clear from Fig. 7.13 that the variability in v_s is great. The U.S. lakes, as distinct from impoundments, vary within about one order of magnitude (approximately 3–30 m/yr). This is a sufficiently large range and could markedly influence the calculation of the allowable phosphorus load.

Impoundments, in general, vary about three orders of magnitude with no clear correlation to the hydraulic loading rate. The net settling rate for impoundments is, however, generally greater than that for lakes by a substantial amount. This probably reflects an increase in particulate phosphorus fractions and associated settling velocity.

Figure 7.13 shows the importance of not relying too heavily on the simplified graphical approach of Fig. 7.12. Each individual lake and impoundment will have an associated v_s that should be evaluated on a site specific basis.

However, from the point of view of engineering control, the simple nutrient mass balance equation coupled with the empirical expressions indicated above provide an excellent starting point for assessing the efficacy of proposed nutrient reduction programs for lakes. For other situations where (a) the assumptions used for the simple analysis are no longer sufficient, (b) water bodies such as estuaries and bays must be analyzed, and (c) additional quantitative understanding of the processes of eutrophication is desired, it is important to explore phytoplankton relationships with a simple dynamic modeling framework as discussed in Section 7.6.

7.5.4 Relationships between Total Phosphorus and Lake Trophic Status

In addition to the total phosphorus concentration of a lake, the trophic status is also measured by the chlorophyll concentration, secchi depth, and hypolimnetic oxygen depletion (see Table 7.1). Some expressions relating phosphorus concentration to chlorophyll levels include the following:

Bartsch and Gakstatter (1978)

$$\log_{10}(\text{chl } a) = 0.807 \log_{10}(p) - 0.194 \tag{7.18a}$$

Rast and Lee (1978)

$$\log_{10}(\text{chl } a) = 0.76 \log_{10}(p) - 0.259 \tag{7.18b}$$

Dillon and Rigler (1974)

$$\log_{10}(\text{chl } a) = 1.449 \log_{10}(p_s) - 1.136 \tag{7.18c}$$

Smith and Shapiro (1981a)

$$\log_{10}(\text{chl } a) = 1.55 \log_{10}(p) - b \tag{7.18d}$$

for

$$b = 1.55 \log_{10}\left[\frac{6.404}{0.0204(\text{TN:Tp}) + 0.334}\right]$$

and where chl a and p are in μg/l and TN:Tp $\equiv$ total nitrogen/total phosphorus ratio.

Figure 7.14 shows the relationships between median total phosphorus and mean chlorophyll a concentration for Eq. 7.18a. The notation of trophic status provides an approximate division between the three states of oligotrophy, mesotrophy, and eutrophy.

Figure 7.15 shows the correlation between secchi depth and chlorophyll concentration as given by Rast and Lee (1978). This is a particularly important measure since water clarity is easily perceived by the general public. The increase in water clarity (secchi depth increase) as a result of reductions in chlorophyll concentrations is most readily seen by the public as the response of the lake to nutrient control.

Figure 7.16 is the relationship of Rast and Lee (1978) between the areal hypolimnetic oxygen depletion rate and the ratio of loading to the lake characteristics. The equation is

$$\log_{10} S = 0.467 \log_{10}\left[\frac{W'}{q(1 + \sqrt{t_d})}\right] - 1.07 \tag{7.19}$$

where S is the total hypolimnetic oxygen depletion rate in g O_2/m$^2 \cdot$ day and W' is in mg Tp/m$^2 \cdot$ yr. The oxygen depletion rate is the net observed rate of decline of DO in the hypolimnion and includes phytoplankton respiration and SOD and any uptake of DO to satisfy nonphytoplankton BOD (see Sample Problem 7.2).

Figure 7.14 The relationship between chlorophyll a and total phosphorus concentrations in northeastern U.S. lakes and reservoirs. From Bartsch and Gakstatter (1978).

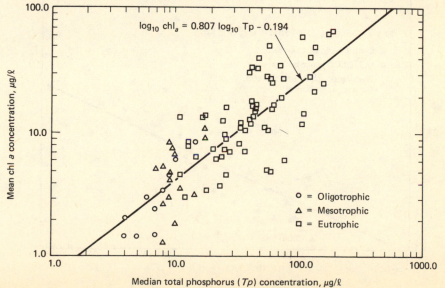

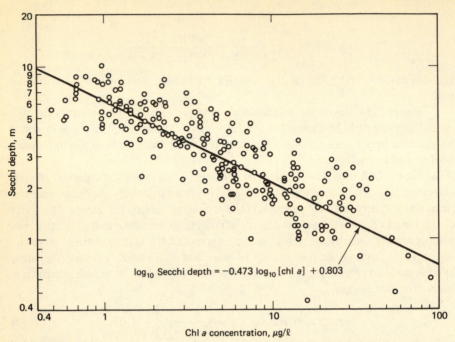

Figure 7.15 Secchi depth and chlorophyll *a* relationship in natural waters. From Rast and Lee (1978).

Figure 7.16 Phosphorus loading characteristics and hypolimnetic oxygen depletion rate in natural waters. From Rast and Lee (1978).

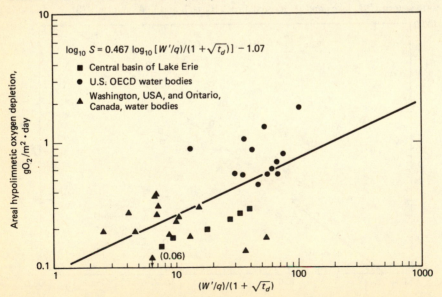

SAMPLE PROBLEM 7.2

DATA

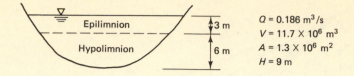

$Q = 0.186$ m³/s
$V = 11.7 \times 10^6$ m³
$A = 1.3 \times 10^6$ m²
$H = 9$ m

period of stratification = 100 days
hypolimnetic DO concentration at beginning of stratification = 9.0 mg/l

PROBLEM

Using the empirical hypolimnetic DO demand rate (S) of Rast and Lee (1978), determine the temporal variation of hypolimetic DO for various total phosphorus loadings.

ANALYSIS

From Eq. 7.19

$$\log_{10} S \text{ (g O}_2 \text{/m}^2\text{·day)} = 0.467 \log_{10} \{W'_{Tp}/[q(1 + \sqrt{t_d})]\} - 1.07$$

where $W'_{Tp} = $ mg Tp /m²·yr, $q = $ m/yr, $t_d = $ yr

$$t_d = \frac{V}{Q} = \frac{11.7 \times 10^6 \text{ m}^3}{0.186 \text{ m}^3/\text{s}} \times \frac{1 \text{ yr}}{3.154 \times 10^7 \text{s}} = 2.0 \text{ yr}$$

$$q = \frac{H}{t_d} = \frac{9 \text{ m}}{2.0 \text{ yr}} = 4.5 \text{ m/yr}$$

The volumetric value for S is:

$$S = \frac{S \text{ (g O}_2/\text{m}^2\text{·day)}}{H(\text{hypolimnetic)m}} = \frac{\text{g O}_2}{\text{m}^3\text{·day}} = \frac{\text{mg O}_2}{\text{l·day}}$$

W'_{Tp} (mg Tp/m²·yr)	$\dfrac{W'_{Tp}}{q(1 + \sqrt{t_d})}$ [mg/m³ $= \mu$g(Tp)/l]	S (g O₂/m²·day)	$\dfrac{S}{H(\text{hypolimnetic})}$ (mg O₂/l·day)	Hypolimnetic DO concentration (mg/l) Time (days) after onset of stratification					
				0	20	40	60	80	100
109	10	0.249	0.0415	9.0	8.17	7.34	6.51	5.68	4.85
218	20	0.345	0.0575	9.0	7.85	6.70	5.55	4.40	3.25
327	30	0.417	0.0695	9.0	7.61	6.22	4.83	3.44	2.05
436	40	0.477	0.0795	9.0	7.41	5.82	4.23	2.64	1.05
545	50	0.529	0.0882	9.0	7.24	5.47	3.71	1.94	0.18
654	60	0.576	0.0960	9.0	7.08	5.16	3.24	1.32	0
763	70	0.619	0.1032	9.0	6.93	4.87	2.81	0.74	0

(continued)

Sample Problem 7.2 (continued)

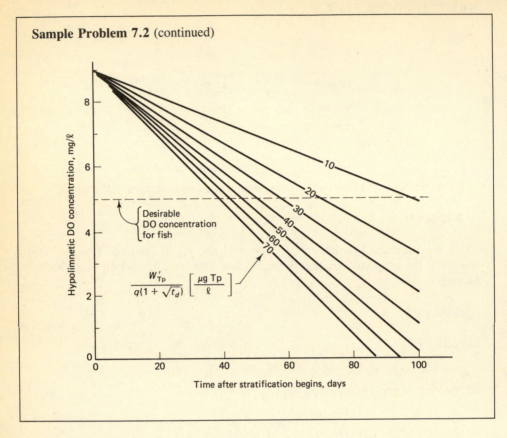

The relationships between total phosphorus as calculated by a mass balance and other measures of eutrophication are particularly useful to provide a more complete picture of lake behavior to changes in external phosphorus inputs.

7.6 PHYTOPLANKTON AND NUTRIENT INTERACTIONS

7.6.1 Phytoplankton Mass Balance

A simple mass balance equation for the phytoplankton can be obtained by following the basic principles of phytoplankton behavior outlined previously in Section 7.2 Considering a completely mixed lake, the basic equation is

$$V \frac{dP}{dt} = V(G_P - D_P)P - Av_P P + Q_{in}P_{in} - QP \qquad (7.20)$$

where P = phytoplankton chlorophyll $[M/L^3, \mu g/l]$
         P_{in} = phytoplankton chlorophyll $[M/L^3]$ in incoming flow

Q_{in} = incoming flow $[L^3/T]$
G_P = growth rate of phytoplankton $[T^{-1}; \text{day}^{-1}]$
D_P = death rate of phytoplankton $[T^{-1}]$
v_P = settling rate of phytoplankton $[L/T, \text{m/day}]$
A = area through which settling is occurring $[L^2]$

The first term on the right-hand side of Eq. 7.19 represents the net growth rate of the phytoplankton, that is, growth minus death. This term is a new term in the mass balance equations for completely mixed systems considered thus far. The second term is the mass of phytoplankton lost to the sediment from net settling. Note that

$$Av_s = \frac{Vv_s}{H} = VK_P$$

where H is the mean depth $[L]$ and K_P $[T^{-1}]$ is the net loss rate = v_P/H.

Estimates of v_P range from about 0.1 to 0.3 m/day and represent the net loss of phytoplankton through the bottom interface of the water body. This net settling depends on phytoplankton species, time of day, and nutritional status of the plankton. In general, active phytoplankton have settling velocities that are considerably less than nonliving particulate matter. Burns and Rosa (1980) carried out a series of experiments using special in situ settling experiments. Table 7.10 summarizes some of their results.

Some phytoplankton such as blue-green algae develop gas vacuoles which result in buoyancy and subsequent aggregation at the water surface. The proliferation of such species is a particular problem since the settling velocity may be zero or negative and phytoplankton tend to remain in the water column or at the surface.

The third term is the mass of phytoplankton chlorophyll entering the lake with the inflow. The last term is the usual mass transport of phytoplankton out of the system due to lake outflow.

TABLE 7.10 SOME RESULTS OF IN SITU SETTLING EXPERIMENTS IN LAKE GEORGE

Particle type	Settling velocity (m/day)
Particulate organic carbon	
1–10 μm	0.2
10–64 μm	1.5
>64 μm	2.3
Phytoplankton	
Scenedesmus acutiforms	0.10
Selenastrum minutum	0.15
Cryptomonas crosa	0.30

Source: From Burns and Rosa (1980). Reprinted by permission of American Society of Limnology and Oceanography.

The growth and death of the phytoplankton represent an important term and will be explored more fully. Consider then the kinetic expression

$$\frac{dP}{dt} = (G_P - D_P)P \tag{7.21}$$

as representative of phytoplankton behavior in a vessel that is kept stirred to eliminate v_P. The solution to this equation for initial condition P_0 is

$$P = P_0 \exp[(G_P - D_P)t] \tag{7.22}$$

Several conditions are now possible, that is,

$G_P > D_P$ (net production of phytoplankton)
$G_P = D_P$ (stationary growth)
$G_P < D_P$ (net decline in phytoplankton)

Figure 7.17 shows the time behavior of these three conditions. Exponential growth of phytoplankton ($G_P > D_P$) would result in massive quantities of biomass if continued unabated for a period of time. For example, if $G_P - D_P$ is $+ 0.3$/day over a 100-day period, then

$$P = P_0 e^{30}$$
$$P = 10^{13} \cdot P_0$$

Clearly since the universe is not completely submerged in phytoplankton, at some point a required nutrient for growth or some other environmental factor limits this growth and $G_P \approx D_P$. At that point, the rate of change of the phytoplankton with time approaches zero and the biomass levels off. Subsequently as temperatures drop or sunlight decreases, $G_P < D_P$ and there is a net decrease in biomass.

The principal dynamics of phytoplankton growth are therefore contained in the behavior of the growth and death of the phytoplankton over the seasons of the year. A more detailed evaluation of the behavior of these kinetic terms is of some importance in understanding the interaction of nutrients and phytoplankton chlorophyll.

7.6.2 Phytoplankton Growth Kinetics

The growth rate of the phytoplankton, as noted earlier, depends on three principal components:

1. Temperature.
2. Solar radiation.
3. Nutrients.

The classical approach is to assume that these effects are multiplicative, although there is no a priori rationale for this assumption. The use of a multiplicative effect of each of these components is, however, somewhat justified from data collected

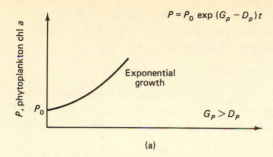

$$P = P_0 \exp(G_p - D_p)t$$

(a)

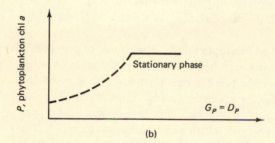

(b)

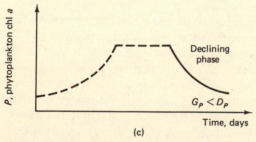

(c)

Figure 7.17 Time variable behavior of phytoplankton for three conditions on net growth. (a) $G_p > D_p$. (b) $G_p = D_p$. (c) $G_p < D_p$.

from laboratory experiments (Di Toro et al., 1971). The phytoplankton growth rate can then be written as:

$$G_P = (\text{temperature effect}) \cdot (\text{light effect}) \cdot (\text{nutrient effect}) \qquad (7.23)$$
$$= G(T) \cdot G(I) \cdot G(N)$$

7.6.2.1 Temperature Effect   Now consider a situation where the available light for growth is at an optimum level and nutrients are plentiful. Then the growth of the phytoplankton will be dependent only on the temperature effect. This relationship of temperature and growth rate can be determined from work by Eppley (1972), among others and is approximated by

$$G(T) = G_{\max}(1.066)^{T-20} \qquad (7.24)$$

where G_{max} is the maximum growth rate of the phytoplankton @ 20°C under optimal light and nutrient conditions. This growth rate is a function of the species of phytoplankton and can vary considerably. To first approximation,

$$G_{max} \approx 1.8/\text{day}$$

This relationship between the growth rate and temperature is shown in Fig. 7.18. Note that the relationship expresses the maximum growth rate as a function of water temperature assuming that no inhibitory effect occurs at high temperatures of greater than 30°C.

The value suggested for G_{max} of 1.8/day is an average condition for a mixed phytoplankton population. Various species of phytoplankton, however, have different growth rates, and, in general, the range of G_{max} is about 1.5–2.5/day.

7.6.2.2 Solar Radiation and Effect on Phytoplankton   The degree of penetration of sunlight into the water column has a significant effect on many areas of water quality including bactericidal effects, impact on the rate of photosynthesis by aquatic plants, and the general clarity, color, or aesthetic quality of the water. The two principal mechanisms for the extinction of solar radiation are *absorption* and *scattering*. In the former, short wave energy is transferred to long wave energy. The presence of particles in the water may also absorb the light. Scattering in water is the effect of reflection and diffraction by particles and in pure water is due to small density fluctuations and other factors.

The degree of solar radiation penetration, therefore, depends on several factors: nonvolatile suspended solids, organic detritus, and living particulates such as phytoplankton. The intensity of solar radiation and its angle with respect to the water surface are also important. Further, different regions of the incoming solar radiation spectrum (i.e., from infrared to ultraviolet) may be selectively absorbed or scattered. This is shown schematically in Fig. 7.19 where the energy spectrum of incoming light in pure water is shifted into the the blue region. For turbid waters, the situation may change significantly and the peak of the energy spectrum may shift to the right toward blue-green and green-yellow.

Figure 7.18   Effect of temperature on growth rate of phytoplankton.

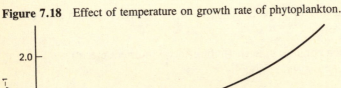

$$G(T) = 1.8(1.066)^{T-20}$$

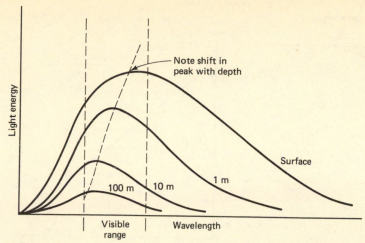

Figure 7.19 Decrease and shift of energy spectrum with depth—pure water.

The penetration (or conversely its extinction) of incoming solar radiation can be described by introducing the extinction coefficient. It has been observed that the extinction of light is proportional to the light at any depth. Therefore, a differential equation that expresses this observation is

$$\frac{dI}{dz} = -K_e I \tag{7.25}$$

$$I = I_0 \ @ \ z = 0 \tag{7.25a}$$

where I is the solar radiation in cal/cm^2 · min, z is the depth in m, K_e is the overall extinction coefficient in m^{-1}, and I_0 is the incoming solar radiation at the surface $z = 0$. The solar radiation reaching the earth's surface is about 0.8 cal/cm^2 · min. The actual amount that arrives at a water surface depends on the duration of sunlight throughout the day and year, the elevation of the sun, cloud cover, and the reflective condition of the water. This is discussed more fully below. Equation 7.25 with boundary condition 7.25a can be solved to give

$$I = I_0 e^{-K_e z} \tag{7.26}$$

or

$$\frac{I}{I_0} = e^{-K_e z}$$

The slope then of $\ln I/I_0$ vs. depth z provides an estimate of K_e. This is shown in Fig. 7.20.

The depth, z, at which 1% of the surface radiation still remains, is of use in eutrophication studies and is given from Eq. 7.26 by

$$z_1 = \frac{4.61}{K_e} \tag{7.27}$$

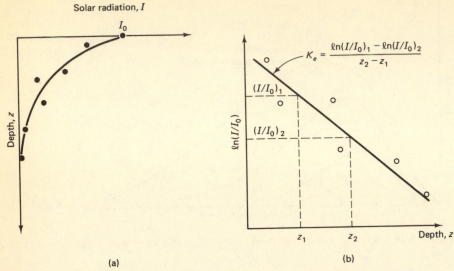

Figure 7.20 (a) Decrease of light with depth. (b) Computation of extinction coefficient.

Measurement of total light penetration into a body of water can be made by use of a pyreheliometer positioned in the boat or other suitable platform in the surface of the water (Fig. 7.21). The pyreheliometer measures the amount of total incoming solar radiation, usually in units of $cal/cm^2 \cdot min$. Simultaneously an underwater photometer is lowered and the radiation recorded at each of a series of depths, throughout the water column. The concurrent measurements provide an estimate of the relative amount of radiation remaining at each depth and the necessary data for computation of K_e as shown in Fig. 7.20(b).

Another more simple method, but less accurate, is to lower a target into the water until, by eye, the target just disappears. The depth at which the target just disappears can then be related to the extinction coefficient. A standardized target used in water quality work is the secchi disk as shown in Fig. 7.21(b). Numerous measurements of "secchi depth" have been made in many water bodies. Sverdrup et al. (1942) and Beeton (1958) and others have developed empirical relationships between the secchi depth, z_s and the extinction coefficient as given by

$$K_e = (1.7 \text{ to } 1.9)/z_s \tag{7.28}$$

Di Toro (1978) has provided a theoretical and empirical basis for estimating the extinction coefficient as a function of nonvolatile suspended solids (N), detritus (D), both in mg/l, and phytoplankton chlorophyll (P) in $\mu g/l$. For a vertical sun angle, his estimate of K_e is

$$K_e = 0.052N + 0.174D + 0.031P \tag{7.29}$$

The nonvolatile suspended solids (the inorganic particulates) both absorb and scatter the light whereas the organic detritus and phytoplankton chlorophyll mainly absorb

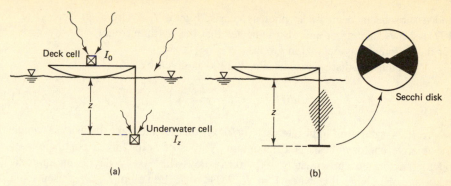

Figure 7.21 (a) Measurement of light extinction with surface and underwater pyrehelio-meters. (b) Measurement with secchi disk.

the light. Di Toro has shown that Eq. 7.29 applies to K_e values of about less than 5.0 m^{-1}. Riley (1956) also related the extinction coefficient to P, the chlorophyll concentration as follows

$$K_e = K_e' + 0.0088P + 0.054P^{2/3} \tag{7.30}$$

where K_e' is the extinction coefficient due to other sources of absorption and scattering. The chlorophyll correction in Eq. 7.29 and that in Eq. 7.30 are virtually identical up to about 15 μg/l. Then Eq. 7.30 appears to be a more suitable correction. Table 7.11 shows some extinction coefficients and depths to 1% surface light for different water bodies. The clarity of some waters, such as Lake Tahoe, is truly remarkable where light penetrates to over 90 m in depth, or a secchi disk can be seen to a depth of about 40 m. It is small wonder then for lakes with this appealing, aesthetically pleasing clarity that great efforts of environmental control are expended to maintain this attribute. The clarity of some of the lakes can be contrasted to the more turbid rivers and estuaries where extinction coefficients are generally one to two orders of magnitude greater.

If now the temperature is held constant and the available light is allowed to vary, it is observed that as the radiation increases, photosynthesis increases up to a maximum level. Further increases in light levels tend to result in photoinhibition

**TABLE 7.11 SOME LIGHT EXTINCTION COEFFICIENTS
FOR VARIOUS WATERS**

Water body and location	K_e (m)$^{-1}$	z_1[a] (m)	Remarks
Tahoe	0.05	92	
Ontario	0.50	9.2	chl = 5 μg/l
Sewage stabilization pond, Kadoka, SD	23.0	0.2	chl = 2.8 mg/l
Delaware @ Trenton, NJ	2.30	2.0	
Delaware @ Philadelphia, PA	6.9	0.7	

[a]z_1 = depth to 1% light penetration. Adapted from Thomann, 1972.

and a subsequent decrease in photosynthesis. Figure 7.22 (from Di Toro et al., 1971) shows this effect and also indicates that the optimum light intensity varies for the different phytoplankton groups.

An equation that represents Fig. 7.22(a) is given by

$$F(I) = \frac{I}{I_s} \exp\left(\frac{-I}{I_s} + 1\right) \tag{7.31}$$

where I_s = saturating light intensity (ly/day), that is, the light intensity at which the relative photosynthesis is a maximum. Values of I_s for mixed populations of phytoplankton are aproximately 100–400 ly/day with 300 ly/day as an approximate average. Note that when $I = I_s$, $F(I) = 1$. Equation 7.31 represents the relative photosynthesis as a function of light intensity at any given point in space

Figure 7.22 Normalized rate of photosynthesis vs incident light intensity. From Di Toro et al. (1971), using theoretical curve from Steele (1965) and data from Ryther (1956). Reprinted with permission from *Nonequilibrium Systems in Natural Water Chemistry*. Copyright 1971, American Chemical Society.

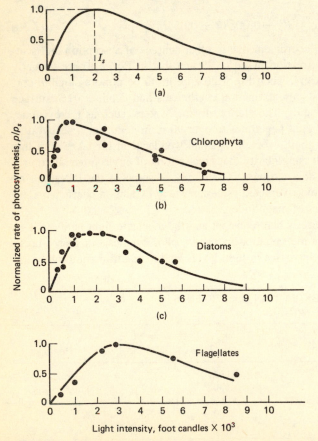

and time. Often, however, the modeling analysis is more tractable by calculating average conditions over a given depth of water and over a fixed interval of time (usually one day). Following Di Toro et al. (1971), and substituting Eq. 7.26 into Eq. 7.31 gives $F(I(z))$. But, the incoming solar radiation, I_0, is a function of the time of day as shown in Fig. 7.23. If the simplifying assumption is made that

$$I_0 = I_a = \frac{I_T}{f} \qquad 0 < t < f \qquad (7.32)$$

$$= 0 \qquad\quad f < t < 1$$

as shown in Fig. 7.23 (where f is the photoperiod and I_T is the total daily solar radiation, as normally reported by meteorological stations), then the average over the depth, H, and over one day is given by

$$G(I_a) = \int_t \int_z F(I(t,z)) \, dt \, dz$$

$$= \frac{2.718f}{K_eH} \left[\exp(-\alpha_1) - \exp(-\alpha_0)\right] \qquad (7.33)$$

where

$$\alpha_1 = \frac{I_a}{I_s} \exp(-K_eH) \qquad (7.33a)$$

$$\alpha_0 = \frac{I_a}{I_s} \qquad (7.33b)$$

It should be recalled here that the extinction coefficient, K_e, depends on the inorganic solids, detrital particles, as well as the phytoplankton levels. Equations 7.29 and 7.30 indicate these relationships.

$G(I)$, the light effect as given by Eq. 7.33, is dimensionless and is used as a multiplier in Eq. 7.23. The behavior of $G(I_a)$ is shown in Fig. 7.24(a). For a

Figure 7.23 (a) Actual variation of solar radiation over one day for a clear day. (b) Approximation for averaging.

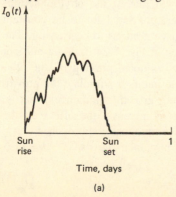

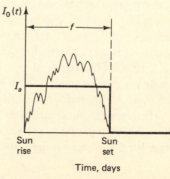

(a)   (b)

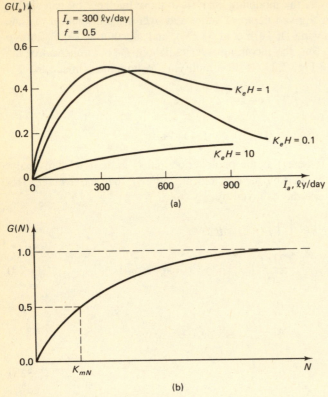

Figure 7.24 Phytoplankton growth rate dependence on (a) light for average depth and across one day and (b) nutrients.

typical range of I_a of 500–1000 (I_T = 250–500) ly/day, the value of $G(I_a)$ ranges from 0.10 to 0.50, that is, the overall effect of nonoptimal light conditions due to light extinction with depth is to reduce the growth rate by about 50–90%.

7.6.2.3 Nutrient Effect   The final effect on the growth that must be evaluated is the impact of varying nutrient levels on the growth rate of the phytoplankton. Therefore, consider again all conditions on temperature and light to be constant and at optimum levels and consider the nutrient concentrations to be varying. The result of this effect is quite significant and is shown in Fig. 7.24(b). At a nutrient level of zero, there is, of course, no growth since, as noted previously, the phytoplankton requires at least some critical nutrients to begin or stimulate the growth of the population. As the nutrient level is increased, growth commences, that is, $G(N) > 0$. However, as nutrient levels continue to increase, the effect on the growth rate of the phytoplankton is reduced and asymptotically approaches unity. An expression for this effect on the growth of the phytoplankton is

$$G(N) = \frac{N}{K_{mN} + N} \qquad (7.34)$$

where N is the concentration $[M/L^3]$ of the nutrient needed for phytoplankton growth (e.g., phosphorus or nitrogen) and K_{mN} is the concentration of the nutrient $[M/L^3]$ at which the nutrient effect $G(N) = \frac{1}{2}$. K_{mN} is called the "Michaelis" constant after the work of Michaelis-Menton on uptake kinetics of organisms on substrates. The implications of this effect are quite profound. Suppose a present nutrient input is such as to result in a concentration of the nutrient that is substantially greater than K_{mN}, that is,

$$N_b >> K_{mN}$$

where N_b is the nutrient concentration before any nutrient reduction. If a nutrient control program is initiated, but the reduction in input load only reduces the nutrient concentration to a level of about two to three times the Michaelis constant, then there will be no effect on the phytoplankton growth. This is equivalent to the notion of the limiting nutrient. Removing a nutrient that is in excess will not have any effect on growth until lower concentrations are reached. The treatment program may then be ineffective. The nutrient effect on phytoplankton, therefore, is a marked contrast to other types of water quality problems where reductions in input load (as in BOD reduction) can generally be considered as being advantageous.

Table 7.12 shows some typical values for the values of K_{mN} for different nutrients. When more than one nutrient may be important (as for example in estuaries), then the nutrient effect is given by

$$G(N) = \min\left\{\frac{N_1}{K_{mN_1} + N_1}; \quad \frac{N_2}{K_{mN_2} + N_2} \quad \cdots \right\} \tag{7.35}$$

where the minimum of the expression in brackets is to be taken as the nutrient effect. That is, the individual nutrient limitation expressions are computed and the minimum value is chosen for $G(N)$.

The full expression for the phytoplankton growth rate is therefore given from Eqs. 7.24, 7.33, and 7.35 as

$$G_P = G_{max}(1.066)^{T-20}\left\{\frac{2.718f}{K_e H}\left[\exp(-\alpha_1) - \exp(-\alpha_0)\right]\right\}$$
$$\cdot \min\left\{\frac{N_1}{K_{mN_1} + N_1}; \quad \frac{N_2}{K_{mN_2} + N_2}; \quad \cdots \right\} \tag{7.36}$$

TABLE 7.12 TYPICAL VALUES FOR K_{mN}. HALF-SATURATION CONSTANTS FOR PHYTOPLANKTON GROWTH

Nutrient	K_{mN}
Nitrogen	10–20 μg N/l
Phosphorus	1–5 μg p/l
Silica (for diatoms only)	20–80 μg Si/l

7.6.3 Phytoplankton Death Kinetics

The second component of phytoplankton kinetics is the overall death or mortality rate of the plankton. The basic components of D_P are:

1. Endogenous respiration rate at which phytoplankton oxidize their organic carbon to CO_2 per unit weight of phytoplankton carbon.
2. Zooplankton predation or grazing of the phytoplankton.

The death rate can, therefore, be written as the sum of these two components, that is,

$$D_P = D_{P1}(T) + D_Z \tag{7.37}$$

where $D_{P1}(T)$ is the endogenous respiration rate as a function of temperature, T, and D_Z is the death rate due to zooplankton grazing.

7.6.3.1 Endogenous Respiration   The endogenous respiration rate is given approximately by

$$D_{P1} = \mu_R(1.08)^{T-20} \tag{7.38}$$

for D_{P1} in day^{-1} and T in °C. Figure 7.25 shows this relationship for μ_R of 0.05 to 0.25/day. To first approximation, a value of 0.15/day is appropriate.

7.6.3.2 Zooplankton Grazing   The loss of phytoplankton due to grazing by herbivorous zooplankton is proportional to the number or concentration of zooplankton that is present. The death rate due to grazing, D_Z, can therefore be written as

$$D_Z = C_g Z \tag{7.39}$$

where C_g is the grazing (filtering) rate of zooplankton (liters/day · mg zoopl C) and Z is the zooplankton concentration in equivalent carbon units (mg C/l). Much work has been done on measurements of the zooplankton grazing rate and Table

Figure 7.25   Phytoplankton respiration rate as a function of water temperature. From USEPA (1983).

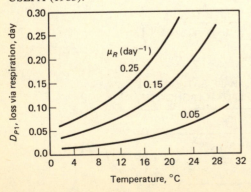

TABLE 7.13   RANGE OF ZOOPLANKTON GRAZING RATES

Organism or location	Grazing rate (l/day · mg zoopl C)
Rotifer (0.1 μg)	1.5–4
Copepod and Cladoceran (1–10 μg)	2–5
Georges Bank	2–3
Lake Ontario	1.5–3

7.13 summarizes some of the data (Di Toro et al., 1971; Thomann et al., 1974). The filtering rate of the zooplankton is, of course, a complicated function of the age, size, temperature, and species.

Equation 7.39 introduces the zooplankton as a new variable. The concentration of the zooplankton may be known or can be estimated. If so, the known concentration can be used in the equation either as an average value during grazing periods or as a time variable value. It is preferable to attempt to calculate the concentration and then compare such a calculated concentration to the observed zooplankton levels. One can, therefore, write an equation for the growth and death of the zooplankton in a manner similar to the phytoplankton (Eq. 7.21). Thus,

$$\frac{dZ}{dt} = (G_Z - D_Z)Z \qquad (7.40)$$

where G_Z and D_Z are the growth and death rates, respectively, of the zooplankton. The growth rate of the zooplankton depends on the rate at which the zooplankton feed on the phytoplankton (the grazing rate, C_g), the amount of phytoplankton available (the chlorophyll level, P), and the efficiency of assimilation or conversion of phytoplankton biomass to zooplankton biomass. The growth rate can then be written as

$$G_Z = a_1' C_g (a_{ZP} P) \qquad (7.41)$$

where a_1' is the efficiency of assimilation (mg zoopl C assimilated/mg zoopl C ingested). A balance of the units is

$$\text{day}^{-1} = \frac{\text{liters}}{\text{day} \cdot \text{mg Z(C)}} \cdot \left[\frac{\text{mg Z(C)}}{\text{mg P(chl)}} \frac{\text{mg P(chl)}}{\text{liter}} \right]$$

The death rate for the zooplankton biomass is simply the sum of the respiration, individual death rate, and predation by the next level in the food chain. Thus,

$$D_Z = K_Z(T) + D_Z' \qquad (7.42)$$

where $K_Z(T)$ is the respiration of the zooplankton and is about 0.02/day at 20°C and D_Z' represents the other losses of zooplankton due to death and higher order predation. This latter coefficient is usually an empirical parameter to be adjusted unless higher order forms are also modeled.

The growth and death of phytoplankton is, therefore, a function of several environmental variables: water temperature, sunlight, turbidity, and zooplankton. From a water quality engineering viewpoint, the time and space distribution of phytoplankton also depends most significantly on the nutrients available for growth. Such dependence is quite nonlinear as shown in Fig. 7.26(b) and, as a result, reductions in ambient nutrient concentrations may not always result in reductions in phytoplankton biomass. It is important then to examine the relationship between phytoplankton and nutrients in some additional detail.

7.6.4 Nutrient Relationships

With the basic principles of phytoplankton growth and death in hand, the interactions of the nutrient and phytoplankton systems can be evaluated by first considering the nutrient mass balance equations with the direct and explicit inclusion of the phytoplankton. It is instructive to begin with a single nutrient that is limiting, say phosphorus, and assume all other nutrients are in excess supply. For phosphorus then, assume that dissolved reactive phosphorus (see Fig. 7.4) is available for uptake by the phtyoplankton as shown in Fig. 7.26. The phosphorus thus absorbed is incorporated into the phytoplankton cellular material. Upon respiration and cell lysis (cell disintegration), the cellular phosphorus of the phytoplankton is released in two principal forms: as particulate organic (detrital) phosphorus and as complexed dissolved organic phosphorus. Both of these forms are grouped under a general heading of "less available" phosphorus. Bacterial degradation and hydrolysis reactions then convert these less available organic forms to the inorganic phosphorus form. Dissolved reactive phosphorus is also released upon cell lysis as shown in Fig. 7.26.

This simplified phosphorus–phytoplankton cycle can now be used to illustrate several important points. Consider this cycle to be operative in a completely mixed

Figure 7.26 A simplified phosphorus–phytoplankton interaction.

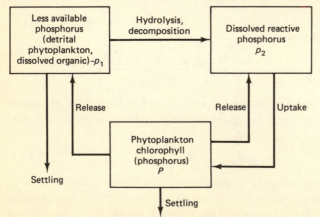

system such as shown in Fig. 7.11. The mass balance equation for each form is given by (where, for simplicity, recycle to p_2 is neglected):

$$V \frac{dp_1}{dt} = W_1 - Qp_1 - K_{11}Vp_1 + D_PVa_pP \qquad (7.43)$$

$$V \frac{dp_2}{dt} = W_2 - Qp_2 + K_{12}Vp_1 - G_PVa_pP \qquad (7.44)$$

$$V \frac{dP}{dt} = W_3 - QP - v_{s3}AP + (G_P - D_P)VP \qquad (7.45)$$

In the equations, p_1 and p_2 are the unavailable and available phosphorus forms respectively, and P is the phytoplankton chlorophyll, all in $[M/L^3]$. W_1 and W_2 are the inputs $[M/T]$ of less available and available phosphorus respectively; W_3 is the input of phytoplankton chlorophyll; K_{12} is the rate $[1/T]$ of conversion of unavailable to available phosphorus; and K_{11} is the loss rate of p_1 due to conversion to available phosphorus as well as any settling of particulate forms, that is,

$$K_{11} = K_{11} + \frac{v_{s1}}{H} \qquad (7.46)$$

for v_{s1} as the effective settling rate $[L/T]$ of particulate forms of less available phosphorus. Similarly, in Eq. 7.45, v_{s3} is the settling rate $[L/T]$ of the phytoplankton and finally, a_p is the phosphorus to chlorophyll ratio of the phytoplankton ($[M\ p/M\ chl]$). The first two of these equations require some additional discussion. The third equation has already been reviewed (see Eq. 7.21).

Equation 7.43, the mass balance equation for the less available phosphorus form includes a source term, D_PVa_pP. This term represents the production of less available phosphorus forms due to the death rate of the phytoplankton, D_P.

Typical units are

$$\begin{array}{ccccc} D_P & V & a_p & P & \\ (\text{day}^{-1}) & (\text{l}) & (\mu g\ p/\mu g\ chl) & (\mu g\ chl/l) & = \mu g\ p/\text{day} \end{array}$$

The second equation (7.44) includes the generation of available phosphorus due to degradation of less available forms in the term $K_{12}Vp_1$. In addition, the uptake of available phosphorus is given by G_PVa_pP. The rate of growth of the phytoplankton and conversion of the inorganic phosphorus to the phytoplankton biomass is accounted for stoichiometrically in the nutrient equation by this term. The production of phytoplankton chlorophyll is, therefore, accompanied by a reduction in the available phosphorus. This was the basis for the discussion on the N/p ratio in Section 7.4. For each $\mu g/l$ of chlorophyll produced, the available phosphorus is reduced stoichiometrically by a_p, the phosphorus to chlorophyll ratio. This can readily be seen by assuming for Eqs. 7.43 to 7.45, a batch reactor vessel that has no external inputs, no flow through the vessel, is kept stirred so the settling velocity is zero and the rate of conversion of less available to available phosphorus forms is fast. Only two variables then need be considered; the available phosphorus and the phytoplankton. Therefore,

$$\frac{dp_2}{dt} = -G_P a_p P + D_P a_p P = -(G_P - D_P)a_p P \tag{7.47}$$

and

$$\frac{dP}{dt} = (G_P - D_P)P \tag{7.48}$$

Now $a_p P$ represents the available phosphorus equivalent of the phytoplankton. Thus, substituting Eq. 7.48 into Eq. 7.47 gives

$$\frac{dp_2}{dt} = -a_P \frac{dP}{dt} \tag{7.49}$$

During the active growth phase of the phytoplankton then, where $G_P > D_P$ (see previous discussion on Eq. 7.21), an equivalent uptake of phosphorus occurs. During the growth phase of the phytoplankton, where, for example, 1 μg chl/l · day were produced, then the product of a_p and this rate is the amount of available phosphorus that had to be supplied.

These nutrient relationships form the basis for controlling the plant biomass in any body of water. Simple calculations can be applied to rivers as noted below. However, the relationship of the plant biomass to the dissolved oxygen is also of importance and is discussed in Section 7.7.

7.6.4.1 Sediment Release of Nutrients

In addition to the external sources of nutrients, the release of nutrients from the sediments may also be important. Such release occurs as a result of a gradient in nutrient concentration between the overlying water and the nutrient in the interstitial water of the sediment. The presence of excessive levels of nutrients in the sediments of water bodies is ultimately traceable back to the external sources. However, for some situations, the problem can be separated and the effect of nutrient release on the phytoplankton can be dealt with as a continuous source. The impact of sediment nutrient release can be significant and result in continuing eutrophication problems even after point sources have substantially reduced. Table 7.14 shows some reported nutrient fluxes from the sediments under both aerobic and anaerobic conditions. As noted, when the overlying water is anaerobic the flux of phosphorus from the sediment increases significantly as a result of increased diffusion between the sediment and the water. Such increased diffusion results from changes in the iron complexes at the water sediment interface.

Sediment nutrient releases can be treated as nutrient sources to a water body to assess the expected responses from reduction in external nutrient inputs.

TABLE 7.14 SOME SEDIMENT NUTRIENT RELEASE RATES

Location	Flux—aerobic conditions (mg/m² · day)		Flux—anaerobic conditions (mg/m² · day)			Reference
	Total dissolved phosphorus	NH₃–N	Total dissolved phosphorus	NH₃–N	Si–silicon	
Muddy River, Boston MA	9.6		(96 Max)			Fillos and Swanson (1975)
Lake Warner, Amherst, MA	1.2		(26 Max)			
Lake Ontario	0.2					Bannerman et al. (1975)
Lake Erie—						Di Toro and Connolly (1980)
Western Basin	6.0	44				
Central Basin	3.0	30				
Eastern Basin	2.0	22				
White Lake, Muskegon Co., MI			34	32	297	Freedman and Canale (1977)
Cape Lookout Bight, NC		40(winter) 325(summer)				Martens et al. (1980)
LaJolla Bight	2.4(–13 to 16)					Hartwig (1975)
Potomac Estuary	1–10					Calendar and Hammond (1982)

7.7 PHYTOPLANKTON–DO RELATIONSHIPS

The relationship between the growth and death of phytoplankton and the DO is a particularly important one. From the DO balance point of view, photosynthesis can result in significant variations in the DO in a water body (see Chapter 6). Also, the production of plant biomass due to the discharge of nutrients can result in the deposition of such biomass to the sediment. The oxidation of this organic material then becomes part of the SOD and can significantly influence the DO. An example is shown in Fig. 7.27 where the DO for the central basin of Lake Erie is shown. The rapid decline of the DO in the hypolimnion in the summer is clearly indicated and is due, in large measure, to the phytoplankton production in the epilimnion. The DO balance equation includes the effect of photosynthesis and respiration by aquatic plants. These terms are given as

$$+ p_a(\text{mg } O_2/\text{l} \cdot \text{day})\text{---photosynthesis}$$
$$- R(\text{mg } O_2/\text{l} \cdot \text{day})\text{---respiration}$$

and represent the direct input or utilization of oxygen through the phytoplankton dynamic of growth and death discussed above. The mass balance equation for DO discussed in Chapter 6 includes these effects as source and sink terms, that is,

$$\frac{dc}{dt} = K_a(c_s - c) - K_d L + p_a - R$$

The connection between the phytoplankton equations given above and the p_a and R in the equation is of particular importance when the DO response due to changes in phytoplankton population must be computed.

It can be recognized from the preceding development that the quantity $G_P P$ is the μg chlorophyll produced/l · day. Now every μg chlorophyll that is produced represents production of phytoplankton carbon to a ratio of about 50–100 μg carbon produced per μg chlorophyll. The stoichiometric oxygen equivalent of this carbon (see Chapter 6) is approximately 2.67 μg oxygen. Therefore, if a_{cP} is the carbon chlorophyll ratio,

$$a_{oP} = 2.67 a_{cP} \tag{7.50}$$

where a_{oP} is the μg DO/μg chl produced. For a_{cP} of 50–100 μg c/μg chl, a_{oP} is 133–266 μg DO/μg chl. The average production of oxygen over a day and over depth is then given by

$$p_a = a_{oP} G_P P \tag{7.51}$$

which can be used as the source term for the DO balance where p_a (mg/l · day) is the average production.

Following the preceding development for the gross growth rate, G_P (Eq. 7.36):

$$p_a = a_{oP}[G_{\max}(1.066)^{T-20} \, G(I)G(N)]P \tag{7.52}$$

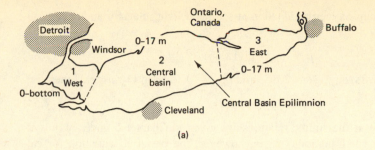

(a)

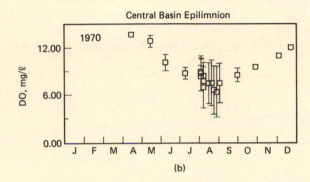

(b)

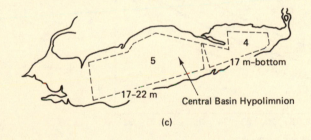

(c)

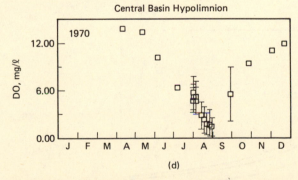

(d)

Figure 7.27 Dissolved oxygen of Lake Erie. (a) Location of epilimnion regions. (b) DO of Central Basin epilimnion. (c) Location of hypolimnion regions. (d) DO of Central Basin hypolimnion. After DiToro and Connolly (1980).

The maximum oxygen production at saturated light intensity and excess nutrients is given by

$$p_s = a_{oP}G_{max}(1.066)^{T-20}P \qquad (7.53)$$

If now $G_{max} = 1.88$/day, $T = 20°C$, and $a_{oP} = 0.133$ mg O_2/μg chl, then

$$p_s = 0.25P \qquad (7.54)$$

which is the same as the simple relationships given in Chapter 6 under photosynthesis (see Eq. 6.42. Equation 7.52 is, however, the preferred equation for calculating the average DO photosynthetic production for use in DO balance equa-

Figure 7.28 Estimating algal productivity from chlorophyll concentrations and water conditions. From USEPA (1983).

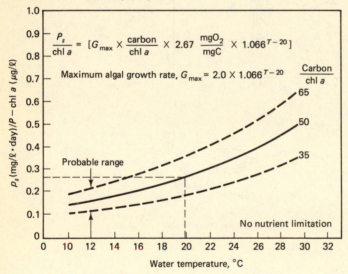

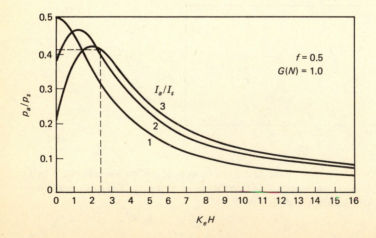

tions. If $G(N) = 1$, then the equation reduces to Eq. 6.40 where the DO production was influenced by light only.

An approximate procedure for estimating p_a under no nutrient limitation is given by Fig. 7.28. The upper panel relates the water temperature and the carbon/chl a ratio to the ratio p_s/P. This latter quantity is the ratio of the maximum productivity at saturated light conditions to the chlorophyll concentration. For a given chlorophyll concentration then, the value of p_s can be estimated.

The lower panel of Fig. 7.28 gives the ratio of the average productivity to maximum productivity (p_a/p_s) as a function of the product of the extinction coefficient and depth (K_eH) for different ratios of the incoming solar radiation to saturated light intensity.

As an example, at 20°C, and carbon/chl of 50, p_s/P is about 0.25. At K_eH of 2.5 and I_a/I_s of 3, p_a/p_s is about 0.4. Therefore,

$$p_a(\text{mg DO/l} \cdot \text{day}) \approx 0.1 P(\mu\text{g chl } a/\text{l}).$$

The uptake of oxygen due to the endogenous respiration of the phytoplankton also follows the preceding development. Thus,

$$R = a_{oP}D_PP \tag{7.55}$$

where it can be recalled that D_PP is the rate of endogenous respiration of the phytoplankton (μg chl/l · day). If $D_P = 0.1$/day and $a_{oP} = 0.25$ mg O_2/μg chl, then

$$R = 0.025P \tag{7.56}$$

Note, however, that Eq. 7.55 is the operative equation and for a range of a_{oP} and D_P that R as a function of chlorophyll may vary considerably from the simple relationship of Eq. 7.56.

The net production of oxygen is then given by the following, using Eqs. 7.52 and 7.55 as

$$p_{\text{net}} = p_a - R$$

7.8 SIMPLIFIED RIVER AND STREAM EUTROPHICATION ANALYSIS— PHYTOPLANKTON

It is becoming increasingly apparent that as more advanced degrees of waste treatment are installed, the water quality of rivers and streams may shift into more difficult problem contexts. For example, when untreated or primary treated effluent is discharged to a stream, the principal water quality effects are sludge deposits, low or anaerobic DO conditions, high turbidity, and an aquatic ecosystem dominated by pollution tolerant organisms. As the treatment level increases, light penetration is enhanced, there is a shift to an aquatic ecosytem that may be dominated by the primary producers—the phytoplankton and the bottom attached algae as well as rooted aquatic macrophytes. The nutrient balance then becomes more important since the biomass may be stimulated by excess nutrient discharges. High concentrations of phytoplankton or attached algae can result in marked diurnal variations in DO. Therefore, as the mean value of DO is improved, the diurnal range may increase due to the stimulated presence of the primary producers. It is important then to analyze the potential impacts of nutrient discharges on resulting plant biomass in rivers and streams. This section explores the behavior of phytoplankton in rivers that is, that component of the aquatic plant system that passively moves downstream at the water column velocity. Attached plants are considered in the next section.

7.8.1 The *N*/*p* Ratio for Streams

At different levels of treatment, the ratio of TN/Tp in the stream at the point of discharge will vary depending on the ratio of river flow to effluent flow. Table 7.8 summarizes this ratio for streams that are dominated by the plant effluent (i.e., approximately 50% or greater of the river flow is effluent flow). Note that under higher rate activated sludge (HRAS) and HRAS plus nitrification, nitrogen would be limiting. But if a biomass problem exists and some nutrient control is necessary, then HRAS and phosphorus removal will shift the system from a nitrogen to a phosphorus limitation. Again, under certain conditions of effluent flow to river flow, phosphorus may be the limiting nutrient under different levels of treatment.

Figure 7.29 shows the relationship between the ratio of TN/Tp for different

Figure 7.29 TN/Tp ratio as a function of the ratio of plant effluent flow/total river flow for different treatment levels.

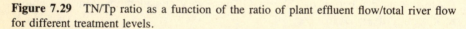
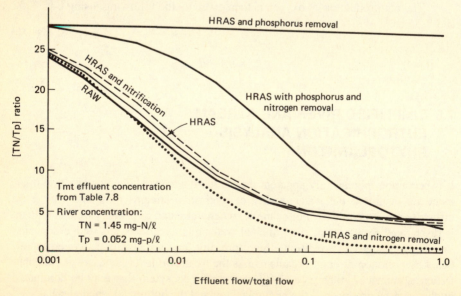

ratios of the plant effluent flow to the total river flow. The upstream flow is assumed to have the following nutrient characteristics:

$$TN = 1.45 \text{ mg N/l}$$
$$Tp = 0.052 \text{ mg p/l}$$
$$TN/Tp = 28$$

Note that in general, runoff from non-point sources will tend to have high TN/Tp ratios indicating that phosphorus is the limiting nutrient for non-point source situations. Figure 7.29 shows that when the effluent flow is less than about 2% of the total river flow, phosphorus would tend to be the limiting nutrient (TN/Tp > 10). For effluent/river flow ratios greater than about 2%, the limiting nutrient depends on the treatment process. For all ratios, HRAS and phosphorus removal results in a phosphorus limited situation.

The N/p ratio is therefore a useful first approximation to determining the significance of each nutrient in controlling phytoplankton biomass. One should not, however, necesssarily conclude that because nitrogen is the limiting nutrient in a given situation that nitrogen is the nutrient to be controlled. As the previous point source discussion indicated when a system was controlled by nitrogen, the installation of HRAS + phosphorus removal would shift the system to a phosphorus limitation.

7.8.2 Simplified Stream Phytoplankton Model

For streams there are several aspects of the above discussion that should be recognized.

1. The distance and hence the time of travel for stream nutrient problem contexts generally tends to be short, perhaps on the order of less than 10 days. This would be equivalent to distances of less than about 100 mi for streams moving at 1 fps. Most stream stretches of interest would be shorter than this so that the travel time of the phytoplankton would be less than the 10-day period. As a result, the phytoplankton chlorophyll biomass may not have enough time to grow to the maximum level calculated from the N/p ratio. The rate of growth of the phytoplankton and the travel time of the stream length are therefore of specific importance.

2. Phytoplankton growth dynamics indicate that the concentrations of nitrogen and phosphorus that limit growth are relatively small compared to normal ambient levels of nutrients in streams subject to point source discharges. This can be seen by considering Eq. 7.29, where the nutrient effect for inorganic nitrogen (IN) and inorganic phosphorus (Ip) is given by:

$$G(N) = \min\left\{ \frac{IN}{K_{mN} + IN}; \frac{Ip}{K_{mp} + Ip} \right\} \tag{7.57}$$

At concentrations of the nutrient greater than about five times the half-saturation concentration, the effect of the nutrient on the growth rate $G(N)$ is less than 20%. That is, at IN and Ip at five times K_{mN} and K_{mp}, $G(N) = 0.83$ and therefore has a relatively small effect on the overall growth rate of phytoplankton as may be seen from Eq. 7.23. Table 7.12 shows that the Michaelis concentration (the concentration at which growth rate is reduced by $\frac{1}{2}$) for nitrogen is about 10–20 μg N/l and for phosphorus is about 1–5 μg p/l. Therefore for IN and Ip greater than about 0.1 mg N/l and 0.025 mg p/l, respectively, these nutrients will not have any effect on the growth rate. Considering Tables 7.7 and 7.8, it is seen that, depending on stream dilution, the concentrations of IN and Ip will generally be significantly above these levels. Figure 7.30 shows this fact for upstream IN and Ip of 0.53 mg N/l and 0.021 mg p/l, respectively, (Omernik, 1977). As indicated, for nitrogen, the concentration at the outfall never reaches limiting levels simply because the upstream IN concentration alone is above the maximum limiting concentration of about 0.1 mg N/l. For phosphorus, when the effluent flows are greater than about 5–10% of the total river flow, then the Ip at the outfall will be above the maximum limiting phosphorus concentration. It should be stressed, however,

Figure 7.30   Inorganic phosphorus and nitrogen at outfall for different ratios of effluent flow to total river flow.

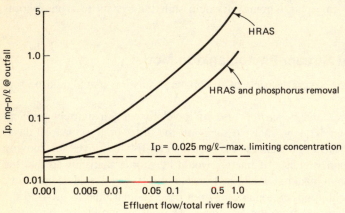

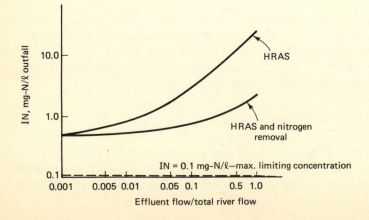

that these considerations are meant as a guide only since limiting nutrient concentrations may be reached at some downstream location from the outfall. As a general rule however, Fig. 7.30 indicates that for ratios of effluent flow to total river flow of greater than about 1%, both phosphorus and nitrogen will tend to be present in excess concentrations. For "small" streams where effluent flow is equal to or greater than upstream flow especially at drought conditions, then it is almost certainly true that the nutrients will be in excess and will not limit phytoplankton growth for a given stretch of river.

   A simple model can now be constructed using the above considerations and the principle of the N/p ratio. Assume that a steady state condition prevails. The analysis is therefore representative of a given flow and temperature regime that has held steady long enough for conditions to have reached an equilibrium state. Consider the phytoplankton equation at steady state as

$$U\frac{dP}{dx} = \frac{dP}{dt^*} = \left(G_P - D_P - \frac{v_s}{H}\right)P = G_n P \qquad (7.58)$$

where as before P is phytoplankton chlorophyll (μ chl/l), and G_P is given by Eq. 7.23. D_P is the respiration rate in day^{-1}, v_s is net settling from the water column in m/day, H is the stream depth, U is the stream velocity, x is distance, and t^* is travel time in days and the net growth rate of the phytoplankton, G_n, is given by

$$G_n = G_P - D_P - \frac{v_s}{H} \qquad (7.59)$$

If $G_P \neq f(\text{IN,Ip})$ then G_n is a function only of temperature, light, extinction coefficient, and net settling velocity. Figure 7.31 shows the behavior of G_n as a function of v_s/H for various values of temperature and light representative of conditions in small shallow streams.

   Thus, the assumption of excess nutrients permits solution of Eq. 7.58 as

$$P = P_0 \exp(G_n)t^* \qquad (7.60)$$

where P_0 is the phytoplankton chlorophyll at the outfall. Figure 7.32 shows the possible forms of Eq. 7.60. Because of light limitations and settling effects, G_n may be less than zero in which case there will be an exponential decrease of phytoplankton with travel time. G_n may also be zero where growth is just balanced by respiration and settling. The biomass then remains constant with distance. Finally, G_n may be positive resulting in exponential growth of the phytoplankton. The ultimate biomass is a hypothetical value which results from conversion of the IN or Ip at the outfall, as discussed previously, which only occurs when the travel time in the stream is long ($t_a^* = t_{\text{ult}}^*$).

   The inorganic nutrient equation can now be examined in the light of the preceding N/p discussion. Assume that a given fraction of IN or Ip will constitute the nutrient pool from which the growth of the phytoplankton will draw nutrients. Let the inorganic pool nitrogen $= N_i$ and the inorganic pool phosphorus $= p_i$. A fraction of 0.75 is a good approximation, that is, 75% of the total inorganic nutrient is available for conversion to phytoplankton. The residual 25% of the nutrient recyles into the detrital pool and less available forms and is also lost by settling.

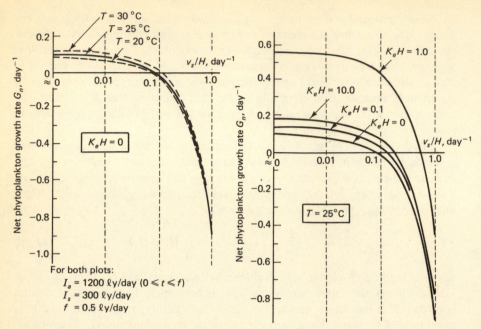

Figure 7.31 Variation of net phytoplankton growth rate with settling for selected temperature and light conditions.

Therefore, the nutrient equation for phosphorus is

$$\frac{dp_i}{dt^*} = -a_p G_P P \qquad (7.61)$$

and for nitrogen,

$$\frac{dN_i}{dt^*} = -a_N G_P P \qquad (7.62)$$

where $a_p = \mu g\ p/\mu g$ chl and $a_N = \mu g\ N/\mu g$ chl. The chlorophyll is given by Eq. 7.60 and substituting into Eqs. 7.61 and 7.62 and solving gives

$$p_i = \frac{a_p G_P P_0}{G_n} (1 - e^{G_n t^*}) + p_{i0} \qquad \text{for}\quad p_i > 25\ \mu g\ p/l \qquad (7.63)$$

and

$$N_i = \frac{a_N G_P P_0}{G_n} (1 - e^{G_n t^*}) + N_{i0} \qquad \text{for}\quad N_i > 100\ \mu g\ N/l \qquad (7.64)$$

where p_{i0} and N_{i0} are the phosphorus and nitrogen concentrations at the outfall. Note that the equations are only applicable for the inorganic nutrients above the maximum limiting concentration of $p_i = 25\ \mu g/l$ and N_i of $100\ \mu g/l$.

If estimates of G_P and G_n are obtained from previous expressions and P_0 and

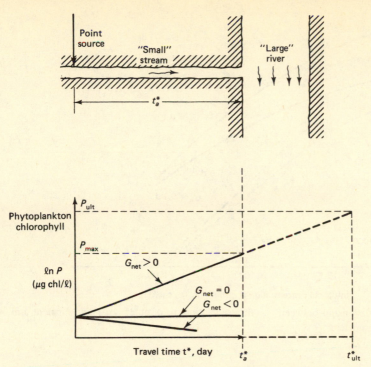

Figure 7.32 Schematic of relationship between phytoplankton chlorophyll, net growth, and travel time.

initial nutrient conditions at the outfall are available, then the travel time to reach the maximum limiting nutrient concentration can be computed. This will then fix the maximum biomass to be expected over that reach. Solving Eq. 7.63 for t_{25}^* at $p_i = 25$ μg p/l gives

$$t^*_{25} = \frac{1}{G_n} \ln \frac{P_0' + p_{0i} - 25}{P_0'} \tag{7.65}$$

where

$$P_0' = \frac{a_p G_P P_0}{G_n}$$

Figure 7.33 shows t_{25}^* for various conditions on the parameters. Using $G_P = 1$/day and $G_n = 0.1$/day, and for p_{0i}, the initial available phosphorus concentration of greater than 100 μg p/l, (see Fig. 7.30), which covers small effluent dominated streams with HRAS and without P removal, Fig. 7.33 indicates that t_{25}^* is generally greater than 5–10 days. Therefore, for small effluent dominated streams, if the study length is less than about 5–10 days, the phosphorus will still be above 25 μg p/l and therefore in excess. At average velocities of 0.5–1 fps, this is equivalent to about 40–160 mi, well within the study reaches usually under investigation.

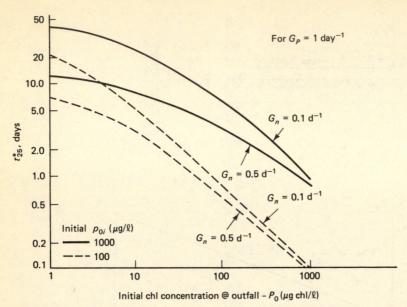

Figure 7.33 Relationship between t_{25}^*, the travel time to p_i of 25 µg/l and nutrient and phytoplankton parameters.

From Fig. 7.33, note that the phytoplankton concentration at the outfall is an important parameter to be specified. At high initial chlorophyll levels at the outfall, the time of travel to reach the ultimate concentration of phytoplankton is markedly reduced since a sufficient "seed" biomass already exists at the outfall. For small streams, however, one would not normally expect upstream concentrations to be greater than 100 µg chl/l.

In this simple model then, the only way in which the maximum phytoplankton biomass will decrease as a function of nutrient removal at a point source is to reduce the time to reach the maximum limiting nutrient concentration. Figure 7.34 shows this effect. If t_{25}^* remains greater than t_a^*, the actual stream length after phosphorus removal, then there will be no reduction in biomass. This is because the stream is simply not long enough for a given net growth rate to result in a nutrient limitation.

To summarize this simple model then, the following procedure is suggested.

1. Compute G_n from Eq. 7.59 and Eqs. 7.36 and 7.37, or estimate G_n from available phytoplankton data on the stream using Eq. 7.60.
2. For a given inorganic phosphorus concentration at the outfall, estimate the percentage available for phytoplankton growth, say 75%.
3. Calculate t_{25}^* from Eq. 7.65 and compare actual stream length t_a^* to t_{25}^*. If t_a^* is less than t_{25}^*, then $P_{\max} = P_0 \exp G_n t_a^*$. If t_a^* is greater than t_{25}^* then the biomass is at least $P_{\max} = P_0 \exp G_n t_{25}^*$ and is probably slightly greater than this amount.

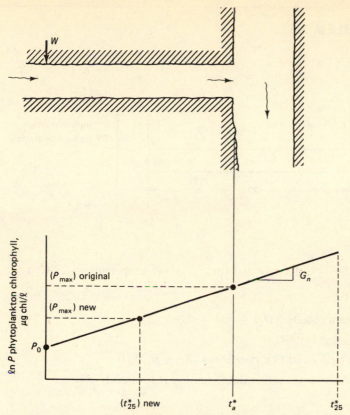

Figure 7.34 Relationship between phytoplankton chlorophyll and travel time showing reduction in biomass due to reduction in time of travel to limiting nutrient.

4. Estimate the new p_{0i} under a phosphorus removal program.
5. Calculate the new t_{25}^* under the removal program.
6. Compare $(t_{25}^*)_{new}$ to t_a^*. If $(t_{25}^*)_{new} > t_a^*$ then no reduction in biomass will occur as a result of the treatment program.

Calculation of G_n and v_s, and application of the above procedure, is illustrated in Sample Problem 7.3.

The case where $G_n = 0$ is an interesting one. For this case, $P = P_0$, a constant. Therefore,

$$\frac{dp_i}{dt^*} = -a_p G_P P_0 \tag{7.66}$$

which is also a constant. The inorganic phosphorus would therefore decline linearly with travel time as

$$p_i = -a_p G_P P_0 t^* + p_{i0} \tag{7.67}$$

SAMPLE PROBLEM 7.3

DATA

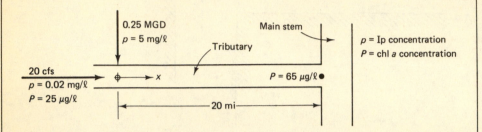

Tributary:

> depth = 3.0 ft,   velocity = 0.5 fps,   water temperature = 23°C

Light:

> daily solar radiation (I_T) = 600 ly/day
>
> photoperiod (f) = 0.5
>
> stream light extinction coefficient (K_e) = 0.33/ft
>
> phytoplankton saturation light intensity (I_s) = 300 ly/day

PROBLEM

Estimate the net phytoplankton growth rate in the stream and the net phytoplankton settling rate. Use μ_R = 0.1/day.

ANALYSIS

Net Phytoplankton Growth Rate

Assuming no nutrient limitation, use an exponential increase of phytoplankton between $x = 0$ and $x = 20$ mi.

$$t_a^* = 20 \text{ mi}/(0.5 \text{ fps} \times 16.4 \text{ mpd/fps}) = 2.44 \text{ days}$$

$$P_{20} = P_0 e^{G_n t_a^*}, \qquad 65 = 25 e^{2.44\, G_n} \qquad\qquad \text{[Eq. (7.60)]}$$

$$\underline{\underline{G_n}} = \frac{\ln(65/25)}{2.44} = \underline{\underline{0.391/\text{day}}}$$

Phytoplankton Growth and Death Rates

$$G(T) = 1.8 \times 1.066^{(23-20)} = 2.18/\text{day} \qquad\qquad \text{[Eq. (7.24)]}$$

Sample Problem 7.3 (continued)

$$\alpha_0 = \frac{I_a}{I_s} = \frac{(600 \text{ ly/day})/0.5}{300 \text{ ly/day}} = 4.00$$

$$\alpha_1 = \alpha_0 e^{-K_e H} = 4.00 e^{-0.33 \times 3.0} = 1.47$$

$$G(I) = \frac{ef}{K_e H} (e^{-\alpha_1} - e^{-\alpha_0}) \qquad \text{[Eq. (7.33)]}$$

$$= \frac{2.718 \times 0.5}{0.33 \times 3.0} \times (e^{-1.47} - e^{-4.00}) = 0.287$$

$$\underline{G_P = G(T) \cdot G(I) = 2.18/\text{day} \times 0.287 = 0.626/\text{day}} \qquad \text{[Eq. (7.23)]}$$

$$\underline{D_P = 0.1 \times 1.08^{(23 \cdot 20)} = 0.126/\text{day}} \qquad \text{[Eq. (7.38)]}$$

Phytoplankton Net Settling Rate

$$\text{with } G_n = G_P - D_P - \frac{v_s}{H} \qquad \text{[Eq. (7.59)]}$$

$$v_s = H(G_P - D_P - G_n) = 3.0(0.626 - 0.126 - 0.371)$$

$$\underline{\underline{v_s = 0.327 \text{ ft/day}}}$$

Check for Nutrient Limitation

$$P_0' = \frac{a_p G_P P_0}{G_n} = \frac{1 \text{ µg } Ip/\text{µg chl } a \times 0.626/\text{day} \times 25 \text{ µg chl } a/\text{l}}{0.391/\text{day}}$$

$$= 40 \text{ µg } Ip/\text{l}$$

$$p_0 = \frac{\sum W}{\sum Q} = \frac{20 \text{ cfs} \times 0.02 \text{ mg/l} + 0.25 \text{ MGD} \times 1.548 \text{ cfs/MGD} \times 5 \text{ mg/l}}{(20 + 0.25 \times 1.548)\text{cfs}}$$

$$p_0 = 0.115 \text{ mg/l} = 115 \text{ µg } Ip/\text{l}$$

$$t_{25}^* = \frac{1}{G_n} \left(\frac{P_0' + p_0 - 25}{P_0'} \right) = \frac{1}{0.391/\text{day}} \ln \left(\frac{40 + 115 - 25}{40} \right) \qquad \text{[Eq. 7.65]}$$

$$= 3.01 \text{ days}$$

Since the actual travel time to $x = 20$ mi, $t_a^* = 2.44$ days, is less than $t_{25}^* = 3.01$ days, no significant inorganic phosphorus nutrient limitation on the growth rate occurs and $G(N) = 1$. Thus, the above analysis is appropriate.

Note: Values of G_P and D_P may be significantly different from those calculated above and sensitivities on v_s should be made for a reasonable range of G_P and D_P values.

The distance to t_{25}^* is therefore

$$t_{25}^* = \frac{p_{i0} - 25}{a_p G_P P_0} \tag{7.68}$$

However, it can be noted that, for this case, reduction of nutrients at the source will not alter the biomass. Again this only applies for the excess nutrient region. If p_{i0} is reduced such that $t_a^* > (t_{25}^*)_{new}$ then it can be presumed that a reduction in biomass will occur but the actual reduction cannot be precisely computed since it will occur in the limiting nutrient region where nonlinear kinetics apply.

7.8.2.1 Relationship to DO   For some stream situations, the phytoplankton concentration itself may not be a matter of great concern in terms of any nuisance condition created by high chlorophyll concentrations. However, the relationship between the resulting phytoplankton biomass and the DO resources may be of significant concern. This is especially true for the diurnal variations in DO occasioned by the phytoplankton. If D_1 is designated as the deficit resulting from photosynthesis, that is,

$$D_1 = c_s - c_P$$

where c_P is the DO resulting from the phytoplankton processes, then

$$\frac{dD_1}{dt^*} = -a_{oP}(G_P - D_P)P(t^*) - K_a D_1 \tag{7.69}$$

Several possibilities now can be examined on $P(t^*)$. These would be cases where data were available on $P(t^*)$ and inferences can be drawn of the effect of $P(t^*)$ on the average DO regime.

Case I. $P(t^*) = $ constant $= P_0$. For this case, $dP/dt^* = 0$ and $(G_p - D_p) = (v_s/H)$. Thus if $v_s/H \neq 0$, then $G_p > D_p$ and there is a positive production of DO given by

$$D_1 = -\frac{a_{oP}(G_P - D_P)P_0}{K_a}(1 - e^{-K_a t^*}) \tag{7.70}$$

If, however, it is determined that $v_s = 0$, then growth just balances respiration (to maintain P_0) and $G_P - D_P = 0$. Therefore, there is no net production of DO. If then from the data, the phytoplankton is constant with distance, then either there is a net production of oxygen for $v_s/H \neq 0$, or the production is zero for $v_s = 0$.

Case II. $P = P_0 \exp G_n t^*$; $G_n > 0$. For this case growth exceeds respiration and settling and there is a net production of oxygen. Substitution into Eq. 7.69 gives

$$\frac{dD_1}{dt^*} = -a_{oP}(G_P - D_P)P_o e^{G_n t^*} - K_a D_1 \tag{7.71}$$

The solution then is

$$D_1 = \frac{a_{oP}(G_P - D_P)P_0}{K_a + G_n} (e^{G_n t^*} - e^{-K_a t^*}) \tag{7.72}$$

which is approximately exponential. It should be recalled that this case can only apply to the "excess nutrient" region where travel times are less than t_{25}^* (for phosphorus) as discussed earlier.

Case III. $P = P_0 \exp G_n t^*$, $G_n < 0$. With net growth less than zero, these data would show phytoplankton declining with distance downstream. In such a case, respiration and net settling exceed growth and there may or may not be a net utilization of oxygen. If $v_s = 0$, then $D_P > G_P$, and a net utilization occurs. For $v_s \neq 0$ however, G_P may still exceed D_P and a net production could occur. This can only be determined by a specific analysis of the actual stream. Substitution into Eq. 7.69 gives

$$\frac{dD_1}{dt^*} = -a_{oP}(G_P - D_P)P_0 e^{-G_n t^*} - K_a D_1 \tag{7.73}$$

and the solution is

$$D_1 = -\frac{a_{oP}(G_P - D_P)}{K_a - G_n} P_0(e^{-G_n t^*} - e^{-K_a t^*}) \tag{7.74}$$

These DO equations, coupled with the simplified phosphorus–phytoplankton model discussed earlier, permits analysis of the response due to nutrient reduction on both the phytoplankton and the DO. Calculation of the average daily DO deficit and associated diurnal range due to phytoplankton is shown in Sample Problem 7.4.

7.9 SIMPLIFIED RIVER AND STREAM EUTROPHICATION ANALYSIS—PERIPHYTON AND ROOTED AQUATIC PLANTS

7.9.1 Introduction

A major problem in the water quality of streams, especially headwater streams, is the impact of nutrient discharges on the growth and downstream distribution of attached organisms, collectively, called the periphyton and rooted aquatic plants. The question of the effect of nutrient reduction on the attached biomass also includes the resulting effect on the diurnal fluctuation of dissolved oxygen. Indeed, in terms of biomass, streams can be extremely productive systems, at least equal to productive mesotrophic and eutrophic lakes. This is partly due to the fixed position of the plant with a continually renewed source of nitrogen and phosphorus passing by a given stream location. Thus, the fixed plants can be contrasted to the plankton which in moving with a parcel of water will, in time, deplete the nutrients

SAMPLE PROBLEM 7.4

DATA

Use the tributary stream of Sample Problem 7.3.

PROBLEM

Estimate the DO deficit on a daily average basis due to the phytoplankton. Also estimate the diurnal variation.

ANALYSIS

It is convenient to segment the stream into five reaches, each of which has a constant value of chlorophyll a, and thus, a constant value of $p_a - R$. The DO deficit is calculated for the end of the first reach, and this is then used as the initial condition of the second reach, which has its specific distributed $p_a - R$ loading. Thus, in step-by-step fashion the DO deficits on a daily averaged basis are estimated.

Reach	chl a (μg/l)	$p_a{}^a$ (mg O_2/l·day)	R^b (mg O_2/l·day)	$p_a - R$ (mg O_2/l·day)	Reach travel time (days)	Daily average DO deficit^a (mg/l)	Diurnal deficit^d (mg/l) Max.	Min.
1	27	2.25	0.45	1.80	0.50	−0.58	0.65	−1.81
2	34	2.83	0.57	2.26	0.50	−0.96	0.27	−2.19
3	41	3.41	0.69	2.72	0.50	−1.25	−0.02	−2.48
4	50	4.16	0.84	3.32	0.50	−1.56	−0.33	−2.79
5	59	4.91	0.99	3.92	0.44	−1.85	−0.62	−3.08

$^a p_a = a_{oP} G_P P = 0.133$ mg O_2/μg chl $a \times 0.626$/day $\times$ chl a μg/l   [Eq. (7.51)]

$^b R = a_{oP} D_P P = 0.133 \times 0.126 \times$ chl a [Eq. (7.55)]

$^c D = D_0 e^{-K_a \Delta t^*} - \dfrac{(p_a - R)}{K_a}(1 - e^{-K_a \Delta t^*})$

$K_a = 12.9 \times \dfrac{(0.5)^{1/2}}{3.0^{3/2}} \times (1.024)^{23-20} = 1.89$/day

^{d}max diurnal deficit = average daily deficit + $\Delta_c/2$

min diurnal deficit = average daily deficit − $\Delta_c/2$

$\Delta_c = p_a/2$ for $K_a < 2$/day   [Eq. (6.47)]

$\Delta_c = 4.91/2 = 2.46$ mg/l

where the maximum value of p_a was used in order to be conservative (cause minimum DO in stream)

Sample Problem 7.4 (continued)

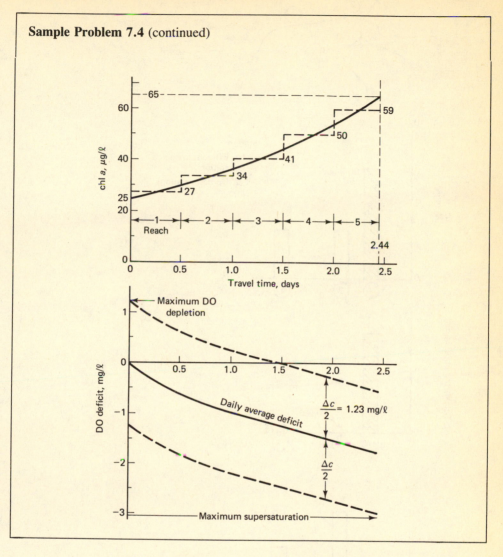

in that parcel. Hynes (1970) provides an excellent overall description of the importance of stream type and the adaptations of plants to running water.

There are numerous examples of the impact of nutrient-rich discharges to streams on the stimulation of fixed plants (e.g., O'Connell and Thomas, 1965; Edwards, 1962; Elwood et al., 1981). Figures 7.35 and 7.36 from Wright and McDonnell (1982) show the stimulation of aquatic plant growth in a small stream receiving the discharge from the Pennsylvania State University sewage treatment plant in comparison to upbasin stations receiving virtually no point discharges. As indicated, the point source dominated effluent stream (characterized by station 1A)

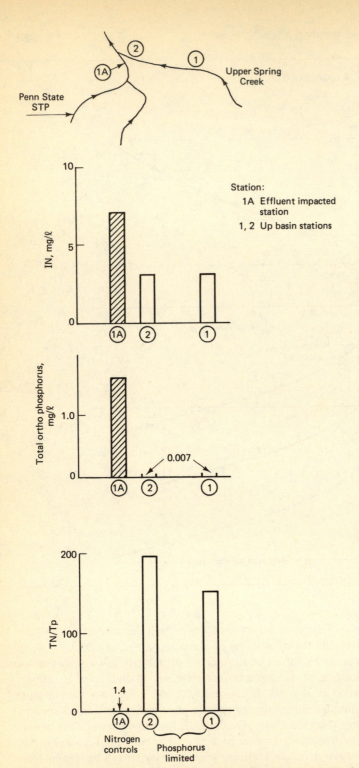

Figure 7.35 Inorganic nitrogen, total orthophosphorus, and *N/p* for a point source dominated stream and an up-basin stream. Plotted from data given in Wright and McDonnell (1982).

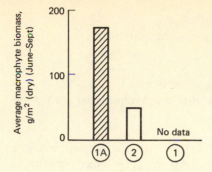

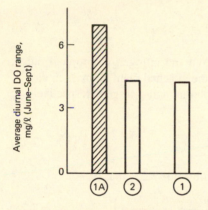

Figure 7.36 Macrophyte biomass and diurnal DO range for stations of Fig. 7.35. Plotted from data given in Wright and McDonnell (1982).

has a low N/p ratio indicating potential nitrogen limitation. In contrast, the upbasin stations (1 and 2) have a very high N/p ratio and indicate phosphorus limitation. The nutrient effect on stimulating the growth of macrophytes is shown in Fig. 7.36 together with the increased diurnal DO range.

In addition to the environmental parameters that affect phytoplankton (i.e., solar radiation, temperature, nutrients), the attached plants are affected by the velocity of the stream and the substrate type. For the former parameter, Horner and Welch (1981) in studies on periphyton development related to current velocity and nutrients indicated that for velocity increases up to about 50 cm/s (1.64 fps) the growth of the periphyton was enhanced. At velocities below about 20 cm/s, production was retarded possibly due to limitations of nutrient diffusion and at velocities above 50 cm/s, the erosive effect of the current physically reduced the biomass.

The purpose of the section is to develop a simple means for rapidly calculating the effect of nutrient reduction on attached plants. Since the objective is to develop a simplified approach, not all of the phenomena that are operative in the production of attached biomass are included and only a part of the overall question is addressed.

7.9.2 The Model

The following assumptions are made for this simplified analysis:

1. Steady state conditions.
2. Only available phosphorus and nitrogen are included.
3. Only attached plants are considered, that is, phytoplankton are not included.

Let p = available inorganic phosphorus in $\mu g/l$, N = available inorganic nitrogen in $\mu g/l$, P = average attached plant biomass in g chl/m^2 or other comparable units. Then for a fixed plant biomass, the nutrient equations are

$$H \frac{dp}{dt^*} = -a_p G_P P = -S_p \tag{7.75}$$

$$H \frac{dN}{dt^*} = -a_N G_P P = -S_N \tag{7.76}$$

where a_p and a_N are the phosphorus/chlorophyll and nitrogen/chlorophyll ratios, respectively, G_P is the growth rate (gross) of the attached plant biomass in day^{-1}, and S_P and S_N are the areal uptake of phosphorus and nitrogen, respectively in mg p, N/m$^2 \cdot$ day. Also,

$$S_N = \frac{a_N}{a_p} S_p \tag{7.77}$$

In general, G_P in Eqs. 7.75 and 7.76 is given by Eq. 7.23, that is,

$$G_P = G(T) \, G(I) \, \min\left(\frac{p}{K_{mp} + p}; \; \frac{N}{K_{mn} + N}\right) \tag{7.78}$$

For this simplified approach, it is assumed that nutrients are in excess, that is, aproximately five times the half-saturation constant. The preceding discussion in Section 7.8 provides justification for this assumption. See, for example, Fig. 7.30. What is sought is the length of stream over which the nutrients have no effect. Thus assume $G_P \neq f(p, N)$. Then for a constant attached plant biomass in a given reach of river, the solutions to Eqs. 7.75 and 7.76 are

$$p = p_0 - \frac{S_p}{H} t^* \qquad \text{for} \quad p \gg K_{mp} \tag{7.79}$$

$$N = N_0 - \frac{S_N}{H} t^* \qquad \text{for} \quad N \gg K_{mN} \tag{7.80}$$

where p_0 and N_0 are the initial concentrations after mixing at the effluent point.

Therefore, the slope of a plot of nutrient vs. time of travel provides a means for estimating the areal uptake of phosphorus and nitrogen for a given stream depth. The task at hand then is to obtain estimates of the nutrient uptake rates either from information on the growth rates or from nutrient profiles.

It should also be noted that the areal nutrient uptake rate can be related to the gross primary production in terms of DO and to the diurnal range of DO. Thus, for p_a = averge daily gross photosynthetic production of oxygen on an areal basis (mg DO/m^2 · day),

$$S_p = \frac{a_{pc}p_a}{2.67} \qquad (7.81)$$

where a_{pc} is the phosphorus carbon ratio and 2.67 is the mg DO/mg C. Note that in terms of the gross primary carbon production, the phosphorus uptake rate is

$$S_p = a_{pc}C_{av} \qquad (7.82)$$

for C_{av} = mg C/m^2 · day. Di Toro (1981) has shown that the diurnal range of DO is related to the average production by Eq. 6.45 and repeated here in areal DO production as

$$p_a = \left\{ \frac{0.5K_a[1 - \exp(-K_a)]}{[1 - \exp(-0.5K_a)]^2} \right\} \Delta_c \cdot H \qquad (7.83)$$

for Δ_c as the diurnal DO range in mg/l, and K_a as the reaeration rate in day^{-1}. Then from Eq. 7.81,

$$S_p = \frac{(a_{pc}FH) \Delta_c}{2.67} \qquad (7.84)$$

where F is the term in braces in Eq. 7.83. Equations 7.81, 7.82, and 7.84 thus permit estimation of the phosphorus uptake rate (or nitrogen rate from Eq. 7.77) given data on either primary production or diurnal DO variation.

Finally, it should be noted that an estimate of p_a, and hence the nutrient uptake, can also be obtained from empirical information on the light energy conversion efficiencies of plants. For example, Edwards (1962) reports that

$$pE = \frac{p_a}{\Sigma I_0} \cdot 100 \qquad (7.85)$$

where pE is the percent photosynthetic efficiency, ΣI_0 is the total incident surface radiation, and assuming that 1 mole of oxygen = 112 kcal, that an average value of 1.5% for pE was representative for aquatic macrophytes (larger aquatic plants) in the River Ivel in England. This is somewhat less than the 7.3% reported by Kevern and Ball (1965) for periphyton in artificial streams. For streams where no data are available, the most approximate estimate of the nutrient uptake can then be obtained by assuming that p_a is about 0.5–5.0% of the total incoming radiation and S_p can be computed from Eq. 7.81.

With estimates of S_p and S_N, Eqs. 7.79 and 7.80 can then be solved for the time of travel from the effluent point to the stream location where nutrients begin to limit (i.e., at $p_c = 5K_{mp}$ or $N_c = 5K_{mN}$). Then for $(t_c^*)_p$ and $(t_c^*)_N$ as the critical times of travel for phosphorus and nitrogen

$$(t_c^*)_p = \frac{p_0 - p_c}{S_p} \qquad (7.86)$$

$$(t_c^*)_N = \frac{N_0 - N_c}{S_N} \qquad (7.87)$$

7.9.2.1 Procedure The above model permits a first estimate to be made of whether a nutrient reduction will have any effect on the attached plants. The steps are as follows:

1. From nutrient profiles, primary productivity or diurnal DO data, estimate S_p and S_N.
2. Choose p_c and N_c, that is, the concentration at which it is estimated that phosphorus or nitrogen would limit the growth of the attached plant biomass.
3. With p_0, and N_0, calculate t_c^* from Eqs. 7.86 and 7.87.
4. Compare $(t_c^*)_p$ and $(t_c^*)_N$ to the actual stream length (t_a^*) to determine whether the stream length is nutrient limited under present conditions.
5. For phosphorus removal, calculate new p_0.
6. Compute new $(t_c^*)_p$ and check to see whether $(t_c^*)_p < (t_a^*)$, that is, that the length of stream with attached biomass will be reduced.

7.9.3 Some Numerical Guidelines and Examples

For a carbon/chlorophyll ratio of 50–100 and a phosphorus/chlorophyll ratio of 1, then the phosphorus/carbon ratio is 0.01–0.02. Letting $F = 3$/day (which is representative for streams with $K_a > 2$/day) and estimating the volumetric phosphorus uptake (S_p' mg p/l · day) from Eq. 7.84 gives

$$\frac{S_p}{H} = S_p' \text{ (mg p/l} \cdot \text{day)} = (0.01 - 0.02) \, \Delta_c \text{(mg DO/l)} \tag{7.88}$$

For a_N/a_p of 7–10, this results in

$$\frac{S_N}{H} = S_N' \text{ (mg N/l} \cdot \text{day)} = (0.07 - 0.20)\Delta_c \tag{7.89}$$

Thus the volumetric phosphorus uptake rate (mg p/l · day) for small streams ($K_a > 2$/day) is about 1–2% of the diurnal DO range. The estimation of the critical nutrient concentrations, p_c and N_c (i.e., the concentrations at which the nutrient begins to limit or about five times the half-saturation constant) can be made to first approximation from the phytoplankton values, although some estimates are available for periphyton. The data of Elwood et al. (1981) in field experiments of the effect of phosphorus enrichment on stream periphyton suggest a p_c of ≤ 60 μg p/l.

If representative phytoplankton values of $p_c = 25$ μg p/l and N_c of 125 μg N/l are used, together with a diurnal range of DO (Δ_c) of 10 mg/l, S_p' at 2% of Δ_c (Eq. 7.88), $S_N' = 7S_p'$, and an effluent dominated stream, then Fig. 7.37 illustrates the behavior of the nutrients with distance downstream.

Figure 7.37 shows that for the effluent dominated stream and no nutrient removal at the effluent, nitrogen would tend to limit the biomass production before phosphorus. The time of travel to the point of nitrogen limitation is about 15 days which is about one-half to one order of magnitude longer than most streams. Therefore, Fig. 7.37(a) indicates that most streams would not experience a nutrient

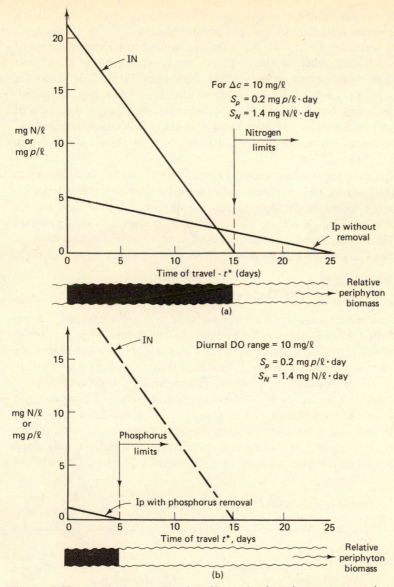

Figure 7.37 (a) Nutrient profiles for effluent dominated stream, no nutrient removal. (b) With phosphorus removal to 1 mg p/1.

limitation in the reach under study and that if a nutrient limitation does exist, nitrogen would limit first. Of course, if the effluent/total river flow decreases, then the effect of upstream dilution flow, which is normally phosphorus limited becomes more significant.

Figure 7.37(b) shows the nutrient behavior under a phosphorus removal program where the mixed phosphorus concentration at the outfall is now 1 mg p/1.

The point of nutrient limitation is now moved upstream to about 5 days (still a considerable stream distance) and now phosphorus controls.

For the preceding conditions, Eq. 7.86 can be used to estimate the time of travel to limiting phosphorus conditions for different concentrations of phosphorus at the outfall after mixing and various diurnal DO ranges. Figure 7.38 shows the result. If the range is known or can be estimated, then Fig. 7.38 may prove useful. As seen, the mixed outfall phosphorus concentration must be on the order of 0.1–0.2 mg p/l for a nutrient limitation to occur in about 1-day travel time or less.

Some nutrient uptake rates by rooted plants and periphyton are shown in Table 7.15.

Although the data of Elliott and McDonnell (1982) (for the same stream as shown in Fig. 7.35) are not a complete data set for the purposes of this analysis, some insight can be gained by applying this simple model to their data and the data on macrophyte biomass obtained from Cole (1973). Figure 7.39 summarizes the results. For a gross average primary production of 15 g/m^2 · day, a depth of 20 cm, the average gross DO production is about 75 mg DO/l · day, rather high. At

$$S_p = 0.0037p_a \quad \text{and} \quad S_N = 0.037p_a$$

the phosphorus and nitrogen uptake rates are 0.28 mg p/l · day and 2.8 mg N/l · day. Cole (1973) indicates a "recovery area" at about 12 km or a travel time of about 0.7 day at an average velocity of 20 cm/s. Figure 7.39 shows the region upstream of the effluent stream to be deficient in phosphorus. Downstream of the effluent stream however, the nitrogen/phosphorus ratio drops considerably. The two calculated slopes are shown, and as indicated, the tendency is for nitrogen to control the area of recovery. At 12 km, phosphorus is calculated at 0.5 mg p/l (using the above uptake rate) which is about 20 times a limiting phosphorus concentration of 0.025 mg p/l. The nitrogen is calculated at 0.4 mg N/l, only about three times a limiting nitrogen condition of 0.125 mg N/l. Indeed if Cole (1973),

TABLE 7.15 PHOSPHORUS AND NITROGEN UPTAKE RATES BY BENTHIC PLANTS

S_p (mg p/m^2 · day)	S_N (mg N/m^2 · day)	S_N/S_p	Remarks	Reference[a]
930	2800	3.0	Treatment ditch by macrophytes	1
87	160	1.8	Quoted in (1)	
146	1100	7.5	Quoted in (1)	
37	238	6.4	Quoted in (1)	
149–381	11,200–17,500	54.0	Quoted in (1)	
—	2300	—	Benthic algae Truckee River	2

[a]References:
(1) Zauke et al. (1982)
(2) O'Connell and Thomas (1965)

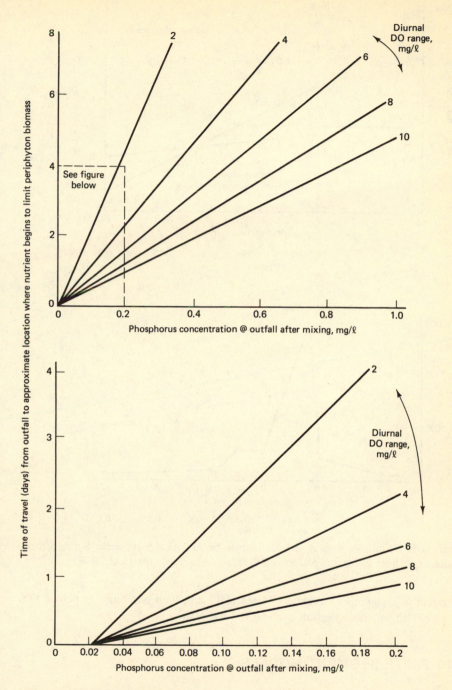

Figure 7.38 Variation of time to travel to limiting phosphorus concentration for various phosphorus concentrations at outfall and diurnal DO ranges.

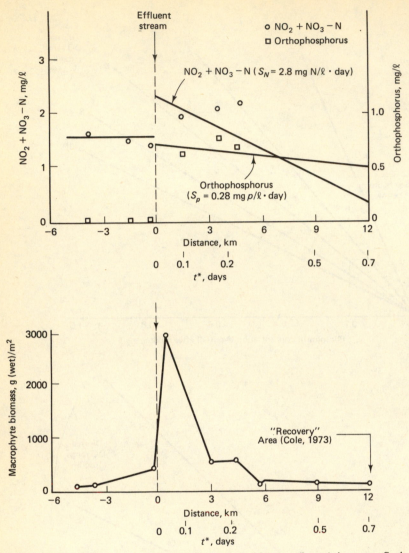

Figure 7.39 Nutrients (upper) and macrophyte biomass (lower) in upper Spring Creek. Nutrient data from Elliott and McDonnell (1982), biomass data from Cole (1973).

is correct in assigning a "recovery" area at 12 km, then one might conclude the limiting nitrogen concentration is about 0.4 mg N/l.

7.10 FINITE SEGMENT MODELS

Finite segment eutrophication models are appropriate for those cases where geometry, advective and dispersive transport, water temperature, and light extinction coefficients vary spatially throughout the water body. These models have been used to calculate phytoplankton concentrations in river, lakes, and estuaries, utilizing varying degrees of kinetic sophistication. One readily available computer program that permits the user to specify desired phytoplankton–nutrient kinetic interactions in a finite difference framework is called WASP, *Water Quality Analysis Simulation Program* (Di Toro et al., 1983). An example of the kinetic interactions employed in an estuarine model (Fig. 7.40a) of the Western Delta–Suisun Bay area of the Sacramento–San Joaquin Rivers in California is shown in Fig. 7.40b (HydroQual, 1981). Eleven systems, or state variables (shown as the square boxes) are modeled, including three organic parameters (carbon, nitrogen, and phosphorus), six inorganic systems (chlorides, dissolved oxygen, ammonia nitrogen, nitrite + nitrate nitrogen, orthophosphate phosphorus, and silica) and two biological compartments (phytoplankton and zooplankton). Direct sources of the systems are indicated by solid arrows, kinetic rates by circles, and directions of reactions are shown by the light arrows. Recycle of dead phytoplankton and zooplankton into the organic carbon, nitrogen, and phosphorus pools is included with provision for sedimentation of fractions of the pool into the bed of the estuary. Four nutrients—ammonia, nitrite + nitrate nitrogen, orthophosphate, and silica—are seen to directly affect the phytoplankton growth rate, in addition to water temperature and solar radiation. Phytoplankton concentrations decrease due to death (D_p) and grazing by zooplankton (C_g). Dissolved oxygen is seen to be utlized by organic carbon (BOD), in the nitrification process (K_N) and by algal respiration (R). It is produced through algal photosynthesis and is affected—through its saturation concentration—by ambient chloride concentrations.

The governing differential equation for each system incorporates advective and dispersive transport terms which inter-relate concentrations in all adjoining segments (see, for example, Eq. 3.83). These transport terms will be represented by $J(s_i)$ subsequently and will be understood to encompass all advective and dispersive transport terms for a segment, for a specific system i. The nonlinear set of differential equations for the eleven systems follows. Segment subscripts are omitted since the kinetic terms only apply to the concentration in the segment for which the equation is being written.

Phytoplankton (s_1)

$$V \frac{ds_1}{dt} = J(s_1) + V(G_P - D_P)s_1 - VC_g s_1 s_2 + W_1 \qquad (7.90)$$

The second term represents the net growth rate of phytoplankton whereas the third term accounts for grazing by the higher order zooplankton, which is dependent on the grazing rate C_g and the ambient levels of zooplankton and phytoplankton. W is a direct source of phytoplankton (e.g., discharge from a treatment plant algal pond).

Zooplankton (s_2)

$$V \frac{ds_2}{dt} = J(s_2) + V(C_g s_1 - D_Z)s_2 \qquad (7.91)$$

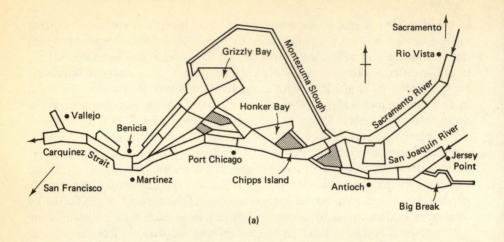

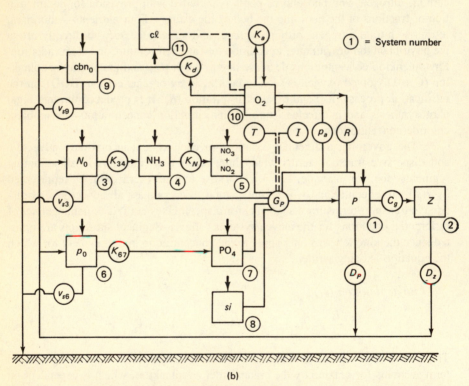

Figure 7.40 (a) Segmentation and (b) kinetics in eutrophication model of Western Delta—Suisun Bay region of the Sacramento-San Joaquin Rivers, California. (a) From Hydroscience, (1974). (b) From HydroQual (1981).

462

Zooplankton grow due to uptake of phytoplankton and decrease due to their own respiration (D_Z). Zooplankton concentrations are represented by the organic carbon equivalence.

Organic Nitrogen (s_3)

$$V \frac{ds_3}{dt} = J(s_3) - VK_{34}s_3 + Va_{13}D_Ps_1 + Va_{23}D_Zs_2 - v_{n3}As_3 + W_3 \quad (7.92)$$

Organic nitrogen decreases as it converts to ammonia (second term) and settles to the estuary sediment (fifth term) and increases due to recycle of dead phytoplankton (third term) and zooplankton (fourth term) and direct sources (W_3). The stoichiometric ratios of cell nitrogen to chlorophyll (phytoplankton) and zooplankton carbon (s_2) are represented by a_{13} and a_{23}.

Ammonia Nitrogen (s_4)

$$V \frac{\partial s_4}{\partial t} = J(s_4) - Va_{14}G_Ps_1 - VK_Ns_4 + W_4 \quad (7.93)$$

The second term accounts for the uptake of inorganic nitrogen by phytoplankton a major fraction of which is assumed to preferentially come from the ammonia nitrogen. Ammonia also decreases due to conversion to nitrate nitrogen and increases due to direct sources.

Nitrite + Nitrate Nitrogen (s_5)

$$V \frac{\partial s_5}{\partial t} = J(s_5) - Va_{15}G_Ps_1 + VK_Ns_4 + W_5 \quad (7.94)$$

Nitrate nitrogen is also used as an inorganic nitrogen source by phytoplankton (second term) and increases due to conversion of ammonia nitrogen (third term). No decay mechanisms (denitrification) of nitrate nitrogen are included in this formulation.

Organic Phosphorus (s_6)

$$V \frac{\partial s_6}{\partial t} = J(s_6) - VK_{67}s_6 + Va_{16}D_Ps_1 + Va_{26}D_Zs_2 - v_{n6}As_6 + W_6 \quad (7.95)$$

Conversion to inorganic phosphorus (second term) and net sedimentation (fifth term) decrease organic phosphorus concentrations whereas recycle of phytoplankton and zooplankton (third and fourth terms) and direct sources cause increases.

Orthophosphate Phosphorus (s_7)

$$V \frac{\partial s_7}{\partial t} = J(s_7) - Va_{17}G_Ps_1 + W_7 \quad (7.96)$$

Silica (s_8)

$$V\frac{\partial s_8}{\partial t} = J(s_8) - Va_{18}G_Ps_1 + W_8 \tag{7.97}$$

Simplified orthophosphate and silica systems, with no removal mechanisms other than algal uptake, are seen to be used.

Organic Carbon (s_9)

$$V\frac{\partial s_9}{\partial t} = J(s_9) - VK_ds_9 + Va_{19}D_Ps_1 + Va_{29}D_Zs_2 - v_{n9}As_9 + W_9 \tag{7.98}$$

Organic carbon (BOD) decreases due to stabilization through uptake of dissolved oxygen (K_d) and sedimentation (fifth term) and is augmented by recylced phytoplankton and zooplankton cell carbon and direct sources. The stoichiometric constant a_{29} is set to unity in this case since zooplankton is expressed as its organic carbon equivalent.

Dissolved Oxygen (s_{10})

$$V\frac{\partial s_{10}}{\partial t} = J(s_{10}) + VK_a(c_s - s_{10}) - Va_{10,9}K_ds_9 - Va_{10,4}K_Ns_4$$
$$+ Va_{10,1}(G_P - D_P)s_1 - S_BA \tag{7.99}$$

DO concentrations are increased by reaeration (second term) and net photosynthesis minus respiration (fifth term) and decrease due to BOD stabilization (K_d), nitrification (K_N), and SOD (S_B). As usual, c_s is the DO saturation concentration and $a_{10,4}$ equals 4.57 mg DO consumed per mg ammonia nitrogen stabilized.

Chlorides (s_{11})

$$V\frac{\partial s_{11}}{\partial t} = J(s_{11}) + W_{11} \tag{7.100}$$

Chloride concentrations are used primarily to validate advective and dispersive transport coefficients as well as establish DO saturation concentrations in conjunction with water temperatures.

Results of a calculation of the annual cycle for all 11 parameters are shown in Fig. 7.41 for one segment (a shallow embayment) of the 37-segment model [Fig. 7.40(a)] of the Western Delta–Suisun Bay (Hydroscience, 1974). Maximum phytoplankton concentrations (s_1) for this single species model are calculated to peak in June–July at approximately 65 μg/l chlorophyll a, whereas observations (indicated by triangles) show a maximum level of approximately 95 μg/l later in the early fall. Zooplankton (s_2) is seen to lag the phytoplankton during the algal high growth period. Ammonia (s_4) and nitrate nitrogen (s_5), as well as silica (s_8), are essentially depleted when algal concentrations are maximum, whereas organic nitrogen (s_3), organic phosphorus (s_6), and organic carbon (s_9) are calculated to increase during this period due to recycle of dead phytoplankton and zooplankton.

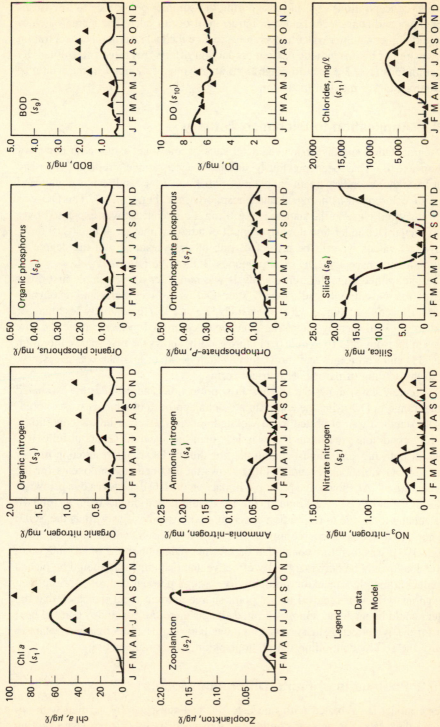

Figure 7.41 Example of eutrophication model calibration—Grizzly Bay, 1970. Western Delta—Suisun Bay region of Sacramento—San Joaquin Rivers, California. From Hydroscience (1974).

Chloride concentrations (s_{11}) increase during the summer period due to low fresh-water flows and consequent landward intrusion of ocean salts. The increased chlo-rides and higher summer water temperatures reduce the DO saturation concentration whereas higher photosynthetic oxygen production offsets this effect. Resulting DO concentrations (s_{10}) are seen to range from approximately 7 mg/l in the winter to approximately 5.5 mg/l in the summer.

7.10.1 Application to Wicomico Estuary

The segmentation and dispersion characteristics of this small estuarine tributary of Chesapeake Bay has been previously discussed (see sample Problem 3.6). The input from the secondary treatment plant of Salisbury, Maryland results in elevated levels of chlorophyll *a*, nitrogen, and phosphorus, see Fig. 7.42(a). The DO was often measured close to the standard of 5.0 mg/l and concern was expressed over the expected DO under low flow river input conditions, that is, at the 7Q10 flow. If low DO values were to be projected, one of the questions was the degree of additional carbonaceous BOD or nitrogenous BOD to be removed.

A simplified finite difference analysis was constructed with simple first-order kinetics for the nutrients and a steady state DO model (see Salas and Thomann, 1975). Inspection of Fig. 7.42(a) shows TN/Tp and IN/Ip generally less than or equal to $7 \rightarrow 10$ thereby indicating a nitrogen controlled system. (Note the low values in IN from mile 10 to mile 0.) Figure 7.42(b) shows the first-stage calibration of the TN and Tp profiles and shows that a small decay or loss rate was necessary, due to the settling of particulate nutrient forms.

Figure 7.43(a) shows the DO low flow projections and includes the estimated diurnal range in DO using the approximate delta approach of Chapter 6. As noted, the DO standard was projected to be violated under 7Q10 river inflow conditions. For July conditions, the mean DO was less than the standard (at about mile 10), while for both July and August conditions, the diurnal DO range resulted in further projected DO violations. Figure 7.43(b) shows the DO deficit components for the July conditions. As shown, the principal impact on the DO deficit was due primarily to the SOD and to $p_a - R$. Indeed, the maximum deficit downstream at mile 10 is calculated to be due to respiration exceeding photosynthesis as well as the SOD, and not due to the point source inputs of CBOD and NBOD which have no effect at mile 10. It was therefore concluded that further reduction in the oxidizable point source load would be only marginally effective and that emphasis should be placed on reducing the phytoplankton productivity through nutrient control. A simplified TN/Tp analysis after removal of the point source nutrient input indicated that the system could be made phosphorus limited. It was therefore concluded that the next step for the Wicomico Estuary was to remove phosphorus at the treatment plant to control the DO by controlling the phytoplankton primary productivity.

7.10.2 Postaudits of Finite Difference Models

After a model is calibrated with one data set, and subsequently verified with ad-ditional data sets, for present conditions, the model is used to project resulting

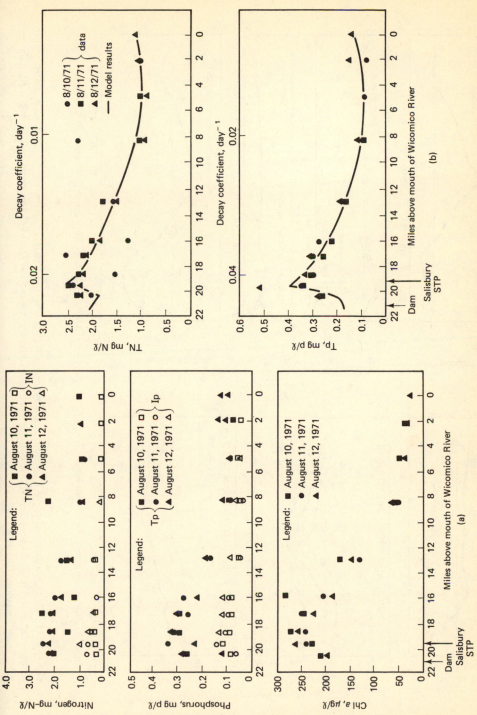

Figure 7.42 (a) Data from state of Maryland on nitrogen, phosphorus, and chlorophyll *a* for Wicomico Estuary. (b) Calibration of TN and Tp using finite difference model. (See Sample Problem 3.6 for segmentation.) From Salas and Thomann (1975).

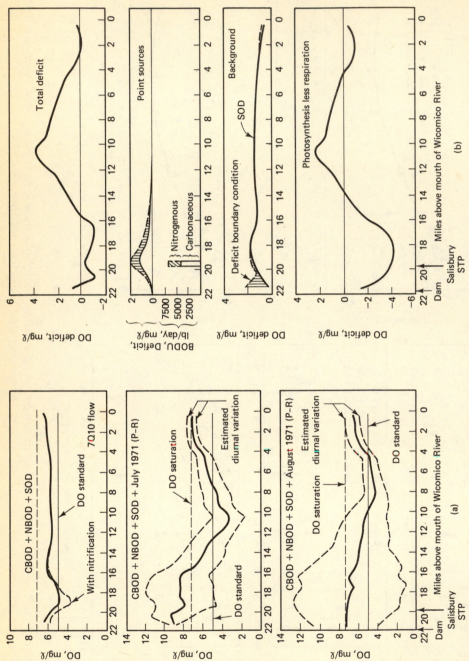

Figure 7.43 (a) DO projections for Wicomico Estuary under 7Q10 (top: effect of CBOD, NBOD, and SOD; middle: above + photosynthesis and respiration of July 1971; bottom: same as middle for August 1971). (b) Components of DO deficit for July 1971. From Salas and Thomann (1975).

water quality for future waste discharges under design low flow conditions. It is a matter of great interest to ascertain the degree with which the model adequately predicts future water quality. Postauditing consists of gathering data sets in the water body, ideally at design flow, after significant changes in the waste discharges have occurred, and comparing these data with the model results obtained using transport and kinetic coefficients determined previously. Postaudits have already been presented for streams in Chapter 6 (Fig. 6.29). Estuarine and lake postaudits will be discussed next.

7.10.2.1 Potomac Estuary A considerable effort has been expended for almost 20 years toward understanding the eutrophication process in the Potomac Estuary occasioned by the discharge of nutrients from point and non-point sources. Freudberg (1985) has provided an excellent review of these efforts and the implications of the modeling work for future management decisions. The effort began in the late 1960s with modeling efforts that included linear nutrient–phytoplankton interactions (Jaworski et al., 1971) followed by dynamic nonlinear phytoplankton modeling (Thomann and Fitzpatrick, 1982). Partly as a result of the early work, a phosphorus reduction program was begun. Figure 7.44(a) shows the reduction in phosphorus point source loadings to the estuary over time. Nitrogen loadings rose to levels of 55,000–60,000 lb TN/day in 1970 and have essentially remained in that range to 1983.

In the late 1970s, as a result of an algal bloom in 1977, an intensive effort was begun to update and expand the modeling effort. The effort was particlarly relevant since the phosphorus removal program was in place although achievement of the effluent allowable phosphorus levels (0.2 mg p/l) had not been fully reached by 1977. The analysis of previous data from the late 1960s (specifically 1966, 1968, 1969, and 1970) was used for calibration of the Potomac Eutrophication Model (PEM). The later data (1977, 1978, 1979) were used for verification (see Fig. 7.45). A full discussion of the PEM is given in Thomann and Fitzpatrick (1982). Analyses of the data indicated that the estuary appeared to be nitrogen limited downstream of the major point source inputs. Phosphorus removal, however, was believed to be feasible to force the estuary to be phosphorus limited by driving the phosphorus low enough so that phosphorus would limit growth instead of nitrogen.

After phosphorus removal was well underway, a major phytoplankton bloom occurred in 1983 causing considerable concern. The 1983 bloom presented, however, from a scientific point of view, an opportunity to postaudit a ''state of the art'' eutrophication model under considerably changed conditions than those used in its development.

Figure 7.44(b) shows the segmentation for the PEM. Sediment segments were included under each water quality segment but the details of sediment chemistry were not included and sediment nutrient fluxes were calibrated to observed in situ flux measurements. The calibration and verification of the PEM was considered suitable for use in analysis of alternatives. Model uncertainty was incorporated using empirical relationships.

Phosphorus removal continued apace and by 1983, point source loadings had

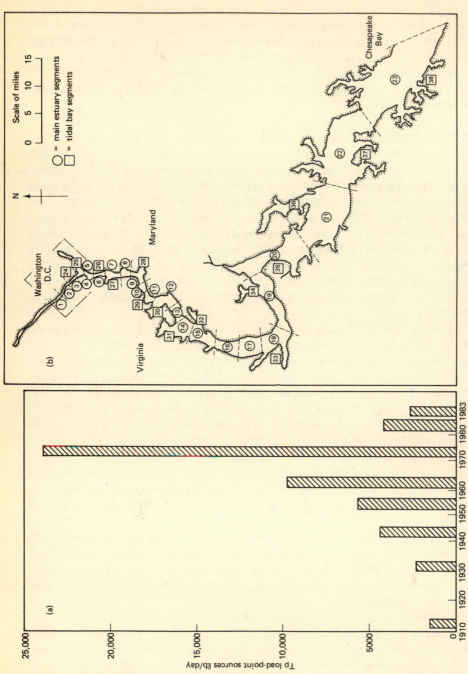

Figure 7.44 (a) Point source total phosphorus loadings for selected years. (b) Segmentation for Potomac eutrophication model (PEM). (a) From Jaworski et al. (1971); Thomann et al. (1985). (b) From Thomann and Fitzpatrick (1982).

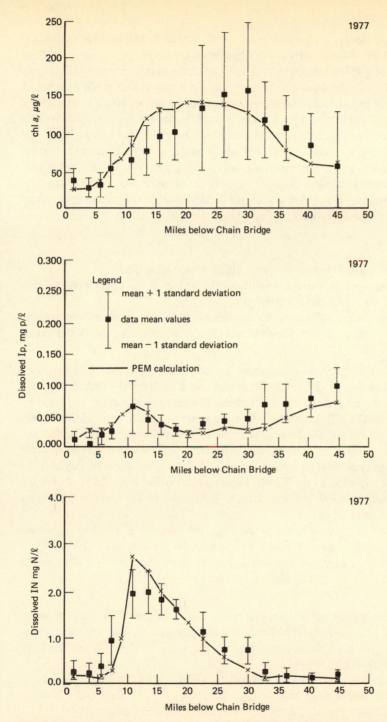

Figure 7.45 Summer mean chlorophyll (top), dissolved inorganic phosphorus (middle), and nitrogen (bottom) in Potomac Estuary. From Thomann and Fitzpatrick (1982).

been reduced to loadings comparable to the 1930s (about 2100 lbs p/day vs. 24,000 lb p/day in 1970). Average effluent concentrations were about 0.4 mg p/l. Then in the summer of 1983, a major bloom developed of the blue-green alga *Microcystis aeruginosa*. *Microcystis* does not fix atmospheric nitrogen and tends to form in scumlike mats at the surface. The bloom extended from about 20 mi in the central estuary and coves and extended for a period from July through September. MWCOG (1984a) have reviewed the 1983 data in detail and showed that chlorophyll *a* levels in the main channel reached about 300 μg/l while in the embayments, levels reached almost 800 μg/l. The water was a brilliant green with some scumming.

The recurrence of these massive blooms of nuisance blue-green algae represented an apparent major setback in the remedial program to enhance water quality of the Potomac Estuary. Some of the specific questions that were raised included:

1. Why did the Potomac Estuary exhibit a bloom in 1983?
2. If there were excessive nutrients in the waters of the upper Potomac Estuary, what was their source?
3. Is there a need to increase the wastewater treatment requirements further?
4. Were existing mathematical models capable of predicting the 1983 bloom?

To help answer the above and other scientific questions, the PEM was used extensively (Thomann et al., 1985). In applying the unchanged PEM to the 1983 conditions, the model tracked the onset of the bloom up to about 100 μg chl/l by the end of July but then failed to reproduce the further intensification of the bloom. The principal reason for the failure of PEM in August–September was that the model ran out of available phosphorus. Sensitivity analyses appeared to indicate that there was an additional phosphorus source of about 4000–8000 lb/day in 1983 which was not included in the PEM. Figure 7.46 (Freudberg, 1985) shows the effect of including a phosphorus source of 4000 lb p/day for mid-July to September in the region of mile 20–40.

As a result of the evaluation of the data, the inability of PEM to reproduce the bloom in August–September and an apparent phosphorus source, a series of hypotheses was evaluated for possible explanations of the observed phenomena (Thomann et al., 1985). It was concluded that:

1. The most likely mechanism that resulted in an increased phosphorus input, thereby increasing the bloom, was an enhanced aerobic sediment phosphorus release brought about by high pH in the overlying water (Di Toro and Fitzpatrick, 1984).
2. Potential contributory mechanisms included:
   a. The longitudinal and vertical transport of particulate and dissolved nutrients resulting from downstream estuarine circulation.
   b. Periodic bioturbation of the sediments with associated enhanced aerobic release of phosphorus.
   c. A reduced settling of particulate phosphorus forms.

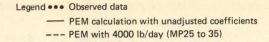

Legend ●●● Observed data
——— PEM calculation with unadjusted coefficients
--- PEM with 4000 lb/day (MP25 to 35)

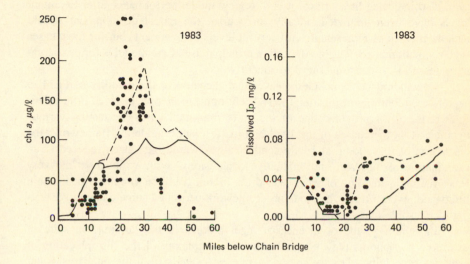

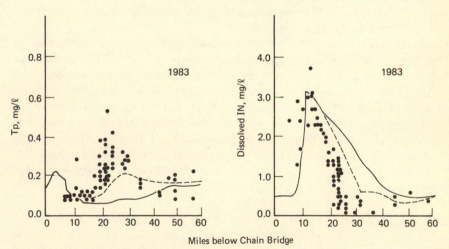

Miles below Chain Bridge

Figure 7.46 August 28, 1983 bloom conditions in Potomac Estuary compared with PEM simulations. From Freudberg (1985).

The failure of PEM to capture the bloom intensification in August and September is a clear example where mechanisms not included in a model formulation apparently become significant after some level of treatment had been installed. The mechanisms that appear to have been most significant included the chemistry of aerobic sediment phosphorus release, the pH–alkalinity–photosynthesis chemistry, and the two-dimensional lower estuary circulation and potential "nutrient trapping" effect.

The entire Potomac estuary analysis from calibration through the 1983 postaudit also indicated how critical it is to review model performance after treatment levels have been upgraded. Ideally, of course, the model development should include conditions representative of several levels of treatment, but for most cases such a situation is not feasible. Indeed, a principal use of the model is for prediction purposes prior to construction of treatment works.

One should not conclude however that, because of the inability of PEM to capture the bloom intensification in 1983, eutrophication models are of little value. Rather, the PEM provided one of the more important bases for testing various hypotheses on the causes of the 1983 bloom. Without a modeling framework, the discussion of that bloom and its potential causes would have been reduced to an examination of the data, and qualitative conclusions. With a model, at the very least, the understanding of what caused the bloom could be circumscribed within limits and, even further, the modeling framework permitted some quantitative testing of plausible hypotheses.

In summary, the occurrence of algal blooms in the Potomac in spite of extensive phosphorus reduction indicates the complexity of the eutrophication of estuaries especially when the region of interest extends from the tidal freshwater portion through a transition zone and into the fully developed estuary. The movement from a phosphorus to nitrogen limited system makes the estuary considerably more complex than the lake environment.

Much was learned from the 1983 experience of the strengths and weaknesses of the PEM, especially since the analysis apparently represented the first time an estuarine eutrophication model has been tested after nutrient removal had been installed. The experience also indicates the need to carefully consider the justification for excluding mechanisms, if they are known, from the model formulation. Finally, the experience also indicates that some mechanisms are "exposed" to view only after treatment levels have been upgraded and, therefore, any modeling framework may still be subject to the uncertainty of some "missed mechanism." The hope is that from experiences such as the 1983 bloom in the Potomac, such possibilities will eventually be eliminated.

7.10.2.2 Lake Erie   Lake Erie was one of the first major water bodies in the United States to be publicly perceived as eutrophied principally through the increase in the area of anoxia in the Central Basin of the lake (see Fig. 7.27). Di Toro and Connolly (1980) reported on the development, calibration, and verification of a detailed eutrophication model of Lake Erie. Figure 7.47 shows the segmentation of the Lake as used in the Di Toro-Connolly model. This time variable model incorporated all of the kinetic interactions discussed previously but, most significantly, also included a sediment computation. The inclusion of the diagenesis mechanisms permitted an internal calculation of the SOD thereby linking the SOD to external phosphorus loads. The model was calibrated to 1970 and 1975 data. A relationship was developed between external total phosphorus load and the DO in the hypolimnion of the Central Basin.

Epilimnion segments

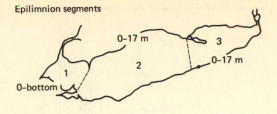

0–17 m

3

2   0–17 m

1

0–bottom

Hypolimnion segments

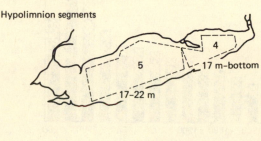

4

5   17 m–bottom

17–22 m

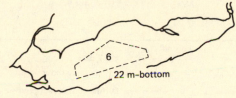

6

22 m–bottom

Sediment segments

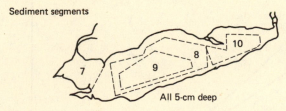

10

8

7   9

All 5-cm deep

Figure 7.47 Segmentation used for Lake Erie eutrophication model. From Di Toro and Connolly (1980).

As a result of extensive investigations by the Great Lakes community, modeling efforts, and implementation of point source nutrient controls, the total phosphorus load to the lake was reduced from a 1972 high of about 25,000 mt/yr to a 1980 load of about 12,000 mt/yr. The International Joint Commission (IJC) target loading to the lake is 11,000 mt/yr. Figure 7.48(a) (from Di Toro et al., 1986) shows the temporal history of the Tp loads from Chapra (1977) and Yaksich et. al. (1982) of the Corps of Engineers.

The response of this large lake to the reduction of Tp input has been a general decline in the area of the anoxia and a general increase in the concentration of the

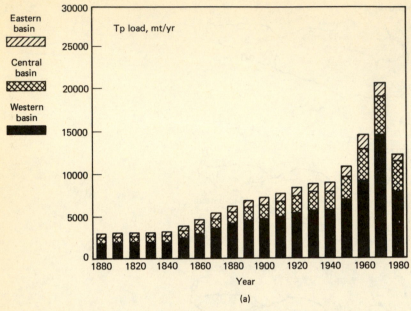

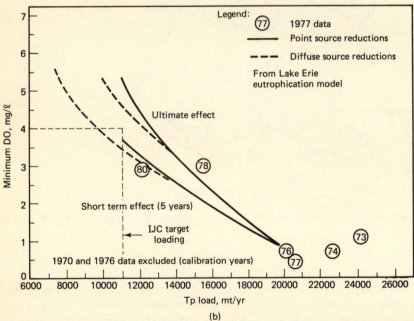

Figure 7.48 (a) Lake Erie total phosphorus loads. (b) Comparison of actual and predicted hypolimnion minimum DO in Central Basin for various total phosphorus external loads. (a) From DiToro et al. (1986), Chapra (1977), Yaksich et al. (1982). (b) From DiToro et al. (1986). Reprinted by permission of International Association for Great Lakes Research.

minimum DO in the hypolimnion of the Central Basin [Fig. 7.48(b)]. Significantly, as summarized by Di Toro et al. (1986), the eutrophication model originally constructed for Lake Erie successfully predicted the area and magnitude of the hypolimnion anoxia as a function of the external total phosphorus load. The original model, calibrated in two years, was subsequently run for 10 years without change of the model structure. The actual loads were used and the results were remarkably representative of the observed data especially the minimum hypolimnion DO. The model success is made especially clear from Fig. 7.48(b) which shows the comparison of the observed DO data for the years when the Tp load was reduced to the calculated trend from the Lake Erie model. It was concluded from this work that the IJC target of 11,000 mt Tp/yr was an appropriate goal in order to reach minimum DO values of about 4 mg/l in the hypolimnion and that the model provided a firm basis for decision making on the reduction of nutrients to Lake Erie.

7.11 EUTROPHICATION CONTROL TECHNIQUES

The control of the eutrophication of various water bodies can take a variety of forms. Perhaps such forms can best be seen by once again reviewing the mass balance equations for phytoplankton chlorophyll and a single nutrient, such as phosphorus. Thus, for a completely mixed system, Eq. 7.20 can be repeated here for the phytoplankton

$$V \frac{dP}{dt} = V(G_P - D_P)P - Av_sP + Q_{in}P_{in} - QP \tag{7.101}$$

and Eq. 7.43 with inputs and flow can be used for the nutrient, that is,

$$V \frac{dp}{dt} = W + Q_{in}p_{in} + W_s - a_p(G_P - D_P)P - v_{s1}Ap - Qp \tag{7.102}$$

where p is representative of the controlling nutrient and W_s represents internal sources of the nutrient such as the nutrient input from the sediment. The engineering control techniques essentially aim at changing the components of these equations in the direction of decreasing plant biomass. As with engineering controls for other water quality problems such as the DO problem, principal interest first centers on control of the inputs, in this case, the nutrient inputs. It can be recalled, however, that there are really three principal control areas:

1. Control of inputs.
2. Alteration of system kinetics.
3. In-stream treatment and flow control.

For the eutrophication problem, then, the engineering techniques include:

1. Reduction or elimination of external nutrient inputs (W).

2. Reduction of internal sources and/or concentrations and/or cycling of nutrients [W_s, v_s, p (and G_P) and D_P].
3. Acceleration of nutrient transport through the system (QP and Qp) and selective withdrawal.
4. Direct biological and chemical control of aquatic plants.

Each of these control techniques have relative advantages and disadvantages when applied to a specific problem context.

7.11.1 Reduction of External Input

There are five means of reducing the direct input of nutrients (Fig. 7.49):

1. Wastewater treatment of municipal and industrial point sources and CSO, including phosphorus and nitrogen removal by physical, chemical, and biological treatment processes.
2. Alteration of non-point nutrient inputs through land conservation practices and changes in agricultural use of nutrients. These practices would include reduction of erosion, testing of soil before fertilizer application, reduction in winter fertilizing, and use of buffer strips.
3. Diversion of a nutrient input to a different water body where the impact is less.
4. Treatment and control of a nutrient at the inflow point including treatment of the entire river inflow and building of upriver detention basins or storage reservoirs.
5. Modification of product to reduce generation of nutrients at the source (e.g., nonphosphate detergents).

Point source phosphorus and nitrogen reduction is accomplished by (a) chemical additions and (b) biological removal or (c) physical processes. Addition of alum [$Al_2(SO_4)_3 \cdot 14H_2O$], for example, will precipitate aluminum phosphate according to

$$Al^{3+} + PO_4^{3-} \rightarrow AlPO_4 \downarrow \qquad (7.103)$$

(Steel and McGhee, 1979). Alum or sodium aluminate ($Na_2O \cdot Al_2O_3 \cdot xH_2O$) may provide the aluminum source. Equation 7.103 indicates a theoretical molar ratio of Al/p of 1 to 1. Other chemicals for phosphorus removal include iron compounds (ferris chloride, pickle liquor) and lime.

Other point source reductions of Tp and TN include physical processes such as NH_3 stripping, biological treatment in stabilization lagoons, treatment by harvesting of algae or rooted plants that have taken up nutrients, and land application. Table 7.16, abstracted from Metcalf and Eddy (1972), shows some ranges of removal efficiencies for different removal processes.

Alteration of non-point nutrient inputs through changes in agricultural practices critically involves obtaining the cooperation of farmers in the control program. Financial incentives and benefits to farm productivity are the keystones for the

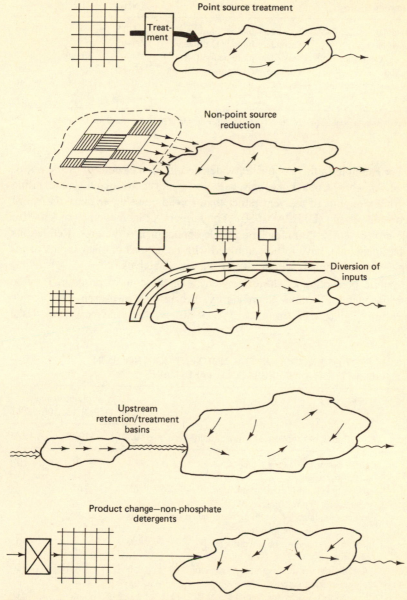

Figure 7.49 Five environmental controls for reducing nutrient inputs.

TABLE 7.16 SOME REMOVAL EFFICIENCIES OF PHOSPHORUS
AND NITROGEN IN POINT SOURCES

Treatment	Wastewater treatment	Percent removal of		
		NH_3	NO_3	PO_4
Chemical precipitation	Secondary biological effluent	5–15	—	90–95
Air stripping of ammonia	Secondary biological effluent	85–98	—	—
Bio nitrification/Denitrification	Secondary biological effluent	—	60–95	—
Land application	Primary and secondary effluent	60–80	5–15	60–90
Stabilization lagoon-harvesting of algae	Secondary biological effluent	50–90	50–90	≈ 50

Source: Abstracted from Metcalf and Eddy, Inc., (1972) with permission of McGraw-Hill Book Company.

program. For example in the Cobbossee watershed in Maine, control of the eutrophication of Annabessacook Lake was partially achieved by manure stockpiling programs for six months of the year rather than spreading of the manure on frozen or snow-covered ground (USEPA, 1980). The farmers' costs ranged from $500 for small scale projects to $20,000 for larger scale containment structures. Reductions in non-point inputs may have to be combined with reductions in other inputs or in-lake treatment to accomplish specific water quality objectives.

The public perception of the effectiveness of the control program is often focused on the increase in lake water clarity. Secchi disk measurements provide such a measure. Table 7.17 from USEPA (1980) for the Annabessacook Lake

TABLE 7.17 EFFECT OF ENVIRONMENTAL NUTRIENT CONTROLS ON
VISIBILITY OF ANNABESSACOOK LAKE, MAINE

Secchi disk depth (m)	Public perception of water quality	Days at given visibility (June 1–Sept 15; 106 days)		
		1972	1977	1979
0–0.9	Gross pollution; lake is totally unusable for recreation	43	10	0
1–1.9	Algae blooms still evident; quality unacceptable for most uses	63	67	0
2–2.9	Some complaints of declining water quality; some impairment of water use	0	28	30
3–3.9	Satisfactory quality; no impairment of water use	0	1	40
4–4.9	Excellent water quality; a positive factor encouraging lake use	0	0	35
5 +	Exceptional quality for this lake	0	0	1

Notes on selected years:
 1972—prior to full diversion of municipal/industrial wastewater
 1977—prior to lake's restoration project.
 1979—after agricultural waste controls and nutrient inactivation treatment by sodium aluminate.
Source: From USEPA (1980)

restoration project shows the increase of secchi depth with implementation of various control measures including treatment, diversion, agricultural control, and in-lake nutrient inactivation by chemical precipitation.

Preimpoundment basins have been used to reduce nutrient inputs especially from difficult to control non-point sources. The retention basins essentially act as settling basins of particulate nutrient forms either in the inorganic or organic phases. Benndorf et al. (1976) in summarizing experience with such basins in the German Democratic Republic indicated removal efficiencies of orthophosphate ranging from 10–30% for detention times of less than 2 days to removal efficiencies of 30–60% for detention times of 2 to 14 days.

7.11.2 Reduction of Internal Sources and/or Cycling

The principal control options under this approach include (Fig. 7.50):

1. Dredging of nutrient rich sediments.
2. Hypolimnetic aeration.
3. Chemical precipitation of nutrients.

In addition, sealing of sediments to reduce nutrient exchange and lake or reservoir drawdown have also been utilized.

Dredging of sediments that accumulated from discharges has had significant

Figure 7.50 Techniques for internal control of eutrophication problems.

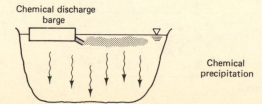

use especially in lake restoration projects. Dredging often accompanies input source control where in some instances even after complete diversion of inputs, the lake did not respond over a sufficient time unless nutrient rich sediments were removed. Peterson (1979) has reviewed the various aspects both pro and con of dredging for lake restoration. Impacts on the bottom sediments are particularly important. A variety of dredging equipment is in use including bucket dredges, drag line dredges, and suction dredges.

A good example of the effect of lake sediment dredging is the case of Lake Trummen in Sweden (see for example, Bengtsson and Gelin, 1975). This lake received wastewater inputs from 1936 to 1958. The wastewater was diverted in 1958 but the lake did not recover due to a 30–40 cm thick sediment layer. Approximately 0.5 m of sediment was removed by suction dredging. Bengtsson and Gelin (1975) indicate that the blue-green biomass declined by about 90% in the two years following the dredging. Secchi depths increased from summer means of 20 to 71 cm. Note the contrast, however, between these depths and those in Table 7.17 for Annabessacook Lake in Maine. Table 7.18 from Peterson (1979) shows the range of costs for dredging of lake sediments.

Aeration of the hypolimnion can be accomplished by aerators that withdraw water from the hypolimnion, mix the water with aerated water, and then release the aerated water back to the hypolimnion. These types of aerators maintain stratification which in some cases is desirable (e.g., maintenance of a stratified fish population). Klapper and Glass (1976) indicate that deep water aerators can obtain a maximum oxygen input to the hypolimnion of about 0.7 kg O_2/hr · kw power. Other fan-type pumps mix the lake contents and induce destratification (e.g., Strecker et al., 1977).

Chemical precipitation of nutrients involves the same principles as chemical removal in wastewater treatment plants. In this case, aluminum sulfate (alum) or other aluminum compounds are dosed directly to the water body. Such actions have apparently been applied exclusively to lakes. In some instances sodium aluminate has been used (USEPA, 1980) to counteract any changes in pH resulting from the use of alum. Jar tests for optimum dosage and bioassays shold be conducted for the specific water body. Table 7.17 shows the improvement in water clarity from sodium aluminate dosing.

TABLE 7.18 MEAN DREDGING COSTS FOR LAKE RESTORATION PROJECTS IN THE UNITED STATES[a]

Geographic location	No. of projects considered	Cost range ($/m^3)	Mean cost ($/m^3)
Great Lakes area	5	0.27– 2.96	1.34
Northwest	3	1.64– 3.16	2.36
Central states	2	2.35– 2.63	2.49
Northeast	7	0.89–15.70	5.63

[a]Costs representative of approximately mid–late 1970s.

Source: From Peterson (1979).

7.11.3 Acceleration of Nutrient Outflow

Selective withdrawal makes use of those strata in a water body with highest quality of water to meet a specific use, such as drinking water. Monitoring of the reservoir quality is coupled to selective withdrawal.

Nutrient outflow and subsequent reduction in phytoplankton growth can be accomplished by the introduction of external sources of dilution water to the eutrophied system. By reducing the detention time and diluting nutrient concentrations, the time for phytoplankton growth is reduced and the amount of nutrients for conversion to phytoplankton is also reduced by dilution. Welch (1981) presents the results of two such dilution control techniques for Green and Moses Lakes in the State of Washington. Table 7.19 shows some of the results from Welch (1981).

7.11.4 Direct Biological and Chemical Control

This category of control aims directly at the nuisance plants through some type of biological control or manipulation or through direct application of weedicides such as copper sulfate. Schuytema (1977) has summarized some of the possible biological controls including:

1. Control by grazing and predation through introduction of predators such as tilapia and carp.
2. Use of plant pathogens.
3. Biomanipulation through changes in grazing, competition, and fish-stocking.

Except for somewhat individualized cases, biological control has not been demonstrated, on any widespread basis, to be a particularly effective eutrophication control approach. The exceptions include use of fish grazers in the tropics and some European countries.

Copper sulfate has been widely used for control of algae and, if properly applied so that dosages are not toxic to indigenous fish, can be effective temporary

TABLE 7.19 EFFECTIVENESS OF USE OF DILUTION FLOW ON EUTROPHICATION OF LAKES

	Secchi depth (m)	Summer chl a (μg/l)	Tp (μg/l)
	Green lake		
Before use of dilution water	1	45	65
After use of dilution water	4	3	20
	Moses lake		
Before use of dilution water	0.9	45	150
After use of dilution water	1.5	21	86

Source: From Welch (1981).

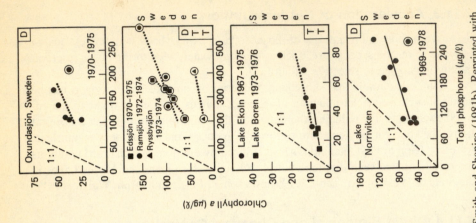

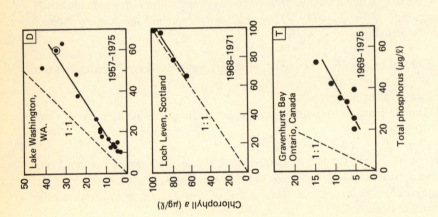

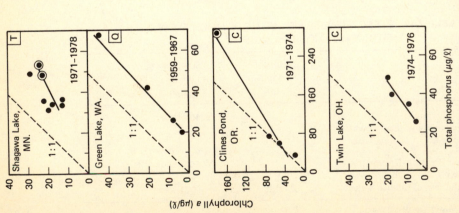

Figure 7.51 Phosphorus dependence on chlorophyll *a* in individual lake restorations. Adapted from Smith and Shapiro (1981b). Reprinted with permission of Environmental Science and Technology. Copyright 1981, American Chemical Society.

means to control algae. Water clarity usually improves rapidly but applications must be repeated at regular intervals. Care must be taken to recognize the impact on DO of the relatively rapid production of organic matter from the dead algae.

The effectiveness of reducing nutrient concentrations in lakes—using several of the techniques discussed above—is demonstrated by the plots in Fig. 7.51 for 13 lakes and one embayment (Smith and Shapiro, 1981b). Chlorophyll *a* concentrations are plotted vs. total phosphorus concentrations, both values representing averages of data gathered during the period of April/June to September/October. Total phosphorus concentrations were reduced by a variety of techniques. In Shagawa Lake, Gravenhurst Bay, Ramsjön, Ryssbysjön, Lake Ekoln, and Lake Boren, wastewater treatment to reduce effluent phosphorus levels was used, whereas wastewater diversion accomplished in-lake total phosphorus reductions for Lake Washington, Oxundasjön, and Lake Norrvikon. Chemical precipitation of phosphorus was effectively implemented in Clines Pond using sodium aluminate or zinconium chloride and in Twin Lake with aluminum sulfate. Increased flushing during the summer in Green Lake was accomplished by augmenting the outflow with low-nutrient water from the Seattle, Washington domestic supply. This resulted in a threefold reduction in the lake detention time (1.2 to 0.4 yr) and, thus, reduced in-lake total phosphorus concentrations.

In all 14 water bodies above, lower lake total phosphorus concentrations were accompanied by lower concentrations of phytoplankton, as measured by chlorophyll *a*. It is interesting to note that this appears to be true even for those lakes where nitrogen limitations are probable (Edssjön, Ramsjön, and Ryssbysjön).

REFERENCES

Bachmann, R. W., and D. E. Canfield, Jr., 1979. Role of Sedimentation in the Phosphorus Budget of Natural and Artificial Iowa Lakes, Project A-063-IA, Iowa State University, Ames, IA, 95 pp.

Bannerman, R. T., D. E. Armstrong, R. F. Harris, and G. C. Holdren, 1975. Phosphorus Uptake and Release by Lake Ontario Sediments, NERC, ORD, USEPA, Corvallis, OR, 51 pp. EPA-660/3-75-006.

Bartsch, A. F., and J. H. Gakstatter, 1978. Management Decisions for Lake Systems on a Survey of Trophic Status, Limiting Nutrients, and Nutrient Landings in American-Soviet Symposium on Use of Mathematical Models to Optimize Water Quality Management, 1975, ERL, ORD, USEPA Gulf Breeze, FL, pp. 371–394. EPA-600/9-78-024.

Beeton, A. M., 1958. Relationship Between Secchi Disk Readings and Light Penetration in Lake Huron, *American Fisheries Society Trans*. **87**:73–79.

Bengtsson, L., and C. Gelin., 1975. Artificial Aeration and Suction Dredging Methods for Controlling Water Quality, Proc. Sym. on the Effects of Storage on Water Quality, Reading University, England, pp. 313–342.

Benndorf, J., K. Putz and W. Kroaty, 1976. On the Function of Preimpoundment Basins, Proc. Eutrosym '76. Int. Symp. on Eutrophication and Rehabilitation of Surface Waters, Inst. of Water Management, GDR, Vol. V. pp 22–38.

Burns, N. M., and F. Rosa, 1980. In Situ Measurement of the Settling Velocity of Organic Carbon Particles and 10 Species of Phytoplankton, *Limnology & Oceanography* **25**(5):855–864.

Calendar, E., and D. E. Hammond, 1982. Nutrient Exchange Across the Sediment-Water Interface in the Potomac River Estuary, *Est. Coast. Shelf Sci.* **15**:395–413.

Chapra, S. C., and S. J. Tarapchak, 1976. A Chlorophyll *a* Model and Its Relationship to Phosphorus Loading Plots for Lakes, *Water Res. Res.*, **12**(6):1260–1264.

Chapra, S. C., 1977. Total Phosphorus Model for the Great Lakes, *J. Env. Eng. Div., Proc. Am. Soc. Civ. Eng.* **103**(EE2):147–161.

Cole, R. A., 1973. Stream Community Response to Nutrient Enrichment, *J. Water Poll. Contr. Fed.* **45**(9):1874–1887.

Dillon, P. J., and F. H. Rigler, 1974. The Phosphorus-Chlorophyll Relationship in Lakes, *Limnol. Oceanogr.* **19**(4):767–773.

Dillon, P. J., and F. H. Rigler, 1975. A Simple Method for Predicting the Capacity of a Lake for Development Based on Lake Trophic Status, *J. Fish. Res. Bd. Can.* **31**(9):1519–1531.

Di Toro, D. M., D. J. O'Connor, and R. V. Thomann, 1971. A Dynamic Model of the Phytoplankton Population in the Sacramento-San Joaquin Delta, Advances in Chemistry, No. 106, American Chemical Society, pp. 131–180.

Di Toro, D. M., 1978. Optics of Turbid Estuarine Waters: Approximations and Applications, *Water Research* **12**:1059–1068.

Di Toro, D. M., and W. F. Matystik, Jr., 1980. Mathematical Models of Water Quality in Large Lakes, Part 1: Lake Huron and Saginaw Bay, ERL, ORD, USEPA, Duluth, MN, 166 pp. EPA-600/3-80-056.

Di Toro, D. M., and J. P. Connolly, 1980. Mathematical Models of Water Quality in Large Lakes, Part 2: Lake Erie, ERL, ORD, USEPA, Duluth, MN 231 pp. EPA-600/3-80-065.

Di Toro, D. M., 1981. Algae and DO. Summer Institute in Water Pollution Control, Manhattan College, pp. 3–99.

Di Toro, D. M., J. J. Fitzpatrick, and R. V. Thomann, 1983. Documentation for Water Quality Analysis Simulation Program (WASP) and Model Verification Program (MVP), Env. Res. Lab, ORD, USEPA, Duluth, MN, 145 pp. EPA-600/3-81-044.

Di Toro, D. M., and J. J. Fitzpatrick, 1984. *A Hypothesis to Account for the 1983 Algae Bloom*, Tech. Memo., HydroQual, Inc., 28 pp.

Di Toro, D. M., N. A. Thomas, C. E. Herdendorf, R. P. Winfield, and J. P. Connolly, 1986. A Post Audit of a Lake Erie Eutrophication Model, *J. Great Lakes Research*. In press.

Edwards, R. W., 1962. Some Effects of Plants and Animals on the Conditions in Freshwater Streams with Particular Reference to Their Oxygen Balance, *Int. J. Air Wat. Poll.* **6**:505–520.

Eisenreich, S. J., P. J. Emmling, and A. M. Beeton, 1977. Atmospheric Loading of Phosphorus and Other Chemicals to Lake Michigan, *J. Great Lakes Research, Inter. Assoc. Great Lakes Research*, **3**(3–4):291–304.

Elliott, J. C., and A. J. McDonnell, 1982. A Conceptual Model of Primary Productivity in Shallow Streams Using Biomass Simulation, Res. Proj. Tech. Comp. Report, OWRT Proj. A-003-PA, Inst. for Res. on Land & Water Res., Penn State University, 129 pp.

Elwood, J. W., J. D. Newbold, A. F. Trimble, and R. W. Stark, 1981. The Limiting Role of Phosphorus in a Woodland Stream Ecosystem: Effects of P Enrichment on Leaf Decomposition and Primary Producers, *Ecology* **62**(1):146–158.

Eppley, R. W., 1972. Temperature and Phytoplankton Growth in the Sea, *Fishery Bulletin*. **70**(4):1063–1085.

Fillos, J., and W. R. Swanson, 1975. The Release Rate of Nutrients from River and Lake Sediments, *J. Water Poll Contr. Fed.* **47**(5):1032–1042.

Freedman, P. L., and R. P. Canale, 1977. Nutrient Release from Anaerobic Sediments, *J. Environ. Eng. Div., Proc. Am. Soc. Civ. Eng.* **103**(EE2):233–244.

Freudberg, S. A., 1985. A Post-Audit of the Potomac Eutrophication Model, Met. Wash. Council of Govts. Paper presented at Amer. Soc. Civil Engr. Water Res. and Planning Div. Conf., Buffalo. NY, 12 pp.

Gakstatter, J. H., M. O. Allum, S. E. Dominguez, and M. R. Crouse, 1978. A Survey of Phosphorus and Nitrogen Levels in Treated Municipal Wastewater, *J. Water Poll Contr. Fed.* **50**:718–722.

Goldman, J. C., D. R. Tenore, and H. I. Stanley, 1973. Inorganic Nitrogen Removal from Wastewater: Effect on Phytoplankton Growth in Coastal Marine Waters, *Science* **180**:955–956.

Hartwig, E. O., 1975. The Impact of Nitrogen and Phosphorus Release from a Siliceous Sediment on the Overlying Water, Third Int. Estuarine Conf., Galveston, TX, Paper No. COO-3279-20, 33 pp.

Herdendorf, C. E., 1983. Lake Erie Water Quality 1970–1982: A Management Assessment, CLEAR Tech. Report No. 279, Ohio State University, Columbus, OH.

Higgins, J. M., and B. R. Kim, 1981. Phosphorus Retention Models for Tennessee Valley Authority Reservoirs, *Water Res. Res.* **17**(3):571–576.

Horner, R. R., and E. B. Welch, 1981. Stream Periphyton Development in Relation to Current Velocity and Nutrients, *Canadian J. Fisheries Aquatic Sciences* **38**(4):449–457.

HydroQual, Inc., 1981. Development and Application of the Western Delta-Suisun Bay Phytoplankton Model, Prepared for Dept. of Water Resources, State of California, June 1981, 5 Chapters.

Hydroscience, Inc., 1974. Western Delta and Suisun Bay Phytoplankton Model-Verifications and Projections, by Di Toro, D. M., D. J. O'Connor, and J. L. Mancini, for the California Department of Water Resources, Oct. 1974.

Hydroscience, Inc., 1977a. The Effects of Forest Management on the Water Quality and Aquatic Biota of Apalachicola Bay, Florida, Prepared for the Buckeye Cellulose Corp., Perry, FL, 6 Chapters + 4 Appendixes.

Hydroscience, Inc., 1977b. National Assessment of Water Quality, Prepared for National Commision on Water Quality, Washington, D. C.

Hynes, H. B. N., 1970. *The Ecology of Running Waters*, University of Toronto Press, R. Clay, Ltd., Bungay, Suffolk, England, 555 pp.

International Joint Commission (IJC), 1978. Environmental Management Strategy for the Great Lakes System, Final Report of Int. Ref. Group on Great Lakes Pollution from Land Use Activities, Windsor, Ontario, 115 pp.

Jaworski, N. A., L. J. Clark, and K. D. Fergnis, 1971. A Water Resources-Water Supply Study of the Potomac Estuary, Tech. Rept. No. 35, EPA Water Quality Office, Middle Atlantic Reg., April 1971.

Kevern, N. R., and R. C. Ball, 1965. Primary Production and Energy Relationships in Artificial Streams, *Limnol. Ocean.* **10**(1):74–87.

Klapper, H., and K. Glass, 1976. Technologies of Deep Water Aeration Developed in the GDR and Experiments in Rehabilitation of Lakes, Proc. Eurosym '76, Int. Symp. on

Eutrophication and Rehabilitation of Surface Waters, Inst. of Water Management, GDR, Vol. V pp. 72–89.

Martens, C. S., G. W. Kipphut, and J. V. Klump, 1980. Sediment-Water Chemical Exchange in the Coastal Zone Traced by In Situ Radon-222 Flux Measurements, *Science* **208**:285–288.

Metcalf and Eddy, Inc., 1972. *Wastewater Engineering*, McGraw-Hill, New York, 782 pp.

MWCOG, 1984a. *The Upper Potomac Estuary: A Report on Water Quality Data for 1983*, Final Draft Report, May 1984, Washington, D.C., 8 Chapters + 3 Appendixes.

National Academy of Science and National Academy of Engineering (NAS/NAE), 1972. Water Quality Criteria, a Report of the Committee on Water Quality.

O'Connell, R. L. and N. A. Thomas, 1965. Effect of Benthic Algae on Stream Dissolved Oxygen, *J. San. Eng. Div., Proc. Am. Soc. Civ. Eng.* **SA3**:1–16.

Omernik, J. M., 1977. Non-Point Source-Stream Nutrient Level Relationships: A Nationwide Study, Corvallis ERL, ORD, USEPA, Corvallis, OR, 151 pp. + Plates. EPA-600/3-77-105.

Peterson, S. A., 1979: Dredging and Lake Restoration, Proc. First EPA National Lake Restoration Conf., Aug 22–24, 1978, Minneapolis, MN, pp. 105–114. EPA-440/5-79-001.

Rast, W., and G. F. Lee, 1978. Summary Analysis of the North American (US Portion) OECD Eutrophication Project: Nutrient Loading-Lake Response Relationships and Trophic State Indices, USEPA, Corvallis Environmental Research Laboratory, Corvallis, OR 454 pp. EPA-600/3-78-008.

Riley, G. A., 1956. Oceanography of Long Island Sound 1952–1954. II. Physical Oceanography, Bulletin Bingham. Oceanog. Collection 15, pp. 15–46.

Ryther, J. H., 1956. Photosynthesis in the Ocean as a Function of Light Intensity, *Limnol. Ocean.* (1):61–70.

Salas, H. S., and R. V. Thomann, 1975. The Chesapeake Bay Waste Load Allocation Study, by Hydroscience, Inc., for the Md. Dept. of Nat. Res. Admin., Annapolis, MD, 287 pp.

Schreiber, J. D., and D. L. Rausch, 1979. Suspended Sediment-Phosphorus Relationships for the Inflow and Outflow of a Flood Detention Reservoir, *J. Envir. Quality* **8**(4):510–514.

Schuytema, G. S., 1977. Biological Control of Aquatic Nuisances—a Review, USEPA, ORD, Corvallis, OR 90 pp. EPA-600/3-77-084.

Smith, V. H., and J. Shapiro, 1981a. A Retrospective Look at the Effects of Phosphorus Removal in Lakes, in Restoration of Lakes and Inland Waters, USEPA, Office of Water Regulations and Standards, Washington, D.C. EPA-440/5-81-010.

Smith, V. H., and J. Shapiro, 1981b. Chlorophyll-Phosphorus Relations in Individual Lakes. Their Importance to Lake Restoration Strategies, *Environ. Sci. Technol.* **15**:444–451.

Steele, J. H., 1965. Notes on Some Theoretical Problems in Production Ecology, in *Primary Production in Aquatic Environments*, Goldman, C. R., Ed., Mem. Inst. Idrobiol., pp. 383–98, 18 Suppl., University of California Press, Berkeley.

Steele, E. W., and T. J. McGhee, 1979. *Water Supply and Sewerage*, McGraw-Hill, New York, 665 pp.

Strecker, R. G., J. M. Steichen, J. E. Garton, and C. E. Rice, 1977. Improving Lake Water Quality by Destratification, *Trans. ASAE* **20**(4):713–720.

Sverdrup, H. U., M. W. Johnson, and R. H. Fleming, 1942. *The Oceans*, Prentice-Hall, Englewood Cliffs, NJ, 1087 pp.

Thomann, R. V., 1972. *Systems Analysis and Water Quality Management,* McGraw-Hill, New York 286 pp. (Reprint: J. Williams Book Co., Oklahoma City, OK.)

Thomann, R. V., R. P. Winfield, and D. M. Di Toro, 1974. Modeling of Phytoplankton in Lake Ontario, (IFYGL) Inter. Assoc. Great Lakes Research, Proceedings 17th Conference, pp. 135–149.

Thomann, R. V., and J. J. Fitzpatrick, 1982. Calibration and Verification of a Model of the Potomac Estuary, HydroQual, Inc., Final Report to D. C. Dept. of Environmental Services, Washington, D.C., 500 pp. + Appendix.

Thomann, R. V., N. J. Jaworski, S. W. Nixon, H. W. Paerl, and J. Taft, 1985. The 1983 Algal Bloom in the Potomac Estuary, Prepared for Potomac Strategy State/EPA Management Committee, Washington, D. C., 7 Chapters.

USEPA, 1974. The Relationships of Phosphorus and Nitrogen to the Trophic State of Northeast and North-Central Lakes and Reservoirs. National Eutrophication Survey Working Paper No. 23.

USEPA, 1980. Lake Restoration in Cobbossee Watershed, Capsule Report. Off. of Water Plan. and Std., Criteria and Stds. Div., ORD, Cinn., OH, 16 pp. EPA-625/2-80-027.

USEPA, 1983. Technical Guidance Manual for Performing Waste Load Allocations, Book II Streams and Rivers, Chapter 2 Nutrient/Eutrophication Impacts, OWRS, Monitoring and Data Support Division, Monitoring Branch, Washington, D.C., Nov. 1983, 6 Chapters.

Vollenweider, R. A., 1968. Scientific Fundamentals of the Eutrophication of Lakes and Flowing Waters, with Particular Reference to Nitrogen and Phosphorus as Factors in Eutrophication, Organ. Econ. Coop. Dev., Paris, Tech. Report No. DAS/CSI/68.27, 159 pp.

Vollenweider, R. A., 1975. Input-Output Models with Special Reference to the Phosphorus Loading Concept in Limnology, *Schweiz. Z. Hydrol.* **37**:53–83.

Vollenweider, R. A., 1976. Advances in Defining Critical Loading Levels for Phosphorus in Lake Eutrophication, *Mem. Ist. Ital. Idrobiol.* **33**:53–83.

Welch, E. B., 1981. The Dilution/Flushing Technique in Lake Restoration, *Water Res. Bull. AWRA* **17**(4):558–564.

Wright, R. M., and A. J. McDonnell, 1982. Stream Rehabilitation Following Treatment for Phosphorus Removal: Pre-treatment Assessment, Res. Proj. Tech. Completion Report OWRT Proj. A-056-PA, Inst. for Res. and Land & Water Res., Penn. State Univ., Univ. Park, PA, 127 pp.

Yaksich, S. M., D. A. Melfi, D. A. Baker, and J. A. Kramer, 1982. Lake Erie, Nutrient Load, 1970–1980, Lake Erie Wastewater Management Study Technical Report, U.S. Army Corps of Engineers, Buffalo, NY, September.

Zauke, G. P., D. Thierfeld, and T. Hopner, 1982. Oxygen Concentrations and Elimination of Inorganic Phosphorus and Nitrogen in an Experimental Watercourse Stocked with Emergent Macrophytes, *Aquatic Botany* **13**:339–350.

PROBLEMS

7.1. In the upstream flow, the phosphorus concentration is 0.01 mg p/l and nitrogen is 1.0 mg N/l. It is required that the nitrogen to phosphorus ratio at *A*, after mixing, be equal to 20. Derive a relationship between the effluent phosphorus concentration and the effluent nitrogen concentration such that the nitrogen to phosphorus ratio of 20 is met after mixing.

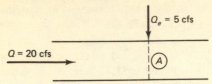

7.2. The Wicomico Estuary in Maryland may be approximated by a constant coefficient estuary with a flow of 40 cfs, cross-sectional area of 4200 ft², and a tidal dispersion coefficient of 0.2 smpd. It receives the effluent from Salisbury, Maryland which has a total nitrogen load of 630 lb/day, a total phosphorus load of 125 lb/day, and a discharge flow of 4 MGD.
   (a) Plot the distribution of nitrogen and phosphorus in the estuary if the nutrients are conservative.
   (b) Repeat (a) for a loss rate of the nutrients equal to 0.01/day.
   (c) Nutrient reduction by treatment at Salisbury could reduce effluent nitrogen and phosphorus concentrations to 2 mg N/l and 0.1 mg p/l. Calculate and plot the response for each case (for $K = 0.01/\text{day}$).
   (d) On the basis of this very simplified analysis, does nitrogen or phosphorus removal seem more appropriate? Why?

7.3. Consider the following:

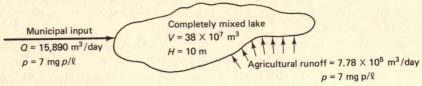

   (a) What is the estimated Tp concentration in the lake in μg/l?
   (b) Using Vollenweider's analysis, is the lake eutrophic?
   (c) What load in g p/day is needed to have the phosphorus concentration equal to 0.01 g/m³?
   (d) If the highest technology of removal for municipal waste results in an effluent of 0.1 mg p/l, can the objective of 0.01 g p/m³ in the lake be achieved by this technology?
   (e) If not, what additional reduction is necessary for the agricultural runoff to achieve the objective?

7.4. A completely mixed lake has the following properties and measured concentration of nutrients:

$$\text{volume} = 1000 \text{ MG}$$
$$\text{flow in} = \text{flow out} = 2 \text{ MGD}$$
$$\text{Tp in incoming flow} = 100 \ \mu\text{g/l}$$
$$\text{Tp in outgoing flow} = 20 \ \mu\text{g/l}$$
$$\text{TN in incoming flow} = 3 \text{ mg/l}$$

   (a) What is the loss rate of total phosphorus (day⁻¹) for this lake?
   (b) Assuming that this rate applies also for nitrogen, what is the total nitrogen concentration in the lake in μg/l?
   (c) Which nutrient would be expected to control the growth of phytoplankton in the lake?

7.5. The chlorophyll level in a lake has been measured at 30 μg/l.

(a) If the net settling rate of total phosphorus is 12.4 m/yr, the volume is 10^{10} m^3, the flow is 10^7 m^3/day and the depth is 10 m, what is the present total phosphorus loading to the lake in kg p/yr?

(b) It is desired to have the chlorophyll level reduced to 5 μg/l. What is the new required total phosphorus loading in kg/yr?

For both parts (a) and (b), use Eqs. 7.18b and 7.18c and compare results.

7.6. A lake has the following geometry:

$$\text{volume} = 1.825 \times 10^9 \text{ m}^3 \qquad \text{depth} = 20 \text{ m}$$
$$\text{flow in} = \text{flow out} = 50 \times 10^5 \text{ m}^3/\text{day}$$

Measurements have indicated that the sediment releases phosphorus on the average over the whole bottom area at a flux of 3 mg/m$^2 \cdot$ day. Also, the external input to the lake is 0.427 mt/yr. The outlet of the lake was measured at 40 μg p/l.

If the external input is eliminated from the lake with no change in inflow, what percent of the bottom area needs to be dredged so that the phosphorus concentration is equal to 10 μg p/l?

7.7. A reservoir has a volume of $2.59 \cdot 10^7$ m^3. The flow into and out of the well-mixed reservoir is 10 m^3/s. The total nitrogen and phosphorus as measured in the incoming flow is 1 mg N/l and 0.05 mg p/l, respectively The average chlorophyll in the reservoir has also been measured at 9.6 μg/l. If the depth of the reservoir is 5 m:

(a) What is the net settling velocity of the phosphorus in m/yr? Use Eq. 7.18b.

(b) It is desired to reduce the chlorophyll in the lake by 50% by addition of chemicals that will increase the phosphorus loss rate. What should the new net settling velocity of phosphorus (m/yr) be so that the chorophyll is reduced by 50%?

7.8. A shallow well-mixed lake received inputs of phosphorus that resulted in a substantial growth of aquatic plants of all types, causing a diurnal DO variation which was unacceptably high. Data obtained are as follows:

$$\text{depth} = 2 \text{ m}$$
$$\text{flow} = 11,000 \text{ m}^3/\text{day}$$
$$\text{total incoming phosphorus load} = 8030 \text{ kg p/yr}$$
$$\text{total phosphorus measured in lake} = 2.0 \text{ mg p/l}$$
$$\text{surface area} = 10^6 \text{ m}^2$$

Data have also been collected on the relationship between the biomass in the lake and the total phosphorus resulting in the following relationship

$$P' = p^{0.8}$$

where p is total phosphorus in μg/l and P' is the areal chlorophyll level in mg chl/m^2. Also, the reaeration rate is 1/day, the carbon to chlorophyll ratio is 100, and the gross growth rate of the plant biomass is 0.2/day. If the allowable diurnal DO variation is restricted to 2 mg/l, what is the allowable input phosphorus load in kg p/yr to meet this restriction?

7.9. (a) Data from the Sacramento–San Joaquin Delta in California indicated a mean depth of 1% light at about 100 cm. What is the extinction coefficient (m^{-1}) for this estuarine system?

(b) Using the relationship between the extinction coefficient and secchi depth, what is the % light penetration at the secchi depth? That is, the secchi depth corresponds to what percent of incoming light at the surface?

7.10. The following data on sunlight penetration were obtained for the Delaware River at Trenton, NJ and for Lake Tahoe.

Delaware River		Lake Tahoe	
Depth (ft)	Relative light intensity (%)	Depth (ft)	Relative (%) light intensity
0	100	0	100
1	42	33	30
2	18.5	60	18
3	11.5	100	9
4	5.1	200	2
5	2.0		
6	0.73		
7	0.27		

(a) Plot the data as a function of depth.

(b) Estimate the extinction coefficients, K_e (to base e) in 1/ft for both sets of data. Also estimate the depth at which 1% of the surface light still exists.

7.11. Given the following conditions for a lake:

$$T = 10°C \qquad I_s = 350 \text{ ly/day} \qquad I_T = 500 \text{ ly/day}$$
$$f = 0.6 \qquad K_e = 0.5/\text{m} \qquad K_{mp} = 2 \ \mu\text{g p/l} \qquad H = 17 \text{ m}$$
$$G_{max} = 1.8/\text{day} \qquad \mu_R = 0.15/\text{day}$$

(a) If nutrients are not limiting, what is the growth rate, death rate (with no zooplankton), and net growth rate of the phytoplankton assuming no settling? If the initial phytoplankton chlorophyll at $t = 0$ is 1.0 μg/l, what would the phytoplankton level be if this net growth rate persisted for 30 days?

(b) Plot the relationship between the growth rate and the available phosphorus concentration.

(c) Repeat (a) if the ambient available phosphorus concentration is 2 μg p/l.

7.12. A waste source with a flow of 10^5 m^3/day, a dissolved phosphorus concentration of 0.1 mg/l, and a dissolved nitrogen concentration of 10 mg/l discharges into a river with an upstream flow of 10^6 m^3/day and upstream concentrations of 0.002 mg p/l and 0.3 mg N/l. The uptake rate of the phosphorus in the river is 0.1/day and of the nitrogen is 0.3/day. Assume that the half-saturation constant for phosphorus is 2 μg p/l and for nitrogen is 10 μg N/l. Which nutrient is controlling the growth rate of phytoplankton at a travel time of 10 days?

7.13. Given the following stream:

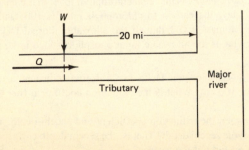

Design Conditions

Flow
    Upsteam   12 cfs
    Plant   0.3 MGD
Stream
    Depth   2.2 ft
    Velocity   0.4 fps
    Temperature   25°C
Sunlight—daily radiation   600 ly/day
    Photoperiod   0.5
    Light extinction in stream   0.33 ft^{-1}
Chlorophyll a—upstream @ $x = 0$ 26 μg/l
    Net settling velocity   0.33 ft/day
Inorganic phosphorus concentration
      upstream of $x = 0$ 0.02 mg/l
Phytoplankton Growth/Death
   $G_{max} = 1.8$/day
   $I_s = 300$ ly/day
   $\mu_R = 0.1$/day

(a) Calculate the net phytoplankton growth rate (day^{-1}) assuming $G(N) = 1.0$.
(b) Calculate the maximum possible phytoplankton concentration (μg/l) in the tributary @ $x = 20$ mi assuming $G(N) = 1.0$.
(c) For an inorganic phosphorus in the effluent of the plant of 1 mg/l, check if phosphorus will limit growth in this 20-mi tributary and, if so, what the phytoplankton concentration would be in that case.

Chapter 8

Toxic Substances

8.1 INTRODUCTION

The issue of the release of chemicals into the environment in concentrations such as to be toxic is an area of intense concern in water quality and ecosystem analyses. Passsage of the Toxic Substances Control Act of 1976 in the United States, unprecedented monetary fines, and continual development of data on lethal and sublethal effects attest to the expansion of control on the production and discharge of such substances. However, as illustrated by pesticides, the ever-present potential for insect and pest infestations with attendant effects on humans and livestock results in a continuing demand for product development. As a result of these competing goals, considerable effort has been devoted in recent years to the development of predictive schemes that would permit an a priori judgment of the fate and effects of a chemical in the environment. This chapter is devoted to an overview of some of these predictive schemes.

The uniqueness of the toxic substances problem lies in the potential transfer of a chemical to humans with possible attendant public health impacts. This transfer can occur through two principal routes:

1. Ingestion of the chemical from the drinking water supply.
2. Ingestion of the chemical from contaminated aquatic foodstuffs (e.g. fish and shellfish) or from food sources that utilize aquatic foodstuffs as a feed.

The concern of the general public lies principally in the apparent unknown long-term effects of such chemical ingestions on the whole person, physical and psychological. Concern over potential cancers, tumors, and birth defects are real and most profound and reflects a certain sense of "mystery" associated with toxic substances. Such substances are therefore unique in water quality in the sense that the potential impact on the public health is perceived to be direct and individualized. This is in contrast to such water quality issues as dissolved oxygen or eutrophication where the impact is primarily on the aquatic ecosystem and the effect on the public health is minimal or nonexistent. The discharge of bacteria and parasites (Chapter 5) is the closest "conventional" issue to the toxic substances concern. However, more is known relatively about waterborne diseases than about the public health implications of toxic substances.

Numerous examples of the interferences with desirable water uses by toxic substances have been documented such as:

1. Metal and organic chemical accumulation in fish resulting in the banning of the fishery for human consumption.
2. Trihalomethane formation in water treatment and in finished water resulting in a potential interference with the use of water for drinking.
3. Leaching and runoff from pesticides in agriculture and subsequent accumulation in the food chain or in water supplies.
4. Leaching and runoff from chemical waste disposal sites resulting in interferences with water supplies and ecosystem accumulation.

In all cases, the interest in the problem is particularly heightened by the potential transfer of chemicals to humans with subsequent short-term or long-term impact on the public health. Particularly extreme cases have occurred such as greatly elevated cadmium and PCB concentrations leading to severe toxic effects in humans.

The toxic substances water quality problem can therefore be summarized as the discharge of chemicals into the aquatic environment which results in concentrations in the water or aquatic food chain at levels that are determined to be toxic, in a public health sense or to the aquatic ecosystem itself, and thus may interfere with the use of the water body for water supply or fishing or contribute to ecosystem instability.

The potentially toxic substances can be grouped into several broad categories:

1. Metals—for example, mercury, cadmium, lead, selenium, resulting from industrial activities such as electroplating, battery manufacturing, mining, smelting, and refining.
2. Industrial chemicals—thousands of chemical compounds for thousands of uses from which several broad subcategories can be identified as:
   a. Plasticizers—for example, alkyl phthalates; used in plastic products.
   b. Solvents—for example, chlorinated benzenes; used in cleaning operations.
   c. Waxes—for example, chlorinated paraffins; used in commercial and residential waxes and cleaners.
   d. Miscellaneous—for example, polychlorinated biphenyls (PCB); originally used in transformers.

3. Hydrocarbons—resulting from oil refining and combustion of fuels; examples include the polycyclic aromatic hydrocarbons (PAH) from coal conversion processes.
4. Agricultural chemicals—a large class of chemicals designed for specific environmental controls such as pesticides, insecticides, herbicides, and weedicides; examples include DDT, malathion, chlordane, atrazine, and heptachlor.
5. Radioactive substances—resulting from production of nuclear energy, nuclear weapons, and production of radioactive materials for industrial use; examples include strontium-90, plutonium-239/240, and cesium-137.
6. Miscellaneous—examples include ammonia and chlorine from municipal and industrial discharges.

The USEPA has listed 129 priority pollutants, not including the radioactive substances and Table 8.1 from Mills et al. (1982) shows the most commonly discharged priority pollutants.

It should, of course, be recognized that a chemical becomes a "toxic substance" when, at some concentration, the chemical has some deleterious impact. The judgment of what constitutes a deleterious impact is usually subject to considerable debate. Nevertheless, the mere presence of a chemical does not constitute the presence of a toxic substance. Therefore, chemicals should really be viewed

TABLE 8.1 MOST COMMONLY DISCHARGED
PRIORITY POLLUTANTS

Pollutant	Percent of samples[a]	Percent of industries[b]
Nonmetals		
bis(2-Ethylhexyl)phthalate	41.9	91
Chloroform	40.2	88
Methylene chloride	34.2	78
Total cyanides	33.4	59
Toluene	29.3	88
Benzene	29.1	78
Phenol	29.1	78
di-*n*-Butyl phthalate	18.9	72
Ethylbenzene	16.7	75
Naphthalene	10.6	56
Phenanthrene and anthracene	10.6	50
Metals		
Copper	55.5	100
Zinc	54.6	100
Chromium	53.7	100
Lead	43.8	96
Nickel	34.7	96

[a]Percent of approximately 2600 samples where compound was found.
[b]Percent of industrial categories and subcategories (32 for nonmetals, 28 for metals).
Source: From Mills et al. (1982).

as "potentially toxic" at some concentration in various sectors of the water system and the effect that the concentration has on the ecoystem or on man (Fig. 8.1). The analysis of the fate and transport is focused on calculating the expected concentration in the water body and its ecosystem and does not make any statements about the significance of the resulting concentration. The analysis of the effects of a chemical usually begins with a given range of concentrations followed by determination of the impact of the concentration range on the ecosystem or a specified organism of the ecosystem. Such impact may be registered in terms of mortality (acute effects) or the reproductive behavior, growth, or other physiologic characteristic of the organism (chronic effects). It is the effects of the chemical that generally forms the basis for the specification of water quality criteria.

8.2 CHEMICAL WATER QUALITY CRITERIA AND STANDARDS

Chemical criteria and guidelines for drinking water quality have been given by the USEPA (U.S. Federal Register, 1980 and 1984) and the World Health Organization (1984a, b). In the former case, risk levels are associated with the chemical where risk is measured in terms of the number of additional cases of cancer in a population of one million. Thus, for carbon tetrachloride, 0.4 µg/l in treated drinking water is proposed for a 10^{-6} risk level (i.e. one additional case per 10^6 people), 0.04 µg/l for 10^{-7} and 4 µg/l for a 10^{-5} risk. Drinking water criteria and guidelines may therefore vary by orders of magnitude depending on the risk assigned to the public health impact. For some chemicals, the drinking water criteria may be more stringent than for protection of the aquatic ecosystem.

Figure 8.1   Interaction of fate and effect of hazardous substances.

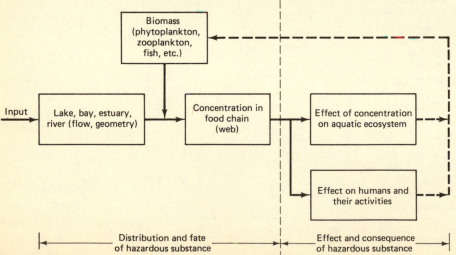

Considering next the focus of the specification of a chemical water quality criterion and standard to be the effect on the aquatic ecosystem, then several factors must be considered:

1. The concentration level that results in acute effects of mortality.
2. The concentration level that shows some specified longer-term chronic effect.
3. The tendency for the chemical to bioaccumulate in the aquatic food chain.
4. The time of exposure of the organism or ecosystem to the chemical at varying concentrations.
5. The return period of the design conditions (e.g., river flow) for use in a chemical waste load allocation.

The questions then are:

1. What are the water quality criteria for a given chemical under both acute and chronic conditions?
2. What should be the target concentration in the water body, that is, the water quality standard or the target concentration of the chemical in organisms of the ecosystem?
3. What river flow, temperature, or other environmental condition (e.g., pH) should be used in arriving at an allowable effluent concentration and mass discharge level?

8.2.1 Acute and Chronic Aquatic Ecosystem Effects

A large body of literature exists on the discipline of aquatic toxicology and only the broad highlights are given here. Sprague (1969, 1970), APHA et al. (1985), and Buikema (1982) provide convenient starting references. Sprague (1969) describes, at the most basic level, acute versus chronic as acute: "coming speedily to a crisis," and chronic: "continuing for a long time, lingering." In general, the measurement of acute toxicity has equated "speedily" to 96 hr or 4 days. Toxicologists have determined that this is approximately the time frame for which test organisms experience acute toxic effects. The range, however, is less than 1 day to greater than 14 days (Sprague, 1969). Nevertheless, because of the large amount of data, acute water quality criteria are now usually (but not always) based on 96-hr exposure times.

The measurement of chronic toxicity has generally equated "a long time, lingering" with periods of up to 30–60 days for which there are no adverse effects on early life stages, partial life cycles (less than 15 months), or whole life cycles.

Figure 8.2 shows the basic components for the determination of the acute toxicity of a chemical, given a chemical at different water concentrations and over varying times of exposure. Buikema et al. (1982) and CDM (1982) discuss these basic components in additional detail. Suitable controls, number of aquatic organisms of a given species (e.g., fish), and concentration ranges are determined from standard procedures. The percent of aquatic organisms that are killed is determined

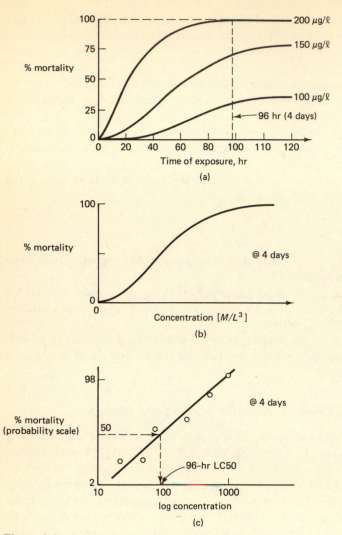

Figure 8.2 Steps in the development of an LC50.

for each concentration and exposure time. As noted above, aquatic toxicologists have generally standardized on the 96-hr exposure time as a reference time for acute toxicity. Figure 8.2(b) then shows the cross plot at 4 days of the percent mortality vs. the concentration of the chemical. Typically a sigmoidal curve results. Data have indicated that plotting the percent mortality on a normal probability scale against the log of the concentration results in a straight line relationship. The median lethal concentration (LC) at which 50% mortality occurs, the LC50, is then estimated from this plot. Thus in Fig. 8.2(c), the 96-hr LC50 is about 90 μg/l of the toxicant.

Table 8.2 (from Thurston et al., 1985) shows the range of LC50 concentrations for different organic chemicals on the same test species, that is, the fathead

TABLE 8.2 SOME LC50 CONCENTRATIONS FOR DIFFERENT CHEMICALS.
AQUATIC TEST ANIMAL: FATHEAD MINNOW

Chemical	Description	LC50[a] (mg/l)
2-(2-Ethoxyethoxy)-ethanol	Narcotic alcohol	9650 (7910–11,800)
2-Methyl-l-propanol	Narcotic alcohol	1510 (1370–1670)
2,4-Pentanedione	Nerve toxin	141 (113–175)
Hexachloroethane	Bioaccumulative narcotic	1.1 (0.967–1.25)
Pentachlorophenol	Uncouples oxidative phosphorolation	0.266 (0.222–0.319)
Endrin	Acetylcholinesterase inhibitor	0.00064 (0.00059–0.00069)

[a]Numbers in ( ) indicate the 95% confidence interval.
Source: Data from Thurston et al. (1985).

minnow. As noted, from a nonspecific chemical such as a narcotic to a more specific mode of toxic action such as endrin occasions, the range in LC50s is over about seven orders of magnitude.

In order to include the relative toxic effect on different aquatic organisms, the approach can be repeated for different aquatic species resulting in responses such as shown in Fig. 8.3. One definition then of an "ecosystem" acute toxicity is the final acute value (FAV) given by the 96-hr LC50 for the species at the 5% level of sensitivity. That is, the FAV is the 96-hr LC50 which is exceeded by 95% of the species in the data base or conversely 5% of the species have 96-hr LC50 concentrations equal to or less than the FAV. Recognizing the uncertainty of the overall approach, the USEPA (*Federal Register*, 1984) has then proposed that the water quality criterion maximum (CM) be determined as

$$CM = \frac{FAV}{2} \qquad (8.1)$$

Testing for chronic effects varies widely and includes plants, micro and macro invertebrates, early life stages of organisms, and fish at a variety of life stages. As noted, exposure times may vary from days for algae to months or more than a year for larger fish. The data from such tests are used to estimate the "safe" concentration at which no defined effect is observed over the life of the test. The resulting

Figure 8.3 Estimating the final acute value (FAV) of a chemical. From CDM, 1982.

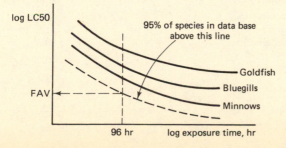

concentration is called the maximum acceptable toxic concentration (MATC). Bui-kema et al. (1982) indicate that the MATC is "empirically determined, and is the highest exposure concentration that does not result in significant harm to the test organism in terms of survival, growth or reproduction, or it is interpolated as the geometric mean of the lowest concentration having an effect and the highest con-centration having no effect." From chronic toxicity data and the resulting MATCs, a criterion average (CA) has been proposed (*Federal Register*, 1984) as a 30-day average concentration with one 96-hr excursion above the average allowed in any 30-day period.

8.2.2 Ecosystem Water Quality Criteria and Action Limits

LC50 values have often been adopted as water quality standards. More recently, attempts have been made to develop the criterion from two numbers: (a) the CM concentration and (b) the CA concentration. Also, since the toxicity of certain chemicals depends on environmental factors such as pH and hardness, such factors are also a part of the criterion specification. Table 8.3 summarizes some proposed criteria (*Federal Register*, 1984) using the CA and CM concentrations. The effect of environmental factors such as pH and hardness can be noted (e.g., on ammonia and cadmium). The chemical is defined in specific terms (e.g., active, total resid-ual). The average is less than the acute and the range across chemicals is large (e.g., mercury CA of 0.20 μg/l and ammonia CA of 90 μg/l).

In most instances, state and local regulatory agencies tend to adopt published

TABLE 8.3   SOME PROPOSED CHEMICAL WATER QUALITY CRITERIA, FRESHWATER AQUATIC LIFE

			Criterion	
Chemical	Unit	Environmental factor/level	Maximum concentration	Average concentration
Ammonia	mg N/l (total NH_3)	$T = 25°C$; pH/7.0	8.48	1.67
		pH/8.0	2.33	0.58
		pH/9.0	0.41	0.09
Cadmium	μg Cd/l "active"[a]	Hardness[b]/50	2.0	2.0
		Hardness/100	4.5	4.5
		Hardness/200	10.0	10.0
Chlorine	μg/l "active"[a]	Hardness[b]/50	8.4	5.8
		Hardness/100	16.0	11.0
		Hardness/200	29.0	20.0
Mercury	μg/l "active"[a] divalent inorganic		1.1	0.20

[a]"Active"—filtration through 0.45-μm filter after acidification to pH 4 with nitric acid.
[b]mg/l $CaCO_3$.
Source: *Federal Register* (1984).

TABLE 8.4 SOME MAXIMUM ALLOWABLE CONCENTRATIONS
IN FISH ("ACTION LIMITS")

Chemical	Maximum concentration in fish (edible portion) (ppm)[a]
Aldrin/dieldrin	0.3
DDT and metababolites	5.0
Endrin	0.3
Heptachlor/heptachlor epoxide	0.3
Kepone	0.3
Mercury	1.0
Mirex	0.1
Polychlorinated biphenyls (PCBs)	5.0[b]
Toxaphene	5.0

[a]ppm = one part chemical per million parts wet weight of fish = $\mu g/g(w)$; g(w) = g wet wt.

[b]2.0 ppm subsequently adopted by USFDA.

Source: USFDA (1978).

chemical water quality criteria as the standards for the chemical. Therefore, the criterion becomes the standard or the target concentration to be used in waste load allocation and permitting requirements. This is not necessarily the best approach since there may be other site specific considerations (such as ecosystem adapatability to a given chemical) which would warrant a more careful analysis of the degree of applicability of the criterion.

For some chemicals, a maximum concentration in edible fish or shellfish is specified. The concentration in the organism itself, in contrast to the water concentration, then becomes the relevant variable that must be controlled. These concentration levels, established in the United States by the U.S. Food and Drug Administration (USFDA) and subsequently adopted by the states with or without modification, are "action" limits. When the limits are exceeded, the USFDA will remove the product from the marketplace. Table 8.4 shows some typical values. These levels are estimated from allowable total intake of the chemical by humans from all sources (i.e., water, meat, fish, eggs, etc.). From the allowable intake, that is, the allowable dose and estimates of the consumption of fish, the allowable concentration is assigned.

Attention can now be directed to the calculation frameworks that can be used to relate the input of a chemical to resulting water concentrations and levels in the food chain.

8.3 PRINCIPAL PHYSIO-CHEMICAL COMPONENTS OF TOXIC SUBSTANCES ANALYSIS

The three features of toxic substances that separate such substances from more conventional pollutants are:

1. The tendency for certain chemicals to sorb to particulates in the water body.

2. The tendency for certain chemicals to be concentrated by aquatic organisms and transferred up the food chain.

3. The tendency for certain chemicals to be toxic at relatively low water concentrations of the $\mu g/l$ or ng/l level.

Figure 8.4 shows the general features of the physio-chemical phases of the transport of a toxic substance in the water body. As with all water quality problems, specification of the inputs of the toxic substance is essential. The inputs include all sources such as municipal and industrial discharges, urban and agricultural runoff, and atmospheric inputs. In the water column and sediment, the principal physical and chemical phenomena to be included are:

1. Sorption and desorption between dissolved and particulate forms in the water column and sediment.

2. Settling and resuspension mechanisms of particulates between the sediment and the water column.

3. Diffusive exchange between the sediment and the water column.

4. Loss of the chemical due to biodegradation, volatilization, photolysis, and other chemical and biochemical reactions.

5. Gain of the chemical due to chemical and biochemical reactions.

6. Transport of the toxicant due to advective flow transport and dispersive mixing.

7. Net deposition and loss of chemical to deep sediments.

Figure 8.4   Schematic of principal features of physical-chemical fate of toxic substances.

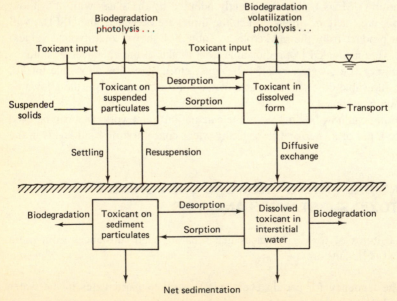

Not all chemicals require consideration of all of these physio-chemical interactions. Some chemicals are more highly volatile and less sorbed to solids while the converse is also true. Mills et al. (1982) estimated that for 103 organic priority pollutants, (a) sorption processes are important for 60 of the chemicals and (b) volatilization is important for 52 chemicals. Also since some chemicals sorb to solids, the sediment of the water body becomes particularly significant as a potential long-term storage reservoir of the chemical. (This is similar to the phosphorus in sediments acting as a long-term source of nutrients in eutrophication.) Before each of the water body types are evaluated, the basic mechanisms of the chemical fate will be presented.

8.3.1 Toxicant Forms

As shown in Fig. 8.4, the toxicant can exist in two basic forms: the toxicant in the dissolved phase and the toxicant on the solids, both for either the water column or sediment. (The toxicant may be selectively sorbed onto certain types of solids such as finely dispersed clays or organic particles, but only one solids category is considered here.) Special care must now be taken in defining the units of the dissolved and particulate forms. Let the dissolved form be specified in some operational way such as all of the chemical that passes through a 0.45-μm filter and let

$$c_d' = \text{dissolved toxicant } [M_T/L_w^3]$$

where M_T is the mass of toxicant and L_w^3 is the volume of water. Let the toxicant in the particulate form be defined by

$$c_p = \text{particulate toxicant } [M_T/L_{s+w}^3]$$

where L_{s+w}^3 is the volume of solids plus water, the bulk volume. The form c_p is therefore the mass of toxicant sorbed on and in the solids relative to the total volume of solids and water. This definition is important in the calculation of sediment toxicant concentrations. In order to properly add these two forms to obtain the total concentration, it is necessary to introduce the volume of water per bulk volume. This ratio is called the porosity (ϕ) and has units $[L_w^3/L_{s+w}^3]$.

The total toxicant concentration anywhere in the water body or sediment is then

$$c_T = c_p + \phi c_d' \tag{8.2}$$

or

$$c_T = c_p + c_d \tag{8.3}$$

where c_d $[M_T/L_{s+w}^3]$ is the porosity corrected dissolved toxicant as

$$c_d = \phi c_d' \tag{8.4}$$

Of course, for the water column, $\phi \approx 1$ since the volume of water is essentially equal to the bulk volume and then $c_d = c_d'$. For the upper strata of the bed sediment, ϕ is generally from about 0.7–0.8. In all subsequent work in this chapter it is assumed that the dissolved form is porosity corrected and all toxicant concentrations are on a bulk volume basis.

The particulate form, c_p, is expressed as a mass of toxicant per bulk volume of solids and water. For a given concentration of solids, c_p can also be expressed as

$$c_p = rm \tag{8.5}$$

where m is the solids concentration $[M_s/L_{s+w}; M_s = \text{mass of solids}]$ and r is the toxicant concentration expressed on a dry weight solids basis $[M_T/M_s]$. Typical units for r are $\mu g \text{ tox}/g$ (dry wt), where g (dry wt) is gram dry weight of solids. Note that a $\mu g/g$ is equivalent to a part per million (ppm). Normally, r is the quantity of the toxicant measured in sediments and then the concentration of the chemical in the particulate form on a bulk volume basis is given from Eq. 8.5.

8.3.2 Sorption–Desorption

Consider now the mechanism of sorption of the chemical to the particulates and desorption from the particulates back into the dissolved phase [Fig. 8.5(a)]. A simple representation of this interaction is given by the following equations where it is assumed that all kinetics are linear.

$$\frac{dc_p}{dt} = k_u m c_d' - K c_p \tag{8.6}$$

$$\phi \frac{dc_d'}{dt} = -k_u m c_d' + K c_p \tag{8.7}$$

where k_u is the sorption rate $[L_w^3/M_s \cdot T]$ and K is the desorption rate $[1/T]$. In terms of the particulate concentration on a solids basis, r, Eq. 8.6 becomes, for constant solids concentration,

$$\frac{dr}{dt} = k_u c_d' - Kr \tag{8.8}$$

In a typical experiment to examine these kinetics, a mass of solids is dispersed into a vessel with a known concentration of the dissolved chemical dispersed in the water phase. Samples are then taken at various intervals, the solids are separated (as with centrifugation), and the chemical on the particles and in the dissolved form is measured. Results are schematically indicated in Fig. 8.5(b). The dissolved concentration drops due to sorption onto the solids which gain in toxicant concentration. Now for a large number of chemicals, the time to the equilibration between the solid and liquid phases is "fast," on the order of minutes to hours. This is fast compared to the kinetics inherent in other mechanisms of the problem. These latter mechanisms include bacterial decay, net loss rates to the sediment, and sedimentation rates that have reaction times on the order of days to years.

8.3.3 Partition Coefficient

The "fast" kinetics of sorption–desorption indicate that for time scales of days to years, there will be a virtually continuous equilibration of the dissolved and par-

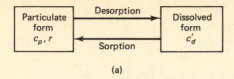

(a)

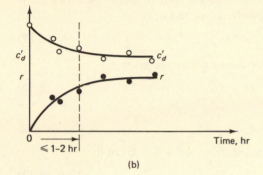

(b)

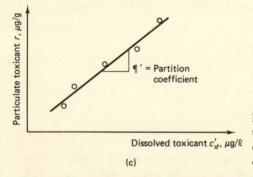

(c)

Figure 8.5 (a) Sorption–desorption of toxicant. (b) "Fast" kinetic relationships of sorption. (c) Partition coefficient as slope of particulate vs. dissolved chemical.

ticulate forms depending on the local solids concentration. This partitioning between the two components permits the specification of the fraction of dissolved and particulate toxicant to the total. The dissolved and particulate toxicant are therefore assumed to be always in a "local equilibrium" with each other.

If the experiment is carried out over a range of dissolved chemical concentrations and the concentrations of both the particulate and dissolved forms are measured at equilibrium, results such as shown in Fig. 8.5(c) are obtained. Assuming that the sorption is reversible (desorption) and that the sorption–desorption kinetics are linear, then a partition coefficient can be defined from the slope of the data of Fig. 8.5(c) as follows:

$$\P' = \frac{r}{c_d'} = \frac{k_u}{K} \tag{8.9}$$

where $\P'$ is the partitioning or distribution of the chemical at equilibrium between the particulate and the dissolved form. The units of $\P'$ are $[M_T/M_s \div M_T/L_w^3]$, as in $\mu g/g$ (dry wt) $\div \mu g/l$. Often the units are also expressed as $\mu g/kg$ (dry wt) $\div \mu g/l$ and since a $\mu g/kg$ and a $\mu g/l$ are both parts per billion parts, the partition coefficient is often expressed in this $(ppb)_{solids}/(ppb)_{water}$ ratio.

From Eq. 8.4,

$$\P = \frac{\P'}{\phi} = \frac{r}{c_d} \tag{8.10}$$

where $\P$ is the porosity corrected partition coefficient $[M_T/M_s \div M_T/L_{s+w}^3]$.

Since the particulate toxicant concentration relative to the bulk volume is given by Eq. 8.5 another relationship between c_p and c_d is

$$c_p = \P m c_d \tag{8.11}$$

Substituting Eq. 8.11 into Eq. 8.3, for the total toxicant gives

$$c_T = (1 + \P m)c_d \tag{8.12}$$

or

$$c_d = f_d c_T$$

where f_d is the fraction of the total that is dissolved and is given by

$$f_d = (1 + \P m)^{-1} \tag{8.13}$$

Also, the particulate toxicant as a fraction of total toxicant is given by

$$c_p = f_p c_T \tag{8.14}$$

for

$$f_p = \frac{\P m}{1 + \P m} \tag{8.15}$$

Also note that

$$f_p + f_d = 1 \tag{8.16}$$

The distribution of the toxicant between the particulate and dissolved phases therefore depends on the partition coefficient and the solids concentration. Table 8.5 shows typical values and Sample Problem 8.1 shows the effect of different orders of the partition coefficient on the fraction particulate and dissolved.

TABLE 8.5 PARTITION COEFFICIENTS FOR
SOME CHEMICALS

Chemical	Partition coefficient[a] $[\mu g/kg \div \mu g/l = l/kg]$
Heavy metals (Cd, Cu, Pb, Zn)	(10^4-10^5)
Benzo(a)pyrene	10^4-10^5
PCB	10^5-10^6
Plutonium-239	10^4-10^5
Methoxychlor	10^4
Napthalene	10^3

[a]To order of magnitude for water column. Some chemicals display a relationship between partition coefficient and solids and organic carbon concentrations (see Fig. 8.6). Sediment partition coefficients may be lower.

SAMPLE PROBLEM 8.1

DATA

A water body has 100 mg/l of suspended solids.

PROBLEM

Calculate the fraction of toxicant in the particulate and dissolved form for:
  (a) PCB, $\P = 100,000$ l/kg.
  (b) Napthalene, $\P = 1000$ l/kg.

ANALYSIS

(a) From Eq. 8.13:

$$(f_d)_{PCB} = (1 + \P m)^{-1}$$
$$= [1 + (10^5 \text{ l/kg} \cdot \text{kg}/10^3\text{g} \cdot 100 \text{ mg/l} \cdot \text{g}/10^3\text{mg})]^{-1}$$
$$= (1 + 10)^{-1}$$
$$= 0.09$$
$$\therefore (f_p)_{PCB} = 1 - (f_d)_{PCB} = 0.91 \qquad\qquad \text{[Eq. (8.16)]}$$

(b) Similarly $(f_d)_{\text{napth.}} = \left(1 + \frac{10^3 \cdot 10^2}{10^3 \cdot 10^3}\right)^{-1}$

$$= (1 + 0.1)^{-1}$$
$$= 0.91$$
$$\therefore (f_p)_{\text{napth.}} = 0.09$$

Summary: For $m = 100$ mg/l

	$\P$(l/kg)	f_d	f_p
PCB	10^5	0.09	0.91
Napthalene	10^3	0.91	0.09

Conclusion: The high partition coefficient for PCB results in 91% of the total concentration in the particulate form whereas the low partition coefficient for napthalene results in the opposite, that is, 91% in the dissolved form.

Note, however, that these relationships assume reversibility and no interaction of the partition coefficient with solids concentration. The "fast" kinetic assumption considers complete reversibility between the solid and liquid phases. These is some considerable evidence that this is not the case for many chemicals (Di Toro et al., 1982a; Karickoff, 1984; Di Toro, 1985).

Di Toro (1985) has described a "resistant" and "reversible" component model to analyze a large set of laboratory data on sorption and desorption of organic chemicals. In the analysis of these data, it was found that there is indeed a fraction of the sorbed chemical which is resistant to desorption, at least for the duration of the normal experiments. It has also been observed (first by O'Connor and Connolly, 1980) that the partition coefficient often displays a relationship to the concentration of suspended solids.

For heavy metals, Thomann (1985) has compiled water column partition coefficients from field data from 15 streams and rivers. The metals copper, zinc, cadmium, chromium, lead, and nickel were combined since there was no systematic difference in ¶ between these metals. The partition coefficient ranges from about 10^2-10^5 1/kg. An appropriate first approximation to the water column partition coefficient for the metals indicated is

$$\P = 250{,}000 \text{ m}^{-1} \tag{8.17}$$

for ¶ in 1/kg and m in mg/l.

With ¶ given by Eq. 8.17 and the fraction dissolved given by Eq. 8.13, f_d will be a constant, that is, $f_d = 0.8$. A range is to be expected however, given the range of ¶ in the data used to obtain Eq. 8.17. Field data indicate then that

$$f_d = 0.8 \pm 0.2 \text{ standard deviation}$$

In summary, to first approximation for studies of the heavy metals, Cu, Cd, Zn, Cr, Pb, and Ni, the partition coefficient depends on the solids concentration and the fraction dissolved can be taken as a constant at about 0.8 (± 0.2) and is approximately independent of the solids concentration.

It should be noted that this relationship ($f_d = 0.8$) is for the water column and the extension of the coefficients to sediment solids concentrations is generally not valid. Although sediment partition coefficients are generally less than in the water column, the rate of decline apparently is reduced at the high bulk density of the sediment.

For organic chemicals, numerous attempts have been made to relate the partitioning of such chemicals to specific properties of the chemical, most notably the octanol–water partition coefficient and the solubility of the chemical in water.

The octanol–water partition coefficient is measured by determining the relative distribution of the chemical between water and octanol and is designated K_{ow}(μg/l octanol $\div$ μg/l water). Karickhoff et al. (1979) for several sediment samples determined that the sorption of organic chemicals to sediment solids is a function of the weight fraction of the organic carbon of the sediment. An organic carbon partition coefficient is, therefore, defined as $\P_{oc}[M_T/M_{\text{org C}} \div M_T/L_w^3$ e.g., μg/g org C $\div$ μg/l]. Karickhoff et al. (1979) obtained a linear relationship between π_{oc} and K_{ow}

$$\pi_{oc} = 0.617 K_{ow} \tag{8.18}$$

Therefore,

$$\pi = \pi_{oc} f_{oc} \tag{8.19}$$

where f_{oc} is the weight fraction of organic carbon of the total solids concentration $[M_{\text{org C}}/M_s]$. This fraction ranges from about 0.001 to about 0.1.

Combining Eqs. 8.18 and 8.19 gives

$$\pi = 0.617 f_{oc} K_{ow} \tag{8.20}$$

Di Toro (1985) as noted above has investigated the effects of resistant and reversible components for organic chemicals and has included the effect of solids concentration on the partition coefficient. His review of experimental data permitted 200 estimates of the "exchangeable" partition coefficient π_x. This is the coefficient that represents the reversible component of the chemical. Di Toro (1985) therefore suggests

$$\pi_x = \frac{f_{oc} \pi_{oc}^x}{1 + m f_{oc} \pi_{oc}^x / 1.4} \tag{8.21}$$

where π_{oc}^x is the exchangeable organic carbon partition coefficient and is given approximately by

$$\pi_{oc}^x \approx K_{ow} \tag{8.22}$$

These expressions permit estimation of π_x for organic chemicals through K_{ow} and environmental conditions through f_{oc} and the solids concentrations, m.

An evaluation of Di Toro's data (Thomann and Salas, 1986) permits an empirical relation between the adsorption partition coefficient and the exchangeable coefficient where to first approximation

$$\pi \approx 2\pi_x \tag{8.23}$$

Figure 8.6 shows the relationship between π (Eqs. 8.21 through 8.23) for different values of the product $f_{oc} K_{ow}$, that is,

$$\pi = \frac{2(f_{oc} K_{ow})}{1 + m(f_{oc} K_{ow})/1.4} \tag{8.24}$$

For high values of $f_{oc} K_{ow}$ (i.e., $\log f_{oc} K_{ow} = 6$), the partition coefficient is inversely proportional to the solids concentration. At $\log f_{oc} K_{ow}$ of about 1 to 3, the partition coefficient is approximately independent of the solids concentration. As with the metals this relationship should not be extended to sediment solids concentrations.

Equation 8.23, using Eqs. 8.21 and 8.22, can be substituted into Eq. 8.13 to determine the fraction dissolved as a function of the solids concentration. Figure 8.7 shows the fraction of the total chemical in the dissolved form using the Di Toro (1985) equations and Eq. 8.23.

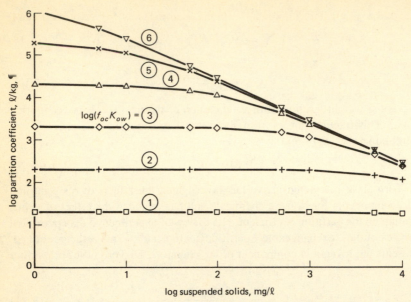

Figure 8.6 Relationships between log partition coefficient (l/kg) and log suspended solids for organic chemicals using Eqs. 8.21 and 8.22 (DiToro, 1985) and Eq. 8.23. Numbers on the curves in the figure are equal to log $(f_{oc}K_{ow})$.

Figure 8.7 Relationships between fraction dissolved of organic chemicals and log suspended solids concentration using Eqs. 8.21 and 8.22 (DiToro, 1985) and Eq. 8.23. Numbers on curves in figure are equal to log ¶ (l/kg).

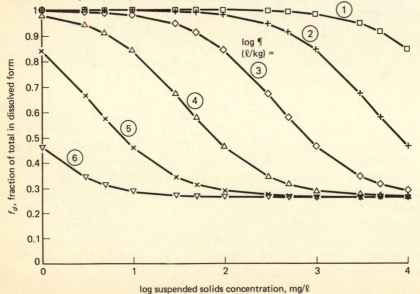

This figure shows that for organic chemicals with a low partition coefficient (e.g., $\P < 100$ l/kg) and a normal range of solids for natural waters of <100–1000 mg/l, the fraction dissolved is greater than about 0.90. This indicates that for organic chemicals with low partition coefficients (< 100 l/kg), the chemical is principally in the dissolved form and to first approximation, the interactions with solids may be neglected. On the other hand, as the partition coefficient increases, Fig. 8.7 indicates that both forms of the particulate and dissolved forms must be included in the analysis. However, at the higher $\P$ values of $>10^5$ l/kg and solids >10 mg/l, the f_d is approximately a constant at about 0.3–0.4 over the entire range of solids. This latter result can be contrasted to Eq. 8.13 which shows $f_d \approx 0(f_p \approx 1)$ for $\P > 10^5$ l/kg.

In each of the expressions above, some estimate must be made of K_{ow}, the octanol–water partition coefficient. K_{ow} coefficients for some frequently encountered chemicals together with other properties of chemicals can be found in several summary references (Mills et al., 1982; Verschueren, 1983; Lyman et al., 1982). K_{ow} values range from about zero for chemicals that do not partition into octanol to 10^6 to 10^7 for chemicals that partition strongly into octanol (see Table 8.6).

K_{ow} may also be estimated from the equations of Chiou et al (1977) and Banerjee et al. (1980) as

$$\log K_{ow} = 5.00 - 0.670 \log c_w \tag{8.25a}$$

$$\log K_{ow} = 6.5 - 0.89 \log c_w - 0.015 \ (mp) \tag{8.25b}$$

where c_w is the chemical solubility in water in μmol/l and mp is the melting point in °C.

Karickhoff (1984) also presents an equation for five chemical families relating $\P_{oc}$ to the water solubility and melting point of the chemical, as follows

$$\log \P_{oc} = -0.83 \log c_w - 0.01(mp - 25) - 0.93 \tag{8.26}$$

where in this equation c_w is the mole fraction solubility (mole fraction = moles of chemical/moles of chemical + moles of water).

The partition coefficient $\P$ also requires an estimate of the fraction of organic carbon in the total solids and the total solids concentration. These data must be obtained from field information or estimated from a general knowledge of the similar types of water bodies.

Finally, these estimates of the partition coefficient do not necessarily apply to the actual bed sediment in the field. The conditions of the sediment that may influence $\P$ include the degree of sediment mixing due to bioturbation (organism burrowing and vertical sediment transport), sediment dissolved organic carbon and humic substances, and sediment particulate organic carbon. At the present time there is no general theory, nor empirical relationships that are available, to estimate stationary bed sediment partition coefficients at solids concentrations of $>100,000$ mg/l.

In the biological sector, the toxicant may be taken up by an aquatic organism directly from the water as well as from contaminated prey. This is discussed in more detail in Section 8.9. The concentration in an organism is designated ν, usually in units of μg chemical/g wt. The details of uptake, depuration, and food

TABLE 8.6 SOME PROPERTIES OF SEVERAL ORGANIC CHEMICALS

Chemical name	Molecular wt (g/mole)	Solubility in water (mg/l)	Log octanol/water partition coefficient K_{ow}	Vapor pressure (atm)	H_e Henry's constant (atm · m³/mole)	Reference[a]
Chlordane	409.6	1.85	2.78	$1.3 \cdot 10^{-8}$	$4.8 \cdot 10^{-5}$	1
Chlorobenzene	112.6	296	2.98	0.016	$3.7 \cdot 10^{-3}$	2, 4
Methyl parathion	263.2	58	2.94	$1.3 \cdot 10^{-8}$	$2.4 \cdot 10^{-7}$	1, 4, 5
PCB-arochlor1242	328.4	0.002	6.72	$1.0 \cdot 10^{-7}$	$8.4 \cdot 10^{-3}$	1
Phenol	94.1	67,000	1.46	$7 \cdot 10^{-4}$	$1.3 \cdot 10^{-6}$	3, 1

[a]References:
(1) USEPA (1980).
(2) Miller et al. (1985).
(3) Veith et al. (1979).
(4) Mackay (1981).
(5) Bowman and Sans (1983).

chain accumulation calculations are discussed later. Here, the basic concepts are presented since the partitioning into aquatic organisms will prove important in the modeling of the fate of the chemical.

If then a fish is exposed to a chemical dissolved in water, the fish will absorb the chemical through its gills and by diffusion into the blood stream, and subsequently into the various organs and components of the fish. The time to equilibrium for fish absorption of the chemical usually takes considerably longer than the equilibrium time for solids sorption, that is, for fish, times of days to weeks to reach a steady state are common. After steady state has been reached and assuming that the fish food is not contaminated, the concentration in the fish is given by

$$\nu = \frac{k_u}{K'} c_d'$$ (8.27)

where k_u is the uptake coefficient $[L_w^3/T \cdot M(w)$ e.g., liter/day $\cdot$ g wet wt] and K' is the sum of the depuration rate of the chemical, $K[1/T; 1/\text{day}]$ and the growth rate of the fish, $G[1/T; 1/\text{day}]$, that is,

$$K' = K + G$$ (8.28)

Thus the concentration in the fish and the dissolved concentration in the water, c_d', are linearly related through a partition coefficient. For the biological system, this coefficient is called the bioconcentration factor (BCF) and is given by

$$N_w = \frac{\nu}{c_d'} = \frac{k_u}{K'}$$ (8.29)

The BCF, N_w, has units $[M_T/M(w) \div M_T/L^3; \mu g/g$ wet wt $\div \mu g/l]$. In a form similar to the solids partition coefficient, the BCF is in typical units of $\mu g/\text{kg}$ wet wt $\div \mu g/l$, that is, $(\text{ppb})_{\text{fish}}/(\text{ppb}_{\text{water}})$. Then

$$N_w' = 1000 N_w$$ (8.30)

A variety of equations have been developed to relate the bioconcentration factor (N_w') of a chemical to a specific property of that chemical such as water solubility or octanol–water partition coefficient (e.g., Chiou et al., 1977; Neeley et al., 1974; Kenaga and Goring, 1978; Veith et al., 1979; Kenaga, 1980). For example, Veith's relationship is

$$N_w' = 0.1995 K_{ow}^{0.85}$$ (8.31)

For the lipid soluble organic chemicals, however, one might expect that the BCF on a lipid basis $[\mu g/\text{kg (lipid)} \div \mu g/l]$ would be equal to K_{ow}, that is,

$$(N_w')_{\text{lip}} = K_{ow}$$ (8.32)

where

$$(N_w')_{\text{lip}} = N_w'/p_i$$ (8.33)

for p_i as the fraction of the wet weight of the organism that is lipid [kg (lipid)/kg (wet)].

TABLE 8.7 SOME BIOCONCENTRATION FACTORS AND EXCRETION RATES

Chemical	Organism	BCF N'_w (l/kg wet wt)	Excretion K (day^{-1})	Reference[a]
Diazinon	Topmouth gudgeon	152	0.79	1, 2
2-Ethylhexl diphenyl phosphate (EHDP)	Rainbow trout fry	1,314	0.72	3
Heptachlor	Spot	4,600	0.08	4
Dieldrin	Yearling lake trout	52,000	0.01	5
Copper	Juvenile rainbow trout	67	0.05	6
Methylmercuric chloride	Brook trout	12,000	0.0[b] 0.0006–0.002[c]	7

[a]References:
(1) Kanazawa (1978).
(2) Kanazawa (1981).
(3) Muir and Grift (1981).
(4) Schimmel et al. (1976).
(5) Reinert et al. (1974).
(6) Dixon and Sprague (1981).
(7) McKim et al. (1976).
[b]After 16 weeks.
[c]As quoted in (7).

Relationships such as these have proved useful (to about 1–2 orders of magnitude) in first-level water quality modeling and assessment of the potential impact of chemicals on the aquatic ecosystem. Specifically, if N'_w can be estimated for a chemical and measurements are made of the concentration of the chemical in the fish then an estimate can be made of the concentration of the chemical in the water via Eq. 8.29. For metals and other nonorganic substances, individual estimates of the BCF must be at hand. Table 8.7 shows some BCF values for various chemicals.

As seen, some chemicals (e.g., dieldrin) are concentrated by aquatic organisms to orders of magnitude above the water concentration. Other chemicals (e.g., copper) are only slightly concentrated. The BCF will prove quite useful since it will be assumed that the concentration in the aquatic organism is linear to the concentration in the water. This assumption (generally a good one) will be used in the next part on the fate of chemicals in the natural water system.

8.4 COMPLETELY MIXED LAKES

Considerable insight into the behavior of a toxicant as schematically shown in Fig. 8.8 can be gained by considering a special case: equilibrium or steady state conditions for a lake that is completely mixed in the water column and interacts with

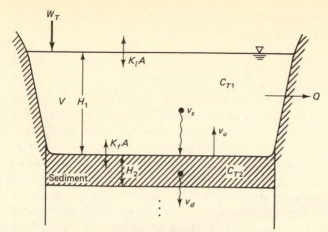

Figure 8.8 Notation for toxicant in completely mixed lake.

one sediment layer. The steady state assumption does not, of course, permit estimation of how long a sediment would take to depurate when the external load is eliminated. The steady state assumption is particularly useful for estimating concentrations for chemicals that will continue to be discharged or for screening chemicals to estimate maximum expected concentrations in water column and sediment.

8.4.1 Suspended Solids Mass Balance

Since as indicated above many chemicals sorb to suspended particulate matter, the first step in the development of a simplified model is the mass balance of suspended solids. A single class of solids is considered and is intended to incorporate inorganic solids and organic particulates (see also Section 4.2.1.1).

For a single completely mixed body of water interacting with a single sediment layer, the mass balance equation for the solids in the water column is given as

$$V \frac{dm}{dt} = W_m - Qm - v_s Am + v_u A m_s \qquad (8.34)$$

where m and m_s are the concentrations of solids in the water column and sediment, respectively, on a mass per bulk volume basis $[M_s/L^3_{s+w}]$, W_m is the mass input of solids $[M_s/T]$, v_s is the overall lake-wide average particle settling velocity $[L/T]$ of the particulates, v_u is the overall lake-wide resuspension velocity $[L/T]$ of the solids from the sediment to the water column and includes parametization of all sediment water particle interactions, and $A[L^2]$ is the interfacial area between the sediment and the water column. Equation 8.34 represents a balance of solids between: (a) input of solids externally (W_m) and internally from the flux due to sediment resuspension ($v_u A m_s$), (b) losses of solids due to flow transport from the lake (Qm) and settling from the water column ($v_s Am$), and (c) the time rate of change of the solids mass in the water column ($V \, dm/dt$). Since Eq. 8.34 depends on an interaction with the sediment, a similar equation must be written for the sediment segment underlying the water column. Thus,

$$V_s \frac{dm_s}{dt} = v_s Am - v_u Am_s - v_d Am_s \tag{8.35}$$

where v_d is the overall lake-wide average net sedimentation velocity of the surface sediment segment $[L/T]$. The equation for the solids in the sediment then is a flux balance between the incoming solids due to settling from the water column, loss due to resuspension, loss due to net sedimentation ($v_d Am_s$), and the time rate of change of solids in the sediment ($V_s \, dm_s/dt$). Note that the sediment is assumed to be stationary and interacting only with the overlying water column. It is assumed now that the solids concentrations are at steady state. Therefore, dm/dt and dm_s/dt in Eqs. 8.34 and 8.35 are zero. Equation 8.35 can then be solved for the sediment concentration and substitution into Eq. 8.34 yields an equation in terms of the water column solids concentration alone, that is,

$$0 = W_m - Qm - v_n Am \tag{8.36}$$

where v_n is the net loss of solids from the water column $[L/T]$ and is given by

$$v_n = \frac{v_s v_d}{v_u + v_d} \tag{8.37}$$

Equation 8.36 can be expressed on an areal loading basis as

$$m = \frac{W'_m}{q + v_n} \tag{8.38}$$

where W'_m is the areal load $[M_s/L^2 \cdot T]$ and q is the ratio of flow to surface area $[L/T]$, the "overflow rate" and is also given by

$$q = \frac{H}{t_d} \tag{8.39}$$

for depth H $[L]$ and hydraulic detention time t_d $[T]$.

With data on input solids, W'_m and observed solids concentration, m, in the water column, Eq. 8.38 then permits direct estimation of the net loss of solids, v_n. A mass balance around the sediment segment yields (on an areal basis)

$$v_n m_1 = v_s m_1 - v_u m_2 \tag{8.40}$$

and also

$$v_d m_2 = v_n m_1 \tag{8.41}$$

Equation 8.40 states that the net solids flux, $v_n m_1$, is the flux out of the water column minus the resuspension flux from the sediment. Equation 8.41 states that the net solids flux into the sediment layer from the water column $[M_s/L^2 \cdot T]$ is balanced by the solids flux leaving the sediment segment due to net sedimentation. Normal solids budget calculation describe only the balance of the fluxes of solids as in Eq. 8.36. For a toxic chemical model, however, it is necessary to separate out the suspended solids concentration from the net sedimentation flux. Also, since the solids concentration in the sediment represents the bulk density, the concentration is given by

$$m_2 = \rho_s(1 - \phi_2) \tag{8.42}$$

$$[M_s/L_{s+w}^3] = [M_s/L_s^3][L_s^3/L_{s+w}^3]$$

or

$$m_2 = \rho_s(1 - \phi_2) \cdot 10^6 \tag{8.43}$$

for ρ_s the density of the solids in g/cm^3 and m_2 in mg/l. If the density of the solids and the average porosity of the sediment are known, the net effective sedimentation velocity can be estimated from Eqs. 8.40 and 8.42 as

$$v_d = \frac{v_n m}{\rho_s(1 - \phi_2)} \tag{8.44}$$

8.4.2 Toxic Substances Model

For a completely mixed lake of volume V (Fig. 8.8), a mass balance of the total toxicant in the water column, c_{T1}, and in the sediment, c_{T2}, recognizing the mechanisms described earlier, is given by

$$V_1 \frac{dc_{T1}}{dt} = W_T - Qc_{T1} + K_f A(f_{d2}c_{T2}/\phi_2 - f_{d1}c_{T1}) - (K_{d1}f_{d1})V_1c_{T1} \tag{8.45}$$

$$+ k_l A \left[(c_g/H_e) - f_{d1}c_{T1}\right] - v_s A f_{p1}c_{T1} + v_u A f_{p2}c_{T2}$$

$$V_2 \frac{dc_{T2}}{dt} = - K_f A(f_{d2}c_{T2}/\phi_2 - f_{d1}c_{T1}) - (K_{d2}f_{d2})V_2c_{T2} \tag{8.46}$$

$$+ v_s A f_{p1}c_{T1} - v_u A f_{p2}c_{T2} - v_d A f p_2 c_{T2}$$

The description of each term in Eq. 8.45 is as follows. The first and second terms on the right side are the input of the total toxicant (W_T) and the outflow of the toxicant from the lake (at a flow Q). The third term is the diffusive exchange of dissolved toxicant between the sediment and the water column (with a diffusion rate of K_f [L/T] and a surface area of A; $\phi_1 = 1$). The fourth term is the degradation of the dissolved form (K_d) due to microbial decay, photolysis, hydrolysis, and so on in the water column (decay of particulate form is assumed zero); the next term is the air–water exchange of the toxicant due to volatilization or gaseous input (at an exchange rate k_l [L/T], vapor phase concentration c_g, and Henry's constant H_e; see Section 8.8.2.1). The next to last term is the settling of the particulate toxicant from the water column to the sediment at velocity v_s and the last term is the resuspension into the water column of the particulate toxicant from the sediment at velocity v_u.

For the sediment, the first term on the right of Eq. 8.46 is the diffusive exchange of dissolved toxicant between water column and the sediment, the second term represents the decay processes in the sediment (again decay of particulate form in the sediment is assumed zero), the third term is the source of toxicant to the sediment from particulate settling from the overlying water column, the next term is the resuspension flux from sediment to water column, and the last term is the loss of toxicant from the sediment due to net sedimentation or burial at a velocity v_d. No exchange is assumed between the surface sediment layer and the deeper sediment. Additional sediment layers can, of course, be added to account for this effect.

Assuming a steady state condition, it can then be shown (Thomann and Di Toro, 1983) that

$$c_{T1} = \frac{W_T'}{q + v_T} \tag{8.47}$$

where W_T' is the areal loading rate of the toxicant $[M_T/L^2 \cdot T]$ and v_T is the overall net loss rate of the chemical from the lake within which is contained all of the sediment–water, air–water, and decay interactions. Note the similarity of Eq. 8.47 to the analysis by Vollenweider of phosphorus in lakes described in Chapter 7 (see Eq. 7.5).

The net loss rate can be expressed as (Thomann and Salas, 1986)

$$v_T = v_{Td} + v_{Ts} \tag{8.48}$$

where

$$v_{Td} = \text{dissolved chemical loss}$$
$$= (K_{d1}H_1 + k_l)f_{d1} \tag{8.49}$$

and

$$v_{Ts} = \text{loss due to sediment interaction}$$
$$= v_n\eta'\left[1 + \frac{m_2(K_{d2}f_{d2}H_2)}{m_1v_n}\right], \qquad \text{for} \quad K_{d2} > 0 \tag{8.50}$$

where

$$\eta' = \frac{v_sf_{p1} + K_ff_{d1}}{v_s + f_{d2}(m_2/m_1)(K_f + K_{d2}H_2)} \tag{8.50a}$$

From a basic mass balance, when $K_{d2} = 0$, $v_{Ts} = v_n\eta'$ and thus, if $v_n = 0$, $v_{Ts} = 0$. It can also be shown (Di Toro et al., 1981, 1982b) that

$$\frac{r_2}{r_1} = \delta = \frac{(v_u + v_d)f_{p2} + K_f(\P_2/\P_1)f_{d2}/\phi_2}{(v_u + v_d)f_{p2} + K_ff_{d2}/\phi_2 + K_{d2}f_{d2}H_2} \tag{8.51}$$

The total toxicant in the sediment on a bulk volume basis is given by

$$c_{T2} = \delta\left[\frac{m_2}{m_1}\frac{f_{p1}}{f_{p2}}\right]c_{T1} \tag{8.52}$$

and the relationship between r_2 and c_{T1} is

$$r_2 = \delta\left(\frac{f_{p1}}{m_1}\right)c_{T1} \tag{8.53}$$

These results show that for steady state, the concentration in the water column given by Eq. 8.47 is identical to the simple water quality constitutent for a completely mixed case (see Chapter 4). Figure 8.9 shows the relationship of Eq. 8.47 for a range of hydraulic loading rates and net loss rates and the resulting ratio of chemical loading to in-lake concentration. For the figure, v_T is assumed as always negative (sink) although it need not be in general. It can be noted that for hydraulic loadings of >100 m/yr, the allowable loading to in-lake concentration ratio is constant over a wide range of v_T, that is, the lake hydraulics dominate the kinetics and "flushing" is more important than all of the chemical interactions. For other water bodies, however, such is not the case, although for lakes with lower overflow rates, there is still a range over which the allowable loading is independent of v_T.

Therefore, for a given v_T and q for a particular toxicant and lake, W_T'/c_{T1} may be calculated. If the water quality standard is in terms of the total chemical, then the discharge load that will result in that concentration at steady state is

$$(W_T')_{\text{allow}} = (q + v_T)c_{T1} \tag{8.54}$$

Note that this relationship is for steady state and therefore represents a conservative estimate of the allowable load since the time to steady state may be long (on the order of years and longer).

The central question of course is: "How does one determine v_T, the net loss of the chemical?" There are two general approaches:

1. From the theoretical formulation of v_T given by Eqs. 8.48 to 8.53 with or without further assumptions.

Figure 8.9 Relationship between hydraulic loading, net loss rate of chemical and chemical loading/in-lake concentration ratio.

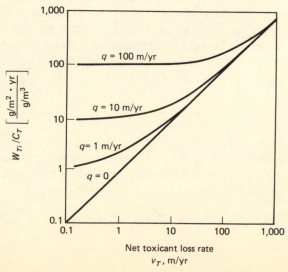

2. From field data on all other terms in Eq. 8.47 and solving for v_T. Thus,

$$v_T = \frac{W_T'}{c_{T1}} - q \tag{8.55}$$

If field data are not available, then the theoretical form of v_T given by Eqs. 8.48 to 8.53 must be used. The calculation of v_T from the equations involves estimates of the various parameters for the dissolved loss rate v_{Td} and the sediment interaction loss rate, v_{Ts}. Before proceeding, however, to the estimates of these parameters for v_T, it is informative to continue and present the model equations for the riverine and finite difference estuarine cases where an interesting result will emerge. At the end of those models, interest will be directed again to the determination of v_T using the theoretical equations 8.48 to 8.53.

8.5 RIVERS AND STREAMS

A simplified approach to describing the distribution of a chemical in a river can be obtained if one is willing to make several key assumptions. Specifically, the assumptions that the river is in a steady state and that the bed sediment is not moving (but may be resuspended) permit substantial simplifications in the analysis. These assumptions indicate that the following analysis is not applicable to the transient events of deposition, scour, and downstream redeposition associated with rapid flow increases. However, for first estimates of the distribution of toxic substances in rivers, it is generally not feasible to carry out detailed time variable sediment transport modeling coupled to a chemical model. Therefore, in spite of the severity of the two aforementioned assumptions, the simplicity of the final results and resulting ease of use urges consideration of the simplified approach for the first-level toxic substances allocations in rivers.

Consider the completely advective stream or river where there is no mixing or dispersion (see Chapter 2 for a discussion of water quality in rivers). For this case, the following conditions apply: an input at $x = 0$, steady-state conditions, constant parameters over a given stretch of river (i.e., solids and reaction rates), and a river flow, Q, with velocity U. Consideration is given first to the solids balance in the river and then to the toxic substance equation.

8.5.1 Solids Balance

As with the lake, the water column solids in a river represent the balance between settling and resuspension. Assuming a reach of the river with constant parameters of settling, resuspension, and deposition, the mass balance equations in the water column and sediment—for the net settling or zero settling case—are:

$$U \frac{dm_1}{dx} = -\frac{v_s}{H_1} m_1 + \frac{v_u}{H_1} m_2 \tag{8.56}$$

and

$$0 = v_s m_1 - v_u m_2 - v_d m_2 \tag{8.57}$$

The equation for the water column solids is then

$$U \frac{dm_1}{dx} = -\frac{v_n}{H_1} m_1(x) \tag{8.58}$$

where v_n is the net solids loss in a reach of the river. This loss is the same expression as the completely mixed lake case (Eq. 8.37), that is,

$$v_n = \frac{v_s v_d}{v_u + v_d}$$

The solution to Eq. 8.58 is

$$m_1 = m_1(0) \exp\left(\frac{v_n}{H_1 U} x\right) \tag{8.59}$$

where $m_1(0)$ is the solids concentration at $x = 0$, the beginning of the reach. For a net loss of solids to the bottom of the river, settling exceeds resuspension and for zero loss of solids, settling is balanced by resuspension. For small streams, $v_n \cong 0$ and the suspended solids behave as a conservative variable. For larger deeper streams, there may be a net deposition of solids and $v_n > 0$.

During certain conditions and reaches of a river, a solids increase may occur in the downstream direction without any external tributaries or drainage input of solids. The solids must then be supplied by the bed sediment through net erosion. This results in $v_n < 0$ and an exponential rise in solids with distance. Alternately, the bed solids concentration can be specified as a spatial constant in the water column equation. Then, following O'Connor (1985) and Delos et al. (1984), the solution for a spatially constant value of m_2 in Eq. 8.56 is

$$m_1 = m_1(0) \exp\left(-\frac{v_s}{H_1 U} x\right) + \frac{v_u m_2}{v_s}\left[1 - \exp\left(-\frac{v_s}{H_1 U} x\right)\right] \tag{8.60}$$

This equation shows the balance between settling (not net settling) and resuspension. If $m_1(0)$ is small then the solids profile builds up to the steady value at $x = \infty$ of

$$m_1(\infty) = \frac{v_u m_2}{v_s} \tag{8.61}$$

and if m_1 is estimated from the solids profile and m_2 and v_s specified, then v_u, the resuspension velocity, can be estimated. O'Connor (1985) suggests using this procedure in the analysis of solids profiles to estimate v_u by extrapolating downstream to $m_1(\infty)$.

The region of net erosion may have to be analyzed in the calibration of a toxics river model since toxicant may have been deposited during a previous depositional low flow period and then be subject to net erosion during a high flow period.

8.5.2 Toxic Substances Model

Assuming a stationary sediment, constant kinetic coefficients, and spatially constant suspended solids in a given reach of a river, the differential equation for the total toxicant concentration in the water column, c_{T1}, is given by

$$U\frac{dc_{T1}}{dx} = K_f A(f_{d2}c_{T2}/\phi_2 - f_{d1}c_{T1}) - K_{d1}f_{d1}c_{T1}$$

$$+ k_l A\left(\frac{c_g}{H_e} - f_{d1}c_{T1}\right) - v_s A f_{p1}c_{T1} + v_u A f_{p2}c_{T2} \qquad (8.62)$$

For the stationary sediment underlying the moving water column, the mass balance equation is

$$0 = -K_f A(f_{d2}c_{T2}/\phi_2 - f_{d1}c_{T1}) - (K_{d2}f_{d2})c_{T2}$$
$$+ v_s A f_{p1}c_{T1} - v_u A f_{p2}c_{T2} - v_d A f_{p2}c_{T2} \qquad (8.63)$$

Notice that the latter equation is an algebraic equation and does not contain a transport derivative, that is $U \, dc/dx$, because the bed is assumed fixed in place. Thus each longitudinal "slice" of the river only sees the water moving past it and does not travel along with the water. As a result, Eq. 8.63 can be solved explicitly for c_{T2} and substituted into Eq. 8.62. After some simplification and assuming no upstream toxic concentration, the solution for the water concentration is given by

$$c_{T1} = \frac{W_T}{Q} \exp\left[-\left(\frac{v_T}{H_1}\right)\left(\frac{x}{U}\right) \right] \qquad (8.64)$$

and the sediment concentration of the toxicant is given by

$$c_{T2} = \delta\left[\frac{m_2}{m_1}\frac{f_{p1}}{f_{p2}}\right]c_{T1} \qquad (8.65)$$

which is the same as for the completely mixed lake (Eq. 8.52). The net loss rate of the toxicant, v_T, is given by Eqs. 8.48 to 8.50 and r_2/r_1 is also as given previously by Eq. 8.51. Thus, the net loss of the chemical in the downstream direction from the input and all the interactive sediment properties are the same as for the completely mixed lake. This initially surprising result is a consequence of the assumption that the sediment bed is not moving.

These simple results show that at steady state: (a) the water column and sediment toxicant concentration decrease exponentially at the same rate, v_T/HU away from the source, (b) the behavior is similar to classical water quality behavior in streams, and (c) the ratio of particulate toxicant concentration in the sediment to that in the water (Eq. 8.51) is identical to that in lakes and is independent of the source strength. Since the sediment concentration on a solids basis is often measured (i.e., r_2 in units of $\mu g/g$ dry wt), then from Eqs. 8.53 and 8.64

$$r_2(x) = \left[\left(\delta\frac{f_{p1}}{m_1}\right)\right]\frac{W_T}{Q} \exp\left(-\frac{v_T}{H_1 U}x\right) \qquad (8.66)$$

Therefore, if data are available on water or sediment concentrations of a toxicant as a function of distance, Eqs. 8.64 and 8.66 indicate that a semilog plot of that data with distance permits estimation of the net loss of the toxicant in the stream. Figure 8.10 shows these relationships. It can be recalled from the lake analysis that the net loss rate v_T embodies all of the complex interactions between the water column and sediment. One of the key parameters then in modeling the behavior of a toxicant in a stream is the net toxicant loss rate. As in classical stream water quality analysis discussed in Chapter 2, estimation of v_T permits calculation of the downstream effect of a discharge.

In Eqs. 8.64 through 8.66, the solids in the water column and bed are assumed spatially constant. Thus, for a stream where solids vary significantly, the above solutions are applicable to a finite length of stream within which the solids may be approximated by a constant concentration.

Reynolds and Gloyna (1964) and Shih and Gloyna (1964) were some of the earliest groups to apply the above basic framework for radionuclides where the principal interest was in the interaction with the sediment due to adsorption and diffusion.

Figure 8.10 Downstream distribution of toxicant in river—loss rates in water column and sediment are equal—no bed sediment movement.

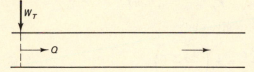

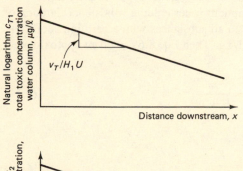

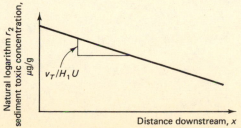

8.5.2.1 Simplification—Heavy Metals and Radionuclides   Assume that heavy metals and long-lived radionuclides neither decay nor volatilize. Then the net loss of the toxicant can be shown from Eqs. 8.48 ff to be

$$v_T = v_n f_{p1} \frac{r_2}{r_1} \tag{8.67}$$

or

$$v_T = v_n m \frac{r_2}{c_{T1}} \tag{8.68}$$

where v_n is the net loss of solids from the water column to the bed sediment in a given reach of the river. This analysis then applies to the river situation where solids are not undergoing net resuspension into the water column. Equation 8.67 assumes also that the change in the fraction particulate in the water column, f_{p1}, is not significant over the reach of the river even though the solids concentration may change from one end to the other due to net deposition to the bed sediment.

Under these conditions (zero decay in the sediment and constant water column solids in the reach), it can be shown from Eq. 8.51 that r_2/r_1, in Eq. 8.67, is given by

$$\frac{r_2}{r_1} = \frac{m}{mv_s + K_f/\P_2} \left(v_s + \frac{K_f}{\P_1 m} \right) \tag{8.69}$$

The second term in brackets is also given by

$$\frac{K_f}{\P_1 m} = K_f \left(\frac{1}{f_{p1}} - 1 \right) \tag{8.70}$$

Now consider those substances (e.g., heavy metals) that have a partition coefficient that results in a fraction particulate, f_{p1}, of approximately equal to or greater than 0.1. For streams with minimum water column solids concentration of about 10 mg/l, this would correspond to all substances with a partition coefficient equal to or greater than about 10,000 l/kg. Then Eq. 8.69 is given approximately by

$$\frac{r_2}{r_1} = \frac{m}{B + m} \tag{8.71}$$

where

$$B = \frac{K_f}{\P_2 v_s} \tag{8.72}$$

Since

$$r_1 = \frac{f_{p1}}{m} c_{T1} \tag{8.73}$$

Equation 8.71 can also be written as

$$\frac{r_2}{c_{T1}} = f_{p1}\left(\frac{1}{B + m}\right) \tag{8.74}$$

or

$$\Pi = \left(\frac{1}{\P_1^{-1} + m}\right)\left(\frac{m}{B + m}\right) \tag{8.75}$$

where $\Pi = r_2/c_{T1}$, a field partition coefficient between sediment and water. Data are sometimes available on this ratio of the sediment toxicant concentration (r_2) to the total water column concentration (c_{T1}). Equations 8.74 and 8.75 indicate that this partitioning should be a function of the solids concentration, the partition coefficient $\P_1$ and the coefficient B representing the relationship between sediment diffusion and settling. These equations can be used in Eq. 8.68 to estimate v_T.

For example, for heavy metals, $\P_1$ is about 10^4 l/kg ($\P_1^{-1} = 100$ mg/l). Analyses of field data indicate B is in the range of 100–300 mg/l. For $m = 10$ mg/l, $B = 200$ mg/l and $\P_1 = 10^4$ l/kg, $\Pi = 0.0004$ l/mg. Equation 8.68 then indicates that $v_T = 0.004\ v_n$.

Laszlo et al. (1975) and Literathy and Laszlo (1977) have presented data on the distribution of heavy metals in the Sajo River in Hungary which permit estimation of the stream net loss rate. Figure 8.11 shows the approximate distribution of cadmium in the sediment and the net loss rate calculated at about 0.12 m/day.

8.6 ESTUARIES

Applying the preceding assumptions of the toxic substances analysis to estuaries, that is, no bed movement, constant solids concentrations, and steady state, then

Figure 8.11 Sediment cadmium concentrations vs. distance. From Thomann (1984); data from Laszlo et al. (1975) and Literathy and Laszlo (1977). Reprinted by permission of Elsevier Science Publishers B.V.

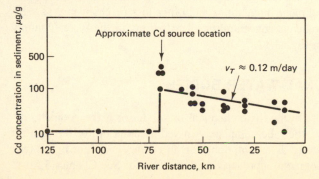

the basic differential equation is

$$0 = -U \frac{dc_{T1}}{dx} + E \frac{d^2 c_{T1}}{dx^2} + K_f A(f_{d2} c_{T2}/\phi_2 - f_{d1} c_{T1})$$

$$- (K_{d1} f_{d1}) c_{T1} + k_l A \left(\frac{c_g}{H_e} - f_{p1} c_{T1} \right)$$

$$- v_s A f_{p1} c_{T1} + v_u A f_{p2} c_{T2} \qquad (8.76)$$

for the water and, for the sediment, the equation is the same as Eq. 8.63. It is obvious by now that the simplified estuary toxic distribution in the water column with a stationary bed sediment will be the same as the distribution of any nonconservative substance (see Chapter 3) with a net loss rate v_T given again by Eqs. 8.48 to 8.51. Thus the solution (see Eqs. 3.8a to 3.8f) for the water column is

$$c_{T1} = c_{T0} \exp(j_1 x) \qquad x \leq 0 \qquad (8.77)$$
$$= c_{T0} \exp(j_2 x) \qquad x \geq 0 \qquad (8.78)$$

where

$$c_{T0} = \frac{W_T}{Q \sqrt{1 + \frac{4(v_T/H_1)E}{U^2}}} \qquad (8.79)$$

and

$$j_1 = \frac{U}{2E} \left[1 + \sqrt{1 + \frac{4(v_T/H_1)E}{U^2}} \right] \qquad (8.80)$$

and

$$j_2 = \frac{U}{2E} \left[1 - \sqrt{1 + \frac{4(v_T/H_1)E}{U^2}} \right] \qquad (8.81)$$

The sediment concentration of the toxicant tracks the water column and is given by the preceding Eqs. 8.52 and 8.53. Thus, to first approximation, the distribution of a toxicant in an estuary behaves in a classical fashion, that is, declining upstream and downstream of the outfall. The rate of decline is a complicated function of the sediment interactions and other loss mechanisms, as well as tidal dispersion. The toxicant concentration distributions in the water column and sediment are shown in Fig. 8.12.

8.7 MULTIDIMENSIONAL WATER BODIES

If significant gradients exist in the toxic chemical or solids, or if the geometry of the water body is complex, then the completely mixed assumption for the lake or the constant geometry and solids assumptions for an estuary may not hold. The finite difference or finite segment approach can then be used. See Chapter 3 for a detailed discussion.

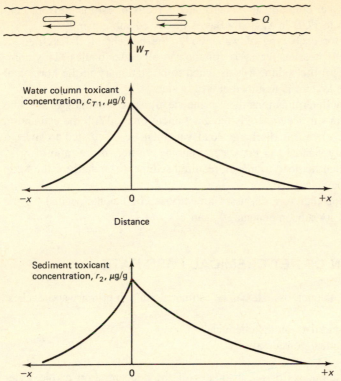

Figure 8.12 Distribution of toxicant in water column and sediment of estuary—stationary bed sediment.

Consider then an estuary divided into five segments with a sediment segment underlying each water column segment. In each segment, the mechanisms included in Eqs. 8.45 and 8.46 apply but now each water column segment is connected to an adjacent segment by flow and horizontal mixing processes. However, it can be shown that the mass balance at steady state for any segment includes the net loss of toxicant v_T in the same form as given in Eqs. 8.48 through 8.50 and that the relations given in Eqs. 8.51 through 8.53 are applicable. Thus for the second segment,

$$0 = W_{T2} + [Q_{12}c_{T1} - Q_{23}c_{T2} + E'_{12}(c_{T1} - c_{T2}) + E'_{23}(c_{T3} - c_{T2})] - v_{T2}A_2c_{T2} \tag{8.82}$$

where the subscripts on Q and E' (the bulk dispersion coefficient, L^3/T) represent the exchanges between adjacent segments and where v_{T2} is the net loss of toxicant from the water column for seg. 2 and is given by Eqs. 8.48 to 8.50 with parameters particular to seg. 2. Since the net loss rate is segment specific with respect to the water column in a multidimensional system, calibration to observed data permits estimates of the net loss of chemical from the water column. Such a net loss includes all interactions with the sediment and atmosphere. Therefore, even for multidimensional systems (with a stationary sediment), the net loss rate of the chemical from the water column can be calculated directly or estimated from field data. An application to PCB in Saginaw Bay is discussed subsequently.

In summary, for each type of water body, the net chemical loss rate v_T appears, where the expression for v_T is common to all water bodies. This commonality is because (a) the sediment is assumed to be stationary in the horizontal direction and (b) the system is assumed at steady state.

One of the key determinations then in the steady state modeling of the fate of toxic substances in surface waters is the assessment of v_T. With the estimation of v_T, an allowable chemical discharge load can then be calculated to meet a required water quality standard in the water column. Further, if the ratio of r_2/r_1 can be determined then the load to meet a required sediment standard for the toxic substance can also be calculated.

The next section therefore discusses the various coefficients needed for the estimation of v_T and its subcomponents v_{Td} and v_{Ts}.

8.8 ESTIMATION OF NET CHEMICAL LOSS RATE

As noted previously, the net loss rate can be estimated via two primary approaches:

1. From the field data on the chemical.
2. From theoretical considerations.

For field measurements of v_T, estimates of the concentration in the water column c_{T1} or in the sediment r_2 are necessary. If data are available for the concentration of the toxicant in a biomonitor such as fish or shellfish, then an estimate can be made of c_{T1} by relationships between v, the concentration in the fish and c_{T1}. This is reviewed in Section 8.9.

8.8.1 Theoretical Estimate of v_T

If data are not available on the toxicant in the field from which v_T can be calculated, for example, from Eq. 8.55, then an estimate can be made from the theoretical form of v_T in Eqs. 8.48 to 8.50. Note from Eq. 8.48 that v_T is composed of two components: the loss of the dissolved fraction due to all decay and exchange processes (e.g., biodegradation, volatilization) and the loss due to interaction with the sediment. The latter loss is shown to depend on the net loss of solids, v_n.

Three levels of complexity on the sediment interaction can then be identified each with successively less restrictive assumptions.

1. Level 1 analysis assumes that:
   a. $\P_2 = \P_1$, that is, the partition coefficient in the sediment is equal to that in the water column.
   b. Sediment decay of the toxicant is zero, that is, $K_{d2} = 0$.
2. Level 2 analysis drops the assumption that $\P_2 = \P_1$, but continues to assume sediment decay $= 0$.
3. Level 3 analysis makes no assumptions about $\P_2$, $\P_1$ and K_{d2}.

Level 1 analysis is useful for rapid assessment since, as will be seen below, the two assumptions result in a particularly simple model framework. Level 2 analysis applies to heavy metals, radionuclides, or recalcitrant toxicants where, to first approximation, it can be assumed that the decay in the sediment is approximately zero. Level 3 analysis is the most complex and requires information about, or estimation of, each of the parameters of v_T. Figure 8.13 shows the various parameters that must be specified at each level of analysis. As noted, only three parameters are needed for the level 1 analysis with an increasing number of parameters required at levels 2 and 3.

8.8.2 Dissolved Chemical Loss Rate

The determination of v_{Td} involves estimation of the following mechanisms:

1. volatilization
2. photolysis
3. hydrolysis
4. biodegradation

Reviews of those mechanisms have been prepared by Di Toro et al. (1981), Mills et al. (1982), Connolly and Winfield (1984), and Delos et al. (1984).

The above losses are summed to obtain the total loss of the dissolved chemical, v_{Td}. Thus following Eq. 8.49,

$$v_{Td} = (K_{d1}H_1 + k_l)f_{d1} \tag{8.83}$$

where

$$K_{d1} = K_p + K_H + K_B \tag{8.84}$$

Figure 8.13 Parameters required for calculation of sediment interaction loss rate of toxicant v_{Ts} for three levels of assumptions. From Thomann and Salas, 1986.

Level 1 $\begin{pmatrix} \P_1 = \P_2 \\ K_{d2} = 0 \end{pmatrix}$ Equal partition coefficients zero sediment decay	Level 2 $\begin{pmatrix} \P_1 \neq \P_2 \\ K_{d2} = 0 \end{pmatrix}$ Unequal partition coefficients zero sediment decay	Level 3 $\P_1 \neq \P_2$ $K_{d2} \neq 0$ Unequal partition coefficients nonzero sediment decay
v_n $\P_1$ m_1	v_n $\P_1$ m_1	v_n $\P_1$ m_1
	$\P_2$ m_2 v_s K_f	$\P_2$ m_2 v_s K_f
		H_2 K_{d2}

Following a discussion of volatilization (k_l) each of the terms on the right side of the equation for K_{dl} are discussed below as K_p = the photolysis rate; K_H = the hydrolysis rate, and K_B = the microbial biodegradation rate. As noted below, the estimation of these rates in the absence of field or experimental data can easily span an order of magnitude for a given organic chemical. Thus, the sensitivity of the final model analysis to this range is necessary for assessment of a waste input or for a calibration and verification analysis.

8.8.2.1 Volatilization The mass exchange of the chemical across the air–water interface can be derived from the difference between exchange from water to air and from air to water. Thus

$$V_1 \frac{dc_{T1}}{dt} = k_l A \left(\frac{c_g}{H_e} - f_{d1} c_{T1} \right) \tag{8.85}$$

where all terms have been defined previously. Note that the dissolved water concentration volatilizes as given by $f_{d1} c_{T1}$ where $\phi_1 = 1$. The equation shows that the flux may be from the air to the water ($c_g/H_e > f_{d1} c_{T1}$) or from the water to the air ($f_{d1} c_{T1} > c_g/H_e$). For many cases, c_g can be assumed zero, so that net volatilization occurs and the toxicant is lost to the atmosphere. The determination of k_l, the overall exchange rate, is then the focus of estimating the importance of the volatilization loss.

As discussed in the above references and as also reviewed by Mackay (1981), the application of the "two film" theory results in the overall volatilization transfer coefficient being given as

$$\frac{1}{k_l} = \frac{1}{K_l} + \frac{1}{K_g H_e} \tag{8.86}$$

where

$$K_l = \text{the liquid film coefficient } [L/T]$$

and

$$K_g = \text{the gas film coefficient } [L/T]$$

As seen from this equation, k_l depends on the chemical (through H_e and also incorporated in K_l and K_g) and through the characteristics of the water body such as velocity (affecting K_l) and wind speed over the water (affecting both K_l and K_g).

The Henry's constant represents a partitioning of the toxicant between the water and atmospheric phases and in general is given by

$$H_e' = \frac{p(@\text{eq})}{c_w(@\text{eq})} \tag{8.87}$$

where $p(@\text{eq})$ and $c_w(@\text{eq})$ are the equilibrium partial pressure in atm and water solubility concentration in mole/m^3 after equilibrium has been reached. Thus in Eq. 8.87, H_e' has the units atm $\cdot$ m^3/mole. The Henry's constant of selected chemicals may be found in Table 8.6. More complete listings of H_e' (and p and c_w) are given in handbooks such as Lyman et al. (1982).

In dimensionless form, H_e in mole/m^3 $\div$ mole/m^3 is

$$H_e = \frac{H_e'}{RT} \tag{8.88}$$

for R, the universal gas constant, $= 8.206 \times 10^{-5}$ atm $\cdot$ m^3/$^\circ$K $\cdot$ mole and T in $^\circ$K. The dimensionless ratio for H_e is represented by

$$H_e = \frac{c_g}{c_w} \tag{8.89}$$

and it is in the dimensionless form that H_e is used in Eq. 8.86.

As can be seen from Eq. 8.86, the overall volatilization rate is approximately equal to the liquid film mass transfer coefficient when Henry's constant is large. The system is then said to be controlled by the liquid phase resistance. This can be seen also from Eq. 8.87 which shows that for high H_e, the chemical partitions relatively easily into the gaseous phase and less so in the liquid phase. There will then be a "resistance" to transfer from the liquid phase into the gaseous phase.

Conversely, if the Henry's constant is "small," then the resistance is in the gas phase and the system is said to be gas phase limited. When both K_l and $K_g H_e$ are of the same magnitude, then both phases contribute to the overall transfer.

The calculation of k_l requires calculation of K_l and K_g. The liquid film coefficient is related to the oxygen reaeration rate. Thus, as reviewed in Mills et al. (1982) and drawing on the work of Smith et al. (1981), the liquid film coefficient is given by

$$K_l = \left[\frac{D_l}{D_L}\right]^n \cdot K_L \tag{8.90}$$

where K_L = oxygen transfer rate at ambient water temperature $[L/T]$
   D_l = diffusivity of toxicant in water $[L^2/T]$
   D_L = diffusivity of oxygen in water $[L^2/T]$
   n = coefficient reflecting turbulent conditions and theoretical approach having a range $0.5 \leq n \leq 1$

Mills et al. suggest n at 0.5 to reflect normal environmental conditions.

This equation is further simplified by Mills et al. (1982) through use of relationships between the diffusivity of a chemical and its molecular weight (M) to the following estimating equation

$$K_l = \left(\frac{32}{M}\right)^{1/4} K_L \tag{8.91}$$

Rathbun and Tai (1981) used this approach to estimate the volatilization rate for four liquid film controlled chemicals (benzene, chloroform, methylene chloride, and toluene) and obtained the following equation

$$K_l = 0.655 K_L \tag{8.92}$$

Equation 8.91 gives a range of 0.7–0.8 for the constant 0.655 for the chemicals used by Rathbun and Tai. Mills et al. then conclude that Eq. 8.91 gives a good estimate of the liquid film volatilization transfer rate. The determination of the liquid film coefficient is then really a problem in determining the reaeration coefficient (see Chapter 6 for discussion). For open bodies of water where wind effects are important, O'Connor (1985) and Di Toro et al. (1981) suggest

$$K_l = 0.17(C_d)\left(\frac{D_l}{\nu_l}\right)U_w \tag{8.93}$$

for C_d equal to the drag coefficient (approximately 0.001), ν_l the kinematic viscosity of water (0.1 cm^2/s), and U_w the wind speed in m/s.

The gas film transfer coefficient can be estimated from O'Connor (1985) and Di Toro et al. (1981) from analysis of the gas film transfer coefficient for water vapor and, including effects of diffusivity, they suggest the following for K_g

$$K_g = 0.001\left(\frac{D_g}{\nu_g}\right)^{0.67} U_w \tag{8.94}$$

where D_g = diffusivity of chemical in atmosphere (cm^2/s)
   ν_g = kinematic viscosity of air (approximately 0.15 cm^2/s)

The diffusivity can be related to the molecular weight of the chemical (M) as (Mills et al., 1982)

$$\frac{D_g}{D_{Mw}} = \left(\frac{18}{M}\right)^{1/2} \tag{8.95}$$

where D_{Mw} = diffusivity of water vapor in air (0.239 cm^2/s).

Mills et al. (1982) following an analysis framework similar to O'Connor combine this diffusivity equation with an equation similar to Eq. 8.94 and suggest

$$K_g = 168\left(\frac{18}{M}\right)^{1/4} U_w \tag{8.96}$$

for K_g in m/day and U_w in m/s. Calculation of the overall exchange rate, k_l, is illustrated in Sample Problem 8.2.

Figure 8.14 shows the behavior of k_l as a function of H_e for different molecular weights, two wind speeds, and two reaeration coefficients using Eqs. 8.91 and 8.96. The plateau reflects liquid film control while the decline indicates increasing gas film resistance. It is seen that k_l is not sensitive to the molecular weight and is generally not sensitive to log H_e over the range 0 to -3.

8.8.2.2 Photolysis   In this decay process, solar energy acts on certain chemical molecules in such a way as to alter the molecular structure. The chemical then degrades or is decomposed by the change in molecular structure and as such this action of solar energy on a molecule is a decay process. The subsequent product of the reaction, however, may still be toxic so that it should not necessarily be assumed that a toxic substance problem has been resolved if a parent chemical is photolyzed. Mills et al. (1982) and Delos et al. (1984) provide reviews of the overall mechanism.

SAMPLE PROBLEM 8.2

DATA

Chlorobenzene is discharged into a stream where the wind speed is 5 m/s and the water temperature is 20°C. The depth of the stream is 0.4 m and the velocity is 0.60 m/s.

PROBLEM

Compute the overall volatilization rate of the chlorobenzene using the following chemical properties:

$$\text{molecular weight, } M = 113 \text{ g/mole}$$
$$\log K_{ow} = 3.0$$
$$\text{Henry's constant, } H_e = 0.0037 \text{ atm·m}^3/\text{mole}$$

ANALYSIS

$$\frac{1}{k_l} = \frac{1}{K_l} + \frac{1}{K_g H_e} \qquad \text{[Eq. (8.86)]}$$

Approximate estimate from Fig. 8.14: Oxygen transfer coefficient can be estimated from Eq. 6.26:

$$K_L = \left(D_L \frac{U}{H} \right)^{1/2} \text{ for } D_L = \text{oxygen diffusivity} = 0.000181 \text{ m}^2/\text{day}$$
$$= (0.000181 \text{ m}^2/\text{day} \cdot 0.60 \text{ m/s} \cdot 86{,}400 \text{ s/day} \div 0.4\text{m})^{1/2}$$
$$= \underline{4.84} \text{ m/day}$$

Figure 8.14 shows that for K_L = 5 m/day, windspeed = 5 m/s and $\log H_e = -2.43$, the overall volatilization rate for chlorobenzene is about 3.5 m/day. Check this result with the analytic solutions below.

Liquid film coefficient:

$$K_l = \left(\frac{32}{M} \right)^{1/4} K_L \qquad \text{[Eq (8.91)]}$$
$$K_l = \left(\frac{32}{113} \right)^{1/4} (4.84 \text{ m/day}) = \underline{3.53} \text{ m/day}$$

Gas film coefficient:

$$K_g = 168 \left(\frac{18}{M} \right)^{1/4} U_w \qquad \text{[Eq. (8.96)]}$$
$$= 168 \left(\frac{18}{113} \right)^{1/4} (5)$$
$$K_g = \underline{531} \text{ m/day}$$

(continued)

Sample Problem 8.2 (continued)

Dimensionless Henry's constant:

$$H_e = \frac{H'_e}{RT} \qquad \text{[Eq. (8.88)]}$$

$$= \frac{0.0037 \text{ atm·m}^3/\text{mole}}{8.206 \cdot 10^{-5} \dfrac{\text{atm·m}^3}{\text{mole·°K}} \cdot 293°\text{K}}$$

$$= \underline{0.154}$$

Substitute into Eq. 8.86 above:

$$\frac{1}{k_l} = \frac{1}{3.53} + \frac{1}{531(0.154)}$$

$$= 0.283 + 0.0122 \text{ (liquid film controlled)}$$

$$\therefore \underline{k_l = 3.38 \text{ m/day}}$$

Briefly, the photolysis rate of decay depends on:

1. The absorption spectrum of the chemical, that is, the structure of the chemical will absorb light from various wavelengths and the degree of energy absorbed may alter a chemical's molecular structure.
2. Incoming solar radiation, which depends on the meteorological and geographical conditions.
3. The subsequent penetration and attenuation of the incoming solar radiation to various depths in the water column; a function of the suspended solids, phytoplankton, and dissolved organic carbon.
4. Quantum yield, ϕ, that is, the fraction of absorbed photons that result in a given reaction; a function of molecular oxygen and chemical speciation.

The estimation of the rate of photolysis is through the approximate first-order reaction

$$\frac{dc_d}{dt} = -K_p c_d \qquad (8.97)$$

where K_p is the overall photolysis rate $[T^{-1}]$ and is given by

$$K_p = K_{dp} + K_{sp} \qquad (8.98)$$

where K_{dp} is the direct photolysis rate and K_{sp} is the sensitized indirect photolysis rate, which results from excess energy from one molecule being transferred to an

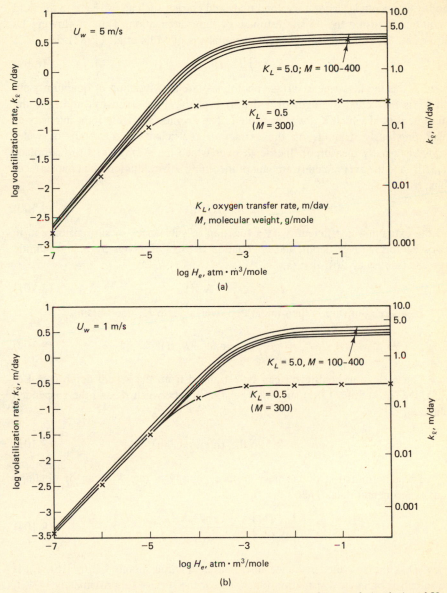

Figure 8.14 Relationship between overall volatilization transfer rate, k_l (m/day) and Henry's constant, H_e (atm-m^3/mole) for different reaeration transfer rates and molecular weights for wind speeds of (a) 5 m/s, (b) 1 m/s. From Thomann and Salas, 1986.

acceptor molecule. All following discussion involves the direct photolysis rate, since a viable theory for sensitized photolysis is not yet available.

The simplest approach is to relate observed experimental estimates of the photolysis rate under laboratory conditions and correct such an estimate to the field conditions by using the on-site estimate of light attenuation. Thus, following Di Toro et al. (1981), and incorporating the analysis of Mills et al. (1982), let

$$K_p = f(K_{d0}, I_0', I_{av}) \tag{8.99}$$

where K_{d0} is the direct near surface photolysis rate as a function of quantum yield and I_{av} is the actual average (over depth) available light intensity, and I_0' is the light intensity at which K_{d0} was measured. The average available light intensity as noted previously depends on the extinction coefficient of the body of water K_e [1/L] (usually measured), the depth over which the averaging is to take place (H_1 [L]), and the average daily amount of incoming solar radiation I_0, at the surface. Thus,

$$I_{av} = I_{av}(K_e, H, I_0) \tag{8.100}$$

and the extinction coefficient K_e is a function of substances in suspension and in dissolved form. Over depth, the attenuation of light is given by a logarithmtic decrease (see Eq. 7.26), that is,

$$I = I_0 \exp(-K_e Z) \tag{8.101}$$

If this relationship is then integrated over a depth H, one obtains

$$I_{av} = \frac{I_0}{K_e H} [1 - \exp(-K_e H)] \tag{8.102}$$

The extinction coefficient can be estimated from the secchi depth (see Eq. 7.28). Di Toro et al. (1981) suggest a relationship between K_e and the suspended solids, m, as

$$\frac{K_e}{m_1} \cong 0.3 \text{ to } 0.8 \text{ (liter/mg} \cdot \text{m)} \tag{8.103}$$

The determination of the near surface photolysis rate K_{d0} can be obtained from experimental data. Then

$$K_p = K_{d0} \frac{I_0}{(I_0)'} \frac{D}{D_0} \left\{ \frac{1 - \exp[-K_e(\lambda_{max}) \cdot H]}{K(\lambda_{max}) \cdot H} \right\} \tag{8.104}$$

where D is the so-called radiance distribution function (about 1.2 to 1.6), D_0 is the so-called radiance distribution function near the surface (approximately = 1.2; therefore D/D_0 is about 1.33), and $K_e(\lambda_{max})$ is the light attenuation coefficient at λ_{max}, the wavelength of the maximum light adsorption.

Values for K_{d0} vary widely depending on the chemical (see e.g. Mills et al., 1982 and Lyman et al., 1982). For example, the K_{d0} for napthalene is 0.23/day and for benzo(a)pyrene is 31/day at I_0' values of 2100 ly/day and λ_{max} of 310 nm and 380 nm, respectively.

8.8.2.3 Hydrolysis In this decay process, the chemical compound reacts with the water (hydrolysis) and a cleavage of a chemical bond occurs. A new compound with either the hydrogen or hydroxyl bond may be formed. Again, the hydrolysis products may not be less toxic than the original compound. In general, the hydrolysis is a second-order reaction because of dependence on the molar concentrations of [H$^+$], [OH$^-$], or water mediators. Since hydrolysis can also be mediated by enzymes (enzymatic hydrolysis), in natural waters hydrolysis may also be a biochemical degradation process. As noted by Delos et al. (1984), the basic problem is the extrapolation of laboratory hydrolysis rates (usually in distilled water) to environmental conditions with associated potential complex interactions with organic chemicals and metals and a natural biota.

The rate equation is

$$\frac{dc_d}{dt} = -K_H c_d \tag{8.105}$$

where

$$K_H = K_n + K_a[\text{H}^+] + K_b[\text{OH}^-] \tag{8.106}$$

for
K_n = neutral hydrolysis rate (day^{-1})
K_a = acid catalyzed hydrolysis rate constant (mole^{-1}day^{-1})
K_b = base catalyzed hydrolysis rate constant (mole^{-1}day^{-1})
[H$^+$] = molar concentration of hydrogen ions
[OH$^-$] = molar concentration of hydroxide ions

Handbooks list various values of K_n, K_a, and K_b (see, e.g., Mills et al., 1982; and Lyman et al., 1982). Overall K_H values range from 10^{-1} to 10^{-7}/day. However, care must be taken in using laboratory measured hydrolysis rates in an actual field situation.

8.8.2.4 Biodegradation In this decay process, the chemical is degraded by bacteria and fungi through metabolic activity. The presence of usually large numbers of such biological processors in natural waters usually means that the rates of biodegradation are one of the more important loss processes of chemicals in aquatic systems.

If the biological community has adapted to a chemical (as presumably during a steady discharge), then the degradation rate of the compound acting as a sole carbon source for the bacteria is given by the Monod equation, that is,

$$\frac{-dc_d}{dt} = \frac{1}{y}\frac{dB}{dt} = \left[\frac{\mu_{max}}{y}\left(\frac{c_d}{K_s + c_d}\right)\right]B \tag{8.107}$$

where B is the bacterial concentration [cells/L^3], y is the bacterial yield coefficient [cells/$L^3 \div M/L^3$], μ_{max} is the maximum specific growth rate [T^{-1}], and K_s [M/L^3] is the half-saturation constant (i.e., the concentration at which the growth rate is one-half of the maximum growth rate). If it is assumed tht $c_d << K_s$, then

$$\frac{dc_d}{dt} = -(K_{B2}B)c_d \tag{8.108}$$

where K_{B2} is the second-order biodegradation rate constant $[(\text{cells}/L^3 \cdot T)^{-1}]$ as

$$K_{B2} = \frac{\mu_{\max}}{yK_s} \qquad (8.109)$$

However, since in the natural environment, other carbon sources may be available for bacterial growth, this equation is usually approximated by a simple first-order kinetic equation as

$$\frac{dc_d}{dt} = -K_B C_d \qquad (8.110)$$

This equation is similar to the familiar BOD decay equation where no particular attempt is made to account for the dependence of the degradation rate on the bacterial biomass. Second-order rates are often reported for laboratory experiments but require estimation of field bacterial concentrations. First-order degradation rates appear more practical and are obtained from experiments under field conditions.

The rates at 20°C should be corrected to the temperature in the field situation by

$$(K_B)_T = (K_B)_{20}\theta^{\,(T-20)} \qquad (8.111)$$

where θ is between 1.04 and 1.095 (Delos et al., 1984).

Mills et al. (1982) in summarizing biodegradation rates indicate a range of K_B from 0.016/day for diazinon to 4.0/day for phenol.

This completes the review of the dissolved chemical loss mechanisms.

8.8.3 Sediment Interaction Loss Rate—Level 1

As noted in Eq. 8.48, the overall loss of the toxicant from the water column v_{T1} depends on the dissolved chemical loss and the sediment interaction chemical loss rate, v_{Ts}. The latter loss represents the interaction with the various mechanisms of the sediment, together with diffusion across the water–sediment interface. Figure 8.13 shows the increase in the number of parameters required to estimate v_{Ts} as the level of analysis increases. This level 1 estimate of v_{Ts} is the simplest approach and requires only two assumptions.

If it is assumed that the partition coefficient in the water column $\P_1$, is equal to that in the bed sediment, $\P_2$, and that the chemical does not decay in the sediment, then Eq. 8.51 shows that

$$r_2 = r_1 \qquad (8.112)$$

which says that the particulate toxicant concentration in the sediment is equal to the particulate concentration in the water. Also, it can then be shown that for this case

$$v_{Ts} = v_n f_{p1} \qquad (8.113)$$

and with

$$v_{Td} = (k_l + K_{d1}H_1)f_{d1} \qquad (8.114)$$

then the total loss rate is

$$v_T = v_{Td} + v_{Ts} = (k_l + K_{d1}H_1)f_{d1} + v_nf_{p1} \qquad (8.115)$$

where v_n is again the net loss of solids.

For the case of zero volatilization and decay (such as the heavy metals and long lived radionuclides), the total net loss rate is given simply by Eq. 8.115 (for $v_{Td} = 0$) as

$$v_T = v_nf_{p1} \qquad (8.116)$$

which says the net loss rate of the chemical is simply the fraction bound to the net loss of the solids. Figure 8.15 shows the behavior of the ratio of the net loss of toxicant to the net loss rate of solids due to sediment interactions from Eq. 8.113. The fraction particulate, therefore can be viewed—for this special case—as the ratio of the sediment interaction loss rate of the chemical to the net loss rate of solids. Sample Problems 8.3 (lake) and 8.4 (river) (see p. 546) are applications of the level 1 analysis.

Also, if an estimate is available at the net sedimentation flux rate of solids, F_s $[M/L^2 \cdot T]$, that is,

$$F_s = v_nm_1 \qquad (8.117)$$

then the loss of chemical due to particulate settling is (from Eq. 8.113)

$$v_{Ts} = \frac{\P_1 F_s}{1 + \P_1m_1} \qquad (8.118)$$

This relationship is also useful if the net particulate loss of a chemical tracer is known (such as a radionuclide) from observed data or from other analyses. Then either the net sedimentation flux F_s—for a known solids concentration in the water column—can be calculated or for known sedimentation flux, the "effective" solids

Figure 8.15 Ratio of net chemical loss to solids loss as a function of suspended solids concentration and partition coefficient; level 1 analysis.

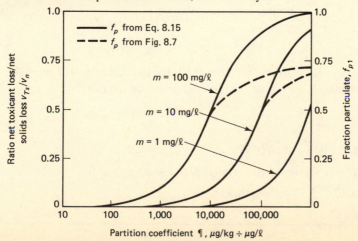

SAMPLE PROBLEM 8.3

DATA

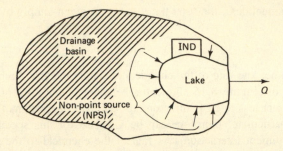

Lake

$$\text{Depth}(H) = 4 \text{ ft}$$

$$\text{Surface area}(A_s) = 81 \times 10^6 \text{ ft}^2$$

$$\text{Outflow}(Q) = 50 \text{ cfs}$$

$$\text{Suspended solids concentration}(m) = 10 \text{ mg/l}$$

$$\text{Net suspended solids loss rate } v_n = 2 \text{ ft/day}$$

Chemical

Sources: NPS $Q = 40$ cfs, $c_T = 2 \ \mu g/l$

IND $Q = 10$ cfs, $c_T = 100 \ \mu g/l$, after "Best Available Treatment" (BAT)

Partition coefficient in water column $= 200{,}000$ l/kg

Chemical does not volatilize or decay

Standards: ≤ 100 ng/l, dissolved, in water column

≤ 10 ppm, total, in sediment

PROBLEM

Using a level 1 analysis, determine whether BAT for this industry will meet the water quality standards. If not, what further reductions are required?

ANALYSIS

$$W_T = W_T(\text{IND}) + W_T(\text{NPS})$$

$$= 10 \text{ cfs} \times (100 \times 10^{-3}) \text{ mg/l} \times 5.4 + 40 \times 2 \times 10^{-3} \times 5.4$$

$$W_T = 5.40 + 0.43 = 5.83 \text{ lb/day}$$

Sample Problem 8.3 (continued)

$$W_T' = \frac{W_T}{A_s} = \frac{5.83 \text{ lb/day} \times 453.6 \text{ g/lb}}{81 \times 10^6 \text{ ft}^2 \times (1\text{m}/3.281 \text{ ft})^2}$$

$$= 3.51 \times 10^{-4} \text{ g/m}^2\text{·day}$$

$$q = \frac{50 \text{ cfs} \times 86400 \text{ sec/day}}{81 \times 10^6 \text{ ft}^2} = 0.0533 \text{ ft/day}$$

$$= 0.01626 \text{ m/day}$$

With $\P m = 200,000 \text{ l/kg} \times 10 \text{ mg/l} \times 1 \text{ kg/}10^6 \text{ mg} = 2$,

$$f_{p1} = \frac{\P m}{1 + \P m} = \frac{2}{1 + 2} = 0.6667$$

$$v_T = v_n f_{p1} = \frac{2 \text{ ft/day} \times 0.6667}{3.281 \text{ ft/m}} = 0.4064 \text{ m/day} \qquad \text{[Eq. (8.116)]}$$

Total Chemical in Water Column (Eq. 8.47)

$$c_T = \frac{W_T'}{q + v_T} = \frac{3.51 \times 10^{-4} \text{ g/m}^2\text{·day}}{(0.01626 + 0.4064)\text{m/day}}$$

$$= 830 \times 10^{-6} \text{ mg/l} = 830 \text{ ng/l}$$

Dissolved Chemical in Water Column

$$f_d = \frac{1}{1 + \P m} = \frac{1}{1 + 2} = 0.333$$

$$c_d = f_d c_T = 0.333 \times 830 = 277 \text{ ng/l}$$

Since $c_d > 100$ ng/l (standard), must reduce loading. With c_d proportional to W_T(total chemical loading),

$$\frac{W_T(\text{allow})}{100 \text{ ng/l}} = \frac{5.83 \text{ lb/day}}{277 \text{ ng/l}}$$

$$W_T(\text{allow}) = 2.10 \text{ lb/day}$$

Assuming NPS is not controllable, the allowable industrial discharge is

$$W_T(\text{allow})|_{\text{IND}} = 2.10 - 0.43 = 1.67 \text{ lb/day}$$

or a 70% reduction in this loading is required.

(continued)

Sample Problem 8.3 (continued)

Sediment Chemical Concentration

For a level 1 analysis, the solid phase concentration in water column and bed are equal, (Eq. 8.112),

$$r_1 = r_2 \qquad (1 = \text{water column}, \ 2 = \text{sediment})$$

with Eqs. 8.5 and 8.14

$$r_1 = \frac{c_{p1}}{m_1} = \frac{f_{p1} c_{T1}}{m_1} = \left(\frac{\P m_1}{1 + \P m_1}\right)\frac{c_{T1}}{m_1} = \left(\frac{2}{1 + 2}\right)\frac{830 \ \text{ng/l}}{10 \ \text{mg/l}}$$

$$r_1 = \frac{553 \ \text{ng/l}}{10 \ \text{mg/l} \times 1 \ \text{g}/10^3 \ \text{mg}} = 55{,}300 \ \text{ng/g}$$

$$\therefore r_2 (= r_1) = 55{,}300 \ \text{ng/g} = 55.3 \ \mu\text{g/g} = 55.3 \ \text{ppm}$$

Since $r_2 = 55.3$ ppm > 10 ppm (standard), this would require a reduction in the loading.

Under the reduced industrial loading, required by the dissolved chemical standard in the water column, the steady state sediment concentration (again, proportional to the total loading) would be

$$\frac{r_2(\text{new})}{\text{total } W_T(\text{new})} = \frac{r_2(\text{present})}{\text{total } W_T(\text{present})}$$

$$r_2(\text{new}) = \frac{2.10 \ \text{lb/day}}{5.83 \ \text{lb/day}} \times 55.3 \ \text{ppm} = 19.9 \ \text{ppm}$$

This value is greater than desired (10 ppm) and, thus, a further reduction in the IND load is required, to obtain $r_2(\text{allow}) = 10$ ppm.

$$\therefore \text{total } W_T(\text{allow}) = \frac{r_2(\text{allow})}{r_2(\text{present})} \text{ total } W_T(\text{present})$$

$$= \frac{10}{55.3} \times 5.83 \ \text{lb/day} = 1.05 \ \text{lb/day}$$

$$\therefore W_T(\text{allow})\big|_{\text{IND}} = 1.05 - 0.43(\text{NPS}) = 0.62 \ \text{lb/day}$$

or a reduction of 90% in present IND load.

Summary

Loading (lb/day)			Water column concentration (ng/l)			Sediment concentration (ppm)
NPS	IND	Total	c_d	c_p	c_T	r_2
0.43	5.40	5.83	277	553	830	55.3
0.43	1.67	2.10	100	199	299	19.9
0.43	0.62	1.05	50	100	150	10.0

concentration can be calculated. This latter quantity can then be used for any other chemical in the particular body of water.

8.8.4 Sediment Interaction Loss Rate—Level 2

At this level of analysis, the assumption that $\P_1 = \P_2$ is eliminated, but sediment decay is still assumed to be zero. The rationale for level 2 is that metals and some recalcitrant organic chemicals (e.g., PCB) appear to decay very slowly in the anaerobic sediment. Further, the decay rate in the sediment is particularly difficult to determine without laboratory or field experiments.

Once the assumption of equal partition coefficients is eliminated, several other parameters must be specified. Equations 8.50 and 8.50a show that for $K_{d2} = 0$,

$$v_{Ts} = v_n \left[\frac{v_s f_{p1} + K_f f_{d1}}{v_s + (f_{d2} m_2 K_f / m_1)} \right] \tag{8.119}$$

The additional parameters that must now be specified are m_2 and $\P_2$ (to obtain f_{d2}), v_s (the solids settling rate, *not* the net solids loss rate), and K_f, the sediment diffusion rate. Note that the sediment depth H_2 does not need to be specified. Note also that if $v_n = 0$, then $v_{Ts} = 0$ even with sediment diffusion and exchange.

8.8.4.1 Settling/Resuspension Velocity
A variety of attempts have been made to characterize the settling velocity (McCave, 1975; Di Toro and Nusser, 1976; Hawley, 1982). Initial efforts focused on the use of Stokes' law, that is,

$$v_s = \frac{g}{18} \left(\frac{\rho_s - \rho}{\mu} \right) d^2 \tag{8.120}$$

where v_s is the settling velocity $[L/T]$, g is the acceleration due to gravity $[L/T^2]$, ρ_s and ρ are the particle and water densities, respectively $[M/L^3]$, d is the particle diameter $[L]$, and μ is the dynamic viscosity $[M/L \cdot T]$. For $g = 981$ cm/s^2, $\mu = 0.014$ g/cm $\cdot$ s, then

$$v_s = 0.033634(\rho_s - \rho) d^2 \tag{8.121}$$

where v_s is in m/day, ρ_s and ρ in g/cm^3, and d in μm. Di Toro and Nusser (1976) in evaluating particle data for the Sacramento-San Joaquin Delta in California determined empirically

$$\rho_s = 2.0 d^{-0.15} \tag{8.122}$$

for ρ_s in g/cm^3 and d in μm. Use of Eqs. 8.121 and 8.122 gives a range of v_s from 0.5 to 9.4 m/day for a particle range of 5–50 μm. Thus, on the basis of Stokes' law for discrete settling, one would expect v_s to be on the order of 0.1 to 10 m/day. Hawley (1982), however, summarized observed data collected for settling velocities of lacustrine and marine particles and showed that such velocities can be up to an order of magnitude higher than the Stokes' velocity.

If v_s is estimated then for a steady solids concentration in a segment of a

SAMPLE PROBLEM 8.4

DATA

Consider again the chlorobenzene discharged into the stream of Sample Problem 8.2 with loading and flow given by:

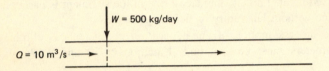

The upstream chlorobenzene concentration is zero and the input waste flow is small compared to the upstream flow. The in-stream suspended solids concentration is a constant at 20 mg/l. The fraction of organic carbon in the water column solids is 0.05.

PROBLEM

Assuming that the only loss is in the dissolved form through volatilization, calculate the downstream profile of chlorobenzene.

ANALYSIS

Since K_{d2} (sediment decay) = 0, this is either a type I or II analysis for sediment interaction.

Calculation of Partition Coefficient

$$\pi_{oc} = 0.617 K_{ow} \qquad\qquad \text{[Eq. (8.18)]}$$

$$= 0.617(1000) = 617 \left(\frac{\mu g/kg \text{ org C}}{\mu g/l} \right)$$

$$\pi = \pi_{oc} f_{oc} \qquad\qquad \text{[Eq. (8.19)]}$$

$$= 617(0.05) = 30.85 \left(\frac{\mu g/kg \text{ dry wt}}{\mu g/l} \right)$$

Compare this value to an estimate from Eqs. 8.21–8.23:

$$\pi = \frac{2(f_{oc} K_{ow})}{1 + m(f_{oc} K_{ow})/1.4} \qquad\qquad \text{[Eq. (8.24)]}$$

Sample Problem 8.4 (continued)

$$\P = \frac{2(0.05 \text{ kgC/kg} \times 1000 \text{ l/kg C}^{\ddagger})}{1 + [20 \text{ mg/l} \times \text{kg}/10^6 \text{mg} \times (0.05 \times 1000)]/1.4}$$

$$\P = 100/(1 + 0.0007) = 99.9 \text{ l/kg dry wt}$$

(Note that the effect of solids in this case is negligible and Eq. 8.24 is simply then $\P = 2f_{oc}K_{ow}$. (See also Fig. 8.6).

The second estimate therefore is 3.24 times the first estimate. Using the latter estimate to calculate the fraction dissolved:

$$f_d = (1 + \P m)^{-1} \qquad\qquad \text{[Eq. (8.13)]}$$
$$= [1 + 100(1/\text{kg}) \, 20(\text{mg/l})(\text{kg}/10^6 \text{ mg})]^{-1}$$
$$= 1.00$$

$\therefore$ All of the chlorobenzene is in the dissolved form. $\therefore f_p = 0.0$

Note that f_d also equals 1.00 for the lower estimate of $\P$.

Downstream Profile Equation

$$c_{T1} = \frac{W_T}{Q} \exp\left[-\left(\frac{v_T}{H_1}\right)\frac{x}{U} \right] \qquad\qquad \text{[Eq. (8.64)]}$$

Since all of chemical is in dissolved form, sediment interaction is not significant. This is also true since $v_n = 0$ (net loss of solids $= 0$) because of the constant solids concentration. (Eq. 8.119 also shows that $v_{Ts} = 0$.)

$$\therefore v_T = v_{Td} + v_{Ts} = (k_l + k_{d1}H_1)f_{d1} + v_n f_{p1} \qquad\qquad \text{[Eq. (8.115)]}$$
$$= (3.38 \text{ m/day}^{\S} + 0)(1.0) + (0)(0)$$
$$= 3.38 \text{ m/day}$$

$\therefore$ Downstream loss rate is all through volatilization.
Equation 8.64 above is then

$$c_{T1} = \frac{500 \text{ kg/day} \cdot 10^6 \text{ mg/kg}}{10 \text{ m}^3/\text{s} \cdot 86\,400 \text{ s/day}} \cdot \frac{\mu\text{g/l}}{\text{mg/kg}} \cdot \exp\left[-\frac{3.38 \text{ m/day} \cdot 10^3 \text{ m/km} \cdot x(\text{km})}{0.4 \text{ m} \cdot 0.6 \text{ m/s} \cdot 86\,400 \text{ s/day}} \right]$$

$$\underline{c_{T1} = 578.7 \exp[-0.163 \cdot x(\text{km})]}$$

(continued)

Sample Problem 8.4 (continued)

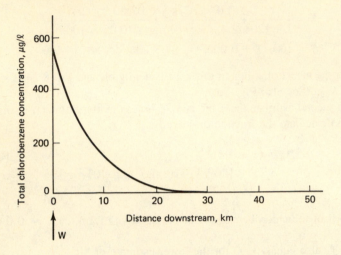

‡Units from Eq. 8.22, $\P_{oc}^X = K_{ow}$

$$\frac{l}{kg\,C} = \frac{l(oct)}{kg\,C} \cdot \frac{l}{l(oct)}$$

§From Sample Problem 8.2; overall volatilization rate = 3.38 m/day.

body of water, the resuspension velocity is given by

$$v_u = v_d\left(\frac{v_s}{v_n} - 1\right) \tag{8.123}$$

Thomann and Di Toro (1984) have used this relationship to estimate v_u in the Great Lakes using values of 2.5 to 5.0 m/day for v_s. (See Section 8.10.3).

8.8.4.2 Sediment Diffusion Rates Di Toro et al. (1981) have summarized some of the data on K_f, the sediment–water diffusive transfer coefficient. As discussed there, the resistance to exchange across the sediment–water interface is essentially all in the sediment since vertical dispersion in the water column is several orders of magnitude higher than that in the sediment. Thus,

$$K_f = \frac{D_2}{\phi_2 \delta_2} \tag{8.124}$$

where D_2 is the interstitial water diffusion coefficient $[L^2/T]$, δ_2 is the characteristic length over which the gradient exists at the sediment–water interface, and ϕ_2 is the sediment porosity. (The porosity appears because the flux of chemical is due to a gradient across a liquid volume, not a bulk volume.) The order of D_2 is at the molecular diffusion level (about 10^{-5} cm²/s or about 1 cm²/day). Using $\delta_2 = 1$ cm, Di Toro et al. (1981) suggest

$$K_f = 19\phi_2(M)^{-2/3} \qquad (8.125)$$

for K_f in cm/day. A typical order of magnitude for K_f is then about 0.1–1.0 cm/day. Sample Problem 8.5 contains a riverine level 2 analysis; a mass balance of the chemical in the sediment is shown in Sample Problem 8.6.

8.8.5 Sediment Interaction Loss Rate—Level 3

The full equations for v_{T_s} (Eqs. 8.50 and 8.50a) now must be used. It can be noted that H_2, the depth of the sediment must also now be specified. This is the first time this parameter must be specified.

Di Toro et al. (1981) indicate that sediment decay rates of chemicals are probably in the range from 0.001 to 0.1/day. Sediment decay rates below about 0.001/day are generally not significant but rates above 0.1/day are significant and must be taken into account. Since the sediment is generally anaerobic, the decay rates for any chemical are reflective of the lower rates in that environment.

The depth of active sediment can be estimated from sediment cores of the chemical, if available, or from sediment tracers. The depth H_2 is assigned to the surface layer of the chemical that appears to be well-mixed or where an approximation can be made to the vertical gradient with an average single layer. Examination of a variety of such ''active'' sediment depths results in a range for H_2 from about 0.1 cm to about 10 cm. For some river systems, H_2 may exceed 10 cm. Sample Problem 8.7 illustrates a level 3 analysis in an estuarine setting.

This completes the modeling framework for the physical-chemical fate of toxic substances. The final part of this chapter is devoted to the area of the accumulation of the chemical in the food chain.

8.9 PRINCIPAL BIOLOGICAL COMPONENTS OF TOXIC SUBSTANCE ANALYSIS

Aquatic organisms may accumulate a chemical from two principal routes:

1. Direct uptake, that is, absorption and/or adsorption from the ''available'' form in the water (''available'' may include the dissolved form and the toxicant adsorbed on microparticulate organic or inorganic particles).
2. Ingestion of the substance through predation of contaminated prey.

It should be noted that the first route, that is, direct uptake of the substance from the water, separates models of the *fate* of toxicants from models of other water quality variables. For example, in analyses of nutrient enrichment problems, it is assumed that upper levels of the food chain receive nutrients only from predation and not directly from the water. As opposed to accumulation, the toxicant may be

SAMPLE PROBLEM 8.5

DATA

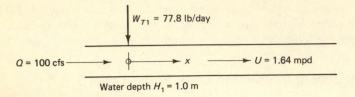

W_{T1} = 77.8 lb/day

Q = 100 cfs ⟶   ○ ⟶ x ⟶ U = 1.64 mpd

Water depth H_1 = 1.0 m

Suspended Solids

Concentration	Water	m_1	10 mg/l
	Bed	m_2	10,000 mg/l
Settling velocity		v_s	0.83 m/day
Sedimentation rate		v_d	0.0 mm/yr

Chemical

Decay rate	Water	v_{Td}	0.1 m/day
	Bed	K_{d2}	0.0 day
Partition coefficient	Water	$\P_1$	50,000 l/kg
	Bed	$\P_2$	10,000 l/kg
Mass discharge rate	Present	W_{T1}	77.8 lb/day
	Future	W_{T2}	
Sediment–water diffusion rate (including bioturbation)		K_f	0.1 m/day

PROBLEM

Determine the allowable discharge rate (W_{T2}) for the stream if the total water column concentration $c_{T1} \leq 50$ μg/l. Display spatial profiles of total, dissolved, and particulate chemical concentrations in the water column and solid phase concentrations in the water column and bed under the allowable discharge rate.

ANALYSIS

Since $\P \neq \P_2$ and $K_{d2} = 0$, this is a Level 2 analysis (see Section 8.8.4).

Sample Problem 8.5 (continued)

Suspended Solids Analysis

Assume m_1 and m_2 spatially constant. With no sedimentation,

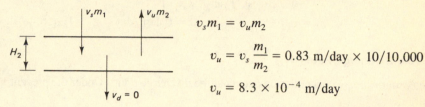

$$v_s m_1 = v_u m_2$$

$$v_u = v_s \frac{m_1}{m_2} = 0.83 \text{ m/day} \times 10/10,000$$

$$v_u = 8.3 \times 10^{-4} \text{ m/day}$$

Since there is no sedimentation of solids, the net loss rate $v_n = 0$. This conclusion follows from inspection of Eq. 8.37 where

$$v_n = \frac{v_s v_d}{v_u + v_d} = \frac{0.83 \times 0}{8.3 \times 10^{-4} + 0} = 0 \qquad [\text{Eq. (8.37)}]$$

Note that Eq. 8.37 is valid for any system whose bed is stationary.

Chemical Parameters

Assume $\phi_2 \approx 1$. (From Eq. 8.43, with $\rho_s = 2.5$, $\phi_2 = 0.996$.)

$$f_{p1} = \frac{\P_1 m_1}{1 + \P_1 m_1} = \frac{50,000 \times 10 \times 10^{-6}}{1 + 50,000 \times 10 \times 10^{-6}} \qquad = 0.333$$

$$f_{d1} = \frac{1}{1 + \P_1 m_1} = 1 - f_{p1} \qquad = 0.667$$

$$f_{p2} = \frac{\P_2 m_2}{1 + \P_2 m_2} = \frac{10,000 \times 10,000 \times 10^{-6}}{1 + 10,000 \times 10,000 \times 10^{-6}} \qquad = 0.990$$

$$f_{d2} = \frac{1}{1 + \P_2 m_2} = 1 - f_{p2} \qquad = 0.010$$

$$\delta = \frac{r_2}{r_1} = \frac{(v_u + v_d)f_{p2} + K_f(\P_2/\P_1)f_{d2}/\phi_2}{(v_u + v_d)f_{p2} + K_f f_{d2}/\phi_2 + K_{d2}f_{d2}H_2} \qquad [\text{Eq. (8.51)}]$$

$$\delta = \frac{r_2}{r_1} = \frac{(8.3 \times 10^{-4} + 0)0.990 + 0.1(10,000/50,000) \times 0.010/1}{(8.3 \times 10^{-4} + 0)0.990 + 0.1 \times 0.010/1 + 0.0}$$

$$\delta = 0.561$$

(continued)

Sample Problem 8.5 (continued)

With $v_n = 0$,

$$v_{Ts} = v_n \left[\frac{v_s f_{p1} + K_f f_{d1}}{v_s + (f_{d2} m_2 K_f / m_1)} \right] = 0. \qquad \text{[Eq. 8.119)]}$$

$$\therefore v_T = v_{Td} + v_{Ts} = 0.1 + 0 = 0.1 \text{ m/day} \qquad \text{[Eq. 8.48)]}$$

Allowable Discharge Rate (W_{T2})

The maximum stream concentration occurs at the discharge point. Under the present discharge,

$$c_{T1}(0) = \frac{W_{T1}}{Q} = \frac{77.8 \text{ lb/day}}{100 \text{ cfs} \times 5.4} \times 10^3 \ \mu g/mg = 144 \ \mu g/l$$

Since $c_{T1} \leq 50 \ \mu g/l$ (required),

$$\frac{W_{T2}}{50 \ \mu g/l} = \frac{W_{T1}}{144 \ \mu g/l}; \quad \underline{\underline{W_{T2} = 77.8 \times \frac{50}{144} = 27.0 \text{ lb/day}}}$$

Water Column Concentrations for W_{T2}

Total Concentration

$$c_{T1} = \left(\frac{W_{T2}}{Q} \right) e^{-(v_T/H_1)x/U} = 50 e^{-0.1x/1.64} (\mu g/l) \qquad \text{[Eq. (8.64)]}$$

Dissolved Concentration

$$c_{d1} = f_{d1} c_{T1} = 0.667 \times 50 e^{-0.1x/1.64} (\mu g/l)$$

Particulate Concentration

$$c_{p1} = f_{P1} c_{T1} = 0.333 \times 50 e^{-0.1x/1.64} (\mu g/l)$$

Solid Phase Concentration

$$r_1 = \frac{c_{p1}}{m_1} = \frac{16.7 e^{-0.0610x} \mu g/l}{10 \text{ mg/l} \times 10^{-3} g/mg} = 1670 e^{-0.0610x} (\mu g/g) \qquad \text{[Eq. (8.5)]}$$

Sediment Concentration under W_{T2}

Solid Phase Concentration

$$r_2 = r_1 \times \delta = 1670 e^{-0.0610x} (0.561) = 937 e^{-0.0610x} (\mu g/g)$$

Plots of the quantities above are shown on p. 553.

Sample Problem 8.5 (continued)

Notes: Since all quantities in the water column and bed decay at the same rate, sediment data may be used to estimate the overall decay rate v_T. It is presumed, of course, that a steady state condition has been reached.

Attainment of the new concentrations in the water column will be more rapid than in the bed, after the discharge has been reduced.

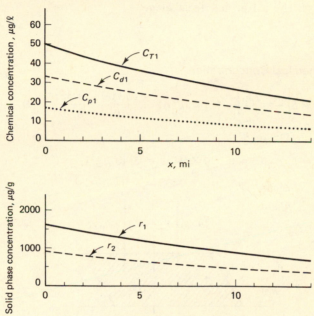

SAMPLE PROBLEM 8.6

DATA

Use the stream of Sample Problem 8.5.

PROBLEM

Demonstrate that a mass balance exists for the sediment located at the discharge point at $x = 0$. Use a 1-m² interfacial area.

ANALYSIS

Sediment Chemical Concentration

$$c_{T2}(0) = \delta\left[\frac{m_2}{m_1} \cdot \frac{f_{p1}}{f_{p2}}\right]c_{T1}(0) = 0.561 \times \frac{10,000}{10} \times \frac{0.333}{0.990} \times 50 \quad \text{[Eq. (8.65)]}$$

$$c_{T2}(0) = 9435 \ \mu g/l$$

$$c_{p2}(0) = f_{p2}c_{T2}(0) = 0.990 \times 9435 = 9340 \ \mu g/l$$

$$c_{d2}(0) = f_{d2}c_{T2}(0) = 0.010 \times 9435 = \quad 94 \ \mu g/l$$

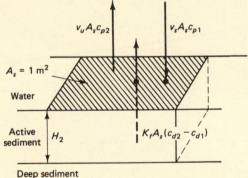

$A_s = 1$ m²

Water

Active sediment H_2

Deep sediment

Into Sediment

$$v_s A_s c_{p1} = 0.83 \ \text{m/day} \times 1 \ \text{m}^2 \times 16.7(\mu g/l = mg/m^3)$$

$$= 13.9 \ \text{mg/day}$$

Out of Sediment

$$v_u A_s c_{p2} = 8.3 \times 10^{-4} \times 1 \times 9340 = 7.8 \ \text{mg/day}$$

$$K_f A_s(c_{d2} - c_{d1}) = 0.1 \times 1 \times (94 - 33.3) = 6.1 \ \text{mg/day}$$

$\therefore M_{in} = M_{out}$, since $13.9 = (7.8 + 6.1 = 13.9)$mg/day.

Note: Since the sediment dissolved concentration ($c_{d2} = 94 \ \mu g/l$) is greater than that in the water column ($c_{d1} = 33.3 \ \mu g/l$), diffusive exchange of the dissolved chemical is *from* the sediment *to* the water column—on a steady state basis. For the model used, diffusive exchange into the deep sediment is assumed negligible.

SAMPLE PROBLEM 8.7

DATA

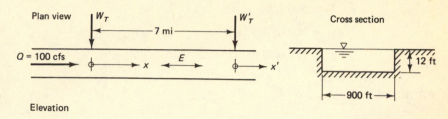

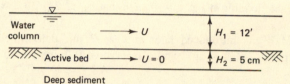

Transport

Freshwater velocity: $U = Q/A = 100 \text{ cfs}/(12 \times 900 \text{ ft}^2)$
$\times 16.4 = 0.152 \text{ mpd}$

Dispersion coefficient: $E = 1.01 \text{ smpd}$

Solids

Water column:

$$m_1 = 50 \text{ mg/l (although decreasing, assume}$$
$$\text{this spatially average value)}$$

$$v_s = 2 \text{ m/day}$$

$$H_1 = 12 \text{ ft } (= 3.66\text{m}) \text{ mean water depth}$$

Stationary active bed:

$$m_2 = 200,000 \text{ mg/l}$$

$$v_d = 5 \text{ mm/yr}$$

$$H_2 = 5 \text{ cm}$$

Chemical

Water Column

$$\P_1 = 8000 \text{ l/kg}$$

$$\frac{v_{Td}}{H_1} = 0.05/\text{day} \begin{cases} \text{Hydrolysis, photolysis, volatilization,} \\ \text{biodegradation, etc.} \end{cases}$$

(continued)

Sample Problem 8.7 (continued)

Stationary Bed

$$\P_2 = 500 \; 1/kg$$

$$K_{d2} = 0.001/day$$

$$K_f = 0.1 \; m/day$$

PROBLEM

(a) Determine allowable loading W_T at $x = 0$ such that $c_{d1} \leq 2.5 \; \mu g/l$ with a reserve of 0.5 $\mu g/l$.

(b) If a second load W_T' is introduced later at $x = 7$ mi, determine its allowable load. Assume that $c_{d1} \leq 2.5 \; \mu g/l$ and a reserve of 0.2 $\mu g/l$ is to be maintained at all locations.

ANALYSIS

Since $\P_1 \neq \P_2$ and $K_{d2} \neq 0$, a Level 3 analysis is required (Section 8.8.5).

Solids Analysis

Mass balance around active stationary sediment

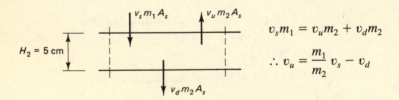

$$v_s m_1 = v_u m_2 + v_d m_2$$

$$\therefore v_u = \frac{m_1}{m_2} v_s - v_d$$

With $v_d = 5 \; mm/yr \times 10^{-3} \; m/mm \times 1 \; yr/365 \; days = 0.137 \times 10^{-4} \; m/day$

$$v_u = \left(\frac{50 \; mg/l}{200,000 \; mg/l} \right) 2 \; m/day - 0.137 \times 10^{-4} \; m/day$$

$$v_u = 5 \times 10^{-4} - 0.137 \times 10^{-4} = \underline{\underline{4.863 \times 10^{-4} \; m/day}}$$

Note:

$$v_n \; (\text{net loss rate of solids from water column})$$

$$v_n m_1 = v_d m_2$$

Sample Problem 8.7 (continued)

$$v_n = 0.137 \times 10^{-4} \times 200,000/50 = 0.0548 \text{ m/day}$$
$$v_n = 20 \text{ m/yr}$$

Chemical Analysis

Water Column

$$\P_1 m_1 = 8000 \times 50 \times 10^{-6} = 0.4$$

$$f_{d1} = \frac{1}{1 + 0.4} = 0.714$$

$$f_{p1} = \frac{0.4}{1.4} = 0.286$$

$$c_{d1} = f_{d1}c_{T1}; \qquad c_{T1} = \frac{c_{d1}}{(f_{d1} = 0.714)} = 1.40c_{d1}$$

$$r_1 = \frac{c_{p1}}{m_1} = \frac{f_{p1}c_{T1}}{m_1} = \frac{0.286c_{T1}(\mu g/l)}{50 \text{ mg/l} \times 10^{-3} \text{ g/mg}} = 5.72c_{T1} \ (\mu g/g)$$

Active Sediment

$$\P_2 m_2 = 500 \times 200,000 \times 10^{-6} = 100$$

$$f_{d2} = \frac{1}{1 + 100} = 0.0099$$

$$f_{p2} = \frac{100}{1 + 100} = 0.9901$$

Overall Toxicant Loss Rate

Assuming $\rho_s = 2.5$,

$$200,000 = 2.5(1 - \phi_2)10^6, \phi_2 = 0.92 \qquad \text{[Eq. (8.43)]}$$

$$\eta' = \frac{2 \times 0.286 + 0.1 \times 0.714}{2 + 0.0099 \times (200,000/50)(0.1 + 0.001 \times 0.05)} \qquad \text{[Eq. (8.50a)]}$$

$$= 0.1079$$

$$v_{Ts} = 0.0548(\text{m/day}) \times 0.1079 \qquad \text{[Eq. (8.50)]}$$

$$\times \left[1 + \frac{200,000(0.001 \times 0.0099 \times 0.05 \text{ m})}{50 \times 0.0548} \right]$$

$$= 0.0061(\text{m/day})$$

$$v_T = v_{Td} + v_{Ts} = 0.1828 + 0.0061 = 0.1889 \text{ m/day} \qquad \text{[Eq. (8.48)]}$$

(continued)

Sample Problem 8.7 (continued)

Note that the sediment interaction effect on v_T is not significant $(0.0061 \ll 0.1828)$.

$$\underline{K_T = \frac{v_T}{H_1} = \frac{0.1889}{3.66} = 0.0516/\text{day}}$$

Also,

$$\delta = \frac{r_2}{r_1} = \frac{\begin{array}{l}(4.863 \times 10^{-4} + 0.137 \times 10^{-4})0.9901 + \\ (10/100 \text{ m/day})(500/8000)(0.0099/0.92)\end{array}}{\begin{array}{l}(4.863 \times 10^{-4} + 0.137 \times 10^{-4})0.9901 + \\ 0.1 \times 0.0099/0.92 + 0.001 \times 0.0099 \times 0.05\end{array}} \quad [\text{Eq. (8.51)}]$$

$$\delta = 0.358$$

Estuary Toxicant Concentrations

$$\alpha = \sqrt{1 + \frac{4K_T E}{U^2}} = \sqrt{1 + \frac{4 \times 0.0516 \times 1.01}{(0.152)^2}} = 3.166$$

(a) For a single point source, maximum concentration is at outfall $(x = 0)$ where $c_{T0} = W_T/(Q\alpha)$.

$$c_{d1}(\text{allow}) = 2.5 \ \mu\text{g/l} - 0.5 \ \mu\text{g/l}(\text{reserve}) = 2.0 \ \mu\text{g/l}$$

$$c_{T1}(\text{allow}) = 1.40c_{d1}(\text{allow}) = 2.80 \ \mu\text{g/l}$$

Since

$$c_{T0} = c_{T1}(\text{allow}), \text{ where } c_{T1}(\text{allow}) = 2.8 \times 10^{-3} \text{mg/l}$$

$$W_T = c_{T1}(\text{allow}) \times Q(\text{cfs}) \times \alpha \times 5.4$$

$$W_T = 2.8 \times 10^{-3} \times 100 \times 3.166 \times 5.4$$

$$\underline{\underline{W_T = 4.79 \text{ lb/day}}}$$

Spatial Distributions

$$j_1 = \frac{U}{2E}(1 + \alpha) = \frac{0.152}{2 \times 1.01}(1 + 3.166) = 0.313/\text{mi}$$

$$j_2 = \frac{U}{2E}(1 - \alpha) = \frac{0.152}{2 \times 1.01}(1 - 3.166) = -0.163/\text{mi}$$

Water Column

Total $c_{T1} = 2.80e^{0.313x}, \qquad x \leq 0$ $c_{T1} = \mu\text{g/l}$

concentration $c_{T1} = 2.80e^{-0.163x}, \qquad x \geq 0$ $= \text{ppb}$

Sample Problem 8.7 (continued)

Dissolved	$c_{d1} = 2.00e^{0.313x},$	$x \le 0$		$c_{d1} = \mu g/l$	
concentration	$c_{d2} = 2.00e^{-0.163x},$	$x \ge 0$		$= $ ppb	

Solid phase concentration
$$r_1 = 5.72(2.80e^{0.313x})$$
$$= 16.0e^{0.313x}, \qquad x \le 0 \qquad\qquad r_1 = \mu g/g$$
$$r_1 = 16.0e^{-0.163x}, \qquad x \ge 0 \qquad\qquad = \text{ppm}$$

Active Sediment

Solid phase concentration
$$r_2/r_1 = 0.358$$
$$r_2 = 5.73e^{0.313x}, \qquad x \le 0 \qquad\qquad r_2 = \mu g/g$$
$$= 5.73e^{-0.163x}, \qquad x \ge 0 \qquad\qquad = \text{ppm}$$

(b) For $W_T = 4.79$ lb/day $+ W_T'$ at $x = 7$ mi $(x' = 0)$

At $x' = 0$, $c_{d1}(\text{allow}) = 2.5 - 0.2(\text{reserve}) = 2.30 \ \mu g/l$
$$c_{T1}(\text{allow}) = 1.40 \times 2.30 = 3.22 \ \mu g/l$$
Due to $W_T = 4.79$ lb/day,
$$c_{T1}(x = 7) = 2.80 \times e^{-0.163 \times 7} = 0.89 \ \mu g/l$$
∴ Allowable $c_{T1}(x = 7)$ due to $W_T' = 3.22 - 0.89 = 2.33 \ \mu g/l.$
$$W_T' = 2.33 \times 10^{-3} \text{ mg/l} \times 100 \text{ cfs} \times 3.166 \times 5.4$$
$$\underline{W_T' = 3.98 \text{ lb/day}}$$

Check spatial distributions to ensure that $c_{d1} \le 2.30 \ \mu g/l$
$$c_{T1} \le 3.22 \ \mu g/l$$

Water Column

Total concentration
$$c_{T1} = 2.80e^{0.313x} + 2.33e^{0.313(x-7)} \qquad x \le 0$$
$$c_{T1} = 2.80e^{-0.163x} + 2.33e^{0.313(x-7)} \qquad 0 \le x \le 7$$
$$c_{T1} = 2.80e^{-0.163x} + 2.33e^{-0.163(x-7)} \qquad x \ge 7$$

Dissolved $\quad c_{d1} = 5.72c_{T1}$

Solid phase $\quad r_1 = 5.72c_{T1}$

Active Sediment

$$r_2 = r_1 \times \left(\frac{r_2}{r_1}\right) = 5.72c_{T1} \times (0.358) = 2.05c_{T1}$$

See spatial distributions on p. 560 which confirm that no violations occur.

(continued)

Sample Problem 8.7 (continued)

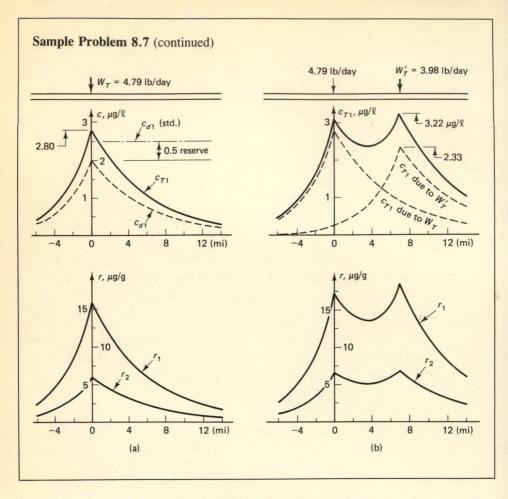

(a)   (b)

excreted from the food chain by physiological processes, released upon death, egested as uneaten or undigested food, or metabolized as part of the chemical processing by the organism.

Data from laboratory and controlled field experiments can be used with subsystem compartment models to obtain estimates of kinetic behavior including values of uptake and clearance, or transfer, between trophic levels. These subsystem models are generally composed of single or 4–5 or more discrete compartments, that is, discrete units of the aquatic ecosystem such as phytoplankton, zooplankton, and fish. Analyses which use compartments of the ecosystem permit the determination of bioconcentration factors, that is, the ratio of the concentration of the substance in the organism to that in the water. Accordingly, it is well to begin the analysis of the fate of a toxicant by a simple mass balance of a single ecological compartment (say, fish) exposed to a controlled environment in laboratory aquarium.

8.9.1 Bioconcentration and Depuration

The uptake of a chemical directly from water through transfer across the gills, as in fish, or through surface sorption and subsequent cellular incorporation, as in phytoplankton, is an important route for transfer of toxicants. This uptake is often measured by laboratory experiments where test organisms are placed in aquaria with known (and constant) water concentrations of the chemical. The accumulation of the chemical over time is then measured and the resulting equilibrium concentration in the organism divided by the water concentration is termed the bioconcentration factor (BCF). A simple representation of this mechanism is given by a mass balance equation around a given organism. Thus,

$$\frac{dv'}{dt} = k_u wc - Kv' \tag{8.126}$$

where v' is the whole body burden of the chemical in μg, k_u is the uptake sorption and/or transfer rate in liters/day $\cdot$ g wet wt, w is the weight of the organism in g wet wt, c is the available water concentration in $\mu g/l$ generally assumed equal to the dissolved concentration c'_d, K is the desorption and excretion rate in day^{-1} and t is time. This equation indicates that if the mass input (μg/day) of toxicant given by $k_u wc$ is greater than the mass lost due to depuration (μg/day) given by Kv', then the chemical will accumulate in the organism. At equal mass rates of uptake and depuration, the chemical will have reached an equilibrium level. It is assumed in the mass balance that the rate of uptake of the chemical is directly proportional to the concentration of the chemical in the water. For most chemicals at normally encountered concentrations in water this is a good assumption. For example, the studies of PCB by Vreeland (1974) for oysters and Hansen et al. (1974) for pinfish, show that the resultant PCB concentration in the test animal was linear to the exposure water concentration.

　　The whole body burden v' is given by

$$v' = vw \tag{8.127}$$

where v is the concentration of the chemical in $\mu g/g$ wet wt in the organism. Substitution of Eq. 8.127 into 8.126 gives

$$\frac{dvw}{dt} = k_u wc - Kvw \tag{8.128}$$

Expanding the derivative and grouping terms yields

$$\frac{dv}{dt} = k_u c - \left(\frac{1}{w} \cdot \frac{dw}{dt} + K\right)v \tag{8.129}$$

Letting

$$\frac{1}{w} \cdot \frac{dw}{dt} = G, \tag{8.130}$$

where $G(\text{day}^{-1})$ is the net growth rate of the organism, and with

$$K' = K + G \tag{8.131}$$

then

$$\frac{dv}{dt} = k_u c - K' v \tag{8.132}$$

It is seen that the loss term on the concentration includes the loss (or gain) due to the changing weight of the organism during the test and may therefore be termed an apparent depuration. The solution to Eq. 8.132 is

$$v = \frac{k_u c}{K'} [1 - \exp(-K't)] + v_0 \exp(-K')t \tag{8.133}$$

where v_0 is the initial concentration of the chemical in the test organism. Note that this expression indicates that the rate of accumulation is a function of K', the sum of the depuration and net growth rates. At equilibrium or steady state, Eq. 8.27 is obtained (with $c \equiv c'_d$), that is,

$$v = \frac{k_u c}{K'} \tag{8.134}$$

and the BCF is given by Eq. 8.29, that is,

$$N_w = \frac{v}{c} = \frac{k_u}{K + G} \tag{8.135}$$

The BCF is defined from Eq. 8.135 as the ratio of the uptake rate, k_u to the sum of the depuration and growth rates and represents the ratio of the equilibrium steady-state chemical concentration in the organism to the dissolved water concentration. It is evident that BCF tests extending over time periods during which the weight of the organism changes are affected by such weight changes. This effect should be incorporated in analyses of the test data through Eq. 8.135. This is especially true for analyses of chemical concentration time history data using independent information on uptake or depuration rates.

For fish, the primary mechanism for uptake directly from water is through transfer of the chemical across the gill surface. The rate of transfer can be calculated from the rate of transfer of oxygen from the water to the blood of the fish.

The rate of mass transport of a chemical is given by

$$M = \frac{DA}{\delta} c \tag{8.136}$$

where M is the mass transport rate in $\mu g/day$, D is the effective diffusivity of the substance in cm^2/day, A is the effective surface area of the gill for transfer in cm^2, δ is the effective thickness of the gill in cm, and c is the concentration of the chemical in the water in $\mu g/l$. It is assumed in Eq. 8.136 that the concentration of the toxicant in the blood is zero. The effective diffusivity incorporates the tendency for some chemicals to be more easily transported across a gill membrane than certain other chemicals. Factors such as the size of organic chemical molecules may influence such transfer. If it is assumed that the mechanism for uptake of the chemical is related to oxygen uptake (i.e., the respiration of the fish) then

$$\frac{M}{M_{O_2}} = \frac{D}{D_{O_2}} \frac{c}{c_{O_2}} \qquad (8.137)$$

where the subscript O_2 refers to the dissolved oxygen (DO). From Eq. 8.137,

$$M = \left(\beta \frac{M_{O_2}}{c_{O_2}} \right) c \qquad (8.138)$$

where $\beta = D/D_{O_2}$, the ratio of the effective diffusivity of chemical to that of oxygen. From Eq. 8.138,

$$M = k_u' c \qquad (8.139)$$

where

$$k_u' = \frac{\beta M_{O_2}}{c_{O_2}} \qquad (8.140)$$

The quantity k_u' represents the mass uptake for the whole fish and has units liters/day. Dividing k_u' by the fish weight gives the uptake rate per unit weight, that is,

$$k_u = \frac{k_u'}{w} = \beta \frac{M_{O_2}/w}{c_{O_2}} \qquad (8.141)$$

The quantity M_{O_2}/w is the respiration rate, r, of the fish, that is,

$$r = M_{O_2}/w \qquad (8.142)$$

where r has units g O_2/g wet wt $\cdot$ day. The uptake rate for the chemical is therefore related to the respiration rate of the organism by

$$k_u = \beta \frac{r}{c_{O_2}} \qquad (8.143)$$

The respiration is often expressed as g wet wt respired/g wet wt of organism per day and is then given by

$$r' = \frac{R_{wd} r}{2.67 R_c} \qquad (8.144)$$

where R_{wd} is the wet weight to dry weight ratio (about 5–10 for lower levels of the food chain, i.e., about 80–90% water, and about 4 for the upper levels, i.e., about 75% water), R_c is the g (carbon) to g dry wt ratio, about 0.40, and r' is in g/g $\cdot$ day. Thus,

$$k_u = \beta' r' \qquad (8.145)$$

where

$$\beta' = \frac{\beta\, 2.67\, R_c}{R_{wd}\, c_{O_2}} \tag{8.146}$$

and β' has units of liters/g wet wt.

Finally, the respiration rate is related approximately to the wet weight of the organism by

$$r' = a' w^{-0.2} \tag{8.147}$$

Therefore, the uptake rate of the chemical can in a general way be related to the weight of the organism by

$$k_u = a_k w^{-0.2} \tag{8.148}$$

where a_k is given by $a'\beta'$ and therefore depends on the chemical diffusivity across the gill, the trophic position of the organisms and the DO of the water.

In depuration experiments, the organism is transferred to a tank with zero toxicant concentration and the time history of the loss of the chemical is measured. With $c = 0$, Eq. 8.133 then becomes

$$\nu = \nu_0 \exp(-K')t \tag{8.149}$$

or in terms of whole body burden

$$\nu' = \nu_0' \exp(-K')t \tag{8.150}$$

Again, the weight change of the organism must be properly taken into account. Equation 8.149 can be used to estimate K' and, with the net growth rate known, the depuration rate K can be obtained. Alternately, Eq. 8.150 can be used directly where the whole body burden is plotted and the depuration rate obtained from such data.

Two examples of uptake and loss due to excretion are shown in Fig. 8.16. For the PCB case, the bioconcentration factor (for pinfish), N_w, is 17 μg/gm ÷ μg/l ($N_w' = 17{,}000$). If the excretion rate is estimated from the uptake, it is found to be 0.045/day or about four times the depuration rate of 0.012/day. Growth of the pinfish and storage in various body compartments may account for the difference. For malathion in carp, $N_w = 0.006$ ($N_w' = 6$), an indication of the lower tendency for malathion to be accumulated. Depuration is rapid at an excretion rate of 1.23/day or about two orders of magnitude faster than the PCB. The excretion rate calculated from the uptake experiment is about 0.6/day or about one-half that calculated from the depuration experiment. Results of the type shown in Figs. 8.16(a) and (b) provide estimates of the key rate parameters and equilibrium conditions for a variety of organisms and substances. A summary of some parameters for fish is given in Table 8.7 (see Sample Problem 8.8 for an example of bioconcentration and depuration).

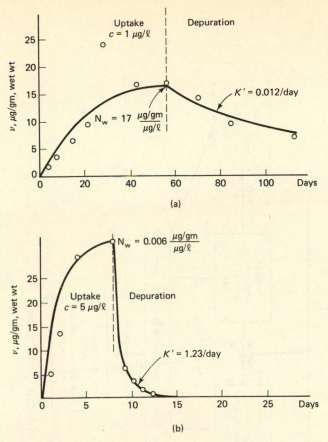

Figure 8.16 Bioconcentration and depuration for: (a) PCB (from Hansen et al., 1974) and (b) malathion (from Bender, 1969).

8.9.2 Food Chain Biomagnification

As noted previously, an aquatic organism can receive a chemical by direct uptake from the water or by consuming prey that are contaminated with the chemical. This latter route through the food chain may result in the biomagnification of chemical concentration as one proceeds up the food chain to top predators. In order to provide a first level of understanding of the operative mechanisms, a simple four level food chain model may be considered. Figure 8.17 shows the schematic of the compartments:

1. Phytoplankton
2. Zooplankton
3. Small fish (approximately less than 300 mm).
4. Large fish (top predator of size 0.1–1.0 m)

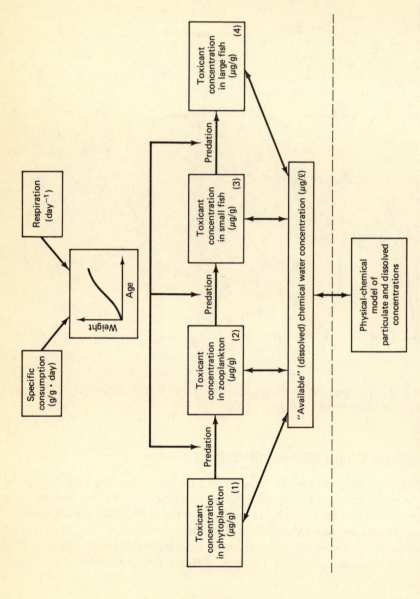

Figure 8.17 Schematic of simplified food chain toxicant transfer.

SAMPLE PROBLEM 8.8

DATA

Fish were exposed to 2 μg/l pentachlorophenol (log octanol/water partition coefficient = 5.0) for 50 days. The fish were than transferred to a clean water tank and analyzed for an additional 20 days during their depuration of the chemical. The data obtained were as follows.

Time (days)		Weight of fish (g wet wt)	Concentration in fish whole body (μg/g wet wt)
0		100	0.0
1		100	0.7
5		102	2.5
10		105	4.7
15		108	5.6
25		116	7.4
30		115	7.2
40		125	7.9
(0) 50		127	8.2
(5)	55	129	5.1
(10)	60	132	3.5
(20)	70	141	1.4

PROBLEM

(a) What is the bioconcentration factor (BCF) in μg/g $\div$ μg/l?
(b) What is the depuration rate in day^{-1}?
(c) What is the uptake rate in liters/day $\cdot$ g(wet wt)?

ANALYSIS

$$N_w = \frac{\nu}{c} = \frac{k_u}{K'} \qquad : \text{BCF} \qquad \text{[Eq. (8.135)]}$$

$$\nu(\mu g/g) = \frac{k_u c}{K'}(1 - e^{-K't}): \begin{array}{l} \text{Uptake} \\ \text{phase} \end{array} \qquad \text{[Eq. (8.133)]}$$

$$\nu = \nu_0 e^{-K't} \qquad : \begin{array}{l} \text{Depuration} \\ \text{phase} \end{array} \qquad \text{[Eq. (8.142)]}$$

$$K' = K + G \qquad \text{[Eq. (8.131)]}$$

$$w = w_0 e^{Gt}$$

(continued)

Sample Problem 8.8 (continued)

A semilog plot of the average weight of the fish gives an estimate of the net growth rate of the fish, $G(1/day)$:

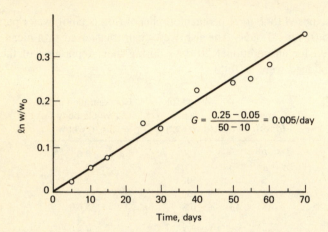

$$G = \frac{0.25 - 0.05}{50 - 10} = 0.005/day$$

(a) A plot of the pentachlorophenol concentration in the fish (see below) gives an estimated equilibrium value of 8 $\mu g/g$(wet wt) . Then,

$$BCF = N_w = \frac{8\,\mu g/g}{2\,\mu g/l} = 4\,\frac{\mu g/g}{\mu g/l}$$

or

$$N'_w = 4000\,\frac{\mu g/kg}{\mu g/l}\ (= 1/kg)$$

(b) By trial and error using uptake data, $K' = 0.085/day$. Check with depuration data.

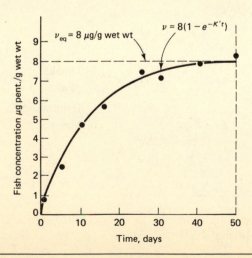

$$v_{eq} = 8\ \mu g/g\ wet\ wt$$

$$v = 8(1 - e^{-K't})$$

Sample Problem 8.8 (continued)

From a semilog plot of depuration data, obtain $K' = 0.085/\text{day}$.

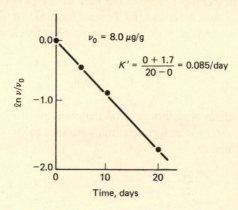

Since

$$K' = K + G \qquad\qquad \text{[Eq. (8.131)]}$$
$$K = K' - G$$

and

$$K = 0.085 - 0.005 = \underline{\underline{0.08/\text{day}}} \qquad \text{depuration rate}$$

(c) Uptake rate:

$$N_w = \frac{k_u}{K'} \qquad\qquad \text{[Eq. (8.135)]}$$

$$k_u = K' \cdot N_w$$

$$= 0.085(1/\text{day}) \cdot 4(1/\text{g(wet wt)})$$

$$\underline{\underline{k_u = 0.34 \ 1/\text{g(wet wt)} \cdot \text{day}}} \qquad \text{uptake rate}$$

The mechanisms included in the model are (Thomann, 1981):

1. Sorption and direct uptake of the chemical from the water.
2. Desorption and excretion of the chemical to the water.
3. Consumption of contaminated food through predation.
4. Respiration and growth rate of organism.

Mechanisms (3) and (4) contribute to the weight change of the organism as a function of age. In the construction of the food chain model, a mass balance equation is written around each of the compartments incorporating the preceding mechanisms. The mass balance equation represents the flows of the chemical into and out of the average organism. Care must be taken to include the weight change of the organism since such changes affect the chemical concentration.

Consider the first compartment as the base of the food chain and composed of the phytoplankton and other detrital organic material of size <100 μm. An equation for this compartment is given by a simple reversible sorption–desorption linear equation as

$$\frac{d\nu_1}{dt} = k_{u1}c - K_1\nu_1 \tag{8.151}$$

For any food chain compartment that may accumulate toxicant via the food route as well as the water route, a general mass balance equation for the food chain in Fig. 8.17 is obtained as

$$\frac{d\nu_i'}{dt} = \frac{d(\nu w)_i}{dt} = \frac{w_i d\nu_i}{dt} + \frac{\nu_i dw_i}{dt} = k_{ui}w_ic - K_i\nu_i' + \alpha_{i,i-1}C_{i,i-1}w_i \tag{8.152}$$

$$i = 2, 3, 4$$

where ν_i' is the chemical whole body burden in μg and is given by Eq. 8.127, $\alpha_{i,i-1}$ is the chemical assimilation efficiency (μg chemical absorbed/μg chemical ingested), C is the specific consumption of organism i on $i - 1$(g wet wt prey/g wet wt predator · day), and t in the equation is interpreted as the age of the ith organism. The chemical assimilation efficiency, α, represents that portion of the ingested chemical mass that is absorbed and incorporated in organism tissue. The quantity $1 - \alpha$ then represents that portion of ingested chemical which is immediately or rapidly egested by the organism and never accumulated in body tissue. The excretion rate therefore in Eq. 8.152 is interpreted as excretion of the chemical that has been absorbed into the organism components.

An equation for the individual organism weight is

$$\frac{dw_i}{dt} = (a_{i,i-1}C_{i-1} - r_i)w_i, \qquad i = 2, 3, 4 \tag{8.153}$$

where $a_{i,i-1}$ is the biomass conversion efficiency (g wet wt predator/g wet wt prey) and r_i is the weight loss in day^{-1} due to routine metabolism, swimming, and other activities. The weight change of Eq. 8.130 is therefore

$$G_i = \frac{dw_i}{dt} \cdot \frac{1}{w_i} = a_{i,i-1}C_{i,i-1} - r_i \tag{8.154}$$

Equation 8.152 can therefore be written as

$$\frac{d\nu_i}{dt} = k_{ui}c + \alpha_{i,i-1}C_{i,i-1}\nu_{i-1} - K_i'\nu_i \qquad i = 2, 3, 4 \tag{8.155}$$

Equation 8.155 forms the basic mass balance chemical equation. The first term of Eq. 8.155 represents the direct uptake of the chemical by the organism from the water. The second term represents the flux of the chemical into the organism through predation. Note that the units of this term are gram of chemical assimilated per gram predator per day. The third term is the loss of chemical due to depuration from body tissue at a rate K_i plus the change in concentration due to the rate of change of the individual's weight. This is reflective of the changing volume available through the size of the organism to "dilute" the concentration of the chemical. In order to obtain additional insight into the food chain transfer process, it is informative to consider a steady state analysis of Eq. 8.155.

8.9.2.1 Steady State Model   Steady state in the context of this analysis is considered as the dynamic balance between the weight change of an individual organism and the total body burden of the chemical. It is viewed as the average toxicant concentration one would observe across an ensemble of specified compartment characteristics, such as age, net growth, and feeding behavior.

In Eqs. 8.151 nd 8.155 if $dv_i/dt = 0$, the steady state equation for the second (zooplankton) compartment is

$$v_2 = \frac{k_{u2}c + \alpha_{21}C_{21}v_1}{K_2'} \tag{8.156}$$

and upon substitution of Eq. 8.151

$$v_2 = \frac{k_{u2}c}{K_2'} + \frac{\alpha_{21}C_{21}}{K_2'} \cdot \frac{k_{u1}c}{K_1} \tag{8.157}$$

or from Eq. 8.135

$$v_2 = (N_{2w} + f_{21}N_{1w})c \tag{8.158}$$

where N_{2w} is the BCF for the zooplankton and f_{21} is the biomagnification ratio due to food chain transfer, and is given by

$$f_{21} = \frac{\alpha_{21}C_{21}}{K_2'} \tag{8.159}$$

The total zooplankton chemical concentration is thus composed of two additive parts: the BCF directly from the water and the chemical accumulation due to predation of contaminated phytoplankton.

Following a development similar to the zooplankton, the steady state equation for the third compartment (small fish) is

$$v_3 = \frac{k_{u3}c + \alpha_{32}C_{32}v_2}{K_3'} \tag{8.160}$$

Substituting Eq. 8.157, the chemical concentration in the small fish compartment is

$$v_3 = \left(N_{3w} + \frac{\alpha_{32}C_{32}N_{2w}}{K_3'} + \frac{\alpha_{32}C_{32}}{K_3'} \frac{\alpha_{21}C_{21}N_{1w}}{K_2'} \right)c \tag{8.161}$$

or

$$v_3 = N_{3w} + f_{32}N_{2w} + f_{32}f_{21}N_{1w})c \tag{8.162}$$

where $N_{3w} = k_{u3}/K_3'$, the bioconcentration factor for uptake of the chemical from the water only by the small fish. At this third level, the concentration is now seen to depend on the two lower levels through successive predation.

For the top carnivore compartment, the procedure is now direct. The steady state toxicant concentration is

$$v_4 = \frac{k_{u4}c + \alpha_{43}C_{43}v_3}{K_4'} \tag{8.163}$$

When successive substitutions are made for each of the lower trophic levels, the resulting concentration factor for the large fish is

$$v_4 = \left(N_{4w} + \frac{\alpha_{43}C_{43}N_{3w}}{K_4'} + \frac{\alpha_{43}C_{43}}{K_4'} \frac{\alpha_{32}C_{32}N_{2w}}{K_3'} \right. \tag{8.164}$$

$$\left. + \frac{\alpha_{43}C_{43}}{K_4'} \frac{\alpha_{32}C_{32}}{K_3'} \frac{\alpha_{21}C_{21}N_{1w}}{K_2'} N_{1w} \right)c$$

or

$$v_4 = (N_{4w} + f_{43}N_{3w} + f_{43}f_{32}N_{2w} + f_{43}f_{32}f_{21}N_{1w})c \tag{8.165}$$

Equation 8.165 can also be written as

$$v_4 = \left[N_{4w} + \sum_{j-1}^{n-1} \left(\prod_{i=j+1}^{n} f_{i,j-1} \right) \right] N_{iw})c \tag{8.166}$$

$$= N_4 c \tag{8.167}$$

where N_4 is the overall accumulation factor. In this model then, there will always be an increase in chemical concentration in a predator due to food chain transfer. The question is the relative significance of this route compared to the water route. Thomann (1981) and Thomann and Connolly (1982) show that if food chain transfer is considered significant at 20% then all chemicals for which the ratio of chemical concentration in the predator to that in the prey in feeding experiments is about 0.2, the food chain route is important. Sample Problem 8.9 contains computations for a four level food chain.

Equation 8.167 together with Eq. 8.47 provide the relationship between the concentration of toxicant in the fish (as the top predator), the external input of the chemical, and the characteristics of the lake, that is,

$$v_4 = \left(\frac{N_4 f_d}{q + v_T} \right) W_T' \tag{8.168}$$

This expression can also be used to estimate the within-lake net loss rate, v_T, if data are available on v_4 and the loading rate W_T'.

For a riverine situation, if it is assumed that fish accumulate the toxicant in a small area of a constant reach of the stream and do not migrate into or out of

SAMPLE PROBLEM 8.9

DATA

Suppose a food chain consists of three levels above the phytoplankton as shown:

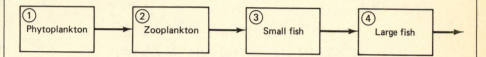

The phytoplankton are in reversible equilibrium with the water and the three trophic levels prey successively on each level below. The characteristics of the three levels are:

Trophic level	Respiration rate coefficient a' (Eq. 8.147)	Net growth rate G (1/day)	Food assimilation efficiency a	Weight of organism w (g (wet wt))
2	0.015	0.025	0.3	0.01
3	0.03	0.0063	0.8	10.0
4	0.03	0.0025	0.8	1000.0

A toxic substance in the water is at a dissolved water concentration of 750 ng/l. Laboratory experiments have indicated the following data:

Trophic level	BCF N'_w (l/kg (wet wt))	Depuration rate K (1/day)	Toxic assimilation efficiency α
2	10^4	0.05	0.5
3	10^4	0.01	0.5
4	10^4	0.005	0.5

PROBLEM

(a) Calculate the toxicant concentration (μg/g(wet wt)) in each trophic level.

(b) Calculate the required dissolved water concentration (μg/l) so that the concentration is less than or equal to 5 μg/g(wet wt) in the food chain.

ANALYSIS

First, calculate respiration (Eq. 8.147) and specific food consumption (Eq. 8.154):

$$r'_i = a'_i w_i^{-0.2}$$

$$C_{i,i-1} = \frac{G_i + r'_i}{a_i}$$

(continued)

Sample Problem 8.9 (continued)

Level 2

$$r_2' = 0.015(0.01)^{-0.2} = 0.0377 \text{ g/g} \cdot \text{day}$$

$$C_{21} = \frac{0.025 + 0.0377}{0.3} = 0.209 \text{ g/g} \cdot \text{day}$$

Levels 3 & 4

$$r_3' = 0.0189 \text{ g/g} \cdot \text{day}; \ C_{32} = 0.0315 \text{ g/g} \cdot \text{day}$$

$$r_4' = 0.0075 \text{ g/g} \cdot \text{day}; \ C_{43} = 0.0125 \text{ g/g} \cdot \text{day}$$

Second, calculate food chain ratios, $f_{i,i-1}$ (e.g., Eq. 8.159).

Level 2 on 1: $f_{21} = \dfrac{\alpha C_{21}}{K_2 + G_2} = \dfrac{0.5(0.209)}{0.5 + 0.025}$

$$= \underline{1.393} \left[\frac{\mu\text{g tox assim/g(wet wt)zoop}}{\mu\text{g tox ingest/g(wet wt)phyto}} \right]$$

Level 3 on 2: $f_{32} = \dfrac{\alpha C_{32}}{K_3 + G_3} = \dfrac{0.5(0.0315)}{0.01 + 0.0063}$

$$= \underline{0.966} \left[\frac{\mu\text{g tox assim/g(wet wt)sm fish}}{\mu\text{g tox ingest/g(wet wt)zoop}} \right]$$

Level 4 on 3: $f_{43} = \dfrac{\alpha C_{43}}{K_4 + G_4} = \dfrac{0.5(0.0125)}{0.005 + 0.0025}$

$$= \underline{0.833} \left[\frac{\mu\text{g tox assim/g(wet wt)lrg fish}}{\mu\text{g tox ingest/g(wet wt)zoop}} \right]$$

Now, calculate concentrations at each trophic level:

Level 2 (Eq. 8.158)

$$\nu_2 = (N_{2w} + f_{21}N_{1w})c; \ N = N'/1000 \qquad \text{[Eq. (8.136)]}$$

$$= [10 + 1.393(10)] \cdot 750 \text{ ng/l} \cdot \mu\text{g/1000 ng}$$

$$\frac{\mu\text{g/g}}{\mu\text{g/l}}$$

$$\underline{\nu_2 = 17.95 \ \mu\text{g/g(wet wt)}}$$

Level 3 (Eq. 8.162)

$$\nu_3 = (N_{3w} + f_{32}N_{2w} + f_{32}f_{21}N_{1w})c$$

$$= 10[1 + 0.966 + 0.966(1.393)]0.75$$

$$\underline{\nu_3 = 24.83 \ \mu\text{g/g(wet wt)}}$$

Sample Problem 8.9 (continued)

Level 4 (Eq. 8.165)

$$\nu_4 = (N_{4w} + f_{43}N_{3w} + f_{43}f_{32}N_{2w} + f_{43}f_{32}f_{21}N_{1w})c$$

$$= 10[1 + 0.833 + (0.833)(0.966) + (0.833)(0.966)(1.393)]0.75$$

$$\underline{\nu_4 = 28.19 \ \mu g/g(\text{wet wt})}$$

Note successive effect of lower levels in transferring toxic to top trophic level.

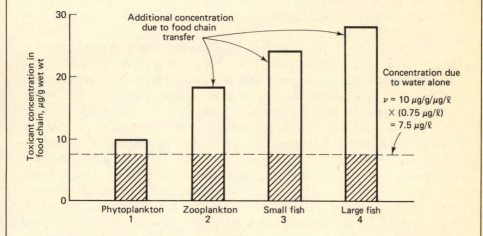

(b) Since maximum concentration is at top level, if reduce top trophic level to 5 μg/g(wet wt), all levels will meet requirement.

Also note that the concentration in large fish is linear to the water concentration:

$$\therefore (c)_{\text{reqd}} = \frac{(\nu_4)_{\text{reqd}}}{(\nu_4)_{\text{present}}} \cdot (c)_{\text{present}}$$

$$= \frac{5}{28.19} (0.75)$$

$c_{\text{reqd}} = 0.13 \ \mu g/l = 130 \ \text{ng/l}$ or a 83% reduction in water concentration and hence in incoming toxicant load.

the small area then, following a similar development as for the completely mixed lake, the chemical concentration in the fish is

$$\nu_4 = N_4 f_d \frac{W_T}{Q} \exp\left[-\left(\frac{v_T}{H_1}\right) \frac{x}{U} \right] \qquad (8.169)$$

Therefore, if data are available on water or fish concentrations of a toxicant as a function of distance, Eqs. 8.64 and 8.169 indicate that a semilog plot of that data with distance permits estimation of the net loss of the toxicant in the stream. It can be recalled (see Eq. 8.48 and following) that the net loss rate, v_T embodies all of the complex interactions between the water column and sediment. For some purposes, where water concentrations are low and difficult to measure or erratic over time, Eq. 8.169 may give better estimates of v_T since the fish will tend to accumulate low water concentrations to measurable levels and average out short term transient fluctuations.

Data on the PCB concentration in the Shiawassee River, a small stream in Michigan, provide an example (Env. Res. Group, 1981). This stream received a discharge of PCB from an industrial user and subsequent to the cessation of the discharge, data were collected on the PCB in the sediments and in the fish. Figure 8.18(a) shows the fish PCB concentrations and Fig. 8.18(b) shows the sediment data. The data on the fish generally follow the theoretical distribution given by Eq. 8.169 which indicates a logarithmic decrease in the concentration with distance. Using a velocity of 0.33 m/s, an overall loss rate of PCB for the Shiawassee is calculated at 3.0/day. This estimate is supported by the analysis of the sediment PCB data shown in Fig. 8.18(b) where a similar number is calculated. The theoretical development discussed previously indicated that the net loss rate of the chemical at steady state should be equal among the water column, sediment, and fish concentrations. This analysis of the PCB fish and sediment data therefore tends to confirm the theoretical formulation for the Shiawassee River.

A similar logarithmic variation in the copper concentration in caged mussels is shown in Fig. 8.19 where again such data can be used to estimate the net loss of the chemical in the region downstream of a discharge.

8.10 FINITE DIFFERENCE APPLICATIONS

All of the preceding kinetic and transport mechanisms as noted in 8.7 are applicable to the finite difference approach discussed previously (Chapter 3) and shown in previous applications. Both steady and time variable toxic substance models are used with the finite difference grid.

8.10.1 Steady State—PCB in Saginaw Bay

The segmentation for Saginaw Bay on Lake Huron, as shown in Fig. 8.20(a), is from Richardson et al. (1983) who have also summarized the data base for PCB concentrations and the overall properties and circulation of the bay. The resulting

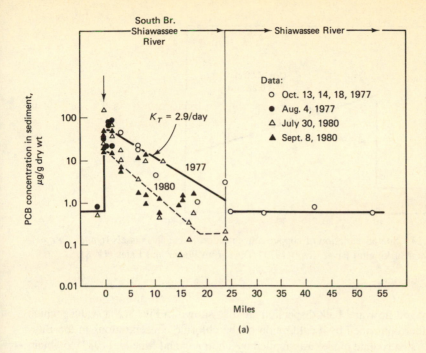

(a)

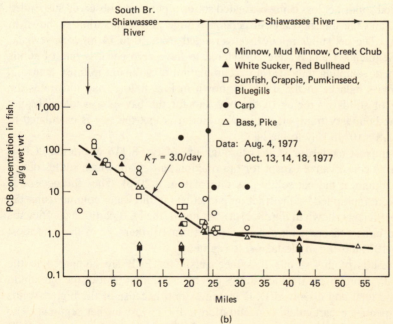

(b)

Figure 8.18 (a) Sediment PCB concentration vs. distance, South Branch Shiawassee River. (b) PCB concentration in fish, whole body vs. distance. Arrow at mile zero location of source. Data from Env. Res. Group (1981). Reprinted by permission of Elsevier Science Publishers B.V.

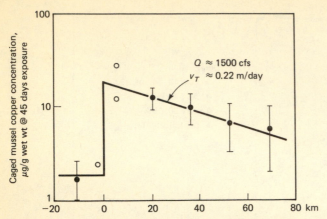

Figure 8.19 Spatial variation of copper concentration in caged mussels (*Quadrula quadrula*) in the Muskingum River, Sept. 1973. (Data from Foster and Bates (1978).

time averaged flow and bulk dispersion are also shown in Fig. 8.20a, values which were further confirmed by a calibration of the chloride concentrations in the Bay.

The steady state model was applied by Thomann and Mueller (1983) to obtain the suspended solids net loss using estimated solids input loads, observed suspended solids data, and the transport and dispersion regime. The calibration is shown in Fig. 8.20(b). The net solids removal rates ranged from 10 to 14 m/yr in segs. 2, 4, and 5. Segments 1 and 3 were considered to have zero net loss rate of solids since sediment sampling indicated no accumulation of sediment in those areas.

The mass balance of the solids is shown in Fig. 8.20(c) and illustrates the net settling of solids by region of the bay and for the bay as a whole. The net effect of the boundary transport with Lake Huron is that the bay is calculated to add about $1.8 \cdot 10^6$ lb/day to the lake.

The simplest model for PCB is through use of Eq. 8.116 where the PCB is assumed to be conservative except for the fraction particulate which settles out of the water column at the net solids loss rate of Fig. 8.20(b). Thus the net solids loss rates were multiplied by the fraction particulate in the water column using the partition coefficients shown in Fig. 8.21(a). As noted, the PCB partition coefficients were varied in accordance with observed data from 10^4 liters/kg in the innermost segment to 10^5 liters/kg in the lake-side segments.

The results of this simple loss mechanism for PCB are compared to the observed data in Fig. 8.21(a) which shows a good calibration. Notice the maximum values of the total and dissolved PCB in seg. 1 but, because of the higher solids in that segment, the particulate concentration is the lowest in that segment. The mass balance for PCB in the bay is shown in Fig. 8.21(b). The net loss of PCB to the sediment is shown as 0.7 lb/day or about one-third of the total input. The other two-thirds of the external input is exchanged on a net basis with Lake Huron.

This simple calculation permits estimation of the steady state response of PCB concentration in the bay to be expected if the external load were reduced or

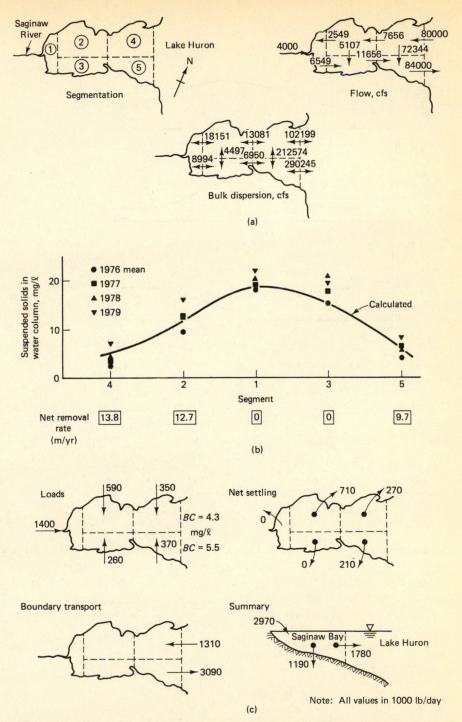

Figure 8.20 Suspended solids balance for Saginaw Bay, MI. (a) Segmentation of bay, net flow and dispersion between segments. (b) Calibration of suspended solids using indicated net loss rates of solids. (c) Mass balance of solids. From Thomann and Mueller (1983). Reprinted by permission of Butterworth Publishers (for Ann Arbor Book Co.).

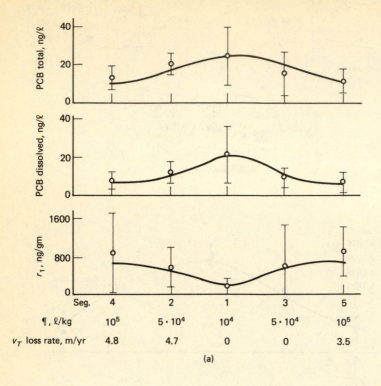

(a)

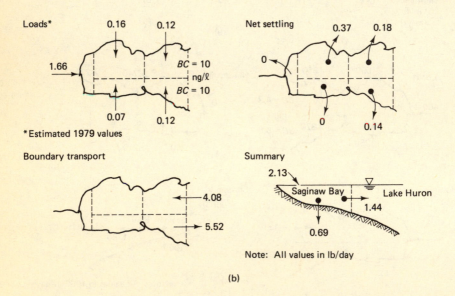

*Estimated 1979 values

(b)

Figure 8.21 Saginaw Bay PCB calibration and analysis. (a) Calibration of water column PCB—total, dissolved, and particulate—for 1979. (b) Long-term mass balance components for PCB. From Thomann and Mueller (1983). Reprinted by permission of Butterworth Publishers (for Ann Arbor Book Co.).

580

increased. However, since the calculation is steady state, it is not possible to describe time variable behavior of such systems. Such a model is illustrated below. The use of the steady state modeling framework, however, permits the generation of steady state response matrices for toxics in the water column and sediment.

8.10.2 Steady State Toxic Substances Response Matrices—Great Lakes

Using Eq. 8.82 for the chemical balance equation in the water column, the set of matrix coefficients for Chapter 3 (Eqs. 3.32) and the relationship between water column and sediment as in Eq. 8.53, it is possible to generate a marix inverse reflecting the steady state response in each water column and sediment segment due to unit inputs of chemical loading. Thus, the matrix equation is

$$[\mathbf{A}](c_T) = (W) \tag{8.170}$$

where the coefficients of $[\mathbf{A}]$ include the flow and dispersion as before but where $K_i \equiv v_{Ti}/H_i$, the net chemical loss rate calculated for each segment. The elements of (c_T) are the water column concentrations in each water segment. Equation 8.53 or subsequent simplifications in Eq. 8.112 can then be used to calculate the sediment concentration. Once calculated, the response matrices can be displayed in tabular form for subsequent use. The responses due to a series of loads into each segment are calculated simply from the tables by multiplication of the column responses by the actual load and summing over all columns that receive input loads.

For the application to the Great Lakes, the segment geometry used for calculating the steady state chemical concentration response matrices utilizes the representation of the Great Lakes given in Thomann and Di Toro (1983, 1984). The Lakes are considered well mixed, with the exception of Lake Erie, which is divided into three basins, and the inclusion of Saginaw Bay. A total of eight segments is used. Tables 8.8 and 8.9 show the segment geometry and the solids balances. (See also Fig. 8.23 for the segmentation arrangement.)

Tables 8.10 and 8.11 show the water column and sediment responses of a particular chemical with ¶ = 10^5 liters/kg everywhere and zero dissolved loss. Figure 8.22 shows the utility of the tables in rapidly estimating the water column and sediment concentrations at steady state for the loaded lake and all downstream lakes.

8.10.3 Great Lakes Time Variable Toxics Model

This application utilizes the time variable chemical equations (Eqs. 8.45 and 8.46) in a finite difference framework with three sediment layers (Thomann and Di Toro, 1983). The purpose of the model was to provide a framework for assessing the time variable response of toxic inputs to the Great Lakes and also how long it would take for chemical concentration in the lakes to decline to various levels. The emphasis was on the PCBs in the lakes because of the impact of the presence of that chemical on the fisheries, particularly the lake trout of Lake Michigan.

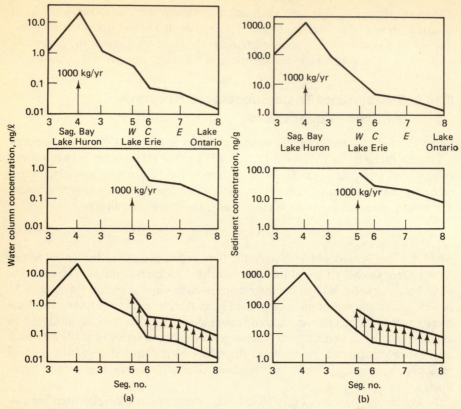

Figure 8.22 (a) Water column concentration of chemical with ¶ = 10⁵ l/kg everywhere in the Great Lakes for 1000 kg/yr into Saginaw Bay (top), Western Lake Erie (middle), and the sum of both inputs (bottom). (b) Similar to (a) but for surface sediment concentration. From Thomann (1986).

TABLE 8.8 GEOMETRIC AND TRANSPORT PARAMETERS OF WATER
COLUMN USED IN GREAT LAKES MODEL

Segment no.	Lake or region	Volume (km³)	Depth (m)	Surface area (km²)	Flow In (m³/s)	Flow Out (m³/s)	Horizontal exchange (m²/s)
1	Superior	11,326	137	82,882	2,033	2,033	—
2	Michigan	4,716	82	57,772	1,470	1,470	—
3	Huron	3,243	56	57,437	3,503	4,967	—
4	Saginaw	25	6	4,222	153	153	1,229
5	Erie, West	23	8	3,026	4,967	5,282	
6	Central	288	18	15,716	5,282	5,517	4,425
7	East	175	28	6,254	5,517	5,692	23,354
8	Ontario	1,666	86	19,485	5,692	6,740	—

Source: From Thomann and Di Toro (1983).

TABLE 8.9   NET SOLIDS LOSS RATES AND SEDIMENTATION VELOCITIES FOR GREAT LAKES[a]

Segment	Lake/ region	Net solids loss rate [m/yr (m/day)]	Net sediment deposition rate $(g/m^2 \cdot yr)$	Net sediment velocity[b] (mm/yr)
1	Superior	195(0.53)	98	0.41
2	Michigan	137(0.38)	69	0.29
3	Huron	213(0.58)	107	0.45
4	Saginaw Bay	8(0.02)	64	0.27
5	Erie, West	87(0.24)	1740	7.24
6	Central	216(0.59)	1080	4.52
7	East	185(0.51)	927	3.87
8	Ontario	447(1.22)	224	0.93

[a]See Table 4.3 for overflow rates, solids loadings, and solids concentrations.
[b]Using porosity of surface sediment = 0.9 through ρ_s = 2.4 and Eq. 8.35; porosity = 0.85 and 0.82 for 2–4 cm and 4–6 cm sediment layers, respectively.
Source: From Thomann and Di Toro (1983).

The model was calibrated using the radionuclide plutonium 239,240 since (a) long-term loading information was available, (b) water column and sediment concentrations were also available for certain years, and (c) plutonium partitions onto solids at about the same order of magnitude as PCBs, and does not volatilize.

Figure 8.23(a) shows the segmentation and Tables 8.8 and 8.9 show the geometry and the solids balances for each segment. Eight water column segments and three sediment segments of 2 cm each were used for a total of 32 segments. Figure 8.23(b) shows the assumed plutonium loading (Wahlgren et al., 1981) and Fig. 8.23(c) shows the result of a 24-year calculation for the plutonium using that loading history. The initial conditions were zero plutonium everywhere and ¶ in the water column was set equal to ¶ in the sediment. Three settling velocity conditions were used to determine the degree of any resuspension. Figure 8.23(c) and Fig. 8.24(a) show that setting $v_s = v_n$ in the time variable calculation did not produce a good comparison to the observed data. Assuming a settling velocity of about 2.5m/day, however, resulted in a much better comparison in all the lakes.

As noted in Eq. 8.40, if the solids settling velocity is greater than the net settling velocity of the solids, then resuspension is implied. It was therefore concluded on the basis of the calibration of the plutonium data as shown in Figs. 8.23(c) and 8.24(a) that some effective resuspension was occurring in the lakes. Table 8.12 shows the effective resuspension velocity resulting from the calibration and shows that the range of v_u is about 1–100 times the net sedimentation rate, v_d.

Figure 8.24(b) shows the calibration for the three sediment layers of Lake Michigan using the data of Edington et al. (1975a, 1975b). It should be noted that the sediment comparison is the result of the sediment buildup of plutonium over time, since the plutonium was intially set at zero concentration. The comparison to sediment data over time therefore represents an additional test of the model.

With a calibrated plutonium model, the framework was applied to PCB using a range of external inputs of PCB load (Thomann and Di Toro, 1983). In order to

TABLE 8.10 GREAT LAKES STEADY STATE RESPONSE MATRIX WATER COLUMN (NG/L). LOADING OF TOXICANT (1000 kg/yr) INTO SEGMENT[a]

	Lake Superior 1	Lake Michigan 2	Lake Huron 3	Saginaw Bay 4	Western Erie 5	Central Erie 6	Eastern Erie 7	Lake Ontario 8
Lake Superior 1	1.1983							
Lake Michigan 2	0.0000	2.3544						
Lake Huron 3	0.1018	0.1466	1.3376	1.1022				
Saginaw Bay 4	0.0746	0.1074	0.9797	19.7086				
Western Erie 5	0.0350	0.0504	0.4598	0.3788	2.1974	0.1837	0.1040	
Central Erie 6	0.0064	0.0092	0.0843	0.0695	0.4028	0.6329	0.3582	
Eastern Erie 7	0.0045	0.0065	0.0590	0.0486	0.2818	0.4427	1.0184	
Lake Ontario 8	0.0013	0.0019	0.0169	0.0139	0.0808	0.1269	0.2920	1.5982

[a]Toxicant properties: zero dissolved loss, ¶ = 10^5 l/kg everywhere

TABLE 8.11 GREAT LAKES STEADY STATE RESPONSE MATRIX SURFACE SEDIMENT (*NG/GM*). LOADING OF TOXICANT (1000 kg/yr) INTO SEGMENT

	Lake Superior 1	Lake Michigan 2	Lake Huron 3	Saginaw Bay 4	Western Erie 5	Central Erie 6	Eastern Erie 7	Lake Ontario 8
Lake Superior 9	114.0782							
Lake Michigan 10	0.0000	224.1389						
Lake Huron 11	9.6914	13.9563	127.3395	104.9294				
Saginaw Bay 12	4.1440	5.9661	54.4223	1094.8127				
Western Erie 13	1.1667	1.6801	15.3274	12.6273	73.2503	6.1236	3.4668	
Central Erie 14	0.4266	0.6133	5.6194	4.6329	26.8506	42.1891	23.8776	
Eastern Erie 15	0.3000	0.4333	3.9329	3.2397	18.7848	29.5104	67.8865	
Lake Ontario 16	0.1238	0.1809	1.6089	1.3233	7.6922	12.0809	27.7984	152.1486

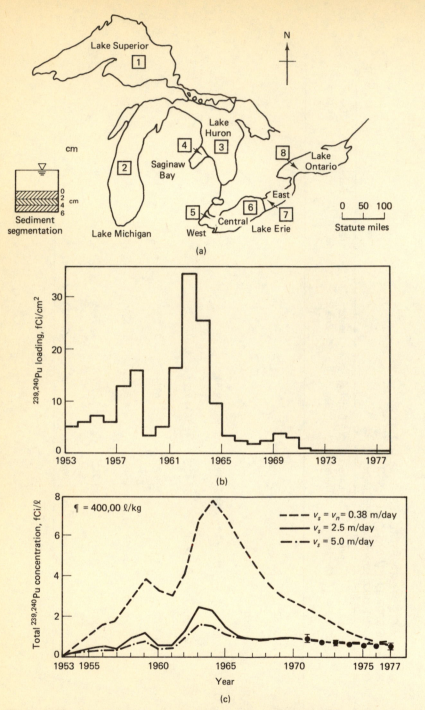

Figure 8.23 Great Lakes toxic substances model. (a) Segmentation. (b) Areal loading of plutonium (from Wahlgren, et al., 1981) with constant loading 1973–1977. (c) Calibration of Lake Michigan plutonium data (Wahlgren et al., 1981) for three conditions of settling velocity v_s. From Thomann and DiToro (1983). Reprinted by permission of International Association for Great Lakes Research.

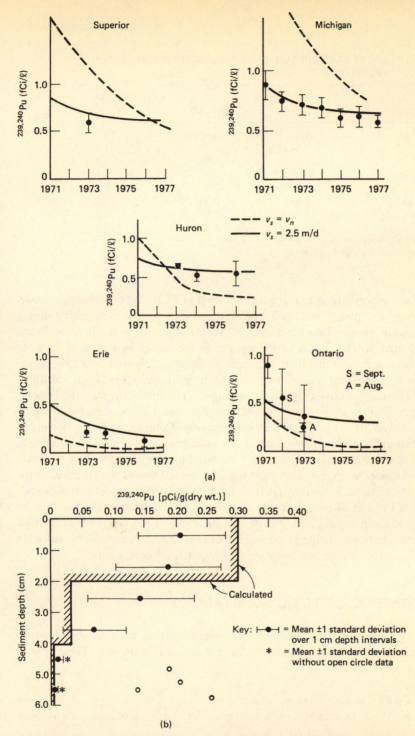

Figure 8.24 (a) Comparison of calculated plutonium concentrations for the Great Lakes for 1971–1977 using the model of Thomann and DiToro (1983) (data from Wahlgren et al., 1981) ¶ = 400,000 l/kg. (b) Comparison to sediment data of 1973–1974, v_s = 2.5 m/day (data from Edington et al., 1975a, 1975b). Reprinted by permission of International Association for Great Lakes Research.

TABLE 8.12 GREAT LAKES RESUSPENSION VELOCITIES FOR SETTLING
VELOCITY OF 2.5 M/DAY

Lake or region	v_u (mm/yr)	v_d (mm/yr)	Ratio v_u/v_d
Superior	1.51	0.41	3.7
Michigan	1.64	0.29	5.6
Huron	1.35	0.56	2.4
Saginaw Bay	30.53	0.27	113.1
Western Lake Erie	68.80	7.25	9.5
Central Lake Erie	14.50	4.50	3.2
Eastern Lake Erie	11.20	2.85	3.9
Ontario	0.89	1.02	0.9

Source: From Thomann and Di Toro (1984).

estimate the response time to reach a given level of PCB, the time variable model
was run with a constant load during buildup and then zero load thereafter during
the depuration phase. Two conditions of volatilization were used: $k_l = 0$ and
$k_l = 0.1$ m/day. The results of the simulation are shown in Fig. 8.25. For the
water column it is seen that there is a rapid initial drop followed by a long period
where the concentration reduces only slightly in the upper lakes. With volatilization, the decrease is obviously even more rapid. The time to reduce the concentration to 50% of the initial values varied from less than 5 years when volatilization
was included to 10–20 years without volatilization. Comparison to the decline of
PCB in fish in Lake Michigan indicated that volatilization is probably occurring at
a rate of about 0.1 m/day. These applications illustrate the utility of the toxic
substances model in approximate estimates of expected fate of the chemical in the
aquatic environment. Sensitivity analysis for the steady-state model aids in bound-
ing expected chemical concentrations in the water and sediment and can then be
used in food chain calculations or toxicity evaluations. Time variable models must
be used for questions relating to the buildup of chemicals to a certain concentration
and a subsequent decline or depuration to a required concentration.

8.11 CONTROL OF TOXIC SUBSTANCES

As reviewed in the previous water quality problem contexts, there are several points
at which control can be exercised to achieve a desirable water use. These control
points include:

1. Control at the source including:
   a. Waste treatment to reduce chemical input using processes such as given
      by (Mulligan et al. 1981):
      i.   Adsorption to activated carbon.
      ii. Chemical oxidation.
      iii. Ultraviolet photolysis and catalyzation.

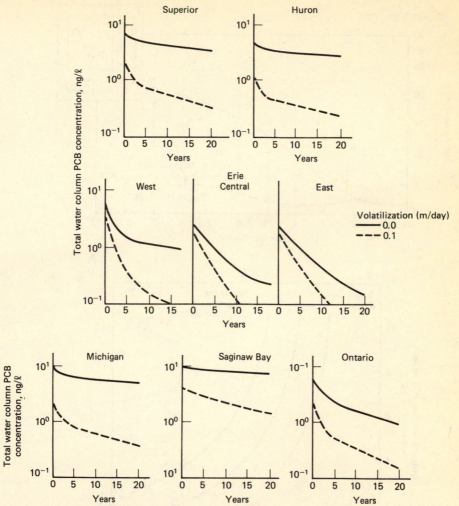

Figure 8.25 Calculated water column PCB concentration response to an instantaneous drop in the PCB external input at $t = 0$. A high load estimate is used, with and without volatilization. From Thomann and DiToro (1983). Reprinted by permission of International Association for Great Lakes Research.

  iv. Air stripping.
  v. Biological treatment.
b. Control of agricultural runoff containing pesticides and urban storm-water runoff containing a variety of chemicals.
c. Modification of the product to:
  i. Increase biodegradation.
  ii. Decrease solids partitioning.
  iii. Decrease bioconcentration and food chain magnification.

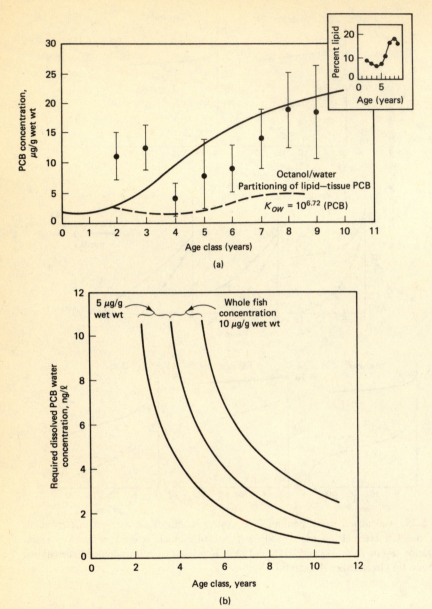

Figure 8.26 Illustration of use of food chain model to estimate allowable age classes for consumption. (a) Calibration of model to PCB in Lake Michigan lake trout. (b) Required water concentration to meet 5–10 μg PCB/g wet wt in lake trout at different age classes. From Thomann and Connolly (1984). Reprinted with permission from Environmental Science and Technology. Copyright 1984, American Chemical Society.

2. Control within the water body including:
   a. Dredging of contaminated sediment with subsequent treatment and disposal of dredge spoils (see Table 7.18 for dredging costs).
   b. Covering or in-place fixation of contaminated sediments.
3. Control at point of use including:
   a. Treatment of municipal water supply before distribution.
   b. Consumption of only certain size or weight class of fish where chemical concentrations are acceptable.

An example of the latter control approach is shown in Fig. 8.26. A model was constructed of the food chain accumulation of a chemical, in this case, PCB. The model was applied to the uptake and transfer of PCB in the lake trout food chain of Lake Michigan (Thomann and Connolly, 1984). Figure 8.26(a) shows the calibration of the observed PCB data in the lake trout to that calculated using age-dependent model equations similar to Eq. 8.155. The early age classes were not fully calibrated due probably to exposure to water concentrations other than the constant 5 ng/l assumed throughout the calculation. Better than 99% of the calculated PCB in the adult lake trout was calculated to be due to food chain transfer. Figure 8.26(a) also shows that a simple partitioning into the lipid phase of the lake trout failed to reproduce the observed data, thereby further supporting the importance of food chain transfer of PCB.

Figure 8.26(b) shows the simulation to meet a level of 5–10 μg/g wet wt PCB concentration in the whole fish at different age classes. As indicated, the older age classes require the lowest dissolved PCB water concentration to meet the 5–10 μg/g wet wt level. Thus, if a level of 2 ng/l were obtained, then whole fish 6 years and older would have concentrations between 5 and 10 μg/g wet wt. Conversely, whole fish less than 6 years old would have PCB concentrations less than 5 μg/g wet wt. Relationships can be developed between whole fish concentrations and the concentrations in the edible portion. It is the latter that is usually subject to regulation (see Table 8.4). It is possible to use a plot such as Fig. 8.26(b) to develop an age-dependent (or more practically a weight or size-dependent) basis for opening up the fishery to consumption. For example, if projections for the next 10 years indicate that the lowest attainable PCB water concentration is 1 ng/l because of diffuse and atmospheric sources, then trout 8 years or older (i.e., 3.5 kg or 70 cm) would be prohibited for consumption. The whole body concentrations in those fish would be expected to be equal to or greater than 5 μg/g wet wt.

REFERENCES

American Public Health Assoc., American Water Works Assoc., and Water Pollution Control Fed., 1985. Standard Methods for the Examination of Water and Wastewater, APHA, Washington, D.C., 874 pp.

Banerjee, S., S. H. Yalkowsky, and S. C. Valvani, 1980. Water Solubility and Octanol/Water Partition Coefficients of Organics. Limitations of the Solubility-Partition Coefficient Correlation, *Env. Sci. Tech* **14**(10):1227–1229.

Bender, M. E., 1969. Uptake and Retention of Malathion by the Carp, *The Progressive Fish-Culturist*, **31**(3): 155–159.

Bowman, B. T., and W. W. Sans, 1983. Determination of Octanol-Water Partitioning Coefficients (K_{ow}) of 61 Organophosphorus and Carbonate Insecticides and Their Relationship to Respective Water Solubility (*S*) Values, *J. Environ. Sci. Health* **B18**(6): 667–683.

Buikema, A. L. Jr., B. R. Niederlehner, and J. Cairns Jr., 1982. Biological Monitoring. Part IV—Toxicity Testing, *Water Research* **16**:239–262.

Camp, Dresser, and McKee (CDM), 1982. Analysis of Data to Determine the Capability of Organics to Withstand Various Concentration Levels of Toxic Pollutants for Short Periods of Time, Report to USEPA, 28 pp.

Chiou, C. T., V. H. Freed, D. W. Schmedding, and R. L. Kohnert, 1977. Partition Coefficient and Bioaccumulation of Selected Organic Chemicals, *Env. Sci. Tech.* **11**(5):475–478.

Connolly, J. P., and R. P. Winfield, 1984. WASTOX, A Framework for Modeling the Fate of Toxic Chemicals in Aquatic Environments. Part 1: Exposure Concentration, Env. Eng. & Sci., Manhattan College, Prepared for USEPA, Coop. Agreement No. R807827/R807853, Large Lakes Res. Station, Grosse Ile, MI, 130 pp.

Delos, C. G., W. L. Richardson, J. V. DePinto, R. B. Ambrose, P. W. Rodgers, K. Rygwelski, J. P. St. John, W. J. Shaughnessy, T. A. Faha, and W. N. Christie, 1984. Technical Guidance Manual for Performing Waste Load Allocations, Book II. Streams and Rivers, Chapter 3, Toxic Substances. Off. of Water Reg. and Stds., Monitoring and Data Support Div., Water Qual. Anal. Br., USEPA, Washington, D.C., 203 pp. + Appendix. EPA-440/4-84-022.

Di Toro, D. M., and J. Nusser, 1976. Particle Size and Settling Velocity Analyses for Bay Area Waters, Hydroscience, Inc., Report prepared for California Dept. of Water Resources.

Di Toro, D. M., D. J. O'Connor, R. V. Thomann, and J. P. St. John, 1981. Analysis of Fate of Chemicals in Receiving Waters, Phase 1. Chemical Manufact. Assoc., Washington, D. C., Prepared by HydroQual, Inc., Mahwah, N. J., 8 Chapters and 4 Appendixes.

Di Toro, D. M., L. M. Horzempa, M. M. Casey, and W. Richardson, 1982a. Reversible and Resistant Components of PCB Adsorption-Desorption: Adsorbent Concentration Effects, *J. Great Lakes Res.* **8**(2):336–349.

Di Toro, D. M., D. J. O'Connor, R. V. Thomann, and J. P. St. John, 1982b. Simplified Model of the Fate of Partitioning Chemicals in Lakes and Streams, K. S. Dickson, A. W. Maki, and J. Cairns, Jr., eds., *Modeling the Fate of Chemicals in the Aquatic Environment*, Ann Arbor Science, Ann Arbor, MI, pp. 165–190.

Di Toro, D. M., 1985. A Particle Interaction Model of Reversible Organic Chemical Sorption, *Chemosphere* **14**(10):1503–1538.

Dixon, D. G., and J. B. Sprague, 1981. Copper Bioaccumulation and Hepotopration Synthesis During Acclimation to Copper by Juvenile Rainbow Trout, *Aquatic Toxicology* **1**:69–81.

Edington, D. N., J. J. Alberts, M. A. Wahlgren, J. O. Karttunen, and C. A. Reeve, 1975a. Plutonium and Americium in Lake Michigan Sediments, IAEA Symposium on Transuranium Nuclides in the Environment, San Francisco, California Argonne Nat. Lab., Argonne, IL, 25 pp.

Edington, D. N., and J. A. Robbins, 1975b. The Behavior of Plutonium and Other Long-Lived Radionuclides in Lake Michigan. II. Patterns of Deposition in the Sediments, Proceedings of International Symposium on Impacts of Nuclear Releases into the Aquatic Environment, Int. Atomic Energy Agency, Vienna, IAEA-SM-198/39:245–260.

Env. Res. Group, 1981. Appendix B (1977 and 1981 data) of Final Report, Problem Definition Survey of South Branch of the Shiawaissee River, Sept. 1980. ERG, Inc., Ann Arbor, MI.

Federal Register, 1980. Part V. EPA Water Quality Criteria Documents; Availability, Vol. 45, No. 231, Nov. 28, 1980, pp. 79318–79379.

Federal Register, 1984. EPA Request for Comments on Nine Documents Containing Proposed Ambient Water Quality Criteria. 49FR4551 Feb. 7, 1984.

Foster, R. B., and J. M. Bates, 1978. Use of Freshwater Mussels to Monitor Point Source Industrial Discharges, *Env. Sci. Tech.* **12**:958–962.

Hansen, D. J., P. R. Parish, and J. Forester, 1974. Aroclor 1016: Toxicity to and Uptake by Estuarine Animals, *Env. Res.* **7**:363–373.

Hawley, N., 1982. Settling Velocity Distribution of Natural Aggregates, *J. Geo. Res.* **87**(C12):9489–9498.

Kanazawa, J., 1978. Bioconcentration Ratio of Diazinon by Freshwater Fish and Snail, *Bull. Env. Contam. Toxicol.* **20**:613–617.

Kanazawa, J., 1981. Measurement of the Bioconcentration Factors of Pesticides by Freshwater Fish and Their Correlation with Physiochemical Properties or Acute Toxicities, *Pestic. Sci.* **12**:417–424.

Karickhoff, S. W., D. S. Brown, and T. A. Scott, 1979. Sorption of Hydrophobic Pollutants on Natural Sediments, *Water Research* **13**:241–248.

Karickhoff, S. W., 1984. Organic Pollution Sorption in Aquatic Systems, Am. Soc. Civ. Eng. J. Hydr. Div., **10**(6):707–735.

Kenaga, E. E., and C. A. I. Goring, 1978. Relationships Between Water Solubility, Soil-Sorption, Octanol-Water Partitioning, and Bioconcentration of Chemicals in Biota, Presented at ASTM Symposium, New Orleans, October, 1978, 63 pp.

Kenaga, E. E., 1980. Correlation of Bioconcentration Factors of Chemicals in Aquatic and Terrestrial Organisms with These Physical and Chemical Properties, *Env. Sci. Tech.* **14**(5):553–556.

Laszlo, F., P. Literathy, and P. Benedek, 1975. Heavy Metals Pollution in the Sajo River, Hungary, In Symposium Proceedings of Int. Conference on Heavy Metals in the Environment, Toronto, Canada, Vol. II, Part 2, pp. 923–931.

Literathy, P., and F. Laszlo, 1977. Uptake and Release of Heavy Metals in the Bottom Silt of Recipients, H. L. Golterman, Ed., *Interactions Between Sediments and Fresh Water*, Junk, B. W. Pub., The Hague, pp. 403–409.

Lyman, W. J., W. F. Reehl, and D. H. Rosenblatt, 1982. *Handbook of Chemical Property Estimation Methods, Environmental Behavior of Organic Compounds*, McGraw-Hill, New York, 26 Sections, 3 Appendixes.

McCave. I. N., 1975. Vertical Flux of Particles in the Ocean, *Deep Sea Res.* **22**:491–502.

McKim, J. M., G. F. Olson, G. W. Holcombe, and E. P. Hunt, 1976. Long-Term Effects of Methylmercuric Chloride on Three Generations of Brook Trout (*Salvelinus fontinalis*): Toxicity, Accumulation, Distribution, and Elimination, *J. Fish. Res. Board Con.* **33**:2726–2739.

Mackay, D., 1981. *Environmental and Laboratory Rates of Volatilization of Toxic Chemicals from Water in Hazard Assessment of Chemicals, Current Developments*, Vol. 1, J. Saxena and F. Fisher, Eds., Academic Press, New York, pp. 303–322.

Miller, M. M., S. P. Wasih, G. L. Huang, W. Y. Shiu, and D. Mackay, 1985. Relationships Between Octanol-Water Partition Coefficient and Aqueous Solubility, *Env. Sci. Tech.* **19**(6):522–529.

Mills, W. B., J. D. Dean, D. B. Porcella, S. A. Gherini, R. J. M. Hudson, W. E. Frick, G. L. Rupp, and G. L. Bowie, 1982. Water Quality Assessment: A Screening Procedure for Toxic and Conventional Pollutants, Part 1. Tetra Tech, Inc., Env. Res. Lab., Office of Res. and Devel., USEPA, Athens, GA, 570 pp. EPA-600/6-82-004a.

Muir, D. C. G., and N. P. Grift, 1981. Environmental Dynamics of Phosphate Esters. II. Uptake and Bioaccumulation of 2-Ethylhexl Diphenyl Phosphate and Diphenyl Phosphate by Fish, *Chemosphere* **10**(8):847–855.

Mulligan, T. J., J. A. Mueller, and O.K. Scheible, 1981. Treatment of Toxic Wastewaters, Presentation at May 1981 Am. Soc. Civ. Eng. Int'l Conv. & Exposition, New York, 12 pp.

Neely, W. B., D. R. Branson, and G. E. Blau, 1974. Partition Coefficient to Measure Bioconcentration Potential or Organic Chemicals in Fish, *Env. Sci. Tech.* **8**:1113–1115.

O'Connor, D. J., and J. P. Connolly, 1980. The Effect of Concentration of Adsorbing Solids on the Partition Coefficient, *Water Research* **14**:1517–1523.

O'Connor, D. J., 1985. Modeling Frameworks, Toxic Substances Notes, Manhattan College Summer Institute in Water Pollution Control.

Rathbun, R. E., and D. Y. Tai, 1981. Technique for Determining the Volatilization Coefficients of Priority Pollutants in Streams, *Water Research* **15**(2):243–250.

Reinert, R., L. J. Stone, and H. L. Bergman, 1974. Dieldrin and DDT: Accumulation from Water and Food by Lake Trout (*Salvelinus namaycush*) in the Laboratory, Proc. 17th Conf. Great Lakes Res., pp. 52–58.

Reynolds, T. D., and E. F. Gloyna, 1964. Uptake and Release of Radionuclides by Stream Sediments, Advances in Water Pollution Research, Proc. Second Int'l Conf., O. Jaag, Ed., Tokyo, Japan, Aug. 1964; Vol 1 pp. 151–164.

Richardson, W. L., V. E. Smith, and R. Wethington, 1983. Dynamic Mass Balance of PCB and Suspended Solids in Saginaw Bay—A Case Study, D. Mackay, S. Paterson, S. J. Eisenreich, and M. S. Simmons, Eds., *Proc. Physical Behavior of PCBs in the Great Lakes*, Ann Arbor Science, Ann Arbor, MI. pp. 329–366.

Schimmel, S. C., J. M. Patrick, Jr., and J. Forester, 1976. Heptachlor: Uptake, Depuration, Retention, and Metabolism by Spot, *Leiostomus xanthurus, J. Toxicol. Env. Health.* **2**:164–178.

Shih, C-S., and E. F., Gloyna, 1969. Influence of Sediments on Transport of Solutes, *Am. Soc. Civ. Eng., J. Hydr. Div.* **95**(HY4):1347–1367.

Smith, J. H., D. C. Bomberger, Jr, and D. L. Haynes, 1981. Volatilization Rates of Intermediate and Low Volatility Chemicals from Water, *Chemosphere* **19**:281–289.

Sprague, J. B., 1969. Review Paper, Measurement of Pollutant Toxicity to Fish. I. Bioassay Methods for Acute Toxicity, *Water Research* **3**:793–821.

Sprague, J. B., 1970. Review Paper, Measurement of Pollutant Toxicity to Fish. II. Utilizing and Applying Bioassay Results, *Water Research* **4**:3–32.

Thomann, R. V., 1981. Equilibrium Model of Fate of Microcontaminants in Diverse Aquatic Food Chains, *Can. J. Fish. Aquatic Sci.* **38**:280–296.

Thomann, R. V., and D. M. Di Toro, 1983. Physico Chemical Model of Toxic Substances in the Great Lakes, *J. Great Lakes Res.* **9**(4):474–496.

Thomann, R. V., and J. A. Mueller, 1983. Steady-State Modeling of Toxic Chemicals— Theory and Application to PCBs in the Great Lakes and Saginaw Bay, Mackay et al., Eds., *Physical Behavior of PCBs in the Great Lakes*, Ann Arbor Science, Ann Arbor, MI, 442 pp.

Thomann, R. V., and J. P. Connolly, 1982. An Age Dependent Model of PCB in a Lake Michigan Food Chain, Large Lakes Research Station, Grosse Ile, MI, Final Report, USEPA, ERL-Duluth, pp. 1–110.

Thomann, R. V., and J. P. Connolly, 1984. Model of PCB in the Lake Michigan Lake Trout Food Chain, *Env. Sci. Tech.* **18**:65–71.

Thomann, R. V., 1984. Physio-Chemical and Ecological Modeling of the Fate of Toxic Substances in Natural Water Systems, *Ecological Modelling* **22**(1983/1984):145–170.

Thomann, R. V., and D. M. Di Toro, 1984. Physico Chemical Model of Toxic Substances in the Great Lakes, Project Report to USEPA, Large Lakes Res. Sta., Grosse Ile, MI, 163 pp.

Thomann, R. V., 1985. A Simplified Heavy Metals Model for Streams, in preparation.

Thomann, R. V., 1986. Steady State Toxic Substances Reponse Tables for the Great Lakes, Section A of Pending Report to USEPA, Grosse Ile, MI, 9 pp., 3 Tables, 14 Figures.

Thomann, R. V., and H. J. Salas, 1986. Manual on Toxic Substances in Surface Waters, Pan American Health Organization, CEPIS, Lima, Peru.

Thurston, R. V., T. A. Gilfoil, E. L. Moyn, R. K. Zajdel, T. I. Aoki, and G. D. Veith, 1985. Comparative Toxicities of Ten Organic Chemicals to Ten Common Aquatic Species, *Water Research* **19**:1145–1155.

USFDA, 1978. Action Levels for Poisonous or Deleterious Substances in Human Food and Animal Feed.

USEPA, 1980. Treatability Manual. Vol., I. Treatability Data, ORD, USEPA, Washington D.C., 15 Chapters. EPA-600/8-80-042a.

Veith, G. D., N. M. Austin, and R. T. Morris, 1979. A Rapid Method for Estimating log P for Organic Chemicals, *Water Research* **13**:43–47.

Verschueren, K., 1983. *Handbook of Environmental Data on Organic Chemicals,* 2nd Ed., Van Nostrand Reinhold Co., New York, 1310 pp.

Vreeland, V., 1974. Uptake of Chlorobiphenyls by Oysters. *Env. Poll.* **6**:135–140.

Wahlgren, M. A., J. A. Robbins, and D. N. Edington, 1981. Plutonium in the Great Lakes, W. C. Hanson, Ed., *Transuranics in the Environment*, U.S. Dept. of Energy (TID-22800), p. 659–683.

World Health Organization, 1984a. *Guidelines for Drinking Water Quality, Vol. 1, Recommendations*, WHO, Geneva, Switzerland, 130 pp.

World Health Organization, 1984b. *Guidelines for Drinking Water Quality, Vol. 2, Health Criteria and Other Supporting Information*, WHO, Geneva, Switzerland, 335 pp.

PROBLEMS

8.1. Consider a large river, 10 m deep, and with a velocity of 0.1 m/s. The wind speed over the river is 5 m/s. An input of methyl parathion enters this river through a tributary. Calculate the overall volatilization rate in m/day for methyl parathion in this river. Use

$$\text{molecular weight} = 263$$
$$\text{solubility} = 25 \, \text{mg/l}$$
$$\text{temperature} = 20°C$$
$$\text{vapor pressure} = 1.3 \cdot 10^{-8} \, \text{atm}$$

8.2.

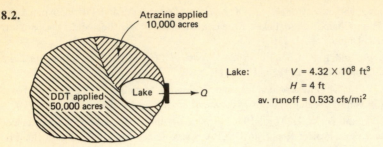

Atrazine applied
10,000 acres

Lake:
$V = 4.32 \times 10^8$ ft^3
$H = 4$ ft
av. runoff = 0.533 cfs/mi^2

Suspended solids: loading = 1.38×10^6 lb/day

lake concentration = 100 mg/l

Pesticide	Runoff Concentration (ug/l)	Hydrolysis (day^{-1})	Photolysis (day^{-1})	$\pi_1 = \pi_2$ (μg/kg)/(μg/l)
Atrazine	630	0.1	0.02	50
DDT	14	2.5×10^{-4}	1.5×10^{-4}	50,000

For a steady state condition and no sediment decay, determine the following:
(a) Lake detention time.
(b) Suspended solids net loss rate.
(c) Lake atrazine concentrations (total, dissolved, particulate).
(d) Lake DDT concentrations (total, dissolved, particulate).

Assume volatilization and biodegradation are negligible.

8.3. A lake with a surface (equal to sediment) area of 10^8 m^2, a detention time of 2 years and a depth of 10 m receives a toxicant load of 10^6 g/yr. The solids concentration in the lake is 10 mg/l. The toxicant has a partition coefficient of 10^4 liters/kg which is equal in both the water column and the sediment. There is no decay of the toxicant in the sediment or in the water column. The sediment concentration of the toxicant has been measured and averages 15.15 μg/g dry wt.
(a) What is the net flux in g/yr of the toxicant to the sediment?
(b) What new load is necessary to result in the dissolved toxicant equal to 0.5 μg/l in the water column?

8.4. A river has a flow of 10 m^3/s, a depth of 3 m, and receives a point input of PCB of 1000 g/day. The partition coefficient is 10^5 liters/kg average solids concentration is 20 mg/l, and net solids settling velocity is 0.2 m/day. The volatilization loss is 1.5 m/d. Using a level 1 analysis,
(a) Compute and plot the longitudinal distribution of the total PCB.
(b) Compute and plot the dissolved form of the PCB as a function of travel time.

8.5. An estuary receives a chemical discharge of 100 lb/day. The properties of the estuary and the chemical are given by:

$Q = 1000$ MGD   Decay rate of dissolved chemical = 0.1/day
$E = 5$ smpd   Partition coefficient = 10,000 liters/kg
$U = 0.2$ mpd   Average solids concentration, m = 50 mg/l
                      Depth = 10 ft
                      Net solids settling velocity = 0.5 ft/day

Using a level 1 analysis, what is the maximum concentration of this chemical in the estuary?

8.6. Reanalyze the lake of Sample Problem 8.3, using a level 2 framework. Assume the following additional sediment parameters:

$$\P_2 = 250 \text{ l/kg}$$
$$m_2 = 300,000 \text{ mg/l}$$
$$v_s = 5 \text{ m/day}$$
$$K_f = 0.5 \text{ cm/day}$$

8.7. With an increased sediment decay rate of $K_{d2} = 0.1/\text{day}$, repeat the level 3 analysis for Sample Problem 8.7 and determine the allowable load at $x = 0$ (one load only).

8.8. Brook trout were exposed to 2.5 μg/l of a chemical in an aquarium and the following results were obtained over time.

Time (days)	Whole fish concentration (μg/g wet wt)
0	0
15	0.2
30	0.3
45	0.4
60	0.46
75	0.52
90	0.50

The exponential growth rate of the trout during this period was equal to 0.005/day. Compute:
(a) The bioconcentration factor (liters/kg wet wt).
(b) The excretion rate (day^{-1}).
(c) The uptake rate (liters/g · day).

8.9. Data from New York State Fisheries Laboratory (New York State Dept. of Environmental Conservation, Dept. of Fish & Wildlife, monthly report, Vol. 2, Report 10, 1/20/79, 9pp.) were obtained on the loss of mirex from brook trout. The experiment began with the concentration of mirex in the trout at 6.34 μg/g wet wt. The trout were then transferred to clean tanks for 385 days. The weight of the fish at the start of the test were 175 g wet wt. At the end of 385 days, the mirex concentration in the trout was 2.12 μg/g wet wt and the weight of the fish was 572 g wet wt. What is the excretion rate (day^{-1}) derived from this experiment?

8.10. Suppose a food chain has the following characteristics:

Level	Net growth rate (day^{-1})	Food assimilation efficiency	Respiration rate (day^{-1})
2	0.009	0.3	0.08
3	0.005	0.8	0.009
4	0.002	0.8	0.006

If a substance in the water is at 10 μg/l, is bioconcentrated uniformly at a BCF of 10^4 (liters/kg), and is excreted uniformly at 0.01/day, what is the toxicant concentration in the top trophic level? Assume a uniform chemical assimilation efficiency of 0.9.

8.11. A lake has the following characteristics:

$$\text{volume} = 10^{12} \text{ m}^3 \qquad\qquad \text{depth} = 20 \text{ m}$$
$$\text{flow} = 2.74 \cdot 10^8 \text{ m}^3/\text{day}$$
$$\text{suspended solids} = 20 \text{ mg/l} \qquad \text{net solids settling velocity} = 0.2 \text{ m/day}$$

It is desired to know the allowable discharge of a toxicant that will not exceed the FDA level of 0.3 ppm in the fish. The properties of the toxicant are:

partition coefficient = 10,000 liters/kg

BCF between dissolved water concentration and fish concentration = 5000 liters/kg

dissolved concentration is conservative (no loss due to volatilization, photolysis, etc.)

What is the allowable discharge of this toxicant in kg/yr (at steady state) so that the concentration in the fish does not exceed 0.3 ppm? Use a level 1 analysis for the water column.

Chapter 9

Temperature

9.1 SIGNIFICANCE OF WATER TEMPERATURE

From a water quality engineering point of view, the temperature of a body of water is of particular significance for three principal reasons: (a) the discharge of excess heat from industrial or municipal effluents may positively or negatively affect the aquatic ecosystem, (b) temperature influences all biological and chemical reactions, and (c) variations in temperature affect the density of water and hence the transport of water. This chapter explores the first area of temperature significance, namely the impact of the discharge of excess heat on a given water system. The influence of temperature on biological and chemical reactions was discussed at appropriate points in the topics of previous chapters. The temperature–density relationships were discussed in Chapter 3.

Water temperature and its variation represents the state of a water body as a result of heat inputs, losses, and exchange. The units of heat used in various studies are summarized in Appendix B, Conversion Factors. Excess heat which may change ambient water temperatures can affect the aquatic ecosystem in several ways including:

1. Direct lethal effect on sensitive plants or animals.
2. Indirect long-term effects on the aquatic ecosystem through effects on growth and/or reproduction.
3. Indirect effects through changes in the species distribution of the system.

9.1.1 Temperature Criteria

The National Academy of Sciences and National Academy of Engineering (1972) summarized data for chinook salmon that indicates the importance of the acclimation temperature, that is, the temperature to which the organism has become accustomed. For example, for chinook acclimatized to 24°C, 50% of the fish die at 26°C (a 2°C rise) in about 40 hr. But for fish acclimatized to 20°C, a rise to 26°C (a 6°C rise) results in 50% mortality after only about 16 hr. Table 9.1 summarizes some results for three temperature levels for fish: the optimum temperature (temperature of maximum physiological "strength"), upper lethal temperature (i.e., the temperature that the organism cannot be acclimated to without causing death), and the maximum weekly average temperature desirable for sustaining a growing and reproducing fish population. As seen, the range in each of these temperature levels can be significant depending on the species of fish. Regulating agencies have generally adopted a variety of water temperature criteria to protect given ecosystems in a specific area. Such criteria for temperature generally involve some allowable increase (e.g., 2°C) above normal conditions. The specification of an allowable increase does not include a designated "mixing zone" where higher temperatures may be permitted.

9.2 EXCESS HEAT INPUTS

The principal sources of excess heat are from electric generating stations and industrial cooling water. Municipal wastewater effluents, while generally at a temperature level between 10 and 20°C, and industrial process water are usually not significant sources of heat. Per capita energy consumption has increased from about $180 \cdot 10^6$ Btu in 1925 to $359 \cdot 10^6$ Btu in 1973 (Hill, 1977). Power plants that are built to meet these rising energy demands can become quite large and generate substantial quantities of waste heat in the cooling water resulting from the condensing process. For example, a 1000-megawatt (MW) plant (i.e., a plant outputting 10^9 W of energy) can supply an average of about 10^5 people. Most plants are 500 MW and above (e.g., 75% of total steam generating capacity in 1970 was from plants of greater than 500 MW). Projected U.S. energy demand is expected

TABLE 9.1 SOME TEMPERATURE LIMITS FOR FISH

	Temperature (°C)		
Fish species	Optimum	Upper lethal	Maximum weekly average
Brook trout	14.5	25.5	18.2
Lake herring	16	27.7	19.2
Bluegill	22	33.8	25.9
Smallmouth bass	27.3	35.0	29.9
Channel catfish	30.0	38.0	32.7

Source: Abstracted from NAS/NAE (1972).

to increase by 60% to about $170 \cdot 10^{15}$ Btu/yr between 1985 and the year 2000 (Hill, 1977). This energy demand is equivalent to almost 6000 power plants of 1000-MW size.

Figure 9.1(a) shows the typical efficiency of power plants. Figure 9.1(b) indicates that as an average, cooling water can be expected to be about 10 to 30°F above the incoming temperature.

The heat input into a body of water is given by

$$W_H = \rho \cdot c_p \cdot Q_e \cdot T \tag{9.1}$$

where ρ is the effluent water density, T is the temperature of the effluent, c_p is the specific heat of water, Q_e is the effluent flow, and W_H is the heat input. Typical units are

$$W_H(10^6 \text{ Btu/day}) = \rho\left(\frac{\text{lb}}{\text{ft}^3}\right) \cdot 1 \frac{\text{Btu}}{\text{lb-°F}} \cdot Q\left(\frac{\text{MG}}{\text{day}}\right) \cdot T(°F) \cdot \frac{0.1337 \text{ ft}^3}{\text{gal}}$$

Figure 9.1 (a) Efficiency of power plants. (b) Cooling water requirements and temperature elevation of 1000 MW plant. Adapted from Jimeson and Adkins (1972).

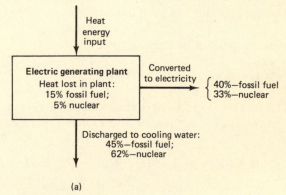

(a)

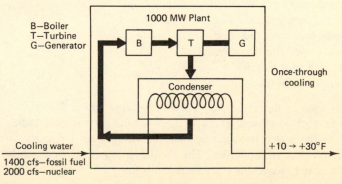

(b)

or

$$W_H(\text{kcal/day}) = \rho\left(\frac{g}{\text{cm}^3}\right) \cdot 1\,\frac{\text{cal}}{g \cdot {}^\circ C} \cdot Q\left(\frac{\text{m}^3}{\text{day}}\right) \cdot T({}^\circ C) \cdot 10^3 \frac{\text{cm}^3 \cdot \text{kcal}}{\text{m}^3 \cdot \text{cal}}$$

9.3 HEAT BALANCE—SOURCES AND SINKS

The temperature of a given water body depends to a considerable extent on the exchange of heat across the air–water interface and the subsequent distribution of that heat throughout the water column. Different sources and sinks of heat therefore must be estimated in order to properly assess a heat balance. Edinger et al. (1974) provide a full and detailed treatment of heat balance, exchange, and resulting temperature.

The principal sources of heat are:

1. Shortwave solar radiation.
2. Longwave atmospheric radiation.
3. Conduction of heat from atmosphere to water.
4. Direct heat inputs from municipal and industrial activities.

The principal sinks (or losses) of heat are:

1. Longwave radiation emitted by water.
2. Evaporation.
3. Conduction from water to atmosphere.

Edinger and Geyer (1965) and Edinger et al. (1968 and 1974) provide a review of each of the processes. The net rate of heat exchange per unit area of air–water interface is:

$$\Delta H = [(H_s - H_{sr}) + (H_a - H_{ar})] - (H_{br} \pm H_c \pm H_e) \qquad (9.2)$$

where ΔH = net heat exchange across the water surface
H_s = shortwave solar radiation
H_{sr} = reflected shortwave radiation
H_a = longwave atmospheric radiation
H_{ar} = reflected longwave radiation
H_{br} = longwave radiation from water
H_c = conductive heat transfer
H_e = evaporative heat transfer

All terms are in units such as cal/cm^2 · day (Btu/ft^2 · day). The first group of terms in brackets on the right-hand side of Eq. 9.2, called the net absorbed radiation, is independent of the water temperature and can be measured directly or computed from meteorological observations. The second group of terms depends in various ways on the water temperature as discussed below.

9.3.1 Net Absorbed Radiation

9.3.1.1 Shortwave Radiation: H_s and H_{sr}   Incoming solar radiation H_s can be measured directly with a pyreheliometer. Some National Weather Service stations record solar radiation as do power stations and many automatic water quality monitoring stations. The reflection of solar radiation depends on the sun's altitude, cloud cover, and water conditions. For an optically smooth surface (not usually met in nature) Fresnel's reflectivity law provides an estimate of the percentage reflection H_{sr}/H_s (Neumann and Pierson, 1966, p. 56) and as shown in Fig. 9.2 for sun altitude angles of greater than 40°, the reflection is only about 2–4%. But, at low sun angles, the percent reflected increases rapidly and at 10° is about 35%.

9.3.1.2 Longwave Radiation: H_a and H_{ar}   The longwave radiation is measured or estimated from

$$H_a = \sigma(T_a + 273)^4(A + 0.031\sqrt{e_a}) \tag{9.3}$$

where σ = Stefan–Boltzmann constant = $11.70 \cdot 10^{-8}$ cal/cm^2 · day · °K^4
        (or $4.15 \cdot 10^{-8}$ Btu/ft^2 · day · °R^4)
   T_a = air temperature (°C)

and

   e_a = air–vapor pressure (mm Hg) = (fraction relative humidity) ×
        $(e_{sat}$ @ $T_a)$
   A = coefficient related to air temperature and ratio of measured solar radiation to clear sky radiation (range 0.5–0.7). This coefficient is estimated from the ratio of solar radiation to clear sky solar radiation. At air temperatures greater than about 20°C, $A \approx 0.7$.
   e_{sat} = saturated vapor pressure of water (mm Hg) at air temperature, T_a

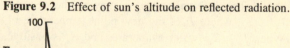

Figure 9.2   Effect of sun's altitude on reflected radiation.

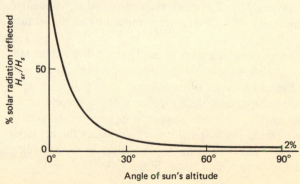

Table 9.2 shows the variation of the saturated vapor pressure with temperature. Figure 9.3 shows the effect of temperature on e_{sat} and, for a given humidity, the relationship between T_a and e_a.

The reflected longwave radiation is generally small; about 3% of incoming longwave radiation.

As seen, then, the net absorbed radiation $(H_s - H_{sr} + H_a - H_{ar})$ is independent of the water temperature and is a function of known or observable meteorological conditions.

9.3.2 Water Temperature-Dependent Heat Terms

The longwave radiation H_{br} emitted by the water follows the Stefan–Boltzmann law for an almost perfect black body and is given by

$$H_{br} = \epsilon\sigma(T_s + 273)^4 \tag{9.4}$$

where ϵ is the emissivity of water (about 0.97) and T_s is the surface water temperature (°C).

The rate of conductive heat transfer H_c depends on the temperature difference between the water and the air as well as the wind speed over the water:

$$H_c = c_1(19.0 + 0.95U_w^2)(T_s - T_a) \tag{9.5}$$

where T_s and T_a are the water surface and air temperature (°C), respectively, and c_1 is Bowen's coefficient, empirically given by 0.47 mm Hg/°C. Consistent with Eq. 9.2, H_c is positive (a net loss from the water) when the water temperature at the surface is greater than the air temperature and is negative when $T_a > T_s$.

The rate of heat loss by evaporation H_e may be estimated from a variety of equations. Edinger et al. (1974) have examined a number of these formulations and have suggested the slightly conservative expression of Brady, Graves, and Geyer (1969):

$$H_e = (19.0 + 0.95\,U_w^2)(e_s - e_a) \tag{9.6}$$

where U_w is the windspeed in m/s measured at a height of 7 m above the water surface and e_s and e_a are in mm Hg. The difference $e_s - e_a$ may be shown to be proportional to the difference $T_s - T_d$, where T_d is the dew point temperature of the air (Edinger et al., 1974). When $T_s > T_d$, water evaporates and H_e is positive (loss of heat from water); conversely, if $T_d > T_s$, water condenses on the surface and H_e is negative (heat gain).

The sum of each of the components can then be used to calculate the net rate of heat exchange in Eq. 9.2. Figure 9.4 shows the various components of the heat balance and the approximate ranges of values of each term.

This net input of heat across the water surface together with any direct sources of heat from industrial discharges (Eq. 9.1) form important inputs into analyses of temperature variations in natural bodies of water. In principle, if ΔH the net input is known, then it assumes a role equivalent to the input load W in the preceding modeling framework for rivers, estuaries, and lakes. In this case

$$W_n = \Delta H A_s \tag{9.7}$$

TABLE 9.2 SATURATED VAPOR PRESURE OF WATER, (mm Hg)

Temperature (°C)	e_{sat} (mm Hg)	Temperature (°C)	e_{sat} (mm Hg)
0	4.579	18	15.477
1	4.926	19	16.477
2	5.294	20	17.535
3	5.685	21	18.650
4	6.101	22	19.827
5	6.543	23	21.068
6	7.013	24	22.377
7	7.513	25	23.756
8	8.045	26	25.209
9	8.609	27	26.739
10	9.209	28	28.349
11	9.844	29	30.043
12	10.518	30	31.824
13	11.231	31	33.695
14	11.987	32	35.663
15	12.788	33	37.729
16	13.634	34	39.898
17	14.530	35	42.175

Source: Abstracted from the *Handbook of Chemistry and Physics*, 1965–1966, by permission of The Chemical Rubber Co.

Figure 9.3 Relationships between temperature, vapor pressure, and relative humidity. Adapted from Edinger et al. (1974), *Heat Exchange and Transport in the Environment*, Report EA-74-049-00-3. Copyright Nov. 1984, Electric Power Research Institute. Reprinted with permission.

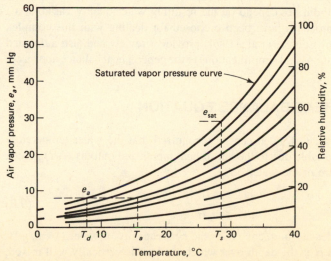

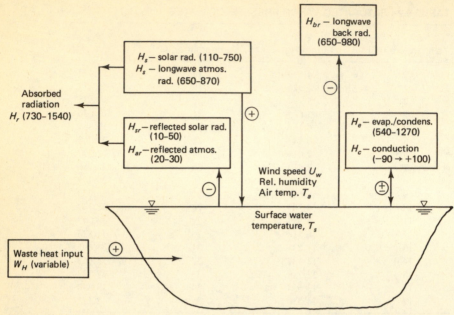

Figure 9.4 Heat balance terms and approximate values in cal/cm^2 · day. After Edinger and Geyer (1965).

where W_n is the net heat input to the water body in k·cal/day or Btu/day and A_s is the surface area over which the analysis is being conducted.

It is noted however that the determination of ΔH involves specification of the water temperature T_s as in Eqs. 9.4 and 9.5. Thus, the specification of the input forcing function becomes dependent on the state variable, water temperature. This is what makes the prediction of the temperature distribution in a water body somewhat different from the preceding modeling frameworks. In those analyses, the input waste load W did not depend on the resulting water quality variable, s.

A variety of approaches have been proposed for dealing with this complex temperature interaction. Edinger et al. (1974) provide a review and also a detailed treatment of approximations and simplifications to the general heat balance approach.

9.4 SIMPLIFIED HEAT BALANCE EQUATION

Following Edinger et al. (1974), it can be recognized that the water temperature is the resulting state variable due to the net of heat inputs and outputs (ΔH). Thus, for T as the temperature of the water body,

$$\frac{dT}{dt} = \frac{\Delta H}{\rho c_p H} \tag{9.8}$$

where ρ is the water density in g/cm^3, c_p is the heat capacity of water (1 cal/g·°C or 4186 J/kg·°C), and H is the depth over which the heat is vertically well mixed.

Edinger et al. (1974) have shown that the net heat input can be represented by

$$\Delta H = K(T_e - T) \tag{9.9}$$

where K is an overall heat exchange coefficient (with units of $W/m^2 \cdot °C$ or $Btu/ft^2 \cdot$ day $\cdot$ °F or $cal/cm^2 \cdot min \cdot °C$) and is a function of water temperature and wind speed and T_e is the equilibrium temperature. This temperature T_e is the temperature that a body of water would reach if all meteorological conditions were constant in time. It is as if all conditions were "frozen" at specific values and the water body is allowed to reach a steady temperature equal to T_e. The equilibrium temperature is therefore that temperature at which $\Delta H = 0$, that is, no net heat exchange across the water surface since under those conditions the water body will approach a water temperature representative of the meteorological conditions. That is, in the sum of all the terms for ΔH, T_e is that temperature where $\Delta H = 0$ when $T = T_e$.

The time rate of change of temperature is given by Eqs. 9.8 and 9.9 as

$$\frac{dT}{dt} = \frac{K(T_e - T)}{\rho c_p H} \tag{9.10}$$

If K and T_e are known or can be estimated, then the rate of heat dissipation in a body of water can be calculated. One procedure is to estimate T_e from the condition $\Delta H = 0$ for each time period and then estimate K from Eq. 9.9. This becomes a complicated interactive procedure. Other approaches taken for example by Edinger et al. (1974) estimate K and T_e with some reasonable approximations.

The exchange coefficient K is therefore a complicated function of the water temperature, windspeed, and other meteorological parameters such as the saturated vapor pressure. Edinger et al. (1974) show that the exchange coefficient in $W/m^2 \cdot °C$ may be determined from

$$K = 4.5 + 0.05T + \beta f(U_w) + 0.47 f(U_w) \tag{9.11}$$

where the wind function $f(U_w)$ is as in Eq. 9.5, that is,

$$f(U_w) = 9.2 + 0.46 U_w^2 \tag{9.12}$$

for $f(U_w)$ in $W/m^2 \cdot mm$ Hg and U_w in m/s. The coefficient β(mm Hg/°C) is given by

$$\beta = 0.35 + 0.015 T_m + 0.0012 T_m^2 \tag{9.13}$$

where

$$T_m = \frac{T + T_d}{2} \tag{9.14}$$

and T_d is the dew point temperature. Figure 9.5 shows this relationship of the heat exchange coefficient K with wind speed and temperature T_m.

The use of this chart requires a first estimate of the water temperature T and the dew point temperature. The solution converges rapidly following the initial estimate. The dew point temperature can be estimated as shown in Fig. 9.3. For air temperature T_a and relative humidity R, e_a is estimated as before. Then T_d is the temperature at the saturated vapor pressure curve of 100% humidity.

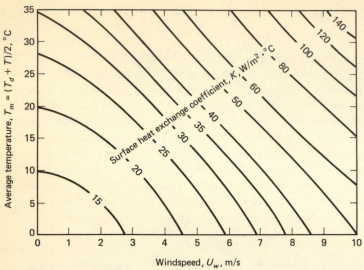

Figure 9.5 Design chart for surface heat exchange coefficient. After Edinger et al. (1974), *Heat Exchange and Transport in the Environment*, Report EA-74-049-00-3. Copyright Nov. 1984, Electric Power Research Institute. Reprinted with permission.

As noted, T_e, the equilibrium temperature, can be estimated for a given set of meteorological conditions by iteration until $\Delta H = 0$. Alternately, under constant coefficients, the equilibrium temperature as given by Edinger et al. (1974), is approximated by the empirical relationship

$$T_e = T_d + \frac{H_s}{K} \tag{9.15}$$

where as before H_s is the shortwave solar radiation. Equation 9.15 should be recognized as an approximate equation only and as such, it may obscure the fundamental nature of T_e. This temperature is really not a function of the variables of Eq. 9.15 because if one waited long enough, under constant coefficients, the equilibrium temperature would be reached regardless of the value of K. Nevertheless, for rapid approximate calculation, Eq. 9.15 together with Fig. 9.5 provide a means for estimating K and T_e for use in Eq. 9.9 (see Sample Problem 9.1).

9.5 TEMPERATURE MODELS

9.5.1 Completely Mixed Lakes

Consider a lake as shown in Fig. 9.6 that is assumed to be completely mixed, that is, the water temperature is equal throughout the water body. (See also Chapter 4 for discussion of completely mixed lakes.) As shown in this figure, the consequence of the completely mixed assumption is that the outflow temperature at the exit of the lake is equal to the temperature in the lake.

SAMPLE PROBLEM 9.1

DATA

Use the surface layer water temperatures from Sample Problem 4.4.

PROBLEM

Estimate the net heat exchange across the surface of Lake Ontario for June 1 and November 1.

ANALYSIS

Quantity	Units	Value on June 1	Value on Nov 1	References
T_s	°C	8.1	10.6	Sample Problem 4.4
T_a	°C	19.2	7.6	Ruffner and Bair (1984)
$I_T{}^a$	cal/cm²·day	610	235	Hutchinson (1957)
U_w	m/s	4.02	4.29	Ruffner and Bair (1984)
Relative humidity	%	68.6	76.4	Ruffner and Bair (1984)
Solar declination[b]	(°)	22.0	− 14.3	NAO (1986)
Solar altitude[c]	(°)	68.5	32.2	

[a]See Section 7.6.2.2.
[b]At noon.
[c]Alt = 90 − latitude (43.5°) + declination.

Shortwave Radiation

$$H_s = I_T = 610 \text{ and } 235 \text{ cal/cm}^2\text{·day for } 6/1 \text{ and } 11/1$$

From Fig. 9.2, using noon solar altitudes,

$$\frac{H_{sr}}{H_s} = 0.02 \text{ and } 0.07 \text{ for } 6/1 \text{ and } 11/1$$

and

$$H_{sr} = 12 \text{ and } 16 \text{ cal/cm}^2\text{·day}$$

Longwave Radiation

$$H_a = 11.70 \times 10^{-8}(T_a + 273)^4(A + 0.031\sqrt{e_a}) \qquad [\text{Eq. (9.3)}]$$

With $T_a = 19.2/7.6$°C and relative humidities of 69 and 76%, $e_a = 11.6$ and 6.0 mm Hg from Fig. 9.3. Using a constant value of 0.7 for A the Brunt coefficient (Edinger et al., 1974 for % clear sky radiation of 70 to 75%)

$$H_a = 687 \text{ and } 563 \text{ cal/cm}^2\text{·day}$$

(continued)

Sample Problem 9.1 (continued)

Using $H_{ar} = 0.03\,H_a$,

$$H_{ar} = 21 \text{ and } 17 \text{ cal/cm}^2\cdot\text{day for } 6/1 \text{ and } 11/1$$

The net absorbed radiation (H_n) is then

$$H_n = H_s - H_{sr} + H_a - H_{ar}$$

$$= \frac{\{610 - 12 + 687 - 21\}}{\{235 - 16 + 563 - 17\}} = \frac{1264 \text{ cal/cm}^2\cdot\text{day} \quad 6/1}{765 \text{ cal/cm}^2\cdot\text{day} \quad 11/1}$$

Longwave Back Radiation

$$H_{br} = 0.97 \times 11.70 \times 10^{-8} \times (T_s + 273)^4 \qquad \text{[Eq. (9.4)]}$$

Substituting $T_s = 8.1/10.6\ °C$ for 6/1 and 11/1,

$$H_{br} = 709 \text{ and } 734 \text{ cal/cm}^2\cdot\text{day}$$

Conduction

$$H_c = 0.47\,(19.0 + 0.95U_w^2)(T_s - T_a) \qquad \text{[Eq. (9.5)]}$$

With

$$U_w = 4.02/4.29 \text{ m/s}, \quad T_s = 8.1/10.6\ °C, \quad \text{and}$$

$$T_a = 19.2/7.6\ °C$$

$$H_c = -179 \text{ cal/cm}^2\cdot\text{day (heat gain) for } 6/1$$

and

$$H_c = +51 \text{ cal/cm}^2\cdot\text{day (heat loss) for } 11/1$$

Evaporation/Condensation

$$H_e = (19.0 + 0.95U_w^2)(e_s - e_a) \qquad \text{[Eq. (9.6)]}$$

From the calculations above

$$e_a = 11.6/6.0 \text{ mm Hg for } 6/1 \text{ and } 11/1$$

The vapor pressure of the water e_s is obtained from Fig. 9.3 using surface water temperatures of $T_s = 8.1$ and $10.6°C$ and the relative humidity of 100%, resulting in

$$e_s = 8.4 \text{ mm Hg for } 6/1$$

and

$$e_s = 9.7 \text{ mm Hg for } 11/1$$

Sample Problem 9.1 (continued)

Substituting the above values for e_s and e_a and $U_w = 4.02$ and 4.29 m/s,

$$H_e = -110 \text{ cal/cm}^2\cdot\text{day for } 6/1$$

and

$$H_e = +135 \text{ cal/cm}^2\cdot\text{day for } 11/1$$

Note that net condensation (heat gain) occurs in June and evaporation (heat loss) in November.

Net Rate of Heat Exchange

From Eq. 9.2,

$$\Delta H = H_n - (H_{br} \pm H_c \pm H_e)$$

$$\Delta H = \begin{cases} 1264 - (709 - 179 - 110) = +844 \\ 765 - (734 + 51 + 135) = -155 \end{cases} \text{cal/cm}^2\cdot\text{day} \quad \begin{matrix} 6/1 \\ 11/1 \end{matrix}$$

Thus, in early June, when the lake water temperature is lower than the air temperature, the atmosphere is a net source of heat to the lake. In November, with air temperatures rapidly cooling and water temperatures higher than those in the air, the lake loses heat to the atmosphere.

Figure 9.6 Heat balance terms for a completely mixed lake.

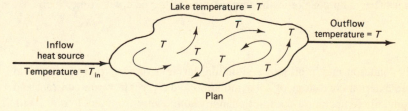

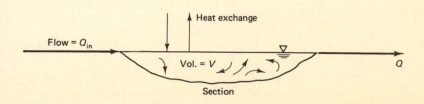

The heat balance is given by

rate of change of temperature in pond = heat in–heat out ± net heat exchange

The rate of change of the temperature, T, in the lake of volume V is given from Eq. 9.9 as

$$\rho c_p V \frac{dT}{dt}$$

with units (g/cm^3 × cal/g·°C × cm^3 × °C/day) = (cal/day). The heat transported into the pond by the inflow Q_{in} $[L^3/T]$ is

$$\rho c_p Q_{in} T_{in}$$

with units (g/cm^3 × cal/g·°C × cm/day × °C) = (cal/day).

The heat transported out of the lake by the lake outflow Q is

$$\rho c_p Q T \text{ (cal/day)}$$

Note that the inflow Q_{in} and outflow Q may be different. The net heat exchange is as given before (Eq. 9.9) with the exchange occurring across the surface area of the lake (A_s); thus the exchange is

$$KA_s(T - T_e) \quad \text{or} \quad \frac{KV}{H}(T - T_e)$$

with units (cal/cm^2·day·°C × cm^2 × °C) = (cal/day). The surface area of the lake is seen to be assumed equal to V/H, for an average depth H of the lake. The complete equation for the heat balance is given by the sum of each of the above components as

$$\rho c_p V \frac{dT}{dt} = \rho c_p Q_{in} T_{in} - \rho c_p Q T - KA_s (T - T_e) \tag{9.16}$$

If it is now assumed that the temperature in the lake is at a temporal steady state, that is, that there is no change in the temperature of the lake with time, then

$$\frac{dT}{dt} = 0$$

This condition may represent a situation during a summer period of several days where external conditions of solar radiation and river inflow are constant and the average daily temperature does not change significantly from day to day. With this assumption and for simplicity letting $Q_{in} = Q$, Eq. 9.16 yields

$$T = \frac{T_{in} + rT_e}{1 + r} \tag{9.17}$$

where

$$r = \frac{KA_s}{\rho c_p Q} \tag{9.18}$$

and is dimensionless (see Sample Problem 9.2). This simple heat balance for a completely mixed lake forms the basis for a first estimate of the required size of a cooling pond to dissipate excess heat. This is discussed below in Section 9.7.

9.5.2 Rivers

In this application, a heated discharge is assumed to be entering the river and elevating the water temperature as shown in Fig. 9.7. The length of the river affected by the discharge is to be estimated. The application incorporates the basic discussion on distribution of water quality in streams and rivers discussed in Chapter 2. Under average steady state conditions, that is, no changes in climatic or environmental variables, Eq. 2.21 becomes for temperature

$$U\frac{dT}{dx} = -\frac{K}{\rho c_p H}(T - T_b) = -K_R(T - T_b) \tag{9.19}$$

where as before T is the average water temperature in °C, T_b is the average background water temperature in the absence of the heated discharge, K is the approximate average heat exchange coefficient in cal/cm^2·min·°C, ρ is the water density in g/cm^3, c_p is the specific heat of water in cal/g·°C, and H is the water depth in cm. The overall exchange coefficient K_R has units day^{-1} and is given by

$$K_R = \frac{K}{\rho c_p H} \tag{9.20}$$

Figure 9.7 Distribution of temperature in a river due to heated discharge.

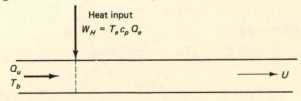

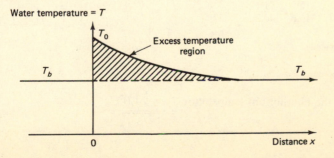

SAMPLE PROBLEM 9.2

DATA

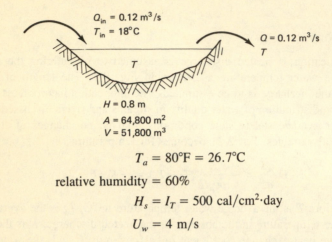

$$Q_{in} = 0.12 \text{ m}^3/\text{s}$$
$$T_{in} = 18°C$$
$$Q = 0.12 \text{ m}^3/\text{s}$$
$$T$$
$$T$$
$$H = 0.8 \text{ m}$$
$$A = 64{,}800 \text{ m}^2$$
$$V = 51{,}800 \text{ m}^3$$

$$T_a = 80°F = 26.7°C$$

$$\text{relative humidity} = 60\%$$

$$H_s = I_T = 500 \text{ cal/cm}^2\cdot\text{day}$$

$$U_w = 4 \text{ m/s}$$

PROBLEM

A shallow natural lake has a reasonably constant temperature for the period of interest. Determine its equilibrium (T_e) and water (T) temperatures. What impact will an excess heat input (W_H) cause if its effluent has a flow of 0.003 m³/s and a temperature of 150°F (65.5°C)?

ANALYSIS

Calculate Equilibrium Temperature

Use trial and error solution. Assume $T = T_a$ for first trial solution.

Trial	Trial T (°C)	$e_a{}^a$ (mm Hg)	$T_d{}^a$ (°C)	$T_m{}^b$ (°C)	K^c (W/m²·°C)	K^c (cal/cm²·day·°C)	H_s/K (°C)	$T_e{}^d$ (°C)
1	26.7	15.7	18.3	22.5	35	72.3	6.9	25.2
2	25.2	14.2	16.9	21.1	33.4	69.0	7.2	24.1
3	24.1	13.4	16.6	20.3	32.5	67.1	7.4	24.0

[a]Figure 9.3 with 60% relative humidity.
[b]Equation 9.14.
[c]Figure 9.5; 1 W/m²·°C = 2.066 cal/cm²·day·°C.
[d]Equation 9.15.

From the above table, the equilibrium temperature $T_e = 24.0°C$.

Sample Problem 9.2 (continued)

Natural Lake Temperature

Assume a well-mixed lake with uniform temperature throughout its volume. Since $Q_{in} = Q_{out}$, use following equations:

$$r = \frac{KA}{\rho c_p Q} = \frac{67.1\,\text{cal/cm}^2\cdot\text{day}\cdot°\text{C} \times 64{,}000\,\text{m}^2 \times 10^4\,\text{cm}^2/\text{m}^2}{1\,\text{g/cm}^3 \times 1\,\text{cal/g}\cdot°\text{C} \times 0.12\,\text{m}^3/\text{s} \times 10^6\text{cm}^3/\text{m}^3 \times 86{,}400\text{s/day}} \quad \text{[Eq. (9.18)]}$$

$$r = \frac{4.294 \times 10^{10}\,\text{cal/day}\cdot°\text{C}}{1.037 \times 10^{10}\,\text{cal/day}\cdot°\text{C}}$$

$r = 4.14$ (dimensionless)

$$T = \frac{T_{in} + rT_e}{1 + r} \quad \text{[Eq. (9.17)]}$$

$$T = \frac{18 + 4.14 \times 24.0}{1 + 4.14} = 22.8°\text{C (73.1°F)}$$

Lake Temperature with Excess Heat Input

Since effluent flow is small, neglect it in flow balance. $\therefore Q_{in} = Q_{out} = Q$. W_H would be added to the right side of Eq. 9.16, resulting in the following:

$$T = \frac{T_{in} + rT_e + W_H/(\rho c_p Q)}{1 + r}$$

with

$$W_H = \rho c_p Q_e T = 1 \times 1 \times 0.003 \times 65.5 = 0.197\,\text{cal/day} \quad \text{[Eq. (9.1)]}$$

$$T = \frac{18 + 4.14 \times 24.0 + 0.197/(1 \times 1 \times 0.12)}{5.14}$$

$$T = 23.1°\text{C}$$

Note the rise of 0.3°C due to the excess heat input.

The boundary condition at the discharge location ($x = 0$) is given by $T = T_0$ where T_0 is the resulting temperature at the outfall after mixing with the upstream river temperature. T_0 is calculated from Eq. 2.18.

The solution then to Eq. 9.19 is

$$T = (T_0 - T_b) \exp\left(-\frac{K_R x}{U}\right) + T_b \quad (9.21)$$

This expression shows that at the outfall the temperature rises to T_0 for a change above the background temperature of $T_0 - T_b$ (see Fig. 9.7). The elevated temperature declines exponentially as a function of the heat exchange coefficient and approaches the background temperature T_b at some distance downstream. The procedure therefore for calculating the downstream temperature distribution is:

1. Estimate upstream background temperature, T_b.
2. Compute mixed temperature at outfall using temperature and flow of discharge and upstream temperature and flow.
3. Estimate heat exchange coefficient from design chart (Fig. 9.5) with an approximate average water temperature over distance.
4. Calculate temperature distribution using field estimate of river depth and velocity.

9.5.3 Other Water Bodies

A procedure similar to the riverine analysis above may be adopted for estuaries to calculate rises ($\Delta T = T - T_b$) above background temperatures (T_b). Equation 9.20 would be used for the overall exchange coefficient and Eqs. 3.8a through 3.8e for the spatial distribution of ΔT.

For complex lake and estuarine settings, finite difference models have been applied (Edinger et al., 1974; Thomann et al., 1975; DiToro and Connolly, 1980). An example of such a calculation is shown in Fig. 9.8 for the New York Harbor complex (Hydroscience, 1975). Twenty-three major thermal loads discharged approximately 66×10^9 Btu/hr (400×10^9 k·cal/day) to the Interstate Sanitation Commission (ISC) waters in 1970, as seen in Fig. 9.8(a). A maximum temperature increase above background of approximately 6°F was calculated in the East River [Fig. 9.8(b)] and 6.5°F in the Arthur Kill [Fig. 9.8(c)].

9.6 REDUCTION OF EXCESS HEAT INPUTS

Excess heat from the generation of electrical power and other heat inputs must usually be dissipated before discharge to a body of water so as not to contravene the applicable temperature standards. The principal means for reducing heat in a discharge are

1. Once-through cooling systems.
2. Cooling ponds.
3. Wet cooling towers and spray ponds.
4. Dry cooling towers.

Krenkel and Novotny (1980), Oleson and Boyle (1972), and Jimeson and Adkins (1972) discuss these heat control methods. Once-through cooling systems as the name implies do not include any external heat dissipation method at the discharge end but rely on a large volume of water for dissipation of the heat. Once-through cooling is therefore simple and economical and there is a minimum amount of water lost (i.e., consumptive use of water). On the other hand, once-through cooling may not be possible because of a lack of available water.

Cooling ponds have the advantages (Oleson and Boyle, 1972) that the costs are generally reasonable, the ponds can act as a settling basin for solids, and may be useful for recreation purposes. Makeup water may not be needed for significant periods of time. The disadvantages of cooling ponds are primarily that such controls require fairly large areas and may result in fogging and icing. Equation 9.17 provides a means for estimating the exit temperature, T for a given size pond. This is the temperature that is normally set by other external conditions such as state or federal water quality objectives for the stream into which the heated discharge is occurring. Equation 9.17 can be re-expressed for that purpose as

$$\frac{T - T_e}{T_{in} - T_e} = \frac{1}{1 + r} \qquad (9.22)$$

The exit temperature can therefore be controlled by increasing the size of the pond, specifically the surface area for dissipation of the heat. In a design situation, the left side of Eq. 9.22 is specified from meteorological conditions. The quantity r can be calculated. For the fixed r, and an estimated heat exchange coefficient at the required exit temperature, the required surface area of the pond can then be computed. Table 9.3 shows cost comparisons between methods.

Wet cooling towers may use mechanical means of maintaining an air flow past the heated water or may use natural draft. These towers can become quite large such as the four 38-story high cooling towers built for a 3100-MW power plant. The principal advantage of wet towers is less land area but the towers tend to be more expensive than cooling ponds. There may also be problems with localized icing or fog caused by the heated moist air discharged from the tower.

TABLE 9.3 UNIT COSTS OF COOLING WATER SYSTEMS FOR STEAM–ELECTRIC PLANTS

Type of system	Investment cost ($/kw)	
	Fossil fueled plant[a]	Nuclear fueled plant[a]
Once-through[b]	2.00–3.00	3.00–5.00
Cooling ponds[c]	4.00–6.00	6.00–9.00
Evaporative cooling towers		
Mechanical draft	5.00–8.00	8.00–11.00
Natural draft	6.00–9.00	9.00–13.00
Nonevaporative cooling towers		
Mechanical draft	18.00–20.00	26.00–28.00
Natural draft	20.00–24.00	28.00–32.00

[a]Based on unit sizes of 600 Mw and larger.

[b]Circulation from lake, stream, or sea and involving no investment in pond or reservoir.

[c]Artificial impoundments designed to dissipate entire heat load to the air. Cost data are for ponds capable of handling 1200 to 2000 MW of generating capacity.

Source: Jimeson, R. M. and G. D. Adkins, Waste Heat Disposal in Power Plants, CEP Tech. Manual: Cooling Towers, 5, 4, 1972. Reprinted by permission of the American Institute of Chemical Engineers.

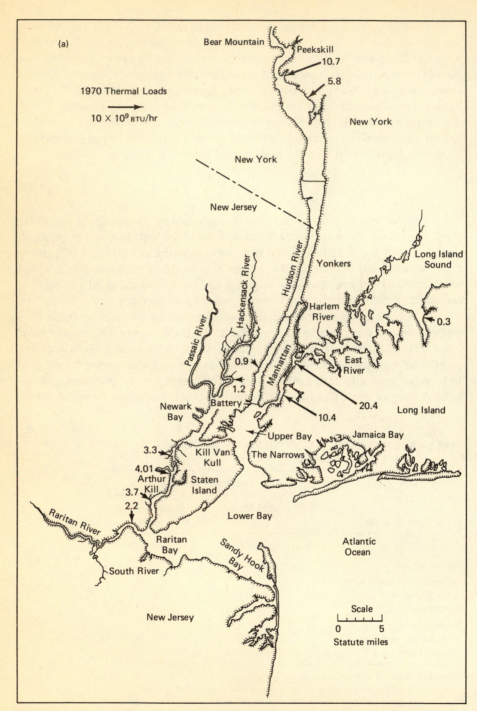

Figure 9.8 Finite difference model calculation of 1970 temperature rise above background due to excess heat inputs to New York Harbor complex. Adapted from Hydroscience (1975).

618

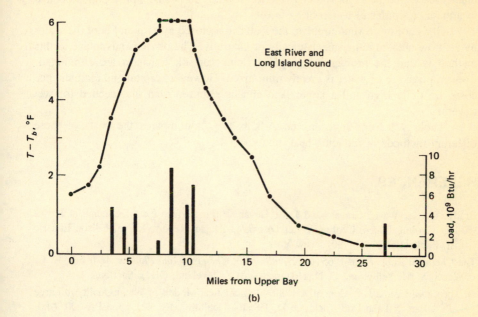

East River and
Long Island Sound

(b)

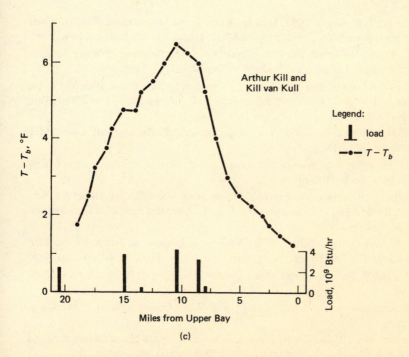

Arthur Kill and
Kill van Kull

Legend:

load

$T - T_b$

(c)

There may also be problems of downwind drift of high salinity water vapor if the plant uses estuarine or brackish water for cooling purposes. Spray ponds behave in a manner similar to wet cooling towers.

Dry cooling towers circulate the water in a closed system and cool the water by air flow created either mechanically or naturally. The obvious advantage to this method is that the problems of fogging, icing, or salt transport are eliminated. Consumptive use of water is greatly minimized. However, capital and maintenance costs are quite high and a larger land area is required than that needed for wet towers.

Table 9.3 from Jimeson and Adkins (1972) compares the costs of these different methods of reducing heat inputs.

REFERENCES

Brady, D. K., W. L. Graves, and J. C. Geyer. 1969. Surface Heat Exchange at Power Plant Cooling Lakes, Cooling Water Discharge Project Report, No. 5, Edison Electric Inst. Publication No. 69-901, New York.

The Chemical Rubber Co. 1965–66. *Handbook of Chemistry and Physics*, 46th Ed., R. C. Weast, S. M. Selby, and C. D. Hodgman, eds., Cleveland, OH, 6 Sections.

Di Toro, D. M., and J. P. Connolly. 1980. Mathematical Models of Water Quality in Large Lakes, Part 2: Lake Erie, ERL, ORD, USEPA, Duluth, MN, 231 pp. EPA-600/3-80-065.

Edinger, J. E., and J. C. Geyer. 1965. Heat Exchange in the Environment, Cooling Water Discharge Project Report No. 2, Publication 65-902, Edison Electric Institute, New York.

Edinger, J. E., D. W. Dutweiler, and J. C. Geyer. 1968. The Response of Water Temperature to Meteorological Conditions, *Water Res. Res.* 4(5):1137–1143.

Edinger, J. E., D. K. Brady, and J. C. Geyer, 1974. Heat Exchange and Transport in the Environment, Report No. 14, Electric Power Res. Inst. Pub. No. EA-74-049-00-3, Palo Alto, CA, Nov. 1974, 125 pp.

Hill, P. G. 1977. *Power Generation; Resources, Hazards, Technology and Costs*, MIT Press, Cambridge, MA, 402 pp.

Hutchinson, G. E., 1957. *A Treatise on Limnology, Vol. I. Geography, Physics and Chemistry*, Wiley, New York, 1015 pp.

Hydroscience, Inc., 1975. Development of a Steady State Water Quality Model for New York Harbor, by St. John, J. P. and Mueller, J. A. for Interstate Sanitation Commission, Oct. 1975, 307 pp. (vol. I).

Jimeson, R. M., and G. D. Adkins. 1972. Waste Heat Disposal in Power Plants, CEP Technical Manual: Cooling Towers, 5, 4, Amer. Inst. Chem. Eng., New York., pp. 1–6.

Krenkel, P. A., and V. Novotny. 1980. *Water Quality Management*, Academic Press, New York, 671 pp.

Nat. Acad. of Sciences and Nat. Acad. of Eng. 1972. Water Quality Criteria, Washington, D.C., 594 pp.

Nautical Almanac Office, 1986. The Astronomical Almanac, by US Naval Observatory and Royal Greenwich Observatory, Superintendent of Documents, Washington, D.C.

Neumann, G., and W. J. Pierson. 1966. *Principles of Physical Oceanography*, Prentice-Hall, Englewood Cliffs, NJ, pp. 41–44.

Oleson, K. A., and R. R. Boyle. 1972. How to Cool Steam-Electric Power Plants, in *Cooling Towers*, Amer. Inst. Chem. Eng., New York, pp. 94–100.

Ruffner, J. A., and Bair, F. E., 1984. The Weather Almanac, Gale Research Company, Detroit, MI.

Thomann, R. V., D. M. DiToro, R. P. Winfield, and D. J. O'Connor, 1975. Mathematical Modeling of Phytoplankton in Lake Ontario, for the USEPA. EPA-660/3-75-005.

PROBLEMS

9.1. The following data for a heat balance study were collected for an area in the southeastern United States:

	Day 1–2 (2000–0800)	Day 2 (0800–2000)
Air temperature (°C)	15.6	26.7
Relative humidity (%)	80	80
Wind speed (m/s)	0.25	1.25
Water temperature (°C)	24.0	24.0
H_s (cal/cm^2·day)	0.0	550

   (a) Estimate the heat exchange coefficient, K in W/m^2·°C and cal/cm^2·day·°C using the simplified approach of Fig. 9.5 for both day and night conditions.

   (b) Estimate the equilibrium temperature (°C) using the simplified approach for both conditions.

   (c) With the daytime estimate of T_e from (b), check whether $\Delta H = 0$ for $T_s = T_e$.

9.2. Suppose for the *average* conditions (day and night) of Problem 9.1 there is a discharge of 3 m^3/s at a temperature of 32°C into a well-mixed pond of 10 ha. The desirable exit temperature from ecological considerations is 25°C.

   (a) Is the size of the present pond sufficient?

   (b) If not, how big (in ha) should it be to meet water quality requirements?

   (c) Assume a source of "cold" water at 20°C is available to mix into the 10-ha pond. How much of this water (m^3/s) is needed to meet the desired 25°C exit temperature?

9.3. A survey of a heat balance of a small impoundment gave the following information:

$$
\begin{aligned}
H_S \text{ (cal/cm}^2\text{·day)} &= 510 \\
\text{air temperature (°C)} &= 30 \\
\text{relative humidity (\%)} &= 90 \\
\text{wind speed (m/s)} &= 1.5 \\
\text{water temperature (°C)} &= 26
\end{aligned}
$$

   (a) Estimate the heat exchange coefficient, K, in W/m^2·°C and cal/cm^2·day·°C using the simplified approach.

   (b) Estimate the equilibrium temperature (°C).

9.4. A heated discharge from an industry has a flow of 3 m^3/s and a temperature of 32°C. An area nearby is available for development as a cooling pond. If it is assumed that $K = 42$ cal/cm^2·day·°C and $T_e = 22.3$°C, develop a plot of area required for the pond (in ha) vs. exit temperatures (°C) for a range of exit temperatures from 20° to 30°C.

9.5. Given the following:

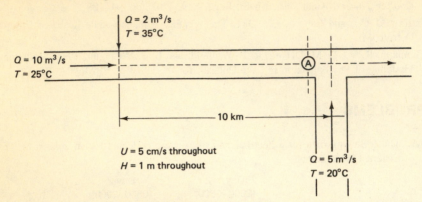

Temperature measurements at point A, just upstream of the incoming tributary, indicated an average temperature of 25.5°C.
- **(a)** What is the heat exchange coefficient (cal/day·cm²·°C) for the first 10-km reach?
- **(b)** Assuming this river heat exchange applies also downstream of the tributary input, what is the temperature of the river 10 km downstream from the tributary?

9.6.

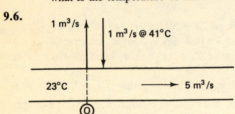

At location O, a plant draws 1 m³/s out of the river at the river temperature of 23°C and returns the flow at a very short distance downstream at 41°C. If $K = 50$ cal/cm²·day·°C, depth = 1 m, U = 10 cm/s, at what distance downstream (in km) will the temperature be within one degree of the background temperature?

9.7. Given the following:

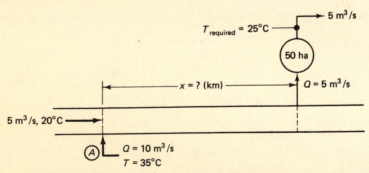

A cooling pond of 50 ha is to be constructed downstream of the heated discharge at location A. The required exit temperature of the pond is to be 25°C. Conditions are as follows:

River	Pond
$K = 62$ cal/cm$^2 \cdot$day$\cdot$°C	$K = 42$ cal/cm$^2 \cdot$day$\cdot$°C
$U = 10$ cm/s	$A = 50$ ha
$h = 0.5$ m	$T_e = 23$°C

What is the distance (in km) from the heated discharge to the withdrawal for the cooling pond such that the required exit temperature of 25°C is met?

9.8. Consider again Problem 9.2. Let $K = 42$ cal/cm$^2 \cdot$day$\cdot$°C for both pond and river, and let $T_e = 23$°C for the pond. A river is near the pond as shown below.

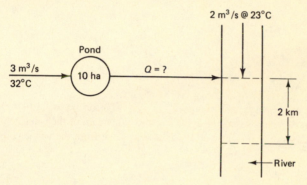

The river has a depth of 1 m and a velocity of 10 cm/s (which is independent of the river flow). How much of the outflow from the pond in m^3/s can be diverted to the river so that the temperature in the river is above 25°C for no more than 2 km?

Nomenclature

Symbol	Definition	Units[a]
a_{cP}	stoichiometric ratio of carbon to chlorophyll a	—
a_p	stoichiometric ratio of phosphorus to chlorophyll a	—
a_N	stoichiometric ratio of nitrogen to chlorophyll a	—
a_{oP}	stoichiometric ratio of oxygen to chlorophyll a	—
A	area, cross-sectional area	L^2
A_s	surface area	L^2
B	width of stream	L
c	concentration of dissolved oxygen (DO), toxicant	M/L^3
c_d'	dissolved toxicant concentration	M_T/L_w^3
c_d	dissolved toxicant concentration, porosity corrected	M_T/L_{s+w}^3
c_g	gas phase concentration	M_T/L_g^3
c_p	particulate toxicant concentration	M_T/L_{s+w}^3
c_T	total toxicant concentration	M_T/L_{s+w}^3
c_s	saturation concentration of dissolved oxygen	M/L^3
c_w	chemical solubility in water	M/L^3
C	weight-specific food consumption by organism	$M(w)/M(w) \cdot T$
	runoff coefficient	—
d	particle diameter	L
D	dissolved oxygen (DO) deficit	M/L^3
	average duration of rainfall event	T
D_o	DO deficit at the outfall location after mixing	M/L^3
D_c	critical (maximum) DO deficit	M/L^3
D_l	diffusivity of toxicant in water	L^2/T
D_L	diffusivity of oxygen in water	L^2/T
D_g	diffusivity of toxicant in gas	L^2/T
D_P	overall death rate of algae	$1/T$

[a]See notes at end of Nomenclature

Symbol	Definition	Units[a]
D_Z	overall death rate of zooplankton	$1/T$
e	base of Naperian logarithms ($= 2.71828$)	—
e_a	air vapor pressure	mm Hg
e_s	saturated vapor pressure of water at air temperature	mm Hg
E	dispersion coefficient	L^2/T
E'	bulk dispersion coefficient	L^3/T
E_{num}	numerical dispersion in finite difference model	L^3/T
E_v	evaporative water loss	L/T
f	ratio CBODU/CBOD5	—
	photoperiod	—
	biomagnification factor	$M_T/M(w) \div M_T/M(w)$
f_d	dissolved toxicant fraction	—
f_{oc}	weight fraction of organic carbon to total solids	M_{orgC}/M_s
f_p	particulate toxicant fraction	—
F_s	mass rate of deposition of particulates	M/T
G	net growth rate of organism (weight basis)	$1/T$
G_n	net phytoplankton growth rate	$1/T$
G_{max}	maximum growth rate of phytoplankton	$1/T$
G_P	gross phytoplankton growth rate	$1/T$
G_Z	gross zooplankton growth rate	$1/T$
H	depth	L
H_a	longwave radiation	$E/L^2 \cdot T$
H_{ar}	reflected longwave radiation	$E/L^2 \cdot T$
H_{br}	longwave radiation emitted by water	$E/L^2 \cdot T$
H_c	rate of conductive heat transfer	$E/L^2 \cdot T$
H_e	evaporative heat loss (gain)	$E/L^2 \cdot T$
	Henry's constant	$M_T/L_g^3 \div M_T/L_w^3$
I	solar radiation	$E/L^2 \cdot T$
	rainfall intensity	L/T
I_a	average solar radiation during photoperiod	$E/L^2 \cdot T$
I_0	solar radiation at water surface	$E/L^2 \cdot T$
I_s	saturating light intensity for species	$E/L^2 \cdot T$
I_T	total daily solar radiation	$E/L^2 \cdot T$
j_1, j_2	estuarine decay coefficients upstream & downstream of discharge	$1/L$
$J(s_i)$	mass flux of substance s_i due to transport terms	M/T
K	first-order decay rate	$1/T$
	depuration rate of chemical from organism	$1/T$
	overall heat exchange coefficient	$E/L^2 \cdot T \cdot (°)$
K_a	reaeration coefficient	$1/T$
K_B	bacterial first-order death rate	$1/T$
	chemical biodegradation rate	$1/T$
K_d	deoxygenation coefficient	$1/T$
	dissolved chemical overall decay rate	$1/T$
K_e	light extinction coefficient	$1/L$
K_f	sediment–water dissolved chemical diffusion rate	L/T
K_g	gas film transfer coefficient	L/T
K_H	chemical decay rate by hydrolysis	$1/T$
k_l	overall volatilization transfer coefficient	L/T
K_l	liquid film transfer coefficient	L/T
K_L	DO transfer coefficient	L/T

[a]See notes at end of Nomenclature

Symbol	Definition	Units[a]
K_{mN}	Michaelis constant	M/L^3
K_N	nitrification rate	$1/T$
K_{ow}	octanol–water partition coefficient	$M_o/L^3 \div M_w/L^3$
K_p	chemical decay rate by photolysis	$1/T$
K_r	gross CBOD decay rate	$1/T$
K_s	net loss rate of solids due to settling	$1/T$
	settling rate of BOD	$1/T$
k_u	uptake rate of toxicant by organism	$L_w^3/M(w) \cdot T$
K_1	laboratory BOD rate	$1/T$
L	CBODU concentration	M/L^3
L_d	dissolved BOD concentration	M/L^3
L_0	CBODU concentration at discharge location	M/L^3
	laboratory CBODU	M/L^3
L_0^N	NBOD concentration at discharge location	M/L^3
L_m	length from discharge to completely mixed concentration	L
L^N	NBOD concentration	M/L^3
L_p	particulate BOD concentration	M/L^3
L_{rd}	distributed source of BOD	$M/L^3 \cdot T$
m	solids concentration	M_s/L_{s+w}^3
M	mass input of substance	M
	molecular weight	—
n	Manning roughness coefficient	$L^{1/6}$
N	number of bacteria per unit volume water	$No./L^3$
	nitrogen concentration	M/L^3
	stability frequency	$1/T^3$
N_w	bioconcentration factor	$M_T/M(w) \div M_T/L_w^3$
p	phosphorus concentration	M/L^3
	partial pressure; barometric pressure	mm Hg
p_a	daily average gross photosynthetic DO production	$M/L^3 \cdot T$
p_m	in-stream maximum daily gross photosynthetic DO production	$M/L^3 \cdot T$
p_{net}	daily average net photosynthetic DO production	$M/L^3 \cdot T$
p_s	saturated (optimum) photosynthetic DO production	$M/L^3 \cdot T$
P	phytoplankton concentration (chlorophyll a)	M/L^3
	wetted perimeter	L
	precipitation	L/T
q	exponent for exponential flow increase	$1/L$
	hydraulic overflow rate	L/T
Q	flow	L^3/T
Q_e	effluent flow	L^3/T
Q_n	net estuarine flow	L^3/T
Q_R	runoff flow	L^3/T
r	relative error of finite difference solution	—
	solid phase toxicant concentration	M_T/M_s
	lake thermal coefficient	—
R	phytoplankton DO respiration rate	$M/L^3 \cdot T$
R_h	hydraulic radius	L
s	concentration	M/L^3
s_0	concentration at discharge location	M/L^3
s_I	concentration in infiltrating flow	M/L^3
s_p	peak concentration	M/L^3

[a]See notes at end of Nomenclature

Symbol	Definition	Units[a]
$\bar{s}$	lake equilibrium concentration	M/L^3
S	hypolimnetic oxygen depletion rate	$M/L^2 \cdot T$
	salinity	M/L^3
	river slope ($= S_0$)	L/L
	average number of storms in interval T	—
S_B	sediment oxygen demand	$M/L^2 \cdot T$
S_D	distributed source or sink (volumetric basis)	$M/L^3 \cdot T$
S_N	areal uptake of nitrogen by benthos	$M/L^2 \cdot T$
S_p	areal uptake of phosphorus by benthos	$M/L^2 \cdot T$
t	time	T
t^*	stream travel time	T
t_c^*	travel time to location of critical DO	T
t_d	lake hydraulic detention time	T
t_p	time to peak concentration	T
t_{90}	time to 90% bacterial mortality	T
T	tidal period	T
	temperature	$(°)$
T_a	air temperature	$(°)$
T_b	natural background temperature	$(°)$
T_e	equilibrium temperature	$(°)$
T_0	temperature at discharge location	$(°)$
T_s	surface water temperature	$(°)$
U	freshwater velocity	L/T
U_T	tidal velocity	L/T
U_w	wind velocity	L/T
U^*	river shear velocity	L/T
$U(t - t_0)$	unit step function	—
V	volume	L^3
v_d	sedimentation velocity	L/T
v_n	net settling velocity of solids	L/T
v_P	phytoplankton net settling rate	L/T
v_s	settling velocity of solids	L/T
v_T	overall net loss rate of chemical	L/T
v_{Ts}	loss of chemical due to sediment interaction	L/T
v_{Td}	dissolved chemical loss rate	L/T
v_u	solids resuspension velocity	L/T
w	weight of organism	$M(w)$
	external distributed input	$M/L \cdot T$
W	mass input rate	M/T
W'	areal mass input rate	$M/L^2 \cdot T$
W_A	long-term average loading rate for storms	M/T
W_H	excess heat input	E/T
W_R	mass discharge due to runoff	M/T
W_T	mass input rate of toxicant	M/T
x	longitudinal distance in direction of flow	L
z	vertical distance	L
z_s	secchi depth	L
Z	zooplankton concentration	M/L^3
α	estuarine coefficient	—
	finite difference weighting value	—
	chemical assimilation efficiency	—

[a]See notes at end of Nomenclature

Symbol	Definition	Units[a]
β	finite difference weighting value ($= 1 - \alpha$)	—
	ratio of effective diffusivity of chemical to diffusivity of oxygen	—
δ	ratio solid phase chemical concentrations in sediment to water	—
$\delta(t - t_0)$	Dirac delta function; unit impulse	$1/T$
Δ	average time between centers of storm events	T
Δ_c	diurnal DO range due to photosynthesis	M/L^3
η	estuary number ($= 4KE/U^2$)	—
ν	toxicant concentration in organism	$M_T/M(w)$
ν'	whole body burden of toxicant	M_T
ν_i	coefficient of variation of rainfall intensity	—
ν_N	coefficient of variation of bacterial concentration	—
ν_q	coefficient of variation of flow	—
ν_W	coefficient of variation of input load	—
ρ	density of water	M/L^3
	lake flushing rate	$1/T$
ρ_s	density of solids	M_s/L_s^3
ϕ	porosity	L_w^3/L_{s+w}^3
¶	toxicant–solids partition coefficient, porosity corrected	$M_T/M_s \div M_T/L_{s+w}^3$
¶'	toxicant–solids partition coefficient	$M_T/M_s \div M_T/L_w^3$
$¶_{oc}$	partition coefficient—organic carbon basis	$M_T/M_{orgC} \div M_T/L_w^3$
exp	$\exp(\ ) \equiv e^{(\ )}$	—
log	common logarithm, base 10	—
ln	natural logarithm, base e	—

Notes: [M] mass; $[M_s]$ mass of solids; $[M_T]$ mass of toxicant; [M(w)] mass of organism, wet weight; [L] length; [E] energy; $[L_{s+w}^3]$ volume of solids and water (bulk volume); $[L_s^3]$ volume of solids; $[L_w^3]$ volume of water; $[L_g^3]$ volume of gas; [T] time; [°] degrees temperature.

Conversion Factors

Length

1 kilometer (km)	= 0.6214 miles (mi)
1 mile (mi)	= 5280 feet (ft)
1 nautical mile	= 6076 feet (ft)
1 meter (m)	= 3.2808 feet (ft)
1 centimeter (cm)	= 0.3937 inches (in)

Area

1 hectare (ha)	= 10000 square meters (m^2)
	= 2.4710 acres
	= 0.003861 square miles (mi^2)
1 square kilometer (km^2)	= 0.3861 square miles (mi^2)

Volume

1 cubic meter (m^3)	= 35.3147 cubic feet (ft^3)
	= 264.172 gallons (gal)
	= 1000 liters (l)
1 cubic foot (ft^3)	= 7.480 gallons (gal)

Velocity

1 meter/second (m/s)	= 3.2808 feet/second (ft/s, fps)
	= 86.40 kilometers/day (km/d)
1 foot/second (ft/s)	= 16.364 miles/day (mi/d, mpd)
1 knot	= 1 nautical mile/hr
	= 1.688 feet/second (ft/s, fps)

Flow

1 cubic meter/second (m^3/s)	=	35.3147 cubic feet/second (ft^3/s, cfs)
	=	22.8245 million gallons/day (MG/d, MGD)
1 million gallons/day (MGD)	=	1.5472 cubic feet/second (ft^3/s, cfs)
1 inch/year	=	0.07367 cubic feet/second · square mile (cfs/mi^2)
1 acre-foot/year	=	0.001381 cubic feet/second (cfs)
1 inch-acre/hour	=	1.008 cubic feet/second (cfs)

Dispersion

1 square mile/day (smpd)	=	29.98 square meters/second (m^2/s)
	=	322.7 square feet/second (ft^2/s)
1 square centimeter/second (cm^2/s)	=	93.00 square feet/day (ft^2/day)

Temperature

Degrees Celsius (°C)	=	(5/9)[degrees Fahrenheit(°F) − 32]
Degrees Fahrenheit (°F)	=	(9/5)[degrees Celsius(°C)] + 32
Degrees Kelvin (°K)	=	degrees Celsius(°C) + 273
Degrees Rankine (°R)	=	degrees Fahrenheit(°F) + 460

Energy

1 langley (ly)	=	1 calorie/square centimeter (cal/cm^2)
1 calorie/square centimeter · day (cal/cm^2 · day)	=	0.4840 watts/square meter (W/m^2)
	=	3.6867 British thermal units/ft^2 · day (Btu/ft^2 · day)
1 joule (J)	=	1 watt-second (W-s)
	=	0.2388 calories (cal)
1 kilojoule (kJ)	=	0.9478 British thermal units (Btu)

Mass

1 pound (lb)	=	453.6 grams (g)
1 kilogram (kg)	=	2.2046 pounds (lb)
1 metric ton (mt)	=	1000 kilograms (kg)

Density (of Pure Water at Atmospheric Pressure)

1 cubic foot	=	62.4 pounds at 60°F
1 cubic centimeter	=	1.00 grams at 4°C

Concentration

$$
\begin{aligned}
1 \text{ milligram/liter (mg/l)} &= 8.34 \text{ pounds/million gallons (lb/MG)} \\
&= 1 \text{ gram/cubic meter (g/m}^3) \\
&= 1 \text{ part/million parts (ppm)}[1] \\
1 \text{ microgram/liter } (\mu\text{g/l}) &= 1 \text{ part/billion parts (ppb)}[1] \\
1 \text{ nanogram/liter (ng/l)} &= 1 \text{ part/trillion parts}[1] \\
1 \text{ part/thousand parts (ppt)} &= 1\text{‰}
\end{aligned}
$$

SI Prefixes

Prefix	Multiplication factor
mega (M)	10^6
kilo (k)	10^3
centi (c)	10^{-2}
milli (m)	10^{-3}
micro (μ)	10^{-6}
nano (n)	10^{-7}
pico (p)	10^{-12}
femto (f)	10^{-15}

[1]For mass density $\rho = 1 \text{ g/cm}^3$.

Appendix C

Solubility of Oxygen in Water Exposed to Water-Saturated Air at Atmospheric Pressure

Temperature (°C)	Oxygen solubility (mg/l)					
	Chlorinity: 0	5.0	10.0	15.0	20.0	25.0
0.0	14.621	13.728	12.888	12.097	11.355	10.657
1.0	14.216	13.356	12.545	11.783	11.066	10.392
2.0	13.829	13.000	12.218	11.483	10.790	10.139
3.0	13.460	12.660	11.906	11.195	10.526	9.897
4.0	13.107	12.335	11.607	10.920	10.273	9.664
5.0	12.770	12.024	11.320	10.656	10.031	9.441
6.0	12.447	11.727	11.046	10.404	9.799	9.228
7.0	12.139	11.442	10.783	10.162	9.576	9.023
8.0	11.843	11.169	10.531	9.930	9.362	8.826
9.0	11.559	10.907	10.290	9.707	9.156	8.636
10.0	11.288	10.656	10.058	9.493	8.959	8.454
11.0	11.027	10.415	9.835	9.287	8.769	8.279
12.0	10.777	10.183	9.621	9.089	8.586	8.111
13.0	10.537	9.961	9.416	8.899	8.411	7.949
14.0	10.306	9.747	9.218	8.716	8.242	7.792
15.0	10.084	9.541	9.027	8.540	8.079	7.642
16.0	9.870	9.344	8.844	8.370	7.922	7.496
17.0	9.665	9.153	8.667	8.207	7.770	7.356
18.0	9.467	8.969	8.497	8.049	7.624	7.221
19.0	9.276	8.792	8.333	7.896	7.483	7.090
20.0	9.092	8.621	8.174	7.749	7.346	6.964
21.0	8.915	8.456	8.021	7.607	7.214	6.842
22.0	8.743	8.297	7.873	7.470	7.087	6.723
23.0	8.578	8.143	7.730	7.337	6.963	6.609

Notes at end

Temperature (°C)	Oxygen solubility (mg/l)					
	Chlorinity: 0	5.0	10.0	15.0	20.0	25.0
24.0	8.418	7.994	7.591	7.208	6.844	6.498
25.0	8.263	7.850	7.457	7.083	6.728	6.390
26.0	8.113	7.711	7.327	6.962	6.615	6.285
27.0	7.968	7.575	7.201	6.845	6.506	6.184
28.0	7.827	7.444	7.079	6.731	6.400	6.085
29.0	7.691	7.317	6.961	6.621	6.297	5.990
30.0	7.559	7.194	6.845	6.513	6.197	5.896
31.0	7.430	7.073	6.733	6.409	6.100	5.806
32.0	7.305	6.957	6.624	6.307	6.005	5.717
33.0	7.183	6.843	6.518	6.208	5.912	5.631
34.0	7.065	6.732	6.415	6.111	5.822	5.546
35.0	6.950	6.624	6.314	6.017	5.734	5.464
36.0	6.837	6.519	6.215	5.925	5.648	5.384
37.0	6.727	6.416	6.119	5.835	5.564	5.305
38.0	6.620	6.316	6.025	5.747	5.481	5.228
39.0	6.515	6.217	5.932	5.660	5.400	5.152
40.0	6.412	6.121	5.842	5.576	5.321	5.078
41.0	6.312	6.026	5.753	5.493	5.243	5.005
42.0	6.213	5.934	5.667	5.411	5.167	4.933
43.0	6.116	5.843	5.581	5.331	5.091	4.862
44.0	6.021	5.753	5.497	5.252	5.017	4.793
45.0	5.927	5.665	5.414	5.174	4.944	4.724
46.0	5.835	5.578	5.333	5.097	4.872	4.656
47.0	5.744	5.493	5.252	5.021	4.801	4.589
48.0	5.654	5.408	5.172	4.947	4.730	4.523
49.0	5.565	5.324	5.094	4.872	4.660	4.457
50.0	5.477	5.242	5.016	4.799	4.591	4.392

Note: Values above calculated from Eqs. 6.14 and 6.15.

Source: APHA (American Public Health Association), 1985. Standard Methods for the Examination of Water and Waste Water, 16th Ed., Washington, D.C., 874 pp. Reprinted by permission of APHA, AWWA (American Water Works Association), and WPCF (Water Pollution Control Federation).

Index

Action limits, for chemicals in fish, 503
Acute toxicity, 499
Aeration
  at hydraulic structures, 374
  effluent, 367
  hypolimnetic, 482
  in-stream, 372
  See also Rearation, atmospheric
Aftergrowth of indicator bacteria, 239
Age-dependent model, lake trout, PCB, 590
Allowable discharge load: *see* Waste load allocation
Anaerobic DO model, river, 313
Annabessacook Lake, ME, nutrient control of, 480
Area, cross-sectional
  average over reach, 47
  estuary, at mean water, 108
  river, 31
Assimilation efficiency
  of chemicals, 570
  of food, 570
  of phytoplankton by zooplankton, 429

Bacteria
  autotrophic, 274
  heterotrophic, 268
  indicator, of disease, 221
  nitrobacter, 274
  nitrosomonas, 274
Benthal oxygen demand: *see* Sediment oxygen demand
Biochemical oxygen demand (BOD)
  effluent variability of, 362

history, 261
sources of, 268
table of inputs, 22
Bioconcentration factor (BCF)
  definition of, 515
  effect of growth rate, 562
  lipid-based, 515
  relation to K_{ow}, 515
Biodegradation of chemicals, 539
Biomagnification ratio, 571
Biomass conversion efficiency, in food chain model, 570
Black River, NY
  DO model, 327
  area–flow, 46
  flow, morphometry, 45
BOD: *see* Biochemical oxygen demand
Boston Harbor, MA
  segments, flow, dispersion, 149
  total coliform bacteria, 251
Bottle BOD rate, 272
Bulk dispersion coefficient, 128

Cadmium, table of inputs, 22
Calibration, model, definition of, 7
Carbonaceous BOD, CBOD
  5-day, CBOD5, 271
  f-ratio, 272
  inputs before treatment, 268
  organic carbon, 273
  oxidation rate, lab, 271
  particulate, settling, 293
  relative to BOD, 268
  river models, 301